Fluxes Between Trophic Levels and Through the Water–Sediment Interface

Developments in Hydrobiology 62

Series editor
H. J. Dumont

Fluxes Between Trophic Levels and Through the Water–Sediment Interface

Proceedings of the Joint Congress of Limnology and Oceanography
held in Marseille, June 26–29, 1989

Edited by
Daniel J. Bonin and Han L. Golterman

Special editor assistant
M.-C. Bonin

Reprinted from Hydrobiologia, vol. 207 (1990)

Kluwer Academic Publishers
Dordrecht / Boston / London

Library of Congress Cataloging-in-Publication Data

CIP

Joint Congress of Limnology and Oceanography (1989 : Marseille, France)
Fluxes between trophic levels and through the water-sediment interface : proceedings of the Joint Congress of Limnology and Oceanography, held in Marseille 26-29 June 1989 / edited by Daniel J. Bonin and Han L. Golterman ; special editor assistant, M.-C. Bonin.
p. cm. -- (Developments in hydrobiology ; 62)

1. Food chains (Ecology)--Congresses. 2. Aquatic ecology--Congresses. 3. Energy transfer--Congresses. I. Bonin, Daniel J. II. Golterman, H. L. III. Title. IV. Series.
QH540.J65 1989
574.5'3--dc20 90-5308

ISBN 978-90-481-4078-7

Published by Kluwer Academic Publishers,
P.O. Box 17, 3300 AA Dordrecht, The Netherlands.

Kluwer Academic Publishers incorporates
the publishing programmes of
D. Reidel, Martinus Nijhoff, Dr W. Junk and MTP Press.

Sold and distributed in the U.S.A. and Canada
by Kluwer Academic Publishers,
101 Philip Drive, Norwell, MA 02061, U.S.A.

In all other countries, sold and distributed
by Kluwer Academic Publishers Group,
P.O. Box 322, 3300 AH Dordrecht, The Netherlands

Printed on acid-free paper

Preface

In this volume we have collected articles presented during the Conference on Limnology and Oceanography held at the Faculty of Sciences in Marseille-Luminy (26–29 June 1989). About 150 communications were presented to nearly 300 participants from 22 countries.

It seemed to the Association Française de Limnologie and Union des Océanographes de France to be a good and new idea to bring together limnologists and oceanographers also in Europe where the two groups of scientists work more separately than in North America.

The conference was organized round two main themes: 1) transfer of energy and matter through food webs in ecosystems of various richness, and 2) exchange between sediments and overlying waters in natural ecosystems. Specialists of all the different disciplines in aquatic sciences (population dynamics, biochemistry, geochemistry, physics and sedimentology) contributed papers concerning experimental or in situ studies, and modelling.

Regarding the main themes, it is clear that some types of ecosystems (e.g. estuaries, lagoons, artificial and natural lakes) are being studied more intensively than others. Few contributions deal with the study of deep ocean waters or rivers and streams, while the exchange between water column and sediment in coastal areas is studied frequently, often over longer periods; this is not surprising, because here access is easy and economic interests play an important role. Similarly, long term studies in several lakes dominate. Eight articles concern the NW Mediterranean coastal waters and four the Lake of Geneva, but other geographical areas have also yielded papers: the Red Sea, Pacific Ocean, St Lawrence (Canada), Bou Regreg (Morocco), the Ria Formosa (S. Portugal) estuaries and several reservoirs. This kind of study corresponds to the scientific demand as expressed in national and international programmes related to environmental protection.

Transition between fresh and marine waters, and between different trophic levels (from oligotrophy to eutrophy) demands serious methodological reflection. Both scientific communities share this interest in methodology and in the functioning of ecosystems in relation to their trophic state. Therefore some papers in this volume deal with the improvement of methods.

The organizers of the symposium and the editors hope that common interest will be the focus of future cooperation and share the wish that this collection of papers may convince the reader of the necessity to assemble the two scientific aquatic communities on a European scale.

The organizing committee would like to thank all public agencies and offices and private companies that contributed to the conference and supported the edition of this volume.

Finally the editors acknowledge the careful reviewing of the manuscripts by the numerous referees and thank Henri Dumont, editor-in-chief of *Hydrobiologia* and Wil Peters of Kluwer Academic Publishers for their interest and assistance in the production of this special issue.

Daniel J. Bonin
Centre d'Océanologie de Marseille
Faculté des Sciences de Luminy
case 901
13288 Marseille cedex 9 – France

Han L. Golterman
Station Biologique de
La Tour du Valat
Le Sambuc
13200 Arles – France

List of participants

ALGERIA
Samson Fatiha, *Alger*

AUSTRIA
Rott Eugen, *Innsbruck*
Wolf Gerard, *Innsbruck*

BELGIUM
Belgrano Andrea, *Ghent*
Billen Gilles, *Bruxelles*
Lancelot Christiane, *Bruxelles*
Servais Pierre, *Bruxelles*

CANADA
Demers Serge, *Mont-Joli*
Desrosiers Gaston, *Rimouski*
Kalff Jacob, *Montreal*
Marsot Pierre, *Rimouski*
Painchaud Jean, *Mont-Joli*
Vincent Bruno, *Rimouski*

EGYPT
Awad Hassan, *Alexandria*
El-Sayed Mahmoud, *Alexandria*

GERMANY
Liebezeit Gerd, *Wilhelmshaven*
Rai Hakumat, *Plön*
Staudinger Barbara, *Bayreuth*
Tilzer Max, *Konstanz*

GREECE
Friligos Nicholas, *Athens*
Souvermezoglou Ekaterini, *Athens*

INDIA
Premchand Kolli, *Cochin*

ISRAEL
Berman Thomas, *Tiberias*

ITALY
Cattaneo-Vietti Ricardo, *Genova*
Cocito Silvia, *La Spezia*
Danovaro Roberto, *Genova*
Morri Carla, *Genova*

MOROCCO
El Mezdi Zohra, *Marrakech*
Foutlane Ali, *Rabat*
Loudiki Mohammed, *Marrakech*
Tifnouti Abdelaziz, *Marrakech*

NORWAY
Steen Harald, *Oslo*

PORTUGAL
Amorim Ferreira Ana, *Lisboa*
Brotas Vanda, *Lisboa*
Cabeçadas Maria Leonor, *Lisboa*
Castro Maria Odete, *Lisboa*
Ova Maria de Luz, *Lisboa*
Sobral Dinah, *Lisboa*
Vale Carlos, *Lisboa*

SPAIN
Borras Calvo Gabriel, *Barcelona*
Cruz Pizarro Luis, *Granada*
Mallo Silvia, *Barcelona*
Margalef Ramon, *Barcelona*
Rodriguez Jaime, *Malaga*
Torrella Francisco, *Murcia*
Vidal Barcelona Montserrat, *Barcelona*
Vives Rego Josep, *Barcelona*

SWEDEN
Enoksson Viveka, *Göteborg*
Selmer Johanne-Sophie, *Göteborg*
Sundbäck Kristina, *Göteborg*

SWITZERLAND
Arbouille Didier, *Versoix*
Bossard Peter, *Kastanienbaum*
Gandais Véronique, *Versoix*
Howa Hélène, *Versoix*
Span Daniel, *Versoix*

THE NETHERLANDS
Duursma Egbert, *Texel*
Vijverberg Jacobus, *Oosterzee*

TUNISIA
Added Ayed, *Tunis*
Kalai Slaheddine, *Tunis*

UNITED KINGDOM
Linton Andrew, *Crewe*

USA
Phelps Harriette, *Washington DC*

USSR
Konovalov Stanislas, *Sevastopol*

FRANCE
Alliot Elisabeth, *Marseille*
Amanieu Michel, *Montpellier*
Amiard Jean-Claude, *Nantes*
André Béatrice, *Pierre-Bénite*
Angeli Nadine, *Thonon-les-Bains*
Aprosi Geneviève, *Chatou*
Arhlac Denise, *Marseille*
Auscher Fabrice, *Caen*
Baleux Bernard, *Montpellier*
Balvay Gérard, *Thonon-les-Bains*
Barthélémy Dominique, *Lyon*
Battini Alain, *Marseille*
Baudin Jean-Pierre, *Cadarache*
Baudinet Dominique, *Marseille*
Bearez Philippe, *Toulouse*
Belamie René, *Lyon*
Bellan Gérard, *Marseille*
Belviso Sauveur, *Gif-sur-Yvette*
Benaïm Jean, *Toulon*
Ben Moussa Mustapha, *Toulouse*
Berland Brigitte, *Marseille*
Berthon Jean-Louis, *Saint-Etienne*
Bertrand Jean-Claude, *Marseille*
Bertru Georges, *Rennes*
Beuffe Henri, *Cestas*
Bidois Laurence, *Paris*
Blanc François, *Marseille*
Blanc-Vernet Laure, *Marseille*
Bonin Daniel, *Marseille*
Bonin Marie-Claude, *Marseille*
Bonin Patricia, *Marseille*
Bonzongo Jean-Claude, *Rennes*
Bosca Catherine, *Toulouse*
Boulanger Bernard, *Orléans*
Boumnich L'Houssaine, *Aubière*
Bourdier Gilles, *Aubière*
Braconnot Jean-Claude, *Villefranche-sur-Mer*
Brun Guy, *Marseille*
Buscail Roselyne, *Perpignan*
Cabannes Jean-Louis, *Marseille*
Cabrera Danielle, *Aix-en-Provence*
Capblancq Jacques, *Toulouse*
Cardinal-Legrand Catherine, *Dijon*
Casabianca Marie-Luce de, *Montpellier*
Caumette Pierre, *Marseille*
Cazaubon Arlette, *Marseille*
Champeau Antoine, *Marseille*
Chappaz Rémi, *Marseille*
Charpy Loïc, *Marseille*
Charry Jean-Paul, *Marseille*
Chikhi Dalila, *Marseille*
Clément René, *Nice*
Collos Yves, *Nieul-sur-Mer*
Coste Bernard, *Marseille*
Coulon Olivier, *Orléans*
Cravinho Arielle, *Aubière*
Daumas Raoul, *Marseille*
Dauta Alain, *Toulouse*
Dauvin Jean-Claude, *Roscoff*
Davoult Dominique, *Wimereux*
Degiovanni Claude, *Marseille*
Denant Véronique, *Paris*
Denis Michel, *Marseille*
Descolas-Gros Chantal, *Banyuls-sur-Mer*
Deslous-Paoli Jean-Marc, *Sète*
Dewarumez Jean-marie, *Wimereux*
Diaz Victoria, *Wimereux*
Ducrotoy Jean-Paul, *St-Valéry-sur-Somme*
Dufour Philippe, *Thonon-les-Bains*
Elkaïm Bernard, *Paris*
Escaravage Vincent, *Arcachon*
Fabre André, *Toulouse*
Faudot Daniel, *Paris*
Fera Philippe, *Orléans*
Fernandez Jean-Michel, *La Seyne-sur-Mer*
Ferra Christiane, *Trévoux*
Ferrara-Guerrero Maria-Jesus, *Marseille*
Ferré Bruno, *Aix-en-Provence*
Feuillade Jacques, *Thonon-les-Bains*
Feuillade Mauricette, *Thonon-les-Bains*
Février Gérard, *La Garde*
Fichez Renaud, *Marseille*
Flatau Gilles, *Nice*
Flesch Anne, *Metz*
Foulquier Luc, *Cadarache*
Frisoni Guy-François, *Montpellier*
Froget Catherine, *Marseille*
Froidefond Jean-Marie, *Talence*
Gadel François, *Perpignan*
Galvin Yannick, *Marseille*
Gambade Gérard, *Dijon*
Garnier Jacqueline, *Cadarache*
Garnier Josette, *Paris*
Gaudy Raymond, *Marseille*
Gentilhomme Valérie, *Marseille*
Gérino Magali, *Marseille*
Gibert Janine, *Villeurbanne*
Gigleux Marc, *Metz*
Gilewicz Michèle, *Marseille*
Ginet René, *Villeurbanne*
Ginsburger-Vogel Thomas, *Nantes*
Girard Dominique, *La Seyne-sur-Mer*
Girard Michelle, *Nieul-sur-Mer*
Giudicelli Jean, *Marseille*
Gnassia-Barelli Mauricette, *Villefranche-sur-Mer*
Golterman Han, *Arles*
Gontier Gilles, *La Seyne-sur-Mer*
Gorsky Gabriel, *Villefranche-sur-Mer*
Gostan Jacques, *Villefranche-sur-Mer*
Goujon Claude, *St-Quentin-en-Yvelines*
Goutx Madeleine, *Marseille*
Grégoire Alain, *Paris*
Grenz Christian, *Marseille*

Groot Cornelis de, *Arles*
Guérin Catherine, *Toulouse*
Gueuné Yvon, *Rennes*
Guidi Laurence, *Banyuls-sur-Mer*
Guilho Marcel, *Rennes*
Guillerm Gérard, *Massy*
Heize René, *Paris*
Helmer Christine, *Toulouse*
Héral Maurice, *La Tremblade*
Herbland Alain, *Nieul-sur-Mer*
Hermin Marie-Noëlle, *Marseille*
Hily Christian, *Brest*
Horn Michel, *Caen*
Jacques Guy, *Banyuls-sur-Mer*
Jamet Jean-Louis, *Aubière*
Julian Robert, *Gap*
Kerambrun Pierre, *Marseille*
Kouwenberg Julia, *Banyuls-sur-Mer*
Lacaze Jean-Claude, *Paris*
Lacroix Georges, *Marseille*
Lafforgue Michel, *Fontainebleau*
Lair Nicole, *Aubière*
Lam Hoai Thong, *Montpellier*
Lampert Denis, *Wissembourg*
Lamy Gérard, *Montpellier*
Lascombe Claude, *Pierre-Bénite*
Laval Philippe, *Villefranche-sur-Mer*
Lavandier Pierre, *Toulouse*
Lebaron Philippe, *Montpellier*
Le Cohu René, *Toulouse*
Lefeuvre Patrick, *Paris*
Legrand Catherine, *Toulouse*
Legrand Claude, *Margency*
Leitao Maria, *Angers*
Le Roux Elisabeth, *Aix-en-Provence*
Leveau Michel, *Marseille*
Lochet Corinne, *Marseille*
Lochet François, *Marseille*
Louis Alain, *Dijon*
Majeed Saad, *Brest*
Mallen Annie, *Marseille*
Martin Claude, *Aix-en-Provence*
Massé Henri, *Marseille*
Masson Jean-Paul, *Besançon*
Mathieu Jean-Louis, *Viroflay*
Mazoyer-Mayère Catherine, *Montpellier*
Meybeck Michel, *Paris*
Minas Hans Joachim, *Marseille*
Minas Monique, *Marseille*
Miramand Pierre, *Cherbourg*
Monaco André, *Perpignan*
Montuelle Bernard, *Lyon*
Neveux Jacques, *Banyuls-sur-Mer*
Nisbet Maud, *Saint-Mandé*
Nourisson Michel, *Metz*
Ouddane Baghdad, *Villeneuve d'Ascq*
Pagano Marc, *Marseille*
Pagès Jean, *Montpellier*
Pattee Eric, *Villeurbanne*
Pavillon Jean-François, *Paris*
Pelletier Jean, *Thonon-les-Bains*
Pena Geneviève, *Montpellier*
Pereira Alice, *Noisy-le-Grand*
Pérès Jean-Marie, *Marseille*
Picot Bernadette, *Montpellier*
Pihan Jean-Claude, *Metz*
Piquemal Françoise, *Toulouse*
Plante Raphaël, *Marseille*
Plante-Cuny Marie-Reine, *Marseille*
Pont Didier, *Arles*
Poulin Michel, *Fontainebleau*
Pourriot Roger, *Paris*
Raimbault Patrick, *Marseille*
Rassoulzadegan Fereidoun, *Villefranche-sur-Mer*
Rebecq Jacques, *Chambéry*
Restituito Françoise, *Le Cendre*
Reyes-Marchant Patricia, *Aubière*
Richou Mireille, *La Garde*
Rivas Joseph, *Pierre-Bénite*
Robert Jean-Michel, *Nantes*
Robert René, *Arcachon*
Robert Serge, *Nieul-sur-Mer*
Rofes Germinal, *Lyon*
Roméo Michèle, *Villefranche-sur-Mer*
Roqueplo Charles, *Cestas*
Saint-Jean Lucien, *Montpellier*
Salen-Picard Chantal, *Marseille*
Saliot Alain, *Paris*
Sauvant-Magnet Philippe, *Meudon-la-Forêt*
Schmidt Sabine, *Gif-sur-Yvette*
Sissakian Chantal, *Marseille*
Slawyk Gerd, *Marseille*
Soriano-Sierra Eduardo, *Talence*
Sornin Jean-Marc, *La Rochelle*
Soto-Mercado Yenise, *Marseille*
Strehmel Gunnar, *Paris*
Szeligiewicz Wojciech, *Fontainebleau*
Taleb Hassan, *Aubière*
Taupier-Letage Isabelle, *La Seyne-sur-Mer*
Téxier Hervé, *Caen*
Thébault Jean-Marc, *Toulouse*
Thimel Anne, *Arcachon*
Travers Marc, *Marseille*
Trousselier Marc, *Montpellier*
Vacelet Eveline, *Marseille*
Valitchek Jean-Claude, *Nanterre*
Vaquer André, *Marseille*
Vaugelas Jean de, *Nice*
Veïn Denis, *Vandoeuvre-les-Nancy*
Videau Christiane, *Paris*
Vidy Guy, *Montpellier*
Vincent Michel, *Marseille*
Vinçon-Leite Brigitte, *Noisy-le-Grand*

Studio Janguy, Marseille

List of sponsors

United Nations Environment Programme (Coordinating Unit for the Mediterranean Action Plan), Athens

Ministère des Affaires Etrangères (DGRCST)
Ministère de la Recherche et de la Technologie (DIST)
Ministère de l'Education Nationale (DAGIC)
Ministère de l'Environnement (Agence de Bassin Rhône-Méditerranée-Corse)
Centre National de la Recherche Scientifique (CNRS-TOAE, INSU, PIREN)
Institut Français de Recherche pour l'Exploitation de la Mer (IFREMER)
Institut Français de Recherche pour le Développement en Coopération (ORSTOM)
Commissariat à l'Energie Atomique (CEA)

Université de Provence (Aix-Marseille I)
Université Aix-Marseille II
Faculté des Sciences de Marseille-Luminy
Centre d'Océanologie de Marseille
Faculté des Sciences et Techniques de Saint-Jérôme (Aix-Marseille III)
Conseil Régional Provence-Alpes-Côte d'Azur
Conseil Général des Bouches-du-Rhône
Conseil Municipal de Marseille

Banque Société Générale (Marseille-Endoume)
Coultronics France
Electricité de France
Europa Scientific
Fondation Océanographique Ricard
Gauthier-Villars (Editions)
Graftek
Hydrodata
Imprimerie Louis-Jean
Isotec
Kluwer Academic Publishers Group (Marketing Department)
Librairie Maupetit
Milton Roy
Oceano Instruments
Ponselle
Sadis-Bruker-Spectrospin
Société des Eaux de Marseille
Wild-Leitz

Hydrobiologia **207**: xiii–xiv, 1990.
D. J. Bonin & H. L. Golterman (eds), Fluxes Between Trophic Levels and Through the Water–Sediment Interface.

List of reviewers

AUSTRALIA
Johannes Robert, *Hobart*
Stephenson W., *Brisbane*
Volkman John, *Hobart*

BELGIUM
Billen Gilles, *Bruxelles*
Lancelot Christiane, *Bruxelles*

CANADA
Bourget Edwin, *Ste Foy*, Québec
Brylinski Michael, *Wolfville*, Nova Scotia
Demers Serge, *Mont-Joli*, Québec
Fréchette Marcel, *Mont-Joli*, Québec
Hynes H. B. N., *Waterloo*, Ontario
Kalff Jacob, *Montreal*, Québec
Kerekes John, *Halifax*, Nova Scotia
Morrison Kenneth A., *Sherbroke*, Québec
Novitsky James, *Halifax*, Nova Scotia
Painchaud Jean, *Mont-Joli*, Québec
Platt Trevor, *Dartmouth*, Nova Scotia
Silverberg Norman, *Rimouski*, Québec
Sundby Bjorn, *Mont-Joli*, Québec

DENMARK
Blackburn T. Henry, *Arrhus*
Jørgensen Bo Baker, *Arrhus*
Kiørboe Thomas, *Charlottenlund*
Nielsen Lars Peter, *Arrhus*
Revsbech Niels Peter, *Arrhus*

FINLAND
Sarvala Jouko, *Turku*

FRANCE
Amanieu Michel, *Montpellier*
Amiard Jean-Claude, *Nantes*
Bellan Gérard, *Marseille*
Benaim Jean, *Toulon*
Boët Philippe, *Paris*
Boucher Guy, *Paris*
Cabioch Louis, *Roscoff*
Champeau Antoine, *Marseille*
Dauta Alain, *Toulouse*
Devaux Jean, *Aubière*
Dufour Philippe, *Thonon-les-Bains*
Elkaïm Bernard, *Paris*
Gerdeaux Daniel, *Thonon-les-Bains*
Herbland Alain, *Nieul-sur-Mer*
Hily Christian, *Brest*
Jacques Guy, *Banuyls-sur-Mer*
Mariotti André, *Paris*
Meybeck Michel, *Paris*
Monaco André, *Perpignan*
Nival Paul, *Villefranche-sur-Mer*
Pattee Eric, *Villeurbanne*
Plante Raphaël, *Marseille*
Pourriot Roger, *Paris*
Relexans Jean-Claude, *Bordeaux*
Saliot Alain, *Paris*
Sarazin Gérard, *Paris*
Sournia Alain, *Paris*

GERMANY
Geller Walter, *Konstanz*
Hoppe Hans Georg, *Kiel*
Liebezeit Gerd, *Wilhelmshaven*
Overbeck Jurgen, *Plön*
Smetacek Victor, *Bremerhaven*
Stabel Hans Henning, *Überlingen-Süssenmühle*
Tilzer Max, *Konstanz*

ISRAEL
Berman Thomas, *Tiberias*
Gophen Moshe, *Tiberias*

MONACO
Fowler Scott, *Monaco*

NORWAY
Sakshaug Egil, *Trondheim*

SPAIN
Margalef Ramon, *Barcelona*
Vives-Rego Josep, *Barcelona*

SWEDEN
Boström Bengt, *Uppsala*
Forsberg Curt, *Uppsala*
Kautsky Nils, *Stockholm*
Tranvik Lars, *Lund*

SWITZERLAND
Bossard Peter, *Kastanienbaum*
Imboden Dieter M., *Dübendorf*

THE NETHERLANDS
Duursma Egbert K., *Texel*
Gulati Ramesh D., *Nieuwersluis*
Helder Wim, *Texel*
de Jonge V. N., *Haren*
Robertson Lesley, *Delft*
Van Bennekom A. J., *Texel*
Vijverberg Jacobus, *Oosterzee*
Zehnder Alexander, *Wageningen*

UNITED KINGDOM
Cranwell Peter, *Ambleside*
Droop Michael, *Argyll*
Hilton John, *Ambleside*
Maker A. F., *Huntingdon*
Massie Lavina C., *Aberdeen*
Pearson Tom, *Argyll*
Williams Peter J. le B., *Menai Bridge*

USA
Azam Farooq, *La Jolla*, CA
Baillie Priscilla W., *Stoors*, CT
Bianchi Thomas S., *Millbrook*, NY
Bienfang Paul, *Honolulu*, HI
Boyer Greg, *Syracuse*, NY
Capone Douglas, *Solomons*, MD
Cloern James E., *Menlo Park*, CA
Dame Richard F., *Conway*, SC
Devol Allan, *Seattle*, WA
Dortch Quay, *Chauvin*, LA
Ducklow Hugh, *Cambridge*, MD
Emery K. O., *Woods Hole*, MA
Eppley Richard W., *La Jolla*, CA
Falkowski Paul G., *Upton*, NY
Fisher Joel, *Fairfax*, VA
Francko David A., *Stillwater*, OK
Fuhrman Jed, *Los Angeles*, CA
Gardner L. R., *Columbia*, SC
Garside Chris, *West Boothbay Harbor*, ME
Glover Hilary, *West Boothbay Harbor*, ME
Goldberg Edward, *La Jolla*, CA
Goldman Charles R., *Davis*, CA
Guckert James L., *Knoxville*, TN
Guillard Robert R. L., *West Boothbay Harbor*, ME
Hopkinson Charles S., *Woods Hole*, MA
Iturriaga Rodolfo, *Los Angeles*, CA
Jassby Alan D., *Davis*, CA
Karl David, *Honolulu*, HI
Kelly John R., *Ithaca*, NY
Landry Michael R., *Honolulu*, HI
Larson Douglas, *Portland*, OR
Leonard Alexander B., *La Jolla*, CA
Malone Thomas, *Cambridge*, MD
Nowicki Barbara, *Narragansett*, RI
Packard Theodore T., *West Boothbay Harbor*, ME
Pomeroy Lawrence R., *Athens*, GA
Poston Ted M., *Richland*, WA
Rhee G. Yull, *Albany*, NY
Rice Donald, *Solomons*, MD
Scavia Donald, *Washington* DC
Sieburth John McNeill, *Narragansett*, RI
Spies Robert, *Livermore*, CA
Tenore Kenneth, *Solomons*, MD
Urban Richard, *Chattanooga*, TN

Contents

PART TWO: WATER–SEDIMENT INTERACTIONS

PART THREE: ROLE OF ORGANISMS

PART FOUR: STUDIES ON METHODS

Hydrobiologia **207**: 1–14, 1990.
D. J. Bonin & H. L. Golterman (eds), Fluxes Between Trophic Levels and Through the Water–Sediment Interface.

Nutrient dynamics and pelagic food web interactions in oligotrophic and eutrophic environments: an overview

Jacques Capblancq
Laboratoire d'Hydrobiologie – Université Toulouse III – 118 Route de Narbonne, 31062 Toulouse Cédex, France

Key words: trophic state, nutrient cycling, phytoplankton dynamics, bacteria, zooplankton, fresh and marine waters

Abstract

The concept of limiting nutrients is a cornerstone of theories concerning the control of production, structure and dynamics of freshwater and marine plankton. The current dogma is that nitrogen is limiting in most marine environments while freshwater ecosystems are mostly phosphorus-limited, although evidence of phytoplankton limitation by either N or P has been found in both environments.

However, the same considerations apply to the availability of phosphorus in freshwater as to nitrogen in oceans. In resource-limited environments the plankton dynamics depend mostly on the internal mechanisms which act to recycle the limiting nutrient many times over within the surface waters. As the overall productivity increases, this dependence on nutrient regeneration decreases.

The relationship between the stock of limiting nutrient, rates of supply and plankton dynamics must therefore be seen in the light of the processes operating within the entire food chain over quite different time scales. There is strong evidence that process-rates are mostly size-dependent and that food web interactions at the microbial level (picophytoplankton, bacteria, microheterotrophs) strongly affect the production of carbon and the regeneration of nutrients in the pelagic zone.

Introduction

The current view in Limnology and Oceanography is that the maximum attainable production at each trophic level in pelagic communities is largely determined by the least available nutrients. Over the last twenty years, this view has gained added acceptance in Limnology through studies of eutrophication of lakes. The crucial role of phosphorus as a controlling factor has been formalized into empirical relationships between external phosphorus supply and lake trophic status. Regression models of chlorophyll concentrations on the total P concentration or P loading as independent variables, of mean zooplankton biomass on phytoplankton biomass, of fish yield on zooplankton biomass or total P (see McQueen *et al.*, 1986; Peters, 1986; Mills & Forney, 1988 for references) and whole lake fertilization experiments (Schindler, 1988) have provided effective evidence of the efficiency of phosphorus control. Nitrogen limitation is considered to have a secondary effect on net production because deficiency may be offset by the growth of heterocystous nitrogen-fixing blue-green algae. However temporary nitrogen limitation of phytoplankton has been recognized in temperate lakes (Smith, 1982; Morris & Lewis, 1988).

In marine waters, the limiting role is generally attributed to nitrogen, although a rigourous demonstration of the extent of N limitation has not been achieved in the marine environment (Smith, 1984; Hecky & Kilham, 1988). The critical role of N is suggested by the fact that, in oceans, the flux from river inputs and from aerobic nitrogen fixing activity is only sufficient to satisfy about one percent of the total photosynthetic flux which is, therefore, very much dependent on recycling *in situ*. Because nitrogen must undergo enzyme-mediated changes in oxidation states during this cycle, one can expect that the velocity of nitrogen cycling may be less than that for phosphorus. Nevertheless, the elemental composition of particulate oceanic material suggests the possible limiting effects of phosphorus in marine environments.

In both marine and fresh waters, eutrophication could lead to Si depletion and the elimination of Diatoms (Officer & Ryther, 1980; Schelske *et al.*, 1986).

The dynamic nature of nutrient cycling in the euphotic zone of marine and freshwaters confounds the design of experiments intended to assess the degree to which nutrient availability can regulate plankton production and determine community structure. A further complication arises from the fact that processes operate over different time scales. On the annual scale, changes in the vertical stability of the water column account for the spring bloom and the following period of phytoplankton growth. At more frequent intervals (a few days) periodic enrichments of surface waters by nutrients may occur during the stratification period due to partial turbulent mixing generated by wind stress at the surface. In the short term, nutrient availability depends on internal processes in the food web which act to recycle the nutrients many times over in surface waters. This recycling takes on added significance when nutrient concentrations and rates of external supply are low. Through feeding and excretion, zooplankton can contribute directly to the cycling of nutrients and to the control of community structure and phytoplankton biomass. Zooplankton species composition and biomass can in turn be controlled from below by the food available and from above by predators. The relative importance of these 'bottom-up' and 'top-down' controls in food web structure and function still remains controversial.

General trends of nutrient-plankton dynamics in fresh and marine waters

Several books and reviews have appeared over the last ten years (Falkowski, 1980; Morris, 1980; Platt, 1981; Pourriot *et al.*, 1982; Reynolds, 1984; Harris, 1986) in which there appear to be a number of similarities between the characteristics of pelagic ecosystems in the oceans and freshwaters which show systematic variations with trophic state.

The pool of dissolved inorganic phosphorus (DIP) in surface waters of oligotrophic and mesotrophic lakes during the summer period is a small percentage of the total P (TP), whereas in more eutrophic waters the DIP pool becomes the dominant fraction of TP. In the same way, dissolved inorganic nitrogen (DIN) in marine waters accounts for a larger proportion of the total N (TN) in more fertile upwelling areas and coastal waters. The near-surface concentrations of DIN in central oceans during summer and the DIP concentrations in the euphotic zone of stratified oligotrophic and mesotrophic lakes are practically undetectable. In both environments, this apparent lack of inorganic nutrients has encouraged research into the role of dissolved organic forms of nitrogen (urea, amino-acids) and phosphorus in the nutrition of phytoplankton.

Rates of turnover of the DIN and DIP pools increase as the pools become smaller, indicating faster internal recycling in oligotrophic waters (Fig. 1). In spite of methodological problems in applying isotope tracer studies as ambient concentrations approach the limits of analytical detection, kinetics of ^{15}N and ^{32}P indicate high turnover rates. In summer the turnover times of the

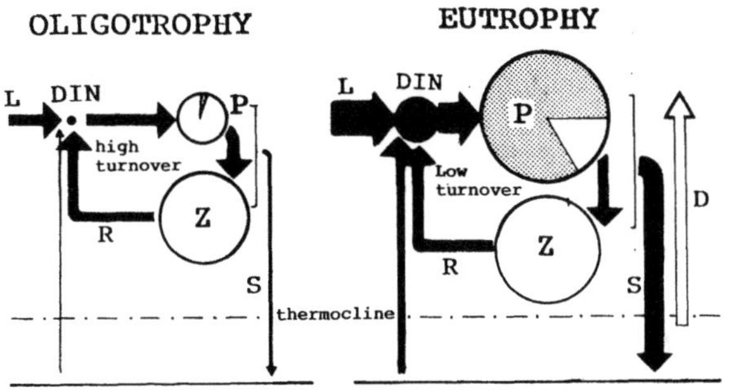

Fig. 1. Comparison between the major pathways of the nitrogen and phosphorus cycle in oligotrophic and eutrophic waters. Increasing nutrient load (L) may increase the biomass of microphytoplankton (P, shaded area) and the ratio of dissolved inorganic nutrients (DIN) to the amount stored in the biomass. In eutrophic waters, the percentage of nutrient regenerated (R) by zooplankton (Z) within surface waters decreases and a relatively greater proportion is temporarily or permanently removed from this cycle by sedimentation (S) and denitrification (D)

DIP pool in epilimnetic water of oligotrophic lakes reduces to minutes. Reported values for NH_4^-N turnover times in marine waters range around 1 day when ambient NH_4^+ concentration is relatively high (1–5 μg-at l^{-1}), and reduce to < 1 hour when ambient NH_4^+ borders on the limit of detection (Glibert *et al.*, 1982). A comparison of the turnover times of DIP in lakes of various trophic states (White *et al.*, 1982) shows that turnover time increases with the size of DIP pool and decreases when algal biomass becomes higher. Examination of data from a number of lakes suggests that there is a systematic change in the ratio of N and P turnover rates in passing from oligotrophy to eutrophy. Figures reported in Harris (1986) indicate that in oligotrophic lakes P is cycled at least twenty times faster than N or C while in eutrophic lakes N, P and C are taken up at the same rate.

The variations of the turnover times of DIN and DIP pools suggest that there is also a relationship between the proportions of 'new' and 'regenerated' production and the trophic state. In the oceans, Harrison (1980) estimates that 20% of the total nitrogen uptake is supplied by NO_3^- originating from deep water (= new production) and 80% by NH_4^+ and urea resulting from *in situ* recycling (= regenerated production). According to different methods of evaluation, the ratio of total production which is new production (the so-called f number) ranges between 0.05 and 0.3. In coastal waters, the f ratio increases asymptotically with an increase in nitrate concentration (Harrison *et al.*, 1987).

As with DIN, the proportion of recycled phosphorus within surface lake waters must be increased in order to supply the needs of algal growth when the pool size of DIP in euphotic layers falls to unmeasurable values. Comparisons of observed rates of change of phytoplankton biomass in the euphotic zone with potential biomass increments, estimated from ^{14}C incubations (Jassby & Goldman, 1976; Tilzer, 1984) and from the frequency of dividing cells (Pollingher & Serruya, 1976; Ramberg, 1980), indicate that rates of production and elimination of phytoplankton carbon in the trophogenic layer are within the same range. In large lakes, losses via sedimentation have been estimated to be about 15%–20% of the newly formed organic carbon and the settling flux of phosphorus is shown to be correlated to the sedimentation rates of particulate organic matter. This means that, as a mean value, 80% of the organic matter formed by the primary production processes is recycled within the euphotic zone (Tilzer, 1984; Stabel, 1985). By examining the relative disappearance of dissolved inorganic C, N and P in the surface waters of lakes during the stratified period, and by measuring the elemental composition of seston, it appears that the rates of removal of C, N and P are close to Redfield ratios in eutrophic waters and that P cycling becomes more and more rapid as the total P content of the water is reduced (Harris, 1986). This is consistent with ^{32}P experiments indicating that turnover rates of DIP depend on the trophic state and time of the year. All these data suggest that, both in lakes and marine environments, there is a shift from new production to regenerated production in passing from eutrophic to oligotrophic waters and from periods of vertical mixing to summer stratification.

Numerous data support the evidence that there are systematic shifts in cell-size distribution of phytoplankton and it is clear that the smaller cells dominate biomass and production in oligotrophic marine and fresh waters. It appears that nanoplankton biomass is less variant that netplankton biomass in most marine waters (Malone, 1980a). Similarly, comparisons of size spectra of plankton communities in lakes of different trophic status (Sprules & Munawar, 1986; Sager *et al.*, 1988) suggest that nanoplankton abundance is similar in all communities. A similar trend was also noted by Stockner (1988) in his review of algal picoplankton data: estimates of algal picoplankton biomass range from 0.05 to 1 mg chl m^{-3} in both fresh and marine waters. Absolute rates of production in the $<20\ \mu m$ size class seem to show little variation along the trophic gradient, but the percentage contribution to total production is considerably higher in oligotrophic environments, where this fraction is responsible for at least 90% of the total ^{14}C, ^{15}N or ^{32}P taken up. Algal picoplankton represent as much as 90% of total carbon production in more oligotrophic regions of the world ocean and 70% in the summer epilimnion of oligotrophic lakes.

Interactions between algal growth, nutrient uptake and regeneration

Due to the very rapid turnover of soluble inorganic nutrient pools in most surface waters, there is often a confusion between the concepts of capacity and intensity (Talling, 1984), i.e. as to whether the nutrients limit the phytoplankton biomass or the rate of growth of individual populations making this biomass. There may be rapid growth associated with a low and stable biomass of phytoplankton, if rapid uptake and growth are balanced by an equally rapid grazing and regeneration of nutrients (Harris, 1986). A recent demonstration was given by Sterner's (1986) experiments: in a natural community with a co-occurring crustacean herbivore (*Daphnia*), the increase in algal reproductive rates caused by nitrogen regeneration from herbivorous *Daphnia* approximately equalled the mortality caused by grazing. This rapid cycling of elemental nutrients between zooplankton and phytoplankton complicated attempts to quantify the internal fluxes. In order to overcome these difficulties, many models on algal growth have been based on cultures.

Nutrition and growth of planktonic algae in cultures and in nature

Algal cultures as models

The nutrition of individual species of marine and freshwater algae and the relationships between the kinetics of nutrient uptake and the growth rate of populations have been the subject of numerous investigations over the last two decades. Such kinetic information on the relationships between nutrient concentration, rates of uptake, internal nutrient storage and growth rate provides the physiological basis for detecting limiting nutrients in waters (Healey & Hendzel, 1979; Maestrini *et al.*, 1984) and for mechanistic models on nutrient limitation (Droop, 1973) and resource competition theory. Several comprehensive reviews have been published (Morris, 1980; Platt, 1981; Bonin, 1982; Tilman *et al.*, 1982; Goldman & Glibert, 1983; Cembella *et al.*, 1984; Ahlgren, 1988; Kilham & Hecky, 1988), and so, rather than enter into a detailed discussion, I will simply state the ecologically significant properties of the phytoplankton (Fig. 2).

As a generalization, the nutrient uptake rate (V) as a function of the nutrient concentration (S) in the culture media follows the rectangular hyperbolic relationship predicted from the Michaelis-Menten enzyme kinetic model: $V = V_{max} \cdot S/(Ks + S)$. The values of the half-saturation coefficient (Ks) for phosphorus seem very similar for different algal classes and are generally $<4\ \mu M$. The Ks values for natural assemblages of phytoplankton generally lie within the lower range of values obtained from laboratory cultures (Cembella *et al.*, 1984). The high affinity of picoplankton to phosphorus could explain this difference. Considering all nitrogen sources, Ks generally ranges between 0,1 and 5 μM.

Algal cells are capable of 'luxury uptake':

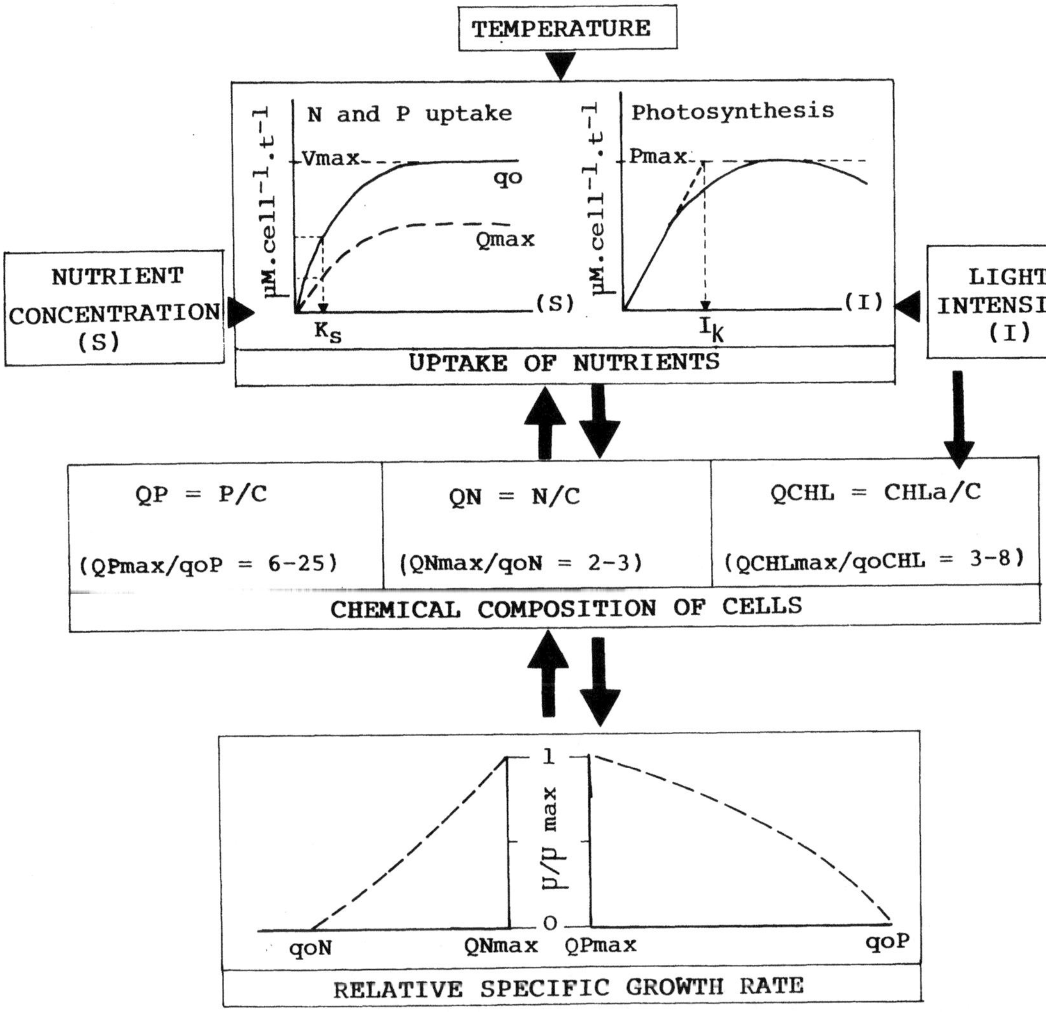

Fig. 2. Schematic representation of the relationships between environmental factors, rates of nutrient uptake, elemental composition of the cells and relative specific growth rates for steady-state populations of algae in cultures. Rates of nutrient uptake (V_{max}), photosynthetic rates (P_{max}), half-saturation constants for nutrient uptake (Ks), cellular subsistence quotas (q_oN, q_oP), chlorophyll content (Q_{CHL}) and potential growth rate (μ_{max}) are correlated with cell size, expressed as volume, or carbon.

nutrients being accumulated in excess of that required for immediate growth. From several studies, it appears that the internal cell quota (Q) of individual species is much more variable for P than for N and it may be quite invariant for C. Diatoms can apparently store an insignificant (<10%) amount of Si beyond that used in cell wall formation (Paasche, 1980), but in some marine species much higher ratios of internal pools to total silicon have been reported.

The uptake capacity (V_{max}) is generally increased by nutrient starvation and negatively correlated with the internal nutrient content of the cell (Q). In this respect, silicon differs from N or P. This reflects the fact that, except at very low external concentrations of Si, uptake is determined by the rate of wall formation and hence by the rate of cell division.

Under constant conditions, the relative growth rates of individual species exhibit a linear relationship with the reciprocal of the cellular content (Q) of the limiting nutrient: $\mu/\mu_{max} = 1 - Q_o/Q$, where Q_o = minimum cell quota at which $\mu = 0$. Because cellular C is quite invariant, the relative growth rate is related to C/N cell content under N limitation and C/P under P limitation. In chemostat experiments with *Scenedesmus*, using a fixed dilution rate and different N/P ratios in the medium, Rhee (1978) demonstrated a very abrupt transition from phosphorus to nitrogen limitation at N:P atomic ratio of about 30. The optimum cellular N:P ratio at which this transition occurs was also found to correspond to the ratio of the minimum cell quotas, Q_o(N):Q_o(P) (Rhee & Gotham, 1980). Most values are, weight by weight, around 10, but interspecific variations from 4 to 40 have been reported (Smith, 1982). When growth rates of algae approach the maximum and hence are not nutrient limited, the elemental composition of phytoplankton is close to the Redfield proportions (Goldman, 1980; Healy & Hendzel, 1979).

The Chlorophyll/Carbon ratio declines with N deficiency and somewhat less with P deficiency. However, the Chlorophyll/Carbon ratio also varies with cell size and light conditions.

Interspecific variations of minimum cell quota (Q_o), saturated uptake rates (V_{max}) and maximum growth rates (μ_{max}) are related to cell size by an allometric equation, showing that smaller cells have both higher nutrient uptake and growth rates than larger ones (Friebele *et al.*, 1978; Shuter, 1978; Malone, 1980b; Smith & Kalff, 1982, 1983). The ratio between the finite maximum level of Q (Q_{max}) and Q_o ranges between 6 and >25 for P, 2 and 3 for N and <2 for Si (Dauta, 1983). Whether this ratio (which corresponds to the number of daughter cells that can arise from the internal store) is size-dependent or not is still unclear.

Algal growth in natural environments

The old question, whether the characteristics of algal growth defined in the idealized steady-state conditions of chemostat cultures apply to natural populations, is still the subject of debate. The difficulty of extrapolating results from simple systems in the laboratory to phytoplankton in nature arises from the fact that most natural planktonic systems are not steady state systems (Harris, 1986). Nutrients in the euphotic zone of lakes and oceans are subject to temporal and spatial variations both in ambient concentrations and rates of supply. In an environment in which nutrients are made available in bursts, at a scale of minutes to a few days, by recycling from grazers and inputs into the euphotic layers by episodic wind-induced turbulence (e.g. Klein & Coste, 1984), the steady-state analogy cannot be applied.

In such variable conditions, storage of nutrients within the algal cells could result in a fairly uniform growth rate over time, depending on the frequency of nutrient encounter, on the cellular storage capacity (Q_{max}/Q_o) for the nutrient and on the light conditions which control growth rate.

Under these transient conditions, short-term measurements of the flux of nutrients into cells may overestimate the actual rate of growth if what is being measured is the temporary replenishment of storage pools (Eppley, 1981). The uncoupling between uptake and growth is the main limit in the interpretation of kinetics measurements with ^{14}C, ^{15}N, or ^{32}P.

The importance of these processes has been sharply focused by the controversy about primary production rates in oligotrophic oceans (Herbland, 1988) and by the fact that phytoplankton appear to grow at near maximal rates despite the apparent absence of a soluble pool of limiting nutrient (Goldman, 1980; Smith & Kalff, 1981). Goldman based his conclusion of high relative growth rates on the observation that atomic C:N:P ratios in oceanic seston were often near the Redfield ratio exhibited by cell cultures growing to their maximum growth rate. Using the same basic approach in combination with bioassays to assess phytoplankton growth rates in lake water in which DIP, DIN and Si are exhausted to very low levels during summer stratification, Sommer (1988) concluded that P and N limitations are weak at the level of community average. However, for short time periods some individual species may be severely N or P limited.

If algae experience non-uniform nutrient supply

in surface waters, then the long standing theory on interspecific competition for limiting nutrients as an important process in determining the phytoplankton species shift during the course of seasonal succession seems questionable. Cultures of natural phytoplankton subjected to steady-state nutrient supply rates exhibit major differences in species composition compared to those receiving the same nutrients in pulsed supplies (Turpin & Harrison, 1979; Sommer, 1985). Moreover, the frequency of pulsed additions of the limiting nutrient seems to modify the species composition of the community. Small phytoplankton species appear to be favoured by frequent additions of small pulses of limiting nutrient, while larger species seem to be favoured when the same amount of nutrient is added in the form of larger and less frequent pulses. This raises the question of the extent of temporal fluctuations in the supply of limiting nutrients in the field. It seems likely that, in waters where the background nutrient concentration is low and phytoplankton are dependent on regeneration processes, diel periodicity of grazing and vertical migration of zooplankton can result in diurnal variations of regenerated nutrient flux. Coupled with light fluctuations and with day-to-day changes in the turbulence structure of the water column, this fluctuating nutrient regime will affect different species in different ways, depending on the relationships between growth rates, uptake and storage capacity and the frequency of disturbance. This leads an increasing number of phytoplankton ecologists (Margalef, 1978; Reynolds, 1984; Harris, 1986) to believe that competitive interaction rarely occurs in frequently disturbed surface waters.

The only groups of phytoplankton for which density-dependent effects can be observed are those, such as blue-greens or dinoflagellates, which grow in a temporally stable environment and exhibit low loss rates. These algae have great bloom potential because cropping by grazers and sinking are minimized; therefore, nutrients can be sequestred in algal biomass for a long enough time to achieve competitive exclusion.

Grazing and nutrient regeneration

Since Gardiner's (1937) indication that nutrients were rapidly released by enclosed zooplankton, several studies have amply demonstrated that nutrient regeneration by grazers might represent a source of inorganic nutrients for phytoplankton. This source is of particular importance in oligotrophic environments.

Lehman (1980a) has estimated that in lake Washington ten times more DIP and three times more DIN are supplied by zooplankton excretion than enter from all other sources during summer stratification. Similarly, studies in marine environments reviewed by Harrison (1980) lead us to estimate that macro- and micro-zooplankton excretion can account for 90% of phytoplankton N requirements and over 100% of the P requirements in surface waters of oligotrophic open oceans.

In general, workers who have examined the release products of single species conclude that NH_3-N and PO_4-P are the dominant forms of excreted nutrients. It seems that, more than any organic P release by zooplankton, the rapid cycling, which results from reabsorption of the excreted PO_4-P by phytoplankton and bacteria with the subsequent reexcretion of some low molecular weight P compounds by the microorganisms (Lean & Nalewajko, 1976), is the process responsible for net accumulation of organic P.

A mean equivalent of 1.5% to 2.5% of the *Daphnia pulex* P content is recycled every hour in *Daphnia pulex* (Lehman, 1980b), meaning that 35–60% of the total P content of the animal turns over each day. This turn-over probably affects the freshly ingested material, of which only a fraction becomes part of the body structure. The rate of N-NH_4 and PO_4-P excretion per unit biomass (Fig. 3) has been demonstrated to be negatively correlated with body size (Ejsmont-Karabin, 1984). This means that Rotifers and Protozoa may be responsible for daily release rates equivalent to more than 100% of their own P content. Some results (e.g. Pace, 1986) suggest that, on average, about 40% of the total zooplankton

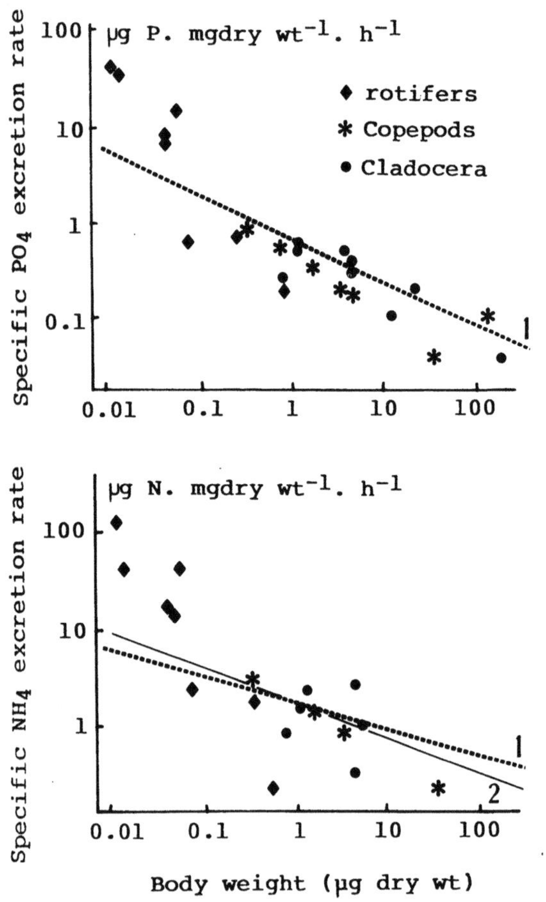

Fig. 3. Relationships between individual body weight of organisms and excretion rates of PO_4 and NH_4. Regression lines calculated by Ejsmont-Karabin 1984 (1) and Harrison 1980 (2) for marine zooplankton, and specific values reported by Ejsmont-Karabin 1984 and Lehman 1980 a & b for particular freshwater species (symbols).

biomass in lakes is accounted for by microzooplankton (Ciliates, Rotifers and Nauplii), so the impact of these small organisms on P cycling is obviously considerable. Experiments performed in lake water enclosures by Mazumder *et al.* (1988) suggest that the turnover of DIP is much shorter when added planktivorous fish reduce the abundance of larger herbivores and their impact on microheterotrophs. In the same way, ingestion of bacteria by heterotrophic nanoflagellates and ciliates strongly increased the remineralization of P from dead algal cells in experiments performed by Bloem *et al.* (1989).

Through grazing and nutrient excretion, zooplankton influence phytoplankton in opposite ways. Size-specific grazing on phytoplankton mainly affects the small algae which have the greatest rates of growth and of nutrient uptake. Concurrently, grazing produces a fertilizing effect due to release of nutrients. Because not all the species of phytoplankton are grazed uniformly, and because not all these species take up nutrients and grow at the same rate (since these rates depends on size and nutritional states of the cells), it is likely that these opposite effects of zooplankton activity act on the shaping of the natural phytoplankton community in quite a complicated way.

Microbial interactions and nutrient cycling

In the original Lindeman trophic hierarchy, bacteria were not assigned a trophic level; they were simply decomposers making inorganic nutrients available for the primary producers through oxidation of organic detritus. Recent work has shown that bacterial production rates can rival the phytoplankton production rates, and that bacteria can be successful competitors with algae for mineral nutrient resources. Both in fresh and marine environments, it has appeared during the last decade that substantial carbon and nutrients flow through the very small organisms ($< 10\ \mu m$) of the so-called 'microbial loop' (Azam *et al.*, 1983). Bacterial, algal and protozoan components of this loop have, in relation with their small size, the potential of extremely rapid growth and turnover rates. Stockner & Antia (1986), Stockner (1988) and Porter *et al.* (1988) have reviewed our present knowledge on the structure and functioning of this microbial assemblage.

Bacteria-phytoplankton competition

The kinetics observed when $^{32}P\text{-}PO_4$ is added to lake water shows that the smaller components of the seston ($< 10\ \mu m$) both take up phosphate and exchange it more rapidly with ambient PO_4-P than microplankton do (Rigler, 1973; Nalewajko & Lean, 1980; Levine *et al.*, 1986). In marine waters, there is also a link between size distribution of plankton and the source of nitrogen used. Recent studies (Malone, 1980b; Wheeler & Kirchman, 1986; Harrison & Wood, 1988) have provided evidence that nanoplankton production depends on 'regenerated' nitrogen i.e. ammonium. Recent emphasis on the picoplankton has revealed that a substantial fraction of the NH_4 uptake may be due to microheterotrophs (Laws *et al.*, 1985; Wheeler & Kirchman, 1986; Harrison & Wood, 1988).

Several papers also suggest that bacteria may not only use inorganic P but may in fact outcompete phytoplankton at ambient PO_4 levels (Currie & Kalff, 1984; Currie *et al.*, 1986; Vadstein *et al.*, 1988). Results reported show that particles smaller than 3 μm were responsible for virtually all the uptake of P-PO_4 (Friebele *et al.*, 1978; Berman, 1985). The proportion of DIP uptake by picoplankton to total uptake seems to decrease in lakes that are not P deficient (Currie *et al.*, 1986). However, in a eutrophic Norwegian Lake, Vadstein *et al.* (1988) estimated that the net consumption of inorganic phosphorus by bacteria is four times higher than that of phytoplankton. From the P:C ratio of phytoplankton (2.5 to 55 μg P (mg C)$^{-1}$, median 5.1), of bacteria (34 to 360 μg P (mg C)$^{-1}$, median 89) and measured production rates, Vadstein *et al.* concluded that about 80% of the P taken up was attributable to the bacteria community. Because bacteria can store P in great excess of their needs, the phosphorus uptake will generally exceed the release and bacteria seem to constitute a P sink.

When heterotrophic and phototrophic plankton are cultivated together (Brown *et al.*, 1981; Jansson, 1988), the dynamic behaviour seems to be the result of competition and commensalism. Because bacteria have a higher P affinity than algae, every 'new' phosphate molecule (e.g. ^{32}P-PO_4 or regenerated phosphate) is most probably taken up by bacterial biomass (competition). The subsequent utilization rate of this phosphate for bacterial growth is controlled by the supply of organic carbon excreted by phytoplankton (commensalism). When the P uptake rate exceeds the utilization rate, bacteria excrete PO_4-P. Consequently, phosphate exchange by bacteria, in combination with the possible carbon limitation of bacterial growth, can explain the partition of P between bacteria and algae in P-limited environments.

According to Porter *et al.* (1988) nanoplankton and microplankton activity seems to be less affected by temperature than picoplankton. This leads these authors to assume that the relative importance of carbon and inorganic nutrient sources in the regulation of autotrophic and heterotrophic production is probably confined to the summer epilimnion in lakes. With the onset of stratification, the spring new production (generally diatoms) sinks, leaving the surface waters with reduced inorganic nutrients and elevated DOM. With the increase in temperature, the epilimnion can subsequently be characterized by the dominance of picoplankton, associated with its predators and with accelerated cycling processes. These conditions prevail until the breakdown of stratification induces an input of nutrients and lowers the temperature, increasing new production and reducing microbial activity.

Predation by flagellates, ciliates, microzooplankton and macrozooplankton on the picoplankton provides another feedback between inorganic nutrient and dissolved organic matter (Fig. 4). Although bacteria can mineralize nutrients from dead phytoplankton, nutrients are often immobilized in bacterial biomass and remineralized only by ingestion of bacteria and subsequent excretion of dissolved P (Bloem *et al.*, 1989). However, the relative contribution of these mechanisms in natural mixed populations of trophically different surface waters remains to be quantified.

Links between microbial web and zooplankton

Bacterial biomass and production rates measured in lake and marine waters indicate that the bacteria generation time is about 0.5–1 day in summer. The same values seem to apply to pico-

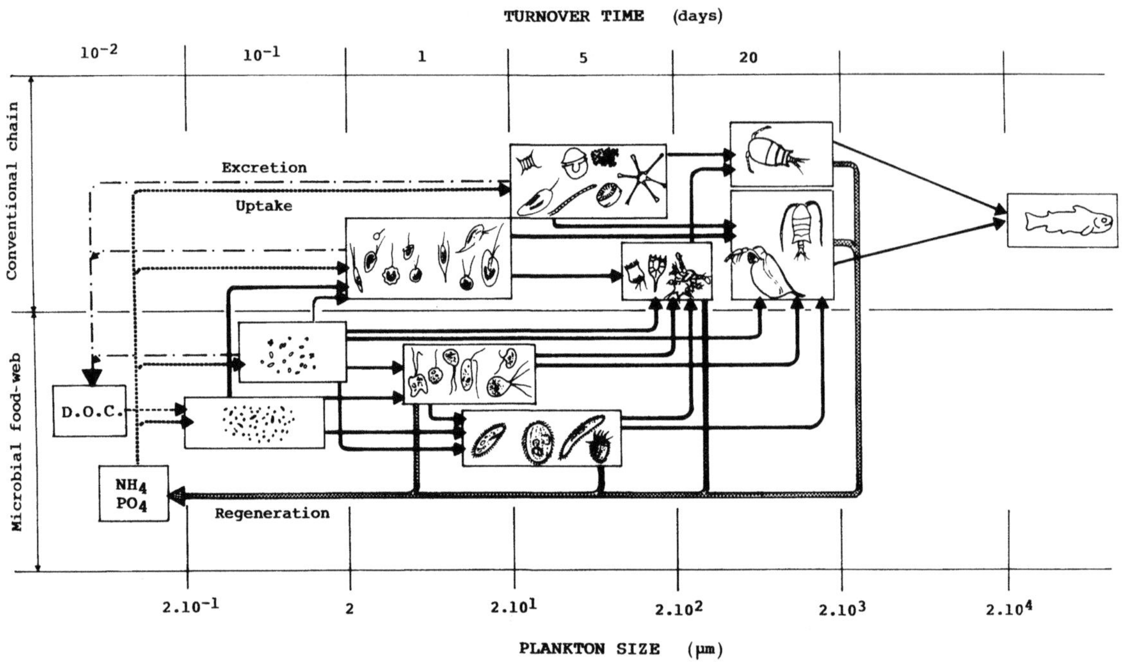

Fig. 4. Schematic representation of trophic interactions in the pelagic lake community on a scale of size of organisms and the related scale of turnover time. The importance of microbial components in the planktonic food-web (bacteria, autotrophic picoplankton, heterotrophic and mixotrophic flagellates, ciliates) is emphasized by their high turnover rates and the variety of trophic linkage with the conventional food chain.

phytoplankton (Stockner, 1988). Despite this high growth potential, abundance of both autotrophic and heterotrophic picoplankton does not show large fluctuation, suggesting a strong regulation by grazing.

The rapid turnover of small cells of the microbial loop can explain the discrepancy observed between short term (2–4 h) and long term (12–24 h) experiments of ^{14}C uptake (Fig. 5). In oligotrophic waters, dominated by picophytoplankton, it is likely that much of the fixed ^{14}C is rapidly cycled by excretion, bacterial uptake, consumption by micrograzers and respiration during relatively long incubations. As a result, the amount of ^{14}C retained on a filter may give apparent rates of uptake 5–15 times lower than short term experiments (Peterson, 1980). In more eutrophic waters with high biomass and larger algae the short and long term incubations give similar estimates of ^{14}C uptake rates.

Microflagellates appear to be the most important grazers of bacterial and algal picoplankton (Stockner & Antia, 1986). Some of these flagellates are obligate heterotrophs, but recent studies from lakes have also shown that several common phototrophic flagellates (*Gymnodinium*, *Cryptomonas*, *Ochromonas*, *Uroglena*, *Dinobryon*) can supplement their nutrition by phagotrophy. Bird & Kalff (1987) estimated that in an oligotrophic lake, the grazing rates of several species of *Dinobryon* are equivalent to those observed for marine heterotrophic microflagellates and have the potential to remove more bacteria from the water column than Ciliates, Rotifers and Crustacean plankton together.

Calanoid Copepods apparently cannot ingest bacteria and freshwater Cladocerans, such as *Daphnia*, appear to be the only large zooplankton that can effectively utilize particles ranging in size from 1 μm to 50 μm. Ciliates and Rotifers can ingest particles with sizes ranging from bacteria to nanoplankton (including some heterotrophic flagellates) providing another link with macrozooplankton.

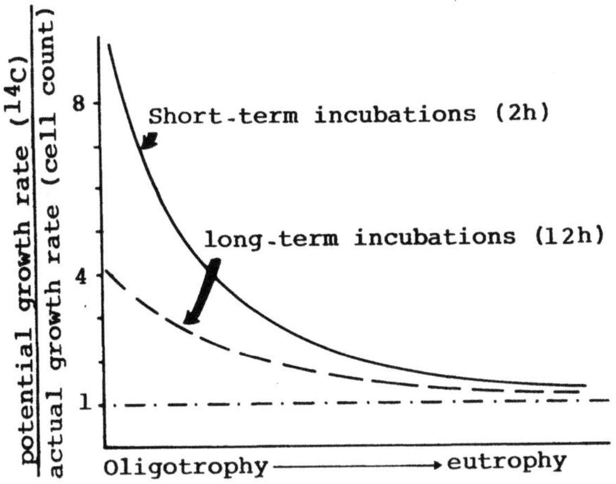

Fig. 5. The ratio of growth rate, calculated from ^{14}C uptake, to the actual growth rate of phytoplankton biomass (determined from cell counts and rates of sedimentation and grazing) as a function of the duration of ^{14}C incubations and the trophic state. High discrepancy between short and long term incubations in oligotrophic waters can arise from consumption of labelled small cells by microheterotrophs and subsequent regeneration of $^{14}CO_2$.

The direct transfer of bacterial production to *Daphnia* was estimated to be about 80–90% in enclosures without fish by Riemann (1985) and 70% in the summer epilimnion of Lake Michigan (Scavia & Fahnenstiel, 1988). These results led to the conclusion that Cladocera are hey organisms in plankton communities and, as opposed to the smaller and more selective Diaptomids, may constitute an efficient link between production at the microbial level and the traditional food chain. In marine environments, Salps and Doliolids possess some of the feeding characteristics of Cladocera and may play the same ecological role (Hobbie, 1988; Lehman, 1988). In lakes with few Cladocera, the bridge between picoplankton and macrozooplankton is provided by heterotrophic flagellates, Ciliates and Rotifers i.e. with additional steps leading to a lowering of the trophic efficiency.

Because the interface between picoplankton, protozoans and macroconsumers is largely controlled by the feeding characteristics of the organisms of the zooplankton, and because the size structure of the zooplankton community can be influenced from above by fish predation, the efficiency of transfer of carbon energy and the cycling of nutrients may equally be influenced by fish manipulation. These 'top-down' effects appear to be well documented (cf. Carpenter, 1988). The regression models of McQueen *et al.* (1986) suggest that top-down effects are not well buffered in oligotrophic lakes, so that zooplankton-phytoplankton interactions would be stronger than in eutrophic lakes. Experimental manipulations suggest, however, that top-down effects can also be very strong in eutrophic systems, but most of the reports come from experiments performed in enclosures or in shallow ponds involving extreme fish manipulations (none vs extremely high fish densities). It is still unclear to what extent the observed 'top-down' effects act in nature.

Conclusion

The recent focus on the role of the microbial food-web in aquatic environments has changed our concepts on the functioning of pelagic ecosystems. In oligotrophic waters, most production is due to organisms <10 μm which convert, both by autotrophy and heterotrophy, soluble organic and inorganic nutrients to particulate biomass.
Important consequences for this microbial loop arise from the ability of bacteria and small algal cells to sequester DIN and DIP at very low concentrations, from active regeneration of nutrients through predation and excretion by flagellates, ciliates, microzooplankton and some Cladocera, and from the fact that most of the organisms of this microbial community do not readily sink, leaving a high percentage of nutrients in the euphotic layer. Owing to the rapid recycling of essential nutrients, the interpretation of data from laboratory cultures and from current methods in the field requires some reexamination. For example, there is little evidence of nutrient-limited growth rate of planktonic algae even in the more oligotrophic areas, where ^{14}C based estimates of primary production have been suspected of grossly underestimating actual production. There are also some reasons to question the

role of algal competition for nutrients in an environment subject to rapid turnover of nutrients and to non-seasonal partial mixing due to frequent meteorological events. The dynamic conception of community structuring is that of flux rates, efficiencies and interaction with predators rather than concentrations and growth rates.

All the data reinforce the conclusion that these interactions between nutrient uptake, growth rate, grazing and nutrient regeneration are basically similar in lakes and in oceans. Many of these interactive processes appear to be, within broad functional groupings, strongly size-dependent. Given the importance of size-structured interactions in planktonic food-webs, the development of other bases for species aggregation than traditional taxons or trophic levels should be encouraged in both theoretical and experimental study of pelagic systems.

References

Ahlgren, G., 1988. Phosphorus as growth-regulating factor relative to other environmental factors in cultured algae. Hydrobiologia 170: 191–210.

Azam, F., T. Fenchel, J. E. Field, J. S. Gray, L. A. Meyer-Rield & F. Thingstad, 1983. The ecological role of water column microbes in the sea. Mar. Ecol. Prog. Ser. 10: 257–263.

Berman, T., 1985. Uptake of (^{32}P) orthophosphate by algae and bacteria in lake Kinneret. J. Plankton Res. 7: 71–84.

Bird, D. F. & J. Kalff, 1987. Algal phagotrophy: Regulating factors and importance relative to photosynthesis in *Dinobryon* (Chrysophyceae). Limnol. Oceanogr. 32: 277–284.

Bloem, J., C. Albert, M. J. Bar-Gillissen, T. Berman & T. E. Cappenberg, 1989. Nutrient cycling through phytoplankton, bacteria and protozoa in selectively filtered lake Vechten water. J. Plankton Res. 11: 119–131.

Bonin, D. J., 1982. Dynamique nutritionnelle du phytoplancton. Oceanis 8: 459–492.

Brown, E. J., D. K. Button & D. S. Lang, 1981. Competition between heterotrophic and autotrophic microplankton for dissolved nutrients. Microb. Ecol. 7: 199–206.

Carpenter, S. R. (ed.), 1988. Complex interactions in lake communities. Springer-Verlag, Berlin, 283 p.

Cembella, A. D., N. J. Antia & P. J. Harrison, 1984. The utilization of inorganic phosphorus compounds as nutrients by eukaryotic microalgae: a multi-disciplinary perspective. C.R.C. Crit. Rev. Microbiol. 10: 317–391.

Currie, D. J. & J. Kalff, 1984. The relative importance of bacterioplankton and phytoplankton in phosphorus uptake in freshwater. Limnol. Oceanogr. 29: 311–321.

Currie, D. J., E. Bentzen & J. Kalff, 1986. Does algal-bacterial phosphorus partitioning vary among lakes? A comparative study of orthophosphate uptake and alkaline phosphatase activity in freshwater. Can. J. Fish. aquat. Sci. 43: 311–318.

Dauta, A., 1983. Conditions de développement du phytoplancton. Etude comparative du comportement de huit espèces en culture. Cinétique d'assimilation et de croissance: étude expérimentale, modélisation appliquée aux cultures et à un milieu naturel (1e Lot). Thèse Doctorat Sci. Univ. Toulouse III, 166 p.

Droop, M. R., 1973. Some thoughts on nutrient limitation in algae. J. Phycol. 9: 264–272.

Ejsmont-Karabin, J., 1984. Phosphorus and nitrogen excretion by lake zooplankton (Rotifers and Crustaceans) in relationship to individual body weights of the animals, ambiant temperature and presence or absence of food. Ekol. Pol. 32: 3–42.

Eppley, R. W., 1981. Relations between nutrient assimilation and growth in phytoplankton with a brief review of estimates of growth rate in the ocean. Can. Bull. Fish. aquat. Sci. 210: 251–263.

Falkowski, P. (ed.), 1980. Primary productivity in the sea. Brookhaven Symposia in Biology 31, Plenum Press, NY., 531 p.

Friebele, E. S., D. L. Correll & M. A. Faust, 1978. Relationship between phytoplankton cell size and the rate of orthophosphate uptake: in situ observations of an estuarine population. Mar. Biol. 45: 39–52.

Gardiner, A. C., 1937. Phosphate production of planktonic animals. J. Cons. int. Explor. Mer 12: 144–146.

Glibert, P., F. Lipschultz, J. J. McCarthy & M. A. Altabet, 1982. Isotope dilution models of uptake and remineralization of ammonium by marine plankton. Limnol. Oceanogr. 27: 639–650.

Goldman, J. C., 1980. Physiological processes, nutrient availability and the concept of relative growth rate in marine phytoplankton ecology. In: P. Falkowski (ed.), Primary productivity in the sea. Brookhaven Symposia in Biology 31, Plenum Press, N.Y.: 179–194.

Goldman, J. C. & P. M. Glibert, 1983. Kinetics of inorganic nitrogen by phytoplankton. In: E. J. Carpenter & D. G. Capone (eds.), Nitrogen in the marine environment – Academic Press, N.Y.: 233–274.

Harris, G. P., 1986. Phytoplankton ecology: structure, function and fluctuation. Chapman and Hall, London, 384 p.

Harrison, W. G., 1980. Nutrient regeneration and primary production in the sea. In: P. Falkowski (ed.), Primary productivity in the sea. Brookhaven Symposia in Biology 31, Plenum Press, N.Y.: 433–460.

Harrison, W. G., T. Platt & M. R. Lewis, 1980. f ratio and its relationship to ambient nitrate concentration in coastal waters. J. Plankton Res. 9: 235–248.

Harrison, W. G. & L. E. Wood, 1988. Inorganic nitrogen

uptake by marine picoplankton: Evidence for size partitioning. Limnol. Oceanogr. 33: 468–475.
Healey, F. P. & L. L. Hendzel, 1979. Indicators of phosphorus and nitrogen deficiency in five algae in culture. J. Fish. Res. Bd Can. 36: 1364–1369.
Hecky, R. E. & P. Kilham, 1988. Nutrient limitation of phytoplankton in freshwater and marine environments: a review of recent evidence on the effects of enrichment. Limnol. Oceanogr. 33: 796–822.
Herbland, A., 1988. Evolution du concept d'oligotrophie au cours des quinze dernières années. Le Courrier de Médiprod (Banyuls-sur-mer, France) 5: 3–7.
Hobbie, J. E., 1988. A comparison of the ecology of planktonic bacteria in fresh and salt water. Limnol. Oceanogr. 33: 750–764.
Jansson, M., 1988. Phosphate uptake and utilization by bacteria and algae. Hydrobiologia 170: 177–189.
Jassby, A. D. & C. R. Goldman, 1974. Loss rates from a lake phytoplankton community. Limnol. Oceanogr. 19: 618–627.
Kilham, P. & R. E. Hecky, 1988. Comparative ecology of marine and freshwater phytoplankton. Limnol. Oceanogr. 33: 776–795.
Klein, P. & B. Coste, 1984. Effects of wind-stress variability on nutrient transports into the mixed layer. Deep-Sea Res. 31: 21–37.
Laws, E. A., W. G. Harrison & G. R. DiTullio, 1985. A comparison of nitrogen assimilation rates based on ^{15}N uptake and autotrophic protein synthesis. Deep-Sea Res. 32: 82–95.
Lean, D. R. S. & C. Nalewajko, 1976. Phosphate exchange and organic phosphorus excretion by freshwater algae. J. Fish. Res. Bd Can. 33: 1312–1323.
Lehman, J. T., 1980a. Nutrient recycling as an interface between algae and grazers in freshwater communities. In: W. C. Kerfoot (ed.), Evolution and ecology of zooplankton communities, University Press of New England, London: 251–263.
Lehman, J. T., 1980b. Release and cycling of nutrients between planktonic algae and herbivores. Limnol. Oceanogr. 25: 620–632.
Lehman, J. T., 1988. Ecological principles affecting community structure and secondary production by zooplankton in marine and freshwater environments. Limnol. Oceanogr. 33: 931–945.
Levine, S. N., M. P. Stainton & D. W. Schindler, 1986. A radiotracer study of phosphorus cycling in a eutrophic Canadian Shield lake, Lake 227, north-western Ontario. Can. J. Fish. aquat. Sci. 43: 366–378.
Maestrini, S. Y., D. J. Bonin & M. R. Droop, 1984. Phytoplankton as indicators of sea water quality: bioassay approaches and protocols. In: L. E. Shubert (ed.), Algae as ecological indicators, Academic Press, London: 71–131.
Malone, T. C., 1980a. Algal size. In: I. Morris (ed.), The physiological ecology of phytoplankton, Blackwell, Oxford: 433–463.
Malone, T. C., 1980b. Size fractionated primary productivity of marine phytoplankton. In: P. Falkowski (ed.), Primary productivity in the sea, Brookhaven Symposia in Biology 31, Plenum Press, N.Y.: 301–319.
Margalef, R., 1978. Life-forms of phytoplankton as survival alternatives in an unstable environment. Oceanol. Acta 1: 493–509.
Mazumder, A., D. J. McQueen, W. D. Taylor & D. R. S. Lean, 1988. Effects of fertilization and planktivorous fish (yellow perch) predation on size distribution of particulate phosphorus and assimilated phosphate: Large enclosure experiments. Limnol. Oceanogr. 33: 421–430.
McQueen, D. J., J. R. Post & E. R. Mills, 1986. Trophic relationships in freshwater pelagic ecosystems. Can. J. Fish. aquat. Sci. 43: 1571–1581.
Mills, E. L. & J. L. Forney, 1988. Trophic dynamics and development of freshwater pelagic food webs. In: S. R. Carpenter (ed.), Complex interactions in lake communities, Springer-Verlag, NY: 12–30.
Morris, D. P. & W. M. Lewis jr, 1988. Phytoplankton nutrient limitation in Colorado mountain lakes. Freshwat. Biol. 20: 315–327.
Morris, I. (ed.), 1980. The physiological ecology of phytoplankton, Blackwell, Oxford, 625 p.
Nalewajko, C. & D. R. S. Lean, 1980. Phosphorus. In: I. Morris (ed.), The physiological ecology of phytoplankton, Blackwell, Oxford: 235–258.
Officer, C. B. & J. H. Ryther, 1980. The possible importance of silicon in marine eutrophication. Mar. Ecol. Prog. Ser. 3: 83–91.
Paasche, E., 1980. Silicon content of five marine plankton diatom species measured with a rapid filter method. Limnol. Oceanogr. 25: 474–480.
Pace, M. L., 1986. An empirical analysis of zooplankton community size structure across lake trophic gradients. Limnol. Oceanogr. 3: 45–55.
Peters, R. H., 1986. The role of prediction in Limnology. Limnol. Oceanogr. 31: 1143–1159.
Peterson, B. J., 1980. Aquatic primary productivity and the ^{14}C-CO_2 method: a history of the productivity problem. Annu. Rev. Ecol. Syst. 11: 359–385.
Platt, T. (ed.), 1980. Physiological bases of phytoplankton ecology. Can. Bull. Fish. aquat. Sci. 210, 346 p.
Pollingher, U. & C. Serruya, 1976. Phased division of *Peridinium cinctum* va *westi* and the development of blooms in lake Kinneret. J. Phycol. 15: 155–162.
Porter, K. G., H. Pearl, R. Hodson, M. Pace, J. Priscu, B. Riemann, D. Scavia & J. Stockner, 1988. Microbial interactions in Lake food webs. In: S. R. Carpenter (ed.), Complex interactions in Lake Communities, Springer-Verlag, Berlin: 209–227.
Pourriot, R., J. Capblancq, P. Champ & J. A. Meyer, 1982. Ecologie du plancton des eaux continentales. Masson, Paris: 198 p.
Ramberg, L., 1980. A population dynamics model for *Oocystis parva* (Chlorophyceae). Arch. Hydrobiol. 89: 118–134.

Reynolds, C. S., 1984. The ecology of freshwater phytoplankton, Cambridge University Press, London, 384 p.
Rhee, G. Y., 1978. Effects of N:P atomic ratios and nitrate limitation on algal growth, cell composition and nitrate uptake. Limnol. Oceanogr. 23: 10–25.
Rhee, G. Y. & I. J. Gotham, 1980. Optimum N:P ratios and coexistence of planktonic algae. J. Phycol. 16: 486–489.
Riemann, B., 1985. Potential influence of fish predation and zooplankton grazing on natural populations of freshwater bacteria. Appl. envir. Microbiol. 50: 187–193.
Rigler, F. H., 1973. A dynamic view of the phosphorus cycle in lakes. In: E. J. Griffith, A. Beeton, J. M. Spencer & D. T. Mitchell (eds.), Environmental phosphorus handbook, Wiley, NY: 539–572.
Sager, P. E., B. Rubio & J. Kirk, 1988. The importance of nanoplankton along the trophic gradient in Green Bay, Lake Michigan. Verh. int. Ver. Limnol. 23: 376–379.
Scavia, D. & G. L. Fahnenstiel, 1988. From picoplankton to fish: complex interactions in the Great Lakes. In: S. R. Carpenter (ed.), Complex interactions in lake communities, Springer-Verlag, Berlin: 85–97.
Schelske, C. L., E. F. Stoermer, G. L. Fahnenstiel & M. Haibach, 1986. Phosphorus enrichment, silica utilization, and biogeochemical silica depletion in the Great Lakes. Can. J. Fish. aquat. Sci. 43: 407–415.
Schindler, D. W., 1988. Experimental studies of chemical stressors on whole lake ecosystems. Verh. int. Ver. Limnol. 23: 11–41.
Shuter, B. J., 1978. Size dependence of phosphorus and nitrogen subsistence quotas in unicellular microorganisms. Limnol. Oceanogr. 23: 1248–1255.
Smith, R. E. H. & J. Kalff, 1981. The effects of phosphorus limitation on algal growth rates: evidence from alkaline phosphatase. Can. J. Fish. aquat. Sci. 38: 1421–1427.
Smith, R. E. H. & J. Kalff, 1982. Size dependent phosphorus uptake kinetics and cell quota in phytoplankton. J. Phycol. 18: 275–284.
Smith, R. E. H. & J. Kalff, 1983. Competition for phosphorus among occurring freshwater phytoplankton. Limnol. Oceanogr. 28: 448–464.
Smith, S. V., 1984. Phosphorus versus nitrogen limitation in the marine environment. Limnol. Oceanogr. 29: 1149–1160.
Smith, V. H., 1982. The nitrogen and phosphorus dependence of algal biomass in lakes: an empirical and theoretical analysis. Limnol. Oceanogr. 27: 1101–1112.
Sommer, U., 1985. Comparison between steady state and non-steady state competition: experiments with natural populations. Limnol. Oceanogr. 30: 335–346.
Sommer, U., 1988. Does nutrient competition among phytoplankton occur in situ Verh. int. Ver. Limnol. 23: 707–712.
Sprules, W. G. & M. Munawar, 1986. Plankton size spectra in relation to ecosystem productivity, size, and perturbation. Can. J. Fish. aquat. Sci. 43: 1789–1794.
Stabel, H. H., 1985. Mechanisms controlling the sedimentation sequence of various elements in prealpine lakes. In: W. Stumm (ed.) Chemical processes in lakes. J. Wiley & Sons, Inc., N.Y.: 143–167.
Sterner, R. W., 1986. Herbivores' direct and indirect effects on algal populations. Science 231: 605–607.
Stockner, J. G., 1988. Phototrophic picoplankton: an overview from marine and freshwater ecosystems. Limnol. Oceanogr. 33: 765–775.
Stockner, J. G. & N. J. Antia, 1986. Algal picoplankton from marine and freshwater ecosystems: a multidisciplinary perspective. Can. J. Fish. aquat. Sci. 43: 2472–2503.
Talling, J. F., 1984. Past and contemporary trends and attitudes in work on primary productivity. J. Plankton. Res. 6: 203–217.
Tilman, D., S. S. Kilham & P. Kilham, 1982. Phytoplankton community ecology: the role of limiting nutrients. Annu. Rev. Ecol. Syst. 13: 349–372.
Tilzer, M. M., 1984. Estimation of phytoplankton loss rates from daily photosynthetic rates and observed biomass changes in Lake Constance. J. Plankton Res. 6: 309–324.
Turpin, D. H. & P. J. Harrison, 1979. Limiting nutrient patchiness and its role in phytoplankton ecology. J. exp. mar. Biol. Ecol. 39: 151–166.
Vadstein, O., A. Jensen, Y. Olsen & H. Reinertsen, 1988. Growth and phosphorus status of limnetic phytoplankton and bacteria. Limnol. Oceanogr. 33: 489–503.
Wheeler, P. A. & D. L. Kirchman, 1986. Utilization of inorganic and organic nitrogen by bacteria in marine systems. Limnol. Oceanogr. 31: 998–1009.
White, E., G. Payne, S. Pickmere & F. R. Pick, 1982. Factors influencing orthophosphate turnover times: a comparison of Canadian and New Zealand lakes. Can. J. Fish. aquat. Sci. 39: 469–474.

Received 20 June 1989; in revised form 24 January 1990; accepted 7 March 1990

Hydrobiologia **207**: 15–29, 1990.
D. J. Bonin & H. L. Golterman (eds), Fluxes Between Trophic Levels and Through the Water–Sediment Interface.

Contribution of the water-sediment interface to the transformation of biogenic substances: application to nitrogen compounds

Raoul Daumas
CNRS, Microbiologie Marine, Faculté des Sciences de Luminy, Case 907, 13288 Marseille Cedex 9, France

Key words: water-sediment interface, microbial activity, bioturbation, nitrogenous substances, freshwater and marine sediments

Abstract

Up to now the water-sediment interface (WSI) has not been considered by limnologists and oceanographers as a theoretical frontier between two phases, but as a variable layer comprising the overlying bottom water, the superficial sediment, and the bioturbed sediment. This heterogeneous benthic boundary layer forms a non-equilibrated assemblage of an aqueous phase, a mineral phase influencing the exchanges, and a dissolved gaseous phase regulating biological activity, detrital organic matter present as dissolved or adsorbed compounds and particles, and a living benthic community dominated by microorganisms. Biological activity and the equilibrium between these components are linked to depth, bottom currents, primary production and allochtonous inputs of organic matter. The settling rate of particulate matter in the water column, which depends on both the size and density of particles, is also an important factor. Settled organic matter is broken down mainly by heterotrophic microorganisms. In the first stage, dissolved or particulate matter is depolymerized by exoenzymes linked to the bacterial membrane. Other heterotrophic organisms, like flagellates and ciliates, contribute to the activity of bacterial populations or to a modification of the physical and chemical characteristics of organic matter like macrofauna. Bacterial activity in the WSI is one to two orders of magnitude higher than in overlying bottom water, depending on the density of the bacterial population, on the amount of available organic matter, and on the presence of predators grazing on bacteria and macrofauna and increasing oxygen and dissolved substrates. The WSI is now considered as a sink for particulate matter and biopolymers, and/or as a sink or source for organic monomers and NH_4^+ depending on their concentrations. As a result of its contribution to mineral and organic cycles, the WSI is a crossroad for exchanges between aqueous and sediment layer. During the last decade, intensive work has been performed in this field, but progress has been slow due to the heterogeneity of the layer and the adsorptive properties of the mineral fraction.

Introduction

Literally, the term ‘interface’ designates a surface forming the common boundary between two parts of matter or space. In science, however, there are as many definitions of interface as there are disciplines. Limnologists and oceanographers consider an interface to be the meeting place of two different layers, in which exchanges and accumulation of material are more intensive than in the bulk of the layer. A number of interfaces can be distinguished in aquatic environments: air-water, particle-water, freshwater-seawater, ice-water, and water-sediment. Although it remains poorly understood, the latter interface, *i.e.* water-sediment, is of particular scientific and technological importance.

The superficial layer of sediment receives a shower of particulate material in the form of 'detritus', *i.e.* organic carbon lost by non-predatory processes from any trophic level (Wetzel *et al.*, 1972). Particle deposit, transport of sediment and considerable biotic activity take place in the water sediment interface (WSI). However the intensity of these activities varies greatly from continental freshwater to seawater areas. Terrestrial inputs increase the complexity of WSI dynamics on the bottom of seas, lakes and brackish ponds. The role of bottom currents in stirring sediments and modifying the exchange across the WSI is greatest in the nearshore regions. Currents determine whether organic carbon mineralization takes place under anaerobic or aerobic conditions. Bacterial and macrofaunal populations also contribute to organic matter (OM) transformation.

The participation of the WSI in mineral and organic exchanges between water and sediment necessitates a knowledge of factors regulating chemical and biological activities.

Description of the water-sediment interface

The WSI can be divided into several different layers: the overlying water layer, the near-sediment layer, the superficial sediment layer, the bioturbed sediment layer, and the compacted sediment layer. Of these layers, the near-sediment layer, the superficial sediment layer, and the bioturbed sediment layer are considered to be actively involved in water-sediment exchanges. These 3 layers form the 'benthic boundary layer', which was defined by the working group of the Nato Conference (Mc Cave, 1976) as including 'the zone near the seabed where there are strong biotic, chemical and physical gradients, which are controlled by the presence of the seabed' and the zone 'within the seabed, the boundary layer of which extends to a depth at which mixing processes are controlled by exchanges or activity occurring at the boundary'. Investigation of the WSI is difficult due to the heterogeneity of this layer and the complexity of the physical and chemical interactions with living components, i.e. the microorganisms.

Water phase: Because it accounts for so large a share of the bulk superficial layer and plays such a crucial role in the biochemical transformation of organic matter (OM), the most important single component in the WSI is the water phase, which can be divided into three compartments according to the mobility of water molecules. 'Free-water' is a circulating phase with an anarchic distribution of ions. The water in this compartment is not bound to sediment particles and separation does not require a pressure system. In the superficial layer, free water represents a fraction of the overlying water layer. 'Interstitial water' forms an intermediary phase between free water and linked water and consequently its definition is related to their properties. Interstitial water can only be extracted from bulk sediment by means of a pressure system or by centrifugation. It contains a large proportion of free ions. Because colloidal particles, between the particle and the bulk dispersion medium, are charged, an electrical potential exists in the 'linked water' compartment and ions of opposite charges are attracted to form a double layer of electrically charged particles (Marshall, 1976). This double layer plays an important part in interactions between bacteria and other solid surfaces. The water phase plays a double role: supplying oxygen and dissolved organic substances to neighboring bacteria and carrying monomeric substances issued from the bacterial breakdown of organic substances.

Mineral particles: Essential for exchanges and biological activity in the layer, their effects depend on their nature and size. The two dominant minerals are clays and quartz, but the distinction between coarse and fine material is more significant. The finest group comprises particles visible by light microscopy whose size is less than 2 μm

(Watling, 1988). Coarse material is made up of small rock fragments or OM fragments, while fine material includes clays and microorganisms. With regard to the influence of mineral particles on the exchanges in the WSI, it is essential to consider the permeability of the layer, which is negatively correlated with size and water content. Clays generally retain about 200% (w/w) of water and form a barrier to exchanges between superficial water and compacted sediment. In contrast, sand, which is porous, retains only a fine water layer around the grain. The existence of a water layer linked to particles is confirmed by the fact that interstitial water can be extracted by means of gas pressure or centrifugation; the percentage of water extracted from the superficial sediment is independent of the water content of the bulk sediment and corresponds maximally to 80% of the water content (Romano *et al.*, 1981).

A major consequence of these differences in permeability involves oxygenation of sediment. In mud, oxygenation is limited to a few millimeters. For sand the depth of oxygenation is several centimeters on beaches and much more in riverbeds. Another parameter linked to the nature of mineral particles is adsorption, which greatly modifies the exchanges of organic or mineral substances, amino-acids and NH_4^+ (Rosenfeld, 1979). Interstitial water carries a dissolved gaseous phase composed mainly of O_2, CO_2, H_2S, CH_4 and N_2. Revsbech *et al.* (1988) used microelectrodes to measure O_2 at the sediment surface, and observed a large difference between an organic-rich, fine grained sediment, in which an oversaturation of oxygen can be detected 0.1 mm below the sediment surface, and a sand-flat sediment, typical of many shallow marine areas, in which the oxygen concentration is notably lower. Photosynthesis of cyanobacterial film contributes to oxygen oversaturation. H_2S and CH_4 result from bacterial transformation of sulfate, acetate or methylotrophe substances. Methane is often undetectable within surface layers of marine sediments where sulfate reduction is most active, whereas in freshwater sediments CH_4 is present.

Organic matter: OM is present in the WSI in the form of particle 'film' adsorbed on mineral fragments and dissolved substances in interstitial water. The ratios between the particulate and dissolved fraction in the overlying water and superficial sediment are opposite. In the water phase, dissolved organic compounds form 96% of the total OM and in the WSI dissolved organic compounds correspond to only 1% of OM (Meyer-Reil *et al.*, 1978). The interaction between organic substances and solid surfaces and the formation of 'organic-mineral aggregates' was noted by Butsch & Stumm (1968). Johnson (1974), showed that polymers extracted from *Aerobacter aerogenes* behaved as bridging polymers between solid surfaces and bacteria. In a recent study of the small-scale features of marine sediments, Watling (1988) demonstrated the presence of a material forming bridges between grains, and showed that this material completely encompassed all grains in muddy sediments. Biddanda (1986) observed that decomposition comprises separate phases, in which detritus aggregates and organo-aggregates are broken down by metazoan consumers or by the dissolving activity of the bacterial population, respectively.

Dissolved organic matter is present in interstitial water at a higher total concentration than in the overlying water. Remarkable gradients of amino-acid concentration have been reported. Daumas *et al.* (1979), Dawson & Liebezeit (1982), Jorgensen *et al.* (1981), Jorgensen (1982), and Stanley *et al.* (1987) measured up to 30-fold higher concentrations of dissolved amino-acids in the sediment surface layer than in the overlying water. However, most of the dissolved OM is present in the high molecular weight fraction (MW > 1000 d) (Krom & Westrich, 1980). Fatty acids are also more abundant in interstitial water, with Saliot (1977) observing an enrichment factor ranging from 10 to 79.

Biota: They represent the active part of the WSI and are composed of microorganisms, meiofauna and macrofauna. 'Microorganisms' are primary microbial decomposers. The important role of the bacterial population in the superficial sediment was observed by early researchers including Russel (1892, 1893), Zobell (1946) and Hayes (1964). However, the contribution of bacteria to OM recycling and nutrient regeneration has been studied by only a few scientists in the last twenty years (review by Fenchel & Blackburn, 1979). The development of new biochemical concepts and techniques over the last ten years has contributed to an increased interest in the metabolic activity of bacteria and their involvement in the biochemical cycle of OM. Focusing his study on the top layer of sediment (< 1 mm), Novitsky (1983) showed that it accounts for only part of the activity of the entire system. The microbiota of freshwater and marine sediments play similar roles in carbon and nutrient regeneration. Distribution and rates of microbial activities are determined by availability of electron acceptors for respiration and metabolizable organic substrates. Dissolved organic matter (DOM) present in overlying water of a stream supports about 55% of the bacterial biomass in surface sediments, while the percentage of bacterial biomass supported in a superficial sediment muffled to remove organic matter is lower (Bott *et al.*, 1984). As a result, sediment organic matter is a source of nutrients as well as a site for colonization. Bacteria decompose particulate and dissolved organic matter including both labile or more refractory compounds (Fig. 1). It has been demonstrated that high molecular weight organic compounds cannot be directly taken up by microorganisms unless they are first hydrolyzed by exoenzymes (Billen, 1984). This preliminary step is the major function of the bacterial population in aquatic systems. The mineralization of organic compounds necessitates the presence of electron-acceptors such as O_2 in the aerobic layer and sulfate or acetate in the reduced sediments. The low availability of nitrate apparently limits the activity of denitrifying bacteria. The relative proportion of methanogenesis and sulfate reduction is a distinguishing feature of freshwater and marine sediments. The sulfate ion concentration in freshwater systems in only one tenth of that in seawater; consequently methanogenesis replaces sulfate reduction as the dominant pathway of

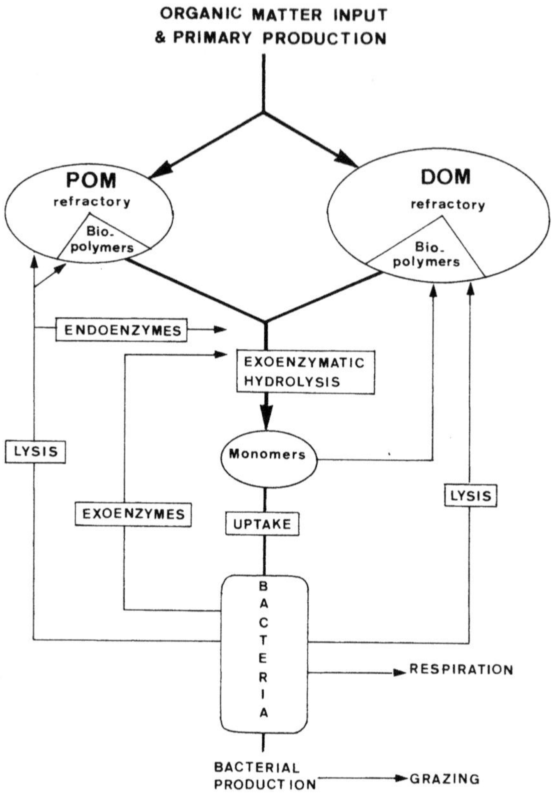

Fig. 1. Role of bacteria in the degradation of organic matter present in the form of particles (POM) or in a dissolved state (DOM), (after Hermin 1989).

terminal carbon metabolism (Ward & Winfrey, 1985). Bacteria are subsequently fed upon by flagellates and ciliates. Grazing by bacterivores maintains the bacterial populations in the vicinity of 10^9–10^{10} bacteria g^{-1} d.w. The majority of them are Gram-negative strains, (Moriarty & Hayward, 1982). These bacteria have complex cell walls with a lipopolysaccharide membrane over a cytoplasmic membrane, and most are embedded in extracellular slime layers.

Recent work on bacteria-protozoan interactions indicates that 'protozoans' are the most important consumers and remineralizers of the bacterial biomass, whereas metazoans do not graze exclusively on bacterial production, (Williams, 1984). The influence of organic and inorganic nitrogen compounds on bacteria-protozoan interactions has been assessed. Phagotrophic protozoans release organic compounds such as metabolic end-products, and bacteria are able to utilize nanomolar concentrations of these compounds, especially dissolved free amino-acids (Rault *et al.*, 1988).

'Meiofauna' are estimated to account for 92% of living biomass in intertidal sediments and 78% in subtidal sediments (Sikora *et al.*, 1977). Gerlach (1972) estimated the biomass of meiofauna to be 3% in comparison with macrofauna in coastal sediment. Schwinghamer & Kepkay (1987) demonstrated that enrichment with *Spartina* detritus has no effect on the size of the microbial and meiofaunal biomass. The 'lack of positive biomass response suggests that the meiofauna was limited not only by the availability of food but also by biotic interaction' (Widbom & Elingren, 1988). 'Macrofauna' can greatly modify biogenic activity both by metabolic processes, such as ingestion of food and excretion of fecal pellets, and by mixing the sediment *i.e.* bioturbation (Aller, 1978). These modifications in the superficial sediment promote development of bacterial activity related to oxygenation of sediment and exchange of organic matter between overlying and interstitial water. The ingestion of sediment by deposit-feeders contributes to a decrease in the bacterial population fixed to the mineral particles, but this mechanism leads not only to fragmentation of particles by the deposit feeders, but also to digestion by the enzymes in the gut (Feral, 1989). Marine and freshwater muds are both dominated by deposit-feeders, but marine animals are restricted by high salinity, whereas freshwater animals are relatively euryhaline. Tentaculate deposit-feeders are found only in marine sediments, while interstitial suspension-feeders are common only in freshwater muds. Most marine animals take up small organic molecules such as amino-acids, whereas freshwater macrofauna exhibit very poor uptake abilities. Preying on bacteria by the bivalve *Ruditapes decussatus*, a filter-feeder catching particles smaller than 1 mm, was evaluated by means of variations in the ATP-biomass of a bacterial population in the water-sediment (Daumas & Bianchi, 1984). After a first period during which it increases rapidly (12 h), the bacterial population decreases as a result of preying by bivalves (12 h). The fact that the biomass at the beginning and end of the experiment is not very different suggests a 'threshold biomass' at which the bivalve intervenes.

Microphytobenthos: It has been reported as playing a role in exchanges in the WSI in nearshore regions. The role also seems to be important in lakes and streams, where the microphytobenthos contribute to regulation of P and N-fluxes between overlying water and interstitial water (Sundbäck & Graneli, 1988). Moreover, microalgae contribute to the bulk OM brought to the WSI. Another contribution to the biological

Table 1. Distribution of nitrogenous forms in different biomasses (from Jocteur-Monrozier, 1984)

Material	Nitrogen fractions (in % of total N)				References
	Non hydrolyzable	α-amine	amino-sugar + NH_4	humic substances	
Beech leaves	5–11	42–81	7–13	0–40	Janel (1978)
Bacteria *(Pseudomonas aeruginosa)*	6.6–9	55.2–83.7	5.1–7.3	0–32.8	Chu & Knowles (1966)
Unicellular algae (benthic diatoms)	0	78	22	0	Roussel (1978)
Protozoan (benthic foraminifera)	0–28	46–74.5	25.5–27	0–27	Pillon (1979)

activity of the WSI is supplied by fungi which have a low C/N and may constitute 0 to 20% of the total microbial biomass (Blum *et al.*, 1988).

Nitrogenous inputs at the water-sediment interface

As shown in previous studies (review in Mann, 1988), the presence of a labile nitrogenous fraction (amino-acids) enhances the biodegradability of OM. When the carbohydrate fraction accounts for the largest proportion, microorganisms and fungi which colonize the particles can supply nitrogen (Marinucci *et al.*, 1983). The nitrogenous material entering the WSI has lost the most labile fractions of OM (intracellular content) because of release into the milieu by autolysis of intracellular enzymes (Gyllenberg & Eklund, 1974). The nitrogen level in fragments of land vegetation is 0.8 to 1.2% (Janel, 1978), and sea material is made up of cell membranes of different thicknesses, chitinous shells and molt envelopes (Rowe & Staresinic, 1979). Nitrogenous forms in these debris are scleroproteins and amino sugars (Table 1). In regions with high sedimentation rates, nitrogen in the particulate material accounts for up to 40% of the total nitrogen. Dissolved organic matter, which represents a large part of the organic material, comes from land and sea vegetation, but condensation and aggregation transform them into 'heteropolycondensates' in which the quantity of degradable nitrogen diminishes with time (Degens, 1970). The 'living' biomass represents only 4% of the nitrogen in land ecosystems and 1% in marine environments. In deeper-regions, fecal pellets account for part of the sedimenting OM (20%), and fine particles ($<20\ \mu m$) account for another part. Because the drop time through the water column is short, fecal pellets constitute the greatest proportion of the labile OM input. Recent observations (Cho & Azam, 1988; Karl *et al.*, 1988) suggest that, as they drop through the water column, fecal pellets undergo desaggregative and reaggregative phases (Biddanda & Pomeroy, 1988) that modify the composition of OM by inclusion of fine particles.

Factors influencing the biological activity at the water-sediment interface

As a result of the complexity of the OM deposited on the sea floor, initial investigations about its degradation dealt only with annual variations of C and N in the sediment and consequently concerned the concentration of nutrients in the interstitial water. With increasing interest in sediment nutrient regeneration, subsequent investigations have focused on sedimentary nitrogen material. With progress in evaluating nitrogen substances, amino-acids and proteins have also been included in research on bacterial activity. Many factors control the input of organic matter to the sediment: primary production, depth of the water column, strength of bottom currents and seasonal variations of inputs all directly or indirectly modify biological activity in the WSI. In addition, exchanges in the WSI also depend on bacterial activity and macrofauna turbation.

In shallow regions, phytoplankton productivity is high and benthic algae increase OM input. On the other hand, currents resuspend superficial sediment contributing to a reoxygenation of the WSI. Large quantities of OM entering the sediment are oxidized by heterotrophic organisms so intensely that the oxygen demand increases greatly (Vosjan, 1987) and the surface area is momentarily anoxic. Absence of oxygen in the WSI is more frequent in lakes because of stratification of water, and Golterman (1984), considering the fate of O_2 in the hypolimnion, raised the question: is O_2 consumed by CH_4 diffusing upwards, or by sediments sinking downwards?

In coastal regions as in lakes, organic material is supplied in waves of variable duration. The composition of OM is complex, ranging from autochthonous substances from planktonic blooms to material from plant debris carried in river flood waters. Deep sea zones are characterized by rain of phytodetritus in the form of fecal pellets, as well as by turbid currents carrying OM from the continental shelf to deeper areas along the slopes and canyons that cut into the shelf. At first glance there would appear to be little difference between the two locations in the quality of organic input, since most OM reaching the bottom is carried at high velocity (100 m/day) by fecal pellets and transfer time is negligible in comparison with the sedimentation rate on the bottom (10^{-1} m/100 y to a 10^{-3} m/1000 y). However, during transfer in the water column, degradation takes place from 0 to 500 m layer (Honjo *et al.*, 1982; Karl *et al.*, 1988), thus suggesting that rapid transfer in the form of fecal pellets is discontinuous and that, during stops, reorganization of organic aggregates occurs (Biddanda, 1986; Biddanda & Pomeroy, 1988). Low bacterial activity in sediment traps (Ducklow *et al.*, 1985) and analysis of OM in the water column (Cho & Azam, 1988) support this view.

The techniques used to study coastal areas are different from those used at deep sea sites. Access to coastal areas is relatively easy and conditions can be readily reproduced in the laboratory; study of deep sea sites requires indirect methods and sophisticated equipment, especially to recreate pressure conditions (Colwell *et al.*, 1980; Cahet *et al.*, 1990). Analysis of plant debris (Trichet, 1967; Olah *et al.*, 1978; Newell *et al.*, 1983; Stuart *et al.*, 1982), bacteria (Taylor & Sullivan, 1984; Novitsky, 1986) or fecal pellets (Hermin, 1989) have provided information about the dynamics of degradation of substrates, but the diversity of sites and ambient conditions in the WSI prohibit any broad generalization about the mechanisms of degradation. Seasonal variations have different effects in freshwater and seawater regions because, although it largely depends on phytoplankton bloom, OM production in lakes and streams is also linked to benthic production and terrestrial leaching related to the rainy period. Leaching leads to a large input of refractory compounds of unknown structure. In the marine environment this phenomenon is less influential than primary production. The settling rate, which determines the lag period between surface production and deposit in the WSI, has been evaluated to be between one and two weeks in the Santa Monica basin where maximum depth is 908 m, (Craven & Carlucci, 1989). In the Kiel Bight, where depth is ca. 18 m, no delay period was observed (Hermin, 1989).

The role of the WSI in the degradation of organic matter (essentially nitrogenous substances)

Bacterial activity in the Kiel Bight (Hermin, 1989): This study was carried out during autumn (13 October – 15 December) in a temperate coastal zone of the Kiel Bight in the Baltic Sea. Samples were taken using a Reineck core sampler at depths of 17 m. In autumn, temperatures vary between 11.5 and 7.6 °C and salinity is around 21 g l^{-1}. Small phytoplankton blooms supply OM to the WSI. Such a bloom occurred at the end of the month of October in the form of a noticeable increase in diatoms and peridinians.

Since deposit of organic material on the bottom is rapid in shallow water, study was focused on variations in the bacterial population, on consequences on exoproteolytic activity (EPA) and on detection of free amino acids (FAA) in interstitial water. The bacterial population at the surface of the sediment (9.10^9 cells cm^{-3} of wet sediment) did not vary significantly during the study; conversely, EPA measured directly in core samples of sediment (Meyer-Reil, 1986) increased concomitantly with the planktonic bloom near the end of October and remained high until the end of November (Fig. 2). Measurement of the EPA/biomass ratio indicated a steady increase in EPA until November 18, followed by a rapid decline until the end of November. Hydrolytic activity of exoenzymes on polymers in OM newly deposited in the WSI led to a release of FAA, as was shown by measuring FAA in interstitial water. Variations in FAA concentrations in the 0–2 cm layer were parallel to exoprotease activity. This *in situ* observation demonstrated the direct link between WSI activity and changes occurring in the euphotic layer. The presence of labile OM in shallow water makes it easier to conceptualize the relationship between the two biotopes. It is likely that sampling at closer intervals in time would have supplied a more explicit picture of the different events.

Effects of organic enrichment in Carteau Bay (northern mediterranean): Addition of labile OM had observable effects on micro-organism popula-

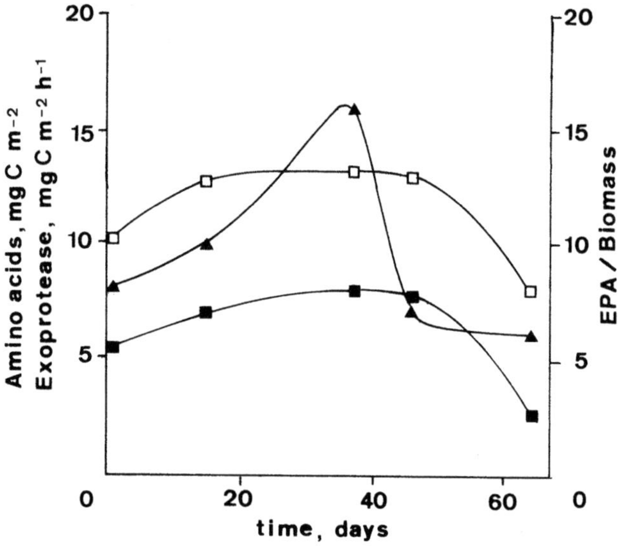

Fig. 2. Exoproteolytic activity, EPA (□), EPA/bacterial biomass ratio (▲) and free amino acids (■) in the top layer of sediment in Kiel Bight (Baltic Sea) between October 13, 1986 (point 0) and December 15, 1986 (point 64), (after Hermin, 1989).

tions. This study was conducted in Carteau Bay, near Marseilles. Fecal pellets collected at a site of intensive mussel farming (*Mytilus galloprovincialis*) were used as the organic material. The *in situ* experiment consisted in spreading freshly collected feces, within a frame, on the bottom and taking sediment samples over two months. The following physical and chemical measurements were made on the sediment: T°, Eh, dry weight, OM content, FAA concentration in interstitial water. To assess the microorganism population, the following parameters were determined: bacterial biomass, exoenzymatic activities, bacterial production. Organic enrichment was carried out on the 4th day of the experiment; it led to a noticeable increase in total nitrogen from 0.06 to 0.16%, while the C:N ratio was only slightly affected. EPA and the bacterial population increased greatly in the two hours following OM supplementation (EPA doubled and bacterial count increased 2.7 fold). The change in EPA was short-lived and, two days later, was no longer detectable. This suggests that enzymatic activity induced by the fecal material was due to the quality of the OM, which was either very labile or low in concentration (Fig. 3). Thereafter, the bacterial population and EPA went through a second growth phase lasting beyond the 20th day. Starting on the 37th day, the two mechanisms dissociated, with the bacterial population stabilizing and EPA steadily decreasing. The behavior of these two parameters is best illustrated by the EPA/bacterial biomass ratio (Fig. 4); agreement with biomass was good except in case of a great increase in biomass; this could alter the relation between EPA and biomass (Fontigny *et al.*, 1987). FAA concentration followed variations in

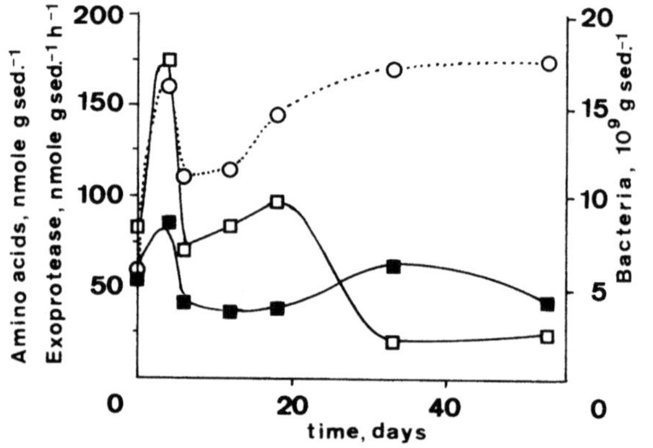

Fig. 3. Bacterial population (○), exoproteolytic activity (□) and free amino acids (■) in the top layer of sediment (0–2 cm) of Carteau Bay (North Mediterranean). Supplementation was performed on day 4 of the experiment (after Hermin, 1989).

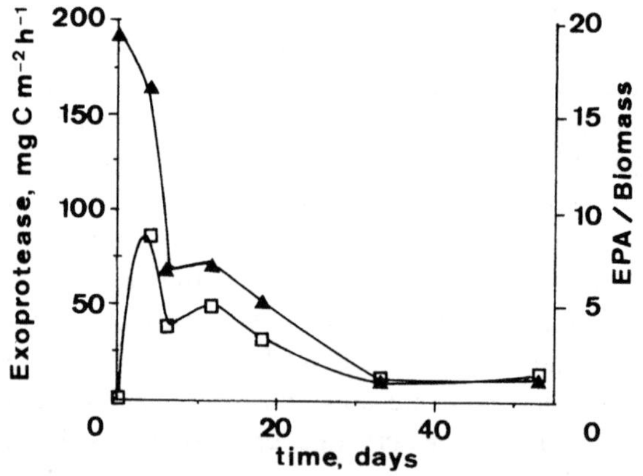

Fig. 4. Comparison of the exoproteolytic activity (□) and the EPA/bacterial biomass ratio (▲) during experimentation in Carteau Bay (after Hermin, 1989).

EPA except with regard to the decrease in EPA, on the 20th day which was not observed for FAA. Measuring uptake of ^{3}H-thymidine in DNA allows assessment of total bacterial production taking into account grazing on the bacterial population by the meio- and macrofauna. Assessment of bacterial production, according to this technique yielded values between 0,3 and 6 mg N m^{-2} d^{-1} (Table 2). The results obtained in this study (Hermin, 1989) showed that the bacterial biomass does not reflect changes in production (Fig. 5).

These two studies performed in the coastal areas using slightly degraded organic material as supplement, rapidly triggered enzymatic activity and bacterial production in the WSI. However, the effects on the biomass were attenuated by predators. On the basis of his *in vitro* study, Mayer (1989) concluded that EPA is closely related to the substrate or number of bacteria, except in the superficial sediment layer. The first stage in the degradation of OM, *i.e.* depolymerization of the substrate, appears essential to understanding OM flux in the underlying water and in interstitial water, as well as uptake by the sediment (King, 1986).

Deep Water Zones: Although organic input is greater in coastal than in deep water areas, the former account for only 14% of the ocean's surface. As along continental edges, settling fecal pellets and turbid currents in deep water are sources of fresh OM and plant debris from dry land. But these areas differ in the rate of sedimentation, the speed of bottom currents and the activity of micro-organisms.

Table 2. Bacterial production in superficial sediments was measured according to the rate of ^{3}H-Thymidine incorporation into DNA. The average cell volume was converted from N content assuming a C content of 22% of wet weight and C/N ratio of 6.25.

Type	Production (mg N m^{-2} d^{-1})	References
Inshore	6	Fallon *et al.* (1983)
Offshore	1	Fallon *et al.* (1983)
Seagrass	6	Moriarty *et al.* (1985)
Inshore	0.3 to 6	Hermin (1989)

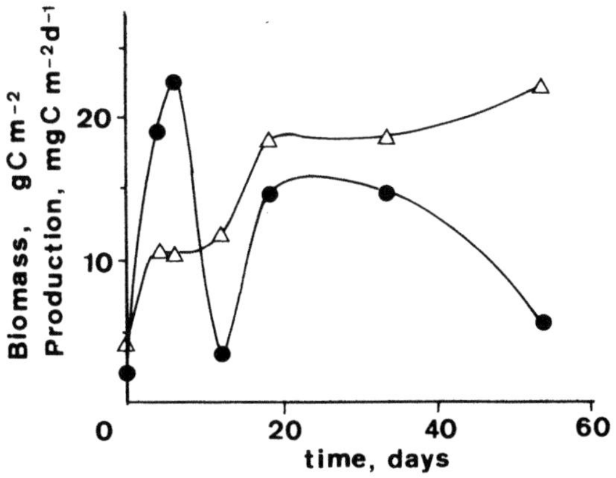

Fig. 5. Biomass (△) and bacterial production (●) in the top layer of sediment in Carteau Bay; evolution during *in situ* experiments between April 21 and June 15, 1988; depth: 4 m (after Hermin, 1989).

Degradation of OM in a western Mediterranean canyon: The purpose of this study (Buscail *et al.*, 1990), which was carried out at 650 m depth in Lacaze-Duthiers Canyon, was to analyze seasonal variations in incoming OM and changes in the WSI, and to diagram the budget of inputs and mineralization in the nitrogenous fraction.

The nitrogen content of suspended particles near the bottom varies from 0.2% in autumn to 4.5% in spring. Comparison with material collected at the surface showed that a great loss of nitrogenous material occurs between the top and the bottom of the water column (Table 3). The residual nitrogen fraction is made up mainly of amino acids (AA) and does not fluctuate in accordance with nitrogen flow. This finding could be explained by resuspension of the top layer of the WSI, release of AA from interstitial water or absorption of AA on clay particles (Daumas *et al.*, 1985; Hedges & Hare, 1987). The high AA content of particles collected in spring and in winter can be attributed to the mechanisms of primary production and to active degradation at these times respectively. The presence of β-glutamic acid, a non-protein AA reported by Henrichs & Farrington (1979) and by Stanley *et al.* (1987) in interstitial water sediment and

Table 3. Comparison of total nitrogen and amino acid content in particles in a sediment trap (600 m) and top sediment in Lacaze-Duthiers canyon (650 m) at different times of the year (from Buscail *et al.*, 1990).

	N %	amino-acids mg/g	C/N
Summer			
Particles	0.40	4.34	4.70
Sediments	0.16	0.87	4.90
Ratio	2.55	5.00	0.96
Autumn			
Particles	0.18	2.27	13.9
Sediments	0.12	1.47	6.90
Ratio	1.50	1.50	2.00
Winter			
Particles	0.20	8.90	8.00
Sediments	0.13	1.82	6.50
Ratio	1.54	4.90	1.20
Spring			
Particles	0.17	16.9	7.60
Sediments	0.13	7.19	6.70
Ratio	1.31	2.30	1.10

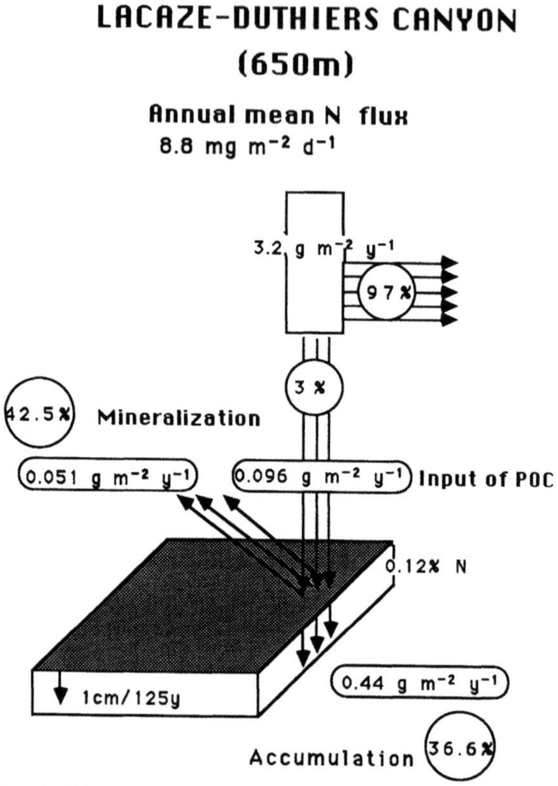

Fig. 6. Diagram of total nitrogen budget in superficial sediments sampled on adjacent slope of the Lacaze-Duthiers canyon at 650 m depth (from Buscail *et al.*, 1990).

sulfato-reducing bacteria, is evidence of intense bacterial activity in the spring.

Values measured in the sediment trap and samples collected in the top layer of sediment allowed estimation of the annual flow of nitrogenous substances on the continental slope (Fig. 6). The flow in the top layers was estimated to be 3.2 g m^{-2} y^{-1}, but only 3% of this amount reach the surface of the sediment. Accumulation of OM in the top layer was estimated to be 36.6% of incoming material. The mineralized fraction deduced from this percentage was 51.5 mg m^{-2} y^{-1} of nitrogen. The rate of accumulation of nitrogenous substances is proportionally higher than that calculated for carbohydrates (45.8% versus 33.3%). This may be due to humification which strengthens the fraction resistant to microorganisms (Jocteur-Monrozier, 1984).

Upwelling zones in Mauritania and in the northern Indian Ocean: Upwelling of bottom water leads to large planktonic blooms in the surface water (Minas *et al.*, 1982). This affects not only the biological chain but also input of phytodetritus on the bottom (Seibold, 1982). When the upwelling area is located off a desert area like Mauritania (northwest Atlantic) and the Arabian coasts (northern Indian Ocean), the OM can be considered to be essentially autochthonous. As part of the 'ORGON' program devoted to the study of the organic geochemistry of deep sea sediments, two missions were carried out in upwelling zones. Particles in the water near the bottom (> 5 m) and in the top layer of sediment were studied. The bathymetric profiles of these two regions (Fig. 7) are similar, and samples could be collected at the top and bottom of the continent slope, as well as from the abyssal plain. In both cases OM was of planktonic origin, but the oxygen deficiency reported in the Indian Ocean (Panakala Rao & Sayraman, 1970; Daumas *et al.*, 1981) was not observed off Mauritania (Daumas *et al.*, 1979).

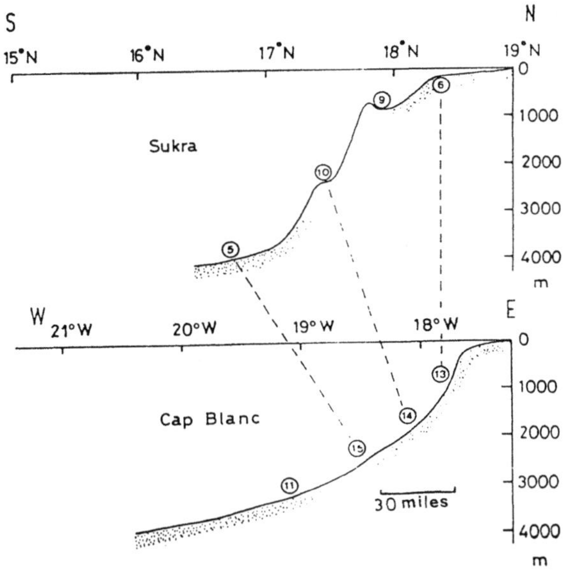

Fig. 7. Bathymetric profiles of Sukra transect (Northern Indian Ocean) and Cap Blanc transect (Northwestern Atlantic Ocean), (from Sautriot 1981).

Comparison of data (Table 4) from Sukra transect (Sta. 6) and Cap Blanc transect (Sta. 13), both of which are on the continental slope, disclosed differences in primary amine content, a parameter characteristic of OM lability (North, 1975; Daumas *et al.*, 1979; Jocteur-Monrozier, 1984). Contents were 2.7 times higher in the Indian Ocean than off Mauritania, even though the levels of organic carbon were almost the same at both sites. The difference was even greater between stations situated at the bottom of the slope (Sta. 10 and 14) or those on the abyssal plain (Sta. 5 and 15). The ratio of soluble carbohydrates to free amines confirms the very labile nature of OM in the top layer of sediment of the Indian Ocean. Degradability can also be estimated by comparing primary amine content in two sediment layers: 0–2 cm and 20 to 22 cm. The greatest decreases were observed at the level of the continental slope (Sta. 6 and 13). Overall, the decrease was greater at stations in the Indian Ocean (68.2%) than off Mauritania (39.1%). These results show that, even in deep water zones, oxygen content in the water column has a direct effect on the quality of OM settling on the bottom. Lower levels of dissolved oxygen seem to enhance conservation of labile nitrogenous fractions in the WSI. This has consequences for diagenesis in top sediment.

Table 4. Oxygen content, organic carbon, total nitrogen and total primary amines in superficial sediment (0–2 cm) collected off Sukra (Indian Ocean) and Cap Blanc (northwest Atlantic), (from Sautriot 1981).

	Sukra				Cap Blanc			
Stations	6	9	10	5	13	14	15	11
Depth (m)	200	830	2390	4010	910	1817	2552	3263
O_2 (mg l^{-1})	3.80	0.52	4.80	3.72	3.30	7.28	7.44	7.52
C (%)	1.32	6.88	2.70	2.06	1.30	2.45	2.20	0.90
N (%)	0.20	0.95	0.35	0.32	0.14	0.39	0.20	0.10
C/N	6.60	7.20	7.70	6.40	9.30	6.30	11.00	9.00
Primary amines (μM g^{-1} eq.leucine)	7.60	41.6	12.4	16.1	2.80	5.80	4.00	3.00
$N(-NH_2)$[a]	5.30	6.10	4.90	7.00	2.80	2.10	2.80	4.30
C(FCH)[b] / $N(-NH_2)$	2.60	1.44	2.45	1.84	4.35	5.33	8.37	3.14

[a] Primary amine nitrogen.
[b] Dissolved carbohydrate carbon: primary amine nitrogen ratios.

Conclusion

A wide variety of factors affect the WSI. I have tried to give an overview and to illustrate the role of the WSI in the transformation of the nitrogenous organic fraction using results of studies performed in coastal and deep water areas. Advances in our knowledge of these mechanisms have been made thanks to collection of samples in deep water areas and to improved 'in situ' experimental techniques for coastal areas. Progress is also linked to the development of new rapid, reliable and sensitive methods that enable monitoring of pre-diagenetic transformations of OM. Unlike old methods focusing on the elementary composition of OM, these new methods monitor variations in different classes of organic components such as carbohydrates, proteins or lipids and even of specific substances such as alanine, glutamic acid, leucine, and glucose. Progress has also been made in the understanding of the enzymatic activities of microorganisms. Finally, experimental investigation has benefited from the expanding use of 'tracers' which constitute the preferential tool for future studies. However, caution is necessary in extrapolating from biological mechanisms in the water column. The heterogeneity of sediment phases contrasts sharply with the relatively uniform conditions in the aqueous phase. Researchers are thus confronted with a difficult problem in choosing sampling points for sediment studies. Is it best to perform extensive sampling in a network of points across the whole area, and limit the number of parameters measured, or to analyze as many biological parameters as possible and limit the number of sampling points? This was well-illustrated in our study of decomposition of OM in the Indian Ocean (Sukra). Discrepancies were noted at a station located on the continental slope, as compared with stations located above and below. The level of organic carbon as well as amino acid and pheopigment contents were abnormally high. The results of stratigraphic and sedimentological studies (Moyes *et al.*, 1981) showed that this could be explained by the presence of a depression filled by accumulated OM. Thus decomposition of OM in the WSI is subject not only to physical, chemical and biological factors, but also to such difficult-to-detect 'sedimentary accidents'. The role of the superficial sediment in cycling nutrient elements has been known for a long time, but in the last ten years greater emphasis has been placed on mineralization rates, bacterial assimilation and sedimenting OM. However, this field of research is at present at an intermediate step between the old concept, in which the WSI is a 'black box' hiding the transformation of OM, and the new attempts to determine the rates at which detritus is metabolized. A comparison between freshwater and seawater WSI brings out some differences with respect to the input of detritus, which is more allochthonous in lakes and streams than in oceans. Another difference involves the role of sulfate-reducing bacteria that is essential in nearshore sediment, but is replaced in lakes by methanogenesis and acetate metabolism; also, macrofauna is more limited in lakes, where cladoceraceae are the main component, than in seashore regions which are inhabited by bacterivores (flagellates and sponges) and detritivores (copepods and euphausiace).

Freshwater and saltwater environments are similar with regard to the participation of microheterotrophs in the transformation of 'detritus'. Insofar as the abundance of bacteria in sediment as compared to overlying water or the role of microorganisms in the oxidation of OM and regeneration of nutrients are concerned, the processes used by microorganisms are identical, leading Wollast (1983) to state that 'a material sedimenting from the same water depth is identical in the two environments'. The WSI can be considered as a sink for particulate matter and biopolymers and as a source for organic monomers, NH_4^+ or nitrate, depending on their respective concentrations. Because of its involvement in mineral and organic cycles, the WSI can be considered as a crossroad for exchanges between aqueous and sediment layers.

The importance of the WSI in the oyster and mussel farming, as well as in extraction of sand and gravels and in fishing (due to its influence on benthic larvae), justifies further investigation of the mineralization of detritus and bacterial production.

Acknowledgements

I would like to thank R. Buscail and M.N. Hermin for the unpublished field and laboratory data cited, and an anonymous reviewer for valuable criticism. I gratefully acknowledge Danièle Mazaudier's help in the preparation of the revised draft of the manuscript.

References

Aller, R. C., 1978. Experimental studies of changes produced by deposit feeders on pore water, sediment and overlying water chemistry. Am. J. Sci. 278: 1185–1243.

Biddanda, B. A., 1986. Structure and function of marine microbial aggregates. Oceanol. Acta. 9: 209–211.

Biddanda, B. A. & L. R. Pomeroy, 1988. Microbial aggregation and degradation of phytoplankton-derived detritus in seawater. I. Microbial succession. Mar. Ecol. Prog. Ser. 42: 79–88.

Billen, G., 1984. Heterotrophic utilization and regeneration of nitrogen. In: J. E. Hobbie & P. J. le B. Williams (eds) Heterotrophic activity in the sea. Plenum Press, Lond.: 313–355.

Blum, L. K., A. L. Mills, J. C. Zieman & R. T. Zieman, 1988. Abundance of bacteria and fungi in seagrass and mangrove detritus. Mar. Ecol. Prog. Ser. 42: 73–78.

Bott, T. L., L. A. Kaplan & F. T. Kusesk, 1984. Benthic bacterial biomass supported by streamwater dissolved organic matter. Microb. Ecol. 10: 335: 344.

Buscail, R., R. Pocklington, R. Daumas & L. Guidi, 1990. Fluxes and budget of organic matter in the benthic boundary layer over the north western Mediterranean margin. Cont. Shelf Res. (in press).

Butsch, P. L. & W. Stumm, 1968. Chemical interaction in the aggregation of bacteria-bio-floculation in waste treatment. Envir. Sci. Technol. 2: 49–53.

Cahet, G., R. Daumas & M. Sibuet, 1990. In situ experimentation at the water-sediment interface in the deep sea: 2. Incorporation of labelled substrates in biopolymers at 2000 m depth in the Bay of Biscay. Progr. Oceanogr. 24 (in press).

Cho, B. C. & F. Azam, 1988. Major role of bacteria in biogeochemical fluxes in the ocean's interior. Nature, Lond. 332: 441–443.

Chu, J. P. H. & R. Knowles, 1966. Mineralization and immobilization of nitrogen in bacterial cells and in certain soil organic fractions. Soil. Sci. Soc. Am. Proc., 30: 210–213.

Colwell, R. R., P. Tabor & J. Deming, 1980. Degradation of organic material and remineralization by microorganisms at the water-sediment interface in deep ocean trenches. In: R. A. Daumas (ed.), Biogéochimie de la matière organique à l'interface eau-sédiment marin. CNRS, Paris: 263–268.

Craven, D. B. & A. F. Carlucci, 1989. Spring and fall microheterotrophic utilization of dissolved amino-acids in a California borderland basin. Mar. Ecol. Prog. Ser. 51: 229–235.

Daumas, R. A. & M. Bianchi, 1984. Bioturbation and microbial activity. Arch. Hydrobiol. Ergebn. Limnol. 19: 284–294.

Daumas, R. A., M.-N. Hermin & N. Drogue, 1985. Dissimulation non-biologique de traceurs organiques par les sédiments marins. Colloque franco-japonais, (Marseille 16–21 Sept. 1984), Oceanis 4: 73–80.

Daumas, R. A., P. Laborde, J. C. Romano & D. Sautriot, 1979. Distribution et évolution des constituants biogéochimiques de la matière organique dans les sédiments. Relations avec la composition des eaux interstitielles. In: M. A. Arnould & R. Pelet (eds.) Géochimie organique des sédiments marins profonds, Orgon III: Mauritanie, Iles du Cap Vert. C.N.R.S., Paris: 67–92.

Daumas, R. A., P. Laborde, J.-C. Romano & D. Sautriot, 1981. Hydrologie et particules en suspension en mer d'Arabie. In: Géochimie organique des sédiments marins profonds. Orgon IV: Golfe d'Aden, mer d'Oman. CNRS, Paris: 71–94.

Dawson, R. & C. Liebezeit, 1982. Biochemical compounds in the pelagic and sedimentary environment of Antarctic waters. In: G.A.B.I.M. (ed.), Indices Biochimiques et milieux marins. C.N.E.X.O., Brest: 67–86.

Degens, E. T., 1970. Molecular nature of nitrogenous compounds in sea water and recent marine sediment. In: D. W. Hood, Organic matter in natural waters. Inst. Mar. Sci., Univ. Alaska, Occ. Publ., 1: 77–106.

Ducklow, H. W., S. M. Hill & W. D. Gardner, 1985. Bacterial growth and the decomposition of particulate organic carbon collected in sediment traps. Cont. Shelf Res. 4: 445–464.

Fallon, R. D., S. Y. Newell & C. S. Hopkinson, 1983. Bacterial production in marine sediments: Will cell-specific measures agree with whole-system metabolism? Mar. Ecol. Prog. Ser. 11: 119–127.

Fenchel, T. & T. H. Blackburn, 1979. Bacteria and mineral cycling. Acad. Press. N.Y., 225 pp.

Feral, J. P., 1989. Activity of the principal digestive enzymes in detrivorous apodous holothuroid *Leptosignapta gallienne*i and two other shallow-water holothuroids. Mar. Biol. 101: 367–379.

Fontigny, A., G. Billen & J. Vives-Rego, 1987. Some kinetic characteristics of exoproteolytic activity in coastal seawater. Estuar. coast. Shelf. Sci. 25: 127–133.

Gerlach, S. A., 1972. Die Produktionleistung der Benthos in der Helgoländer Bucht. Verh. dt. Zool. Ges. 65: 1–13.

Golterman, H. L., 1984. Sediments, modifying and equilibrating factors in the chemistry of freshwater. Verh. int. Ver. Limnol. 22: 23–59.
Gyllenberg, G. & E. Eklund, 1974. Bacteria, In: C. H. Dickinson & G. J. F. Pugh, (eds), Biology of plant litter decomposition, vol. 2, Acad. Press. N.Y. 245–268.
Hayes, F. R., 1964. The mud-water interface. Oceanogr. Mar. Biol. Rev. 2: 121–145.
Hedges, J. J. & P. E. Hare, 1987. Amino acid adsorption by clay minerals in distilled water. Geochim. Cosmochim. Acta. 5: 255–259.
Henrichs, S. M. & J. W. Farrington, 1979. Amino acids in interstitial waters of marine sediments. Nature 279: 319–322.
Hermin, M.-N., 1989. Dégradation microbienne de la matière organique à l'interface eau sédiment en milieu marin. Thèse Doct., Univ. Aix-Marseille II, 202 pp.
Honjo, S., S. J. Manganini & J. J. Cole, 1982. Sedimentation of biogenic matter in the deep ocean. Deep-Sea Res. 29: 609–625.
Janel, Ph., 1978. Etude saisonnière des formes de l'azote organique dans trois types d'humus sous hêtraie: moder, mull acide, mull calcaire; Thèse Doct. 3° cycle, Univ. Nancy I, 63 pp.
Jocteur-Monrozier, L., 1984. Nature et evolution de l'azote organique dans les sols et les sédiments marins récents. Thèse Doct. ès Sci., Univ. Nancy I, 205 pp.
Johnson, R. G., 1974. Particulate matter at the sediment-water interface in coastal environments. J. mar. Res. 32: 313–330.
Jørgensen, N. O. G., 1982. Heterotrophic assimilation and occurrence of dissolved free amino-acids in a shallow estuary. Mar. Ecol. Prog. Ser. 8: 145–159.
Jørgensen, N. O. G., P. Lindroth & K. Mopper, 1981. Extraction and distribution of free amino-acids and ammonium in sediment interstitial waters from the Lingfjord, Denmark. Oceanol. Acta 4: 465–474.
Karl, D. M., G. A. Knauer & J. H. Martin, 1988. Downward flux of particulate organic matter in the ocean: a particle decomposition paradox. Nature, Lond. 332: 438–441.
King, G. M., 1986. Characterization of β-glucosidase activity in intertidal marine sediments. Appl. envir. Microbiol. 51: 373–380.
Krom, M. D. & J. T. Westrich, 1980. Dissolved organic matter in the pore waters of recent marine sediments: a review. In: R. Daumas (ed.), Biogéochimie de la matière organique à l'interface eau-sediment marin. CNRS, Paris: 103–111.
Mc Cave, I. N., 1976. Benthic boundary layer, Plenum press, N-Y: 212 pp.
Mann, K. H., 1988. Production and use of detritus in various freshwater, estuarine, and coastal marine ecosystems. Limnol. Oceanogr. 33: 910–930.
Marinucci, A. C., J. E. Hobbie & J. V. K. Heltrich, 1983. Effects of bitter nitrogen on decomposition and microbial biomass in *Spartina alterniflora*. Microb. Ecol. 9: 27–40.
Marshall, K. C., 1976. Interfaces in microbial ecology. Harvard Univ. Press, Cambridge, Mass., 156 pp.
Mayer, L. M., 1989. Extracellular proteolytic enzyme activity in sediments of an intertidal mudflat. Limnol. Oceanogr. 34: 973–981.
Meyer-Reil, L. A., 1986. Mesurement of hydrolytic activity and incorporation of dissolved organic substrates by microorganisms in marine sediment. Mar. Ecol. Prog. Ser. 31: 143–146.
Meyer-Reil, L. A., R. Dawson, G. Liebezeit & H. Tiedge, 1978. Fluctuations and interactions of bacterial activity in sandy beach sediments and overlying waters. Mar. Biol. 48: 161–171.
Minas, H. J., L. A. Codispoti & R. C. Dugdale, 1982. Nutrients and primary production in the upwelling region off North Western Africa. Rapp. int. Counc. Explor. sea 180: 148–183.
Moriarty, D. J. W. & A. C. Hayward, 1982. Ultrastructure of bacteria and the proportion of Gram-negative bacteria in marine sediments. Microb. Ecol. 8: 1–14.
Moriarty, D. J. W., P. I. Boon, J. A. Hansen, W. G. Hunt, I. R. Poiner, P. C. Pollard, G. W. Skiring & D. C. White, 1985. Microbial biomass and productivity in seagrass beds. Geomicrobiol. J., 4: 21–51.
Moyes, J., J. Duprat, J-C. Faugères, E. Gonthier & C. Pujol, 1981. Etude stratigraphique et sedimentologique. In: Géochimie organique des sédiments marins profonds, Orgon IV: Golfe d'Aden, mer d'Oman, CNRS, Paris: 189–263.
Newell R. C., E. A. S. Linley & M. I. Lucas, 1983. Bacterial production and carbon conversion based on saltmarsh plant debris. Estuar. coast. Shelf Sci. 17: 405–419.
Newell R. C., M. I. Lucas & E. A. S. Linley, 1981. Rate of degradation and efficiency of conversion of phytoplankton debris by marine microorganisms. Mar. Ecol. Prog. Ser. 6: 123–136.
North, B. B., 1975. Primary amines in California coastal waters: utilization by phytoplankton. Limnol. Oceanogr. 20: 20–27.
Novitsky, J. A., 1983. Heterotrophic activity throughout a vertical profile of seawater and sediment in Halifax harbor, Canada. Appl. envir. Microbiol. 45: 1761–1766.
Novitsky, J. A., 1986. Degradation of dead microbial biomass in a marine sediment. Appl. envir. Microbiol. 52: 504–509.
Olah, G. M., O. Reisenberg & G. Kilbertus, 1978. Biodegradation et humification. Atlas ultrastructural. Presses de l'Université Laval (Québec), 331 pp.
Panakala Rao, D. & G. Sayraman, 1970. On the occurrence of oxygen maxima and minima in the upper 500 meters of the North-Western Indian Ocean. Proc. Indian Acad. Sci., (Bengalore) 81: 230–246.
Pillon, P., 1979. Caractérisation organo-minérale dans différents milieux de sédimentation marine. D. E. A. Agro.-Eco.-Pédologie, Univ. Nancy I, 35 p.
Rault, P. L., L. P. Sohier, A. M. Rivier & R. A. Daumas,

1988. Dark bacterial-protozoan culture: acidification, nitrite accumulation, and protozoan inhibition in absence of phytoplankton. J. exp. mar. Biol. Ecol. 116: 273–291.

Revsbech, N. P., J. Nielsen & P. K Hansen, 1988. Benthic primary production and oxygen profile. In: T. H. Blackburn & J. Sørensen (eds), Nitrogen cycling in coastal marine environment. J. Wiley & Sons Ltd, N.Y.: 69–83.

Romano, J. C., D. Sautriot, R. Daumas & P. Laborde, 1981. Les eaux interstitielles: matière organique dissoute et produits de minéralisation. In: Géochimie organique des sédiments marins profonds. Orgon IV: Golfe d'Aden, mer d'Oman. CNRS, Paris: 95–132.

Rosenfeld, J. R., 1979. Ammonium adsorption in nearshore anoxic sediments. Limnol. Oceanogr. 24: 356–364.

Roussel, M., 1978. Etude de la fraction organique associée aux particules de différentes tailles dans des milieux de sédimentation terrestres et marins. D.E.A. Agro-Eco-Pédologie, Univ. Nancy I, 35 p.

Rowe, G. T. & N. Staresinic, 1979. Sources of organic matter to the deep-sea benthos. Ambio, Special Report, 6: 19–23.

Russel, H. L., 1892. Bacterial investigation of the sea and its floor. Bot. Gaz. 17: 312–321.

Russel, H. L., 1893. The bacterial flora of the Atlantic Ocean in the vicinity of Woods Hole, Mass. Bot. Gaz. 18: 383–395.

Saliot, A., 1977. Etude comparative des acides gras de l'eau et de l'eau interstitielle à l'interface eau de mer profonde-sédiment. C.R. Acad. Sci., Paris, 285 (Section D): 857–860.

Sautriot, D., 1981. Composés biogéniques fixés au sédiment. In: Géochimie organique des sédiments marins profonds. Orgon IV: Golfe d'Aden, mer d'Oman. CNRS, Paris: 133–157.

Seibold, E., 1982. Sediments in upwelling areas, particularly off Northwest Africa. Rapp. int. Counc. Explor. Sea 180: 315–322.

Schwinghamer, P. & P. Kepkay, 1987. Effects of experimental enrichment with Spartina detritus on sediment community biomass and metabolism. Biol. Oceanogr. 4: 289–322.

Sikora, J. P., W. B. Sikora, C. W. Erkenbrecher & B. D. Coull, 1977. Significance of ATP, carbon and caloric of meiobenthic nematodes in partitioning benthic biomass. Mar. Biol. 44: 7–14.

Stanley, S. O., K. G. Boto, D. M. Alongi & F. T. Gillan, 1987. Composition and bacterial utilisation of free amino-acids in tropical mangrove sediments. Mar. Chem. 22: 13–30.

Stuart, V., R. C. Newell & M. I. Lucas, 1982. Conversion of kelp debris and faecal material from the mussel *Aulocomya ater* by marine microorganisms. Mar. Ecol. Prog. Ser. 7: 47–57.

Sundbäck, K. & W. Graneli, 1988. Influence of microphytobenthos on the nutrient flux between sediment and water: a laboratory study. Mar. Ecol. Prog. Ser. 43: 63–69.

Taylor, G. T. & C. W. Sullivan, 1984. The use of ^{14}C labeled bacteria as a tracer of ingestion and metabolism of bacterial biomass by microbial grazers. J. microbiol. Methods 3: 101–126.

Trichet, J., 1967. Essai d'explication du dépot d'aragonite sur des substrats organiques. C.R. Acad. Sci. Paris, 265: 1464–1467.

Vosjan, J. H, 1987. A sketchy outline of the fate of organic matter in the Dutch Wadden Sea. Hydrobiol. Bull. 21: 127–132.

Ward, D. M. & M. R. Winfrey, 1985. Interactions between methanogenic and sulfate-reducing bacteria in sediments. In: H. W. Jannasch & P. J. Le B. Williams (eds), Advances in Aquatic Microbiology 3, Acad. Press. Lond.: 142–179.

Watling, L., 1988. Small-scale features of marine sediments and their importance to the study of deposit-feeding. Mar. Ecol. Prog. Ser. 47: 135–144.

Wetzel, R. G., P. H. Rich, M. C. Miller & H. L. Allen, 1972. Metabolism of dissolved and particulate detrital carbon in a temperate, hard-water lake. Mem. Ist. ital. Idrobiol. (Suppl.) 29: 185–243.

Widbom, B. & R. Elingren, 1988. Reponse of the benthic meiofauna to nutrient enrichment of experimental marine ecosystems. Mar. Biol. Prog. Ser. 42: 257–268.

Williams, P. J. le B., 1984. Bacterial production in the marine food chain; the Emperor's new suit of clothes. In: M. J. R. Fasham (ed.) Flows of energy and material in marine ecosystems, theory and practice. Nato Conference Series, Plenum Press. N.Y.: 271–299.

Wollast, R., 1983. Interactions in estuaries and coastal waters. In: B. Bolin & R. B. Cook (eds), The major biogeochimical cycles and their interactions. J. Wiley & Sons Ltd. N.Y.: 385–407.

ZoBell, C. E., 1946. Marine Microbiology. Chronica Botanica Co, Waltham. Mass. 240 pp.

Received 3 August 1989; in revised form 20 April 1990; accepted 15 June 1990

Hydrobiologia **207**: 31–36, 1990.
D. J. Bonin & H. L. Golterman (eds), Fluxes Between Trophic Levels and Through the Water–Sediment Interface.

Long-term changes in zooplankton abundance and water transparency in Lake Geneva

Gérard Balvay
Institut de Limnologie I.N.R.A., B.P. 511, F-74203 Thonon-les-Bains Cedex, France

Key words: Lake Geneva, eutrophication, zooplankton abundance, water transparency, seasonal and annual variations

Abstract

The water quality of Lake Geneva has declined steadily since the 1960s, due to a continuous increase of external phosphorus loading. Average P level in the lake increased steadily to a peak in 1979, and even 1981 in the case of P content in the trophogenic layer. Since then, reduced external inputs related to the delayed effects of phosphorus removal from waste waters initiated many years previously has led to a decrease in P level, and resulted in present stabilization and even improvement in water quality. Long-term changes in zooplankton abundance correspond quite closely to eutrophication level changes. After increasing since the 1960s, maximum zooplankton biomass was recorded for the first time in 1971; a second main peak appeared in 1981 together with the highest eutrophication level. Over the last seven years, zooplankton abundance has decreased continuously, while water transparency has decreased and phytoplankton production has remained at a high level.

Introduction

Eutrophication induces both abiotic and biotic responses, in small water bodies as well as in the largest lakes, and many lakes, such as Lake Constance (Walz *et al.*, 1987), Kasumigaura (Toda *et al.*, 1981), Tahoe (Goldman, 1988) and Piburger (Schaber, 1975; 1976, 1977), are good examples of eutrophication-induced changes. The main biotic responses involve not only changes in species composition and increases in phytoplankton abundance, but also increases in zooplankton biomass (Spodniewska *et al.*, 1973; Skogheim & Rognerud, 1978; Balvay, 1984) and marked changes in zooplankton community structure (Brooks, 1969).

The aim of this work is to present the changes observed in water transparency and zooplankton biomass during the continuous increase and the subsequent reversal of eutrophication in Lake Geneva, one of the largest volumes of fresh water in Europe ($V = 89\ km^3$, $S = 582\ km^2$, $Z_{max} = 309$ m). This lake was subject to accelerated eutrophication from the late 1950s until 1980 due to an increase of nutrient inputs, with maximum total phosphorus concentration in the whole water mass peaking in 1979 (90 $\mu g\ P\ l^{-1}$). Since then, phosphorus concentrations have levelled off, reaching 62 $\mu g\ P\ l^{-1}$ in 1988 (Blanc *et al.*, 1989). In the trophogenic layer (0–20 m), P content increased until 1981 before subsequently declining.

The continuous changes observed in the trophic status of Lake Geneva related to water phosphorus content are accompanied by marked variations in water transparency and zooplankton abundance.

Material and methods

The data presented here are from sampling station SHL 2, located at the centre of the lake, midway between Evian and Lausanne, at a depth of 309 m, sampled monthly in winter, and at least twice a month during the other seasons.

Water transparency was estimated by Secchi disk measurements. Zooplankton abundance was obtained by sedimentation of samples collected by vertical tows from a depth of 50 m to the surface, using a net with a 200 μm mesh. Zooplankton biomass is expressed as fresh weight (FW in g m^{-2}) according to the relation $FW = 0.195\ V^{0.948}$ (Balvay, 1987), where V = volume of settling zooplankton (in ml).

The non-homogeneous distribution of plankton caused by wind-induced dislocations of water bodies has often been observed in Lake Geneva. The use of monthly averages can compensate for spatial heterogeneity to a certain extent and long-term tendencies can be detected despite great year-to-year variability.

Results

Transparency

Highest transparency values (Fig. 1) appear in winter, generally in February, and the lowest in May during the phytoplankton spring-bloom. Water transparency remains quite low in late summer before again increasing in autumn and winter.

Over the last 27 years, transparency trends can be described as follows:

1. a gradual reduction in winter, from a maximum in February 1968 (20.1 m) to a minimum in 1982–1984 (10.2 m). Since then a slight increase took place in 1985 (15.2 m) and 1987 (13.8 m).
2. A constant decline in spring (May) during the period of maximum phytoplankton abundance. Generally greater than 5 m until 1969,

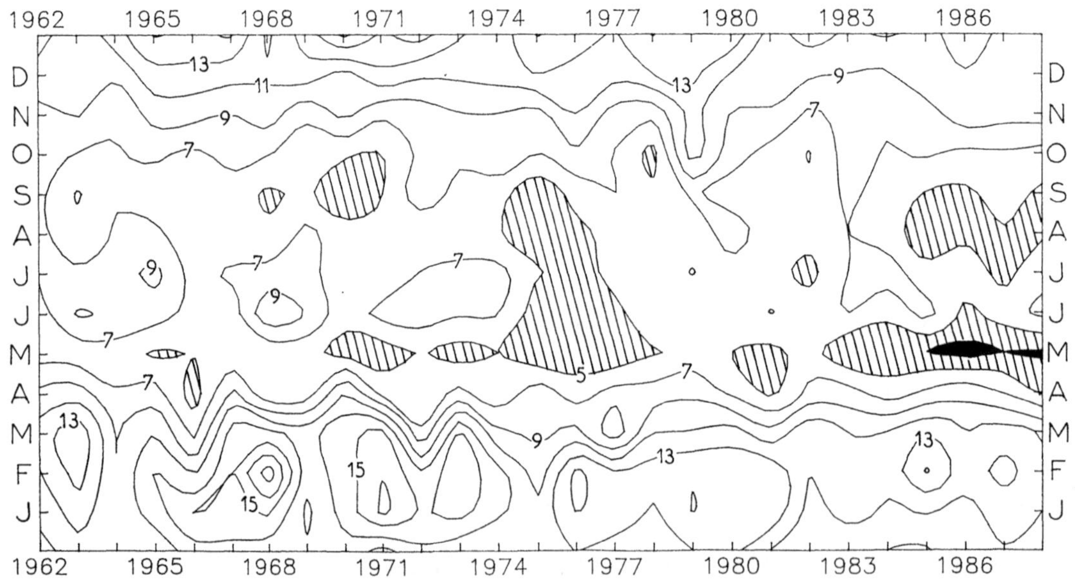

Fig. 1. Annual (X-axis) and seasonal (Y-axis) variations in Secchi disk transparency (m).
shaded areas: transparency <5 m;
black areas: transparency <3 m.

transparency was very often less than 5 m until 1984, and even less than 3 m from 1985 to 1988.

3. Changes in summer (June–September); transparency was temporarily greater than 9 m until 1968 and often greater than 7 m till 1974, then decreased sharply, first in 1969–1970 and more intensively from 1974. After some years of slight improvement, summer transparency has decreased further since 1985, with the appearance of two well-marked periods of low Secchi disk values during this season.

Zooplankton biomass

Zooplankton population is low in winter (Fig. 2), begins to increase in April, and reaches its maximum abundance in June. Zooplankton remain abundant in summer and decrease in autumn. Winter abundance of zooplankton, generally below 10 g m^{-2}, did not show substantial changes from 1962 to 1988.

The most substantial changes were observed in spring and more specially in June. June abundance was generally below 50 g m^{-2} before 1971, but exceeded 50 g m^{-2}, first from 1971 to 1974, and then again from 1977 to 1987, with maximum biomass in 1981 and 1982 (respectively 115 and 120 g m^{-2}). A smaller increase later in the season occurred in 1983 and 1984. In June 1988, the peak decreased again to less than 50 g m^{-2}.

Discussion

Formerly oligotrophic, Lake Geneva was subject to accelerated eutrophication with an increase of nearly one order of magnitude in the average annual concentration of total P in the whole water mass (from about 10 μg P l^{-1} before the sixties, to 90 μg P l^{-1} in 1979). Since then phosphorus concentration continues to decline, to 62 μg P l^{-1} in 1988 (Blanc *et al.*, 1989). In the trophogenic layer (0–20 m) phosphorus content peaked in 1981 (not 1979), and has also tapered off since then (Balvay & Pelletier, 1988). Decrease of P concentrations in lake water from the beginning of the eighties are probably connected with reduced external loading (phosphorus removal in waste

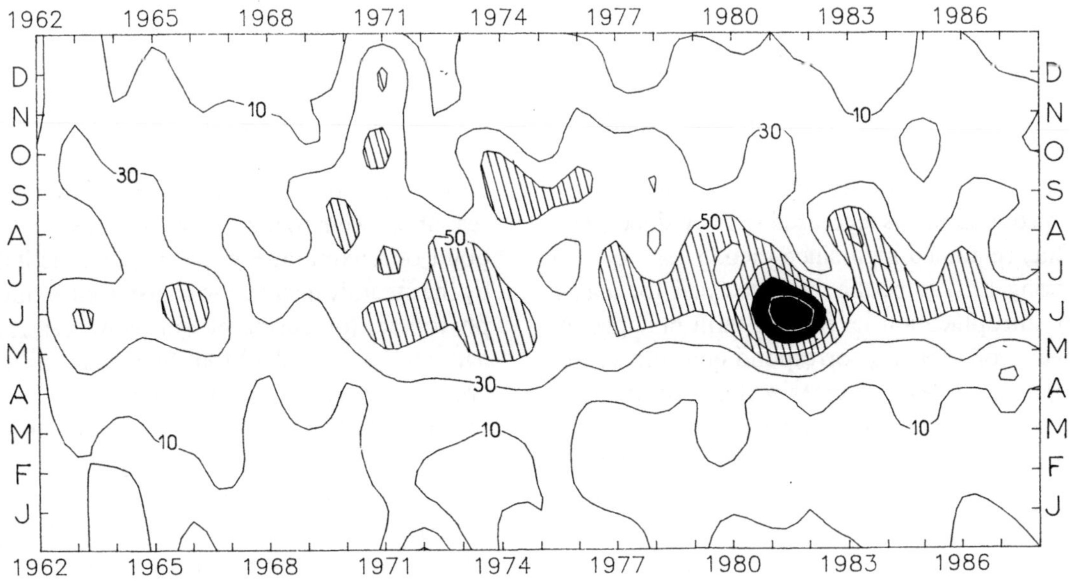

Fig. 2. Annual (X-axis) and seasonal (Y-axis) variations in zooplankton biomass estimated from settling volumes (mg m^{-2}).
shaded areas: zooplankton biomass > 50 mg m^{-2};
black areas: zooplankton biomass > 90 mg m^{-2}.

waters; ban of phosphates in detergents in Switzerland from July 1st 1986).

In the late nineteenth century, measurements of Secchi depth by Forel (1895) demonstrated that seasonal variations of transparency were quite regular, following a bell-shaped curve with the annual minimum in June. Over the last three decades, the seasonal pattern of transparency presents a distinct contrast, with two minima, the first one very pronounced in May and the second in late summer, separated by a June clear-water phase (Balvay *et al.*, 1990). This zooplankton induced clear-water phase does not appear clearly in Fig. 1, due to the use of monthly averages for Secchi disk values.

Transparency is negatively correlated with chlorophyll *a* concentration, and thus gives an estimate of phytoplankton abundance. In Lake Constance pigment concentration changes can explain about 80 per cent of the variations in water transparency (Tilzer, 1988); in Lake Geneva, more than 70 per cent of the variation of transparency can be explained by chlorophyll *a* concentration (Balvay *et al.*, 1990).

Average annual concentration of chlorophyll *a* and annual net primary production exhibited a quasi-cyclical variation with concomitant maxima in 1976 and 1981 (Fig. 3). Moreover, despite a sharp decrease in chlorophyll concentration in 1982, primary production remained at the same high level during that year. Since 1984, chlorophyll *a* concentration has fluctuated from 5.6 to 6.3 mg m^{-3} without a distinct peak, whereas primary production has increased sharply since 1986 and has remained high since then.

Taking 1981 to be the reference year for maximum eutrophication level, long-term changes in zooplankton biomass correspond quite closely to eutrophication level changes. After a continuous increase during the 1960s, maximum zooplankton biomass was first recorded in 1971, and then again in 1981 together with the highest eutrophication level defined by the maximum phosphorus content in the trophogenic layer. Since then zooplankton biomass has steadily decreased, except in 1986 when perch population was very low (Gerdeaux *et al.*, 1989) and the enhancement of primary production probably induced a momentary break in the previously continuous decline in zooplankton abundance. Food resources seem to have increased steadily past the point of maximum P concentration, as indicated by the steadily decreasing transparency during the phytoplankton spring-bloom. However, the average annual zooplankton biomass has not increased in recent years; regression of nanoplankton to the benefit of colonial or large-sized algae is probably one of the causes of this decrease in zooplankton abundance (Druart *et al.*, 1989). On the other hand, the seasonal and long-term changes in N/P and Si/P ratios govern specific composition, seasonal successions and long-term changes in phytoplankton (Gawler *et al.*, 1986), factors which in return act on the zooplankton

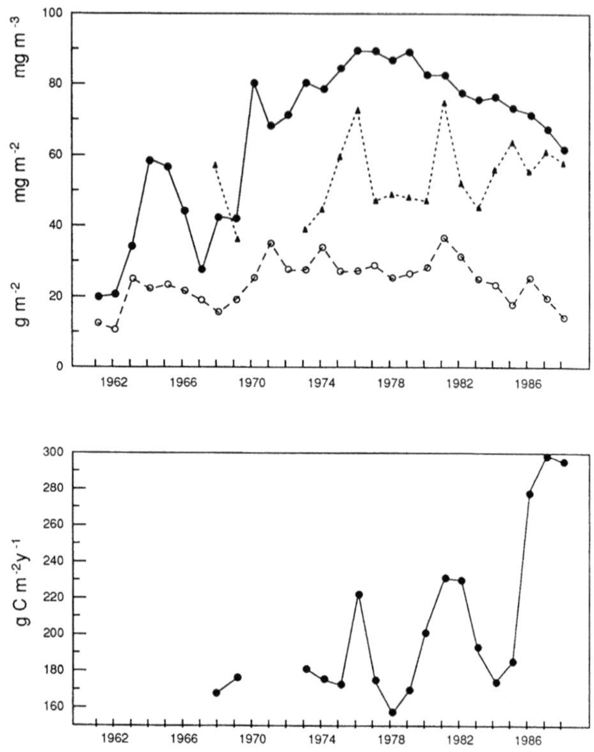

Fig. 3. Changes in average annual chlorophyll *a* concentration (mg m^{-2}), total phosphorus (mg m^{-3}) and average annual zooplankton biomass (fresh weight, g m^{-2}) (upper graph), and net primary production (g C m^{-2} y^{-1}) (lower graph).

(●) total phosphorus; (▲) chlorophyll *a*; (○) zooplankton fresh weight.

community through its food resources. The decline of zooplankton biomass can also be related to increase in planktivorous fish biomass (mainly whitefish and perch) during the last decade (Gerdeaux *et al.*, 1989).

The general increase of zooplankton biomass till the beginning of the eighties can be related to increase of food availability and abundance, as a consequence of an increased state of trophy in Lake Geneva. Such eutrophication-induced changes in zooplankton biomass have been observed many times in many other lakes (De Bernardi & Soldavini, 1976; George & Harris, 1985; Internat. Gewäss. Bodensee, 1987; Einsle, 1988). It can be said that nutrient enrichment, by promoting an increase in standing crop and production of phytoplankton, can result in an increase in zooplankton biomass. However, as shown by the non-homogeneous changes observed in transparency, chlorophyll *a* content, primary production and zooplankton biomass, the return from eutrophication seems to involve a period of imbalance, the importance of which is undoubtedly related to the size of the lake and its residence time (12 years).

Acknowledgements

I thank G. Chapuis et J. P. Moille for typing the manuscript and drawing the diagrams, and N. Angeli and J. Pelletier for their pertinent comments. This study had the support of the 'Institut National de la Recherche Agronomique' and the 'Commission Internationale pour la Protection des Eaux du Léman contre la Pollution'.

References

Balvay, G., 1984. Premières données sur l'évolution quantitative des Entomostracés planctoniques dans le lac Léman. Schweiz. Z. Hydrol. 46: 76–80.

Balvay, G., 1987. Equivalence entre quelques paramètres estimatifs de l'abondance du zooplancton total. Schweiz. Z. Hydrol. 49: 75–84.

Balvay, G. & J. P. Pelletier, 1988. Incidences des mesures de protection sur l'évolution de la qualité des eaux du Léman et du lac d'Annecy. In: G. Balvay (ed.), Eutrophication and lake restoration. Water quality and biological impacts. Thonon-les-Bains, France: 113–128.

Balvay, G., M. Gawler & J. Pelletier, 1990. Lake trophy and the development of the Clear-water Phase in Lake Geneva. In: M. M. Tilzer & C. Serruya (eds.), Ecological structure and function in large lakes. New Monograph Series, Science-Tech Madison: 581–591.

Blanc, P., C. Corvi & F. Rapin, 1989. Evolution physico-chimique des eaux du Léman. Campagne 1988. Rapp. Comm. int. Protection des eaux du Léman contre la pollution. Lausanne, 1989: 23–43.

Brooks, J. L., 1969. Eutrophication and changes in the composition of the zooplankton. In: Nat. Acad. Sci. Washington (ed.), Eutrophication: causes, consequences, correctives. Proc. int. Symp. Madison (Wisc.), 11–15 June 1967: 236–255.

De Bernardi, R. & E. Soldavini, 1976. Long term fluctuations in Lake Mergozzo, Northern Italy. Mem. Ist. ital. Idrobiol. 33: 345–375.

Druart, J. C., G. Balvay, J. Pelletier & R. Revaclier, 1989. Evolution du plancton du Léman, campagne 1988. Rapp. Comm. int. Protection des eaux du Léman contre la pollution, Lausanne, 1989: 51–70.

Einsle, U., 1988. The long-term dynamics of crustacean communities in Lake Constance (Obersee, 1962–1986). Schweiz Z. Hydrol. 50: 136–163.

Forel, F. A., 1895.Le Léman: monographie limnologique. F. Rouge (ed.), Lausanne, 651 p.

Gawler, M., P. Blanc, J. C. Druart & J. P. Pelletier, 1986. Dynamique de quelques populations majeures du phytoplancton printanier du lac Léman en relation avec le broutage et les sels nutritifs. Coll. nation. C.N.R.S. 'Biologie des Populations', Lyon, 4–6 septembre 1986. Univ. Claude Bernard, Lyon, France: 412–419.

George, D. G. & G. P. Harris, 1985. The effect of climate on long-term changes in the crustacean zooplankton biomass of Lake Windermere, UK. Nature 316: 536–539.

Gerdeaux, D., B. Buttiker & D. Pattay, 1989. La pêche et les recherches piscicoles en 1988 sur le Léman. Rapp. Comm. int. Protection des eaux du Léman contre la pollution, Lausanne, 1989: 79–87.

Goldman, C. R., 1988. Primary productivity, nutrients, and transparency during the early onset of eutrophication in ultra-oligotrophic Lake Tahoe, California-Nevada. Limnol. Oceanogr. 33: 1321–1333.

Internationale Gewässerschutzkommission für den Bodensee, 1987. Die Entwicklung des Crustaceen-Planktons im Bodensee-Obersee (1972–1985) und Untersee-Gnadensee und Rheinsee (1974–1985). Ber. Int. Gewässerschutzkomm. 37, 103 p.

Schaber, P., 1975. Rotatorien und Crustaceen des Piburger Sees 1971–1973. Jber. Abt. Limnol. Innsbruck 1: 59–72.

Schaber, P., 1976. Rotatorien und Crustaceen des Piburger Sees 1973–1975. Jber. Abt. Limnol. Innsbruck 2: 78–94.

Schaber, P., 1977. Rotatorien und Crustaceen des Piburger Sees 1976. Jber. Abt. Limnol. Innsbruck 3: 119–132.

Skogheim, O. K. & S. Rognerud, 1978. Recent changes in

plankton communities and present trophic state of Lake Steinfjord. Arch. Hydrobiol. 83: 179–199.
Spodniewska, I., A. Hillbricht-Ilkowska & T. Weglenska, 1973. Long-term changes in the plankton of eutrophic Mikolajskie Lake as an effect of accelerated eutrophication. Bull. Acad. Pol. Sci. 21: 215–221.
Tilzer, M. M., 1988. Secchi disk-chlorophyll relationships in a lake with highly variable phytoplankton biomass. Hydrobiologia 162: 163–171.
Toda, H., A. Hamada & S. Ichimura, 1981. Acceleration of the eutrophication process and community changes in biotic structure due to human activities in Lake Kasumigaura. Verh. int. Ver. Limnol. 21: 646–651.
Walz, N., H. Elster & M. Mezger, 1987. The development of the rotifer community structure in Lake Constance during its eutrophication. Arch. Hydrobiol. suppl. 74: 452–487.

Received 27 June 1989; in revised form 29 January 1990; accepted 26 February 1990

Hydrobiologia **207**: 37–42, 1990.
D. J. Bonin & H. L. Golterman (eds), Fluxes Between Trophic Levels and Through the Water–Sediment Interface.

Dynamics of bacterioplankton in oligotrophic and eutrophic aquatic environments: bottom-up or top-down control?

Gilles Billen, Pierre Servais & Sylvie Becquevort
Groupe de Microbiologie des Milieux Aquatiques, Campus de la Plaine, CP 221, boulevard du Triomphe, 1050 Bruxelles, Belgium

Key words: Bacterioplankton, bacterial growth rate, bacterial mortality, bacterial grazing, oligotrophy, eutrophy

Abstract

Measurements of bacterial biomass, production and mortality have been carried out in a large range of aquatic environments, including eutrophic and oligotrophic ones. The general trends of variations of bacterial biomass, size, specific growth rate and mortality rate in all these environments are examined. The overall flux of bacterial production is taken as an index of the flux of organic matter available to bacteria, thus characterizing the richness of the environment. Bacterial biomass is roughly proportional to richness, while mean cell size increases with it. The turnover rate of biomass, as revealed either by growth or by mortality rates, appears to be fairly independent of richness.

These observations are compatible with a simple resource-limited (bottom-up controlled) model of the dynamics of bacterioplankton. On the other hand, they are in contradiction with the predictions of a predator-controlled (top-down controlled) model.

Introduction

For about ten years, methods have been available for measuring both biomass and production of bacteria in aquatic environment (Daley & Hobbie, 1975; Fuhrman & Azam, 1980, 1982). More recently, methods have been described for evaluating grazing and mortality rates of bacterioplankton (Servais *et al.*, 1985, 1989; McManus & Furhman, 1988). Altogether, these methods provide a good description of the dynamics of heterotrophic bacterioplankton in aquatic environments (see Billen *et al.*, 1988a). Using these methods, the Groupe de Microbiologie des Milieux Aquatiques at the University of Brussels has carried out studies on the microbial ecology of a large number of marine and freshwater aquatic systems, ranging from highly eutrophic (the Scheldt estuary and its tributaries (Billen *et al.*, 1988b); the Meuse river (Servais, 1989)) to mesotrophic (the Belgian coastal zone of the North Sea (Billen & Fontigny, 1987)) and oligotrophic (the Southern Ocean (Billen *et al.*, 1987) or the drinking water from the distribution network in Parisian suburbs (Servais *et al.*, in press)). Because the same methods were used with only minor modifications in all these environments, these studies offer the possibility of comparing bacterial dynamics in systems of very different trophic status.

This paper presents a synthesis of these data. The basic question addressed is that of the control of bacterioplankton and its dependence on the richness of the environment. More specifically: is heterotrophic bacterioplankton biomass controlled by 'bottom-up' effects *i.e.* by the resources in available organic matter, or by 'top-down' effects, *i.e.* by the grazing pressure exerted by predators?

For the needs of the discussion, a suitable index of richness of the environment has to be defined. Richness is a dynamic concept. It is related to the flux of available resources much more than to the standing stock of these resources. For the case of the heterotrophic bacterial community, richness consists in the flux of input of biodegradable organic matter, whatever the origin of this flux (phytoplankton activity or allochtonous inputs) might be. A direct measurement of this flux, including pelagic carbon fixation and additional allochtonous inputs, is not easy to obtain routinely. However, provided the system is not too far away from steady state, organic matter is taken up by bacteria at a rate close to its production rate, maintaining the stock at a low, quasi constant level. Under these conditions, the flux of bacterial utilization of organic matter, or the flux of bacterial production, which, as a first approximation, is proportional to the former, can be considered as a good index of richness of the environment. This is the approach used in this study for comparing spot determination of bacterial biomass and production in a large range of aquatic environments.

Methods

Bacterial numbers were counted by epifluorescence microscopy after acridine orange staining (Daley & Hobbie, 1975). Photographs of about three fields were taken and the bacterial size distribution was measured on enlargements of these photographs. Biomass was calculated from abundance and biovolume distribution using the biovolume dependent conversion factors recently proposed by Simon & Azam (1989).

Bacterial production was estimated by ^{3}H-thymidine incorporation according to the methods described by Fuhrman & Azam (1980, 1982). A conversion factor into cell production was experimentally determined for each environment by the method proposed by Riemann *et al.* (1987). The conversion factors found in the different environments are given in Table 1.

Table 1. Experimentally determined conversion factors relating bacterial cell production to ^{3}H-thymidine incorporation into cold TCA insoluble fraction.

Environment	Cell produced per nmol Thy incorporated into cold TCA insoluble material
Scheldt River and estuary	$1.25\ 10^{9}$
Meuse River	$0.5\ 10^{9}$
North Sea (coastal zone)	$1.7\ 10^{9}$
Prydz Bay (Antarctica)	$5\ 10^{9}$
Drinking water (Parisian suburbs)	$0.5\ 10^{9}$

Bacterial growth rate is calculated as the ratio between cell production and bacterial abundance.

Mortality and grazing rate were determined as described by Servais *et al.* (1985, 1989).

Results

Figure 1 shows the data of bacterial biomass plotted against bacterial production in the different sites considered. As can be judged from the bacterial production values, variations in richness between the environments making up our data base cover 5 orders of magnitude. Biomass values also cover this range of variation. Biomass indeed appears roughly proportional to the richness of the medium (slope close to 1 in log-log plot). Such proportionality would not have been observed with bacterial abundance. Indeed, bacterial numbers and bacterial biomass are not proportional in the whole range of environments considered here, as very different size classes of bacteria dominate in each of them (Fig. 2). Small forms (0.3 to 0.7 μm cocci) dominate in oligotrophic systems, while larger rods are present in eutrophic waters.

The proportionality observed between bacterial biomass and bacterial production (Fig. 1) implies a fairly constant growth rate. As a matter of fact, Fig. 3 shows that the growth rate varies by only 2 orders of magnitude for the 5 orders of magni-

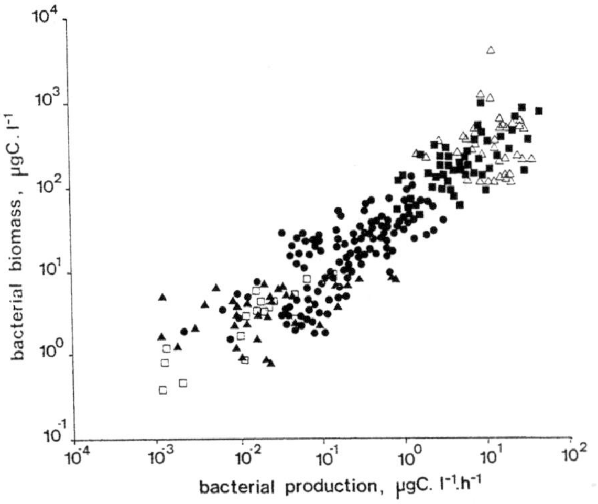

Fig. 1. Bacterial biomass plotted against bacterial production in individual samples from Scheldt River and estuary (△), Meuse River (■), coastal zones of the North Sea (●), Prydz Bay (Antarctica) (▲) and drinking-water distribution network of the Parisian suburbs (□). The flux of bacterial production can be considered here as a general index of the richness of the environment.

log-log linear correlation: log(biomass) = 1.67 + 0.7 log(production) $r = 0.91$ ($n = 288$)

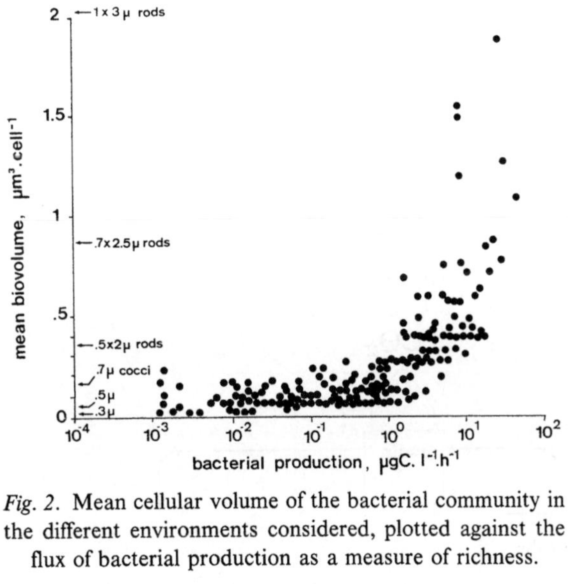

Fig. 2. Mean cellular volume of the bacterial community in the different environments considered, plotted against the flux of bacterial production as a measure of richness.

tude range of richness. Moreover, the lowest values recorded in oligotrophic situations might be underestimated, because of the possible presence of a larger fraction of inactive dormant or resting cells in the biomass there. In meso- and

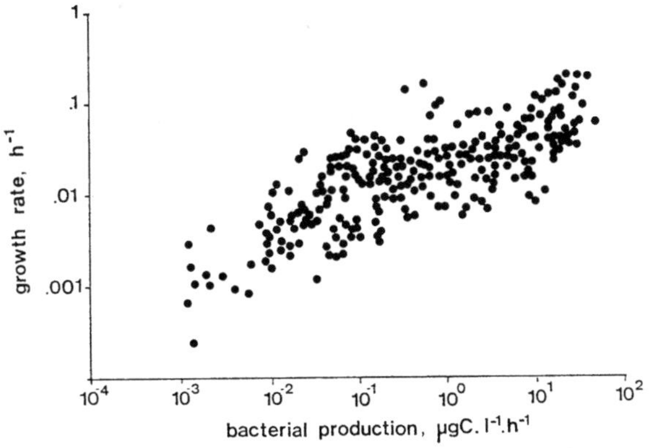

Fig. 3. Mean bacterial growth rate plotted against bacterial production as a measure of richness in the different environments considered.

log-log linear correlation: log(growth rate) = − 1.7 + 0.33 log(production) $r = 0.73$ ($n = 288$)

eutrophic waters, the growth rates fluctuate around a mean value of 0.02 h^{-1}, corresponding to a generation time of about 35 h.

Still more clearly apparent is the lack of important variations in mortality rates between the different environments considered (Fig. 4). Mortality rates are situated at the mid-range of growth rates.

The method used for determining bacterial mortality (Servais *et al.*, 1985) allowed us to distinguish between grazing by protozoans and mor-

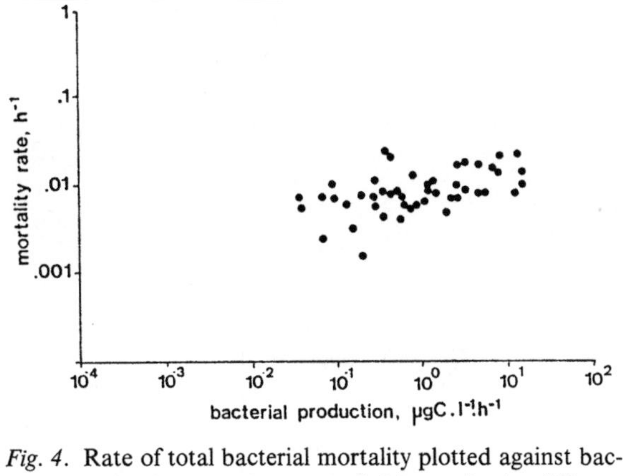

Fig. 4. Rate of total bacterial mortality plotted against bacterial production as a measure of richness in the different environments considered.

log-log linear correlation: log(mortality rate) = − 2.1 + 0.26 log(production) $r = 0.6$ ($n = 49$)

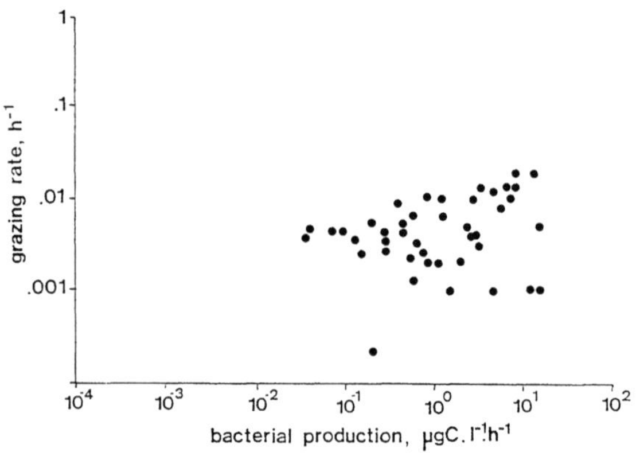

Fig. 5. Rate of bacterial grazing plotted against bacterial production as a measure of richness in the different environments considered.

log-log linear correlation: log(grazing rate) = − 2.4 + 0.21 log(production) $r = 0.35$ ($n = 49$)

tality due to another cause (spontaneous or virus-induced lysis, bacterial predation...). No distinct trend appears for grazing rate. Grazing appears as a variable and not always dominant fraction of total mortality (Fig. 5).

Discussion

The major trends apparent from the data discussed above can be summarized as follows:

1. Bacterial biomass is proportional to the richness of the environment. It reflects directly the trophic status of the system.
2. The turnover of bacterial biomass, as revealed by growth rate or by mortality, appears largely independent of the richness of the environment.

The latter observation might seem paradoxical: it implies that eutrophic systems do not distinguish themselves from oligotrophic ones by a higher growth rate of the bacterial community.

We will show however that this conforms with the predictions of a simplified model of bacterioplankton dynamics discussed elsewhere in another context (Billen *et al.*, 1988a) (Fig. 6). The model assumes that organic substrates (S) are produced at a rate p, which thus represents the richness of the system. The substrates are utilized by bacteria (B) according to Michaelis-Menten kinetics. Growth of bacteria is proportional to the uptake of substrates and their mortality is first-order with respect to their biomass.

The rate of change of S and B can thus be written:

$$\frac{dS}{dt} = p - b_{max} \frac{S}{S + K_s} B \quad (1)$$

$$\frac{dB}{dt} = Yb_{max} \frac{S}{S + K_s} B - k_d B \quad (2)$$

where

p is the input of organic substrates characterizing the richness of the system;

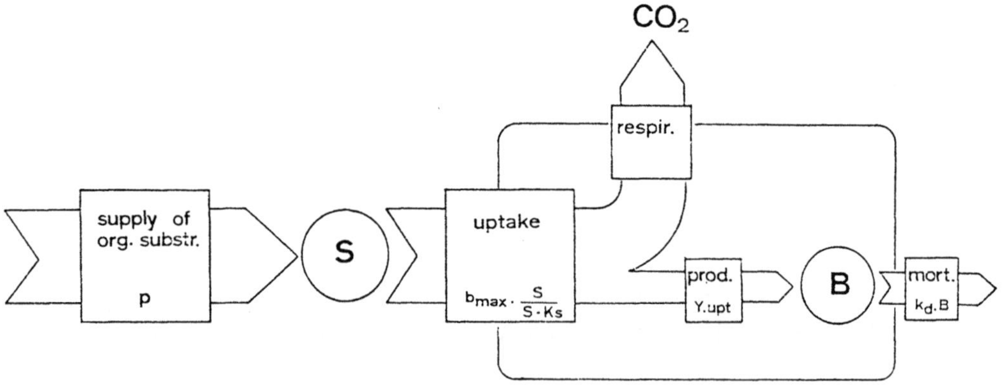

Fig. 6. Approximate representation of the model of bacterioplankton dynamics discussed in the text.

b_{max} is the maximum rate of uptake of organic matter;
K_s is the half-saturation constant of substrate uptake;
Y is the growth yield;
k_d is the first-order mortality rate constant.

At steady state, the model predicts that substrate concentration is maintained at a value independent of p, and dependent only on the physiological characteristics of bacterial uptake:

$$S = \frac{K_s}{\frac{Yb_{max}}{k_d} - 1}$$

Consequently, the growth rate, which depends only on substrate concentration, is constant:

$$\mu = Yb_{max}\frac{S}{S + K_s} = k_d$$

Only the biomass is a direct function of substrate production p:

$$B = \frac{Y}{k_d}p$$

The model thus predicts the major trends observed in our data.

These conclusions, however, are highly dependent on the way the model represents the process of bacterial mortality.

Let us consider another model, including predator (Z)-controlled grazing mortality instead of first-order bacterial mortality. This model represents a top-down control of the bacterial compartment, while the preceding model, ignoring the dynamics of predators, corresponded to simple bottom-up control. The equations can be written:

$$\frac{dS}{dt} = p - b_{max}\frac{S}{S + K_s}B \qquad (1')$$

$$\frac{dB}{dt} = Yb_{max}\frac{S}{S + K_s}B - z_{max}\frac{B}{B + K_B}Z \qquad (2')$$

$$\frac{dZ}{dt} = Y' z_{max}\frac{B}{B + K_b}Z - k_z Z \qquad (3)$$

where

z_{max} is the maximum rate of bacterial grazing per unit predator biomass;
K_b is the half-saturation constant of bacterial grazing;
Y' is the growth yield of the predators;
k_z is the first-order mortality rate of predators.

At steady state this model now predicts that bacterial biomass is independent of the richness of the medium, being determined by the parameters characterizing grazing:

$$B = \frac{K_b}{\frac{Y' z_{max}}{k_z} - 1}$$

Conversely, the growth rate and the mortality rate are predicted to be proportional to p:

$$k_d = \mu = \frac{Y}{K_b}\left(\frac{Y' z_max}{k_z} - 1\right)p$$

The predictions of this latter predator-controlled model are thus contractory with the trends apparent from our experimental observations, where bacterial biomass was found to be proportional to richness, while growth and mortality rates were independent of it. The former resource-controlled model, correctly predicts these trends. This suggests that in the environments studied here, even if grazing undoubtedly contributes to bacterial mortality, the bacterial compartment is not primarily controlled by the dynamics of bacterial grazers, but is controlled more strongly by the amount of resources available.

Basically, the prediction of these trends results from the structure of equation (2), in which neither the term representing bacterial growth, nor that representing bacterial mortality explicitly involves phagotrophic protozoa. However, more subtle interactions than those taken into account in the above models might be operative in controlling bacterial dynamics. For example, Sherr *et al.* (1988) recently reviewed a number of possible mechanisms by which phagotrophic protozoa, beside their direct grazing on bacteria, could also indirectly regulate bacterial growth. These mechanisms all involve mineral nutrient limitation of bacterial growth. Such limitation is highly unlikely, however, in the environments constituting our data base (excepted for some of our North Sea samples). Possibly therefore, our conclusion that in the environments studied here, the bacterial compartment is not primarily controlled by the dynamics of phagotrophic zooplankton, could not apply to situations where, in addition to organic carbon control, mineral nutrient limitation of bacterial growth occurs. It would be interesting to extend the reasoning presented here to broader data bases.

References

Billen, G. & A. Fontigny, 1987. Dynamics of a Phaeocystis-dominated spring bloom in Belgian coastal waters. II. Bacterioplankton dynamics. Mar. Ecol. Prog. Ser. 37: 249–257.

Billen, G., C. Lancelot & S. Mathot, 1987. Ecophysiology of phyto- and bacterioplankton growth in the Prydz Bay area during the Austral summer 1987. Part II: Bacterioplankton activity. Proceedings of the Belgian National Colloquium on Antarctic Research. Prime Minister's Services, Science Policy Office, Brussels.

Billen, G., P. Servais & A. Fontigny, 1988a. Growth and mortality in bacterial population dynamics of aquatic environments. Arch. Hydrobiol. Beih. Ergebn. Limnol. 31: 173–183.

Billen, G., C. Lancelot, E. De Becker & P. Servais, 1988b. Modelling microbial processes (Phyto- and Bacterioplankton) in the Schelde estuary. Hydrobiol. Bull. 22: 43–55.

Daley, R. J. & J. E. Hobbie, 1975. Direct count of aquatic bacteria by a modified epifluorescence technique. Limnol. Oceanogr. 20: 875–882.

Fuhrman, J. A. & F. Azam, 1980. Bacterioplankton secondary production estimates for coastal waters of British Columbia, Antarctica and California. Appl. envir. Microbiol. 39: 1085–1095.

Fuhrman, J. A. & F. Azam, 1982. Thymidine incorporation as a measure of heterotrophic bacterioplankton production in marine surface waters: evaluation and field results. Mar. Biol. 66: 109–120.

McManus, G. B. & J. A. Fuhrman, 1988. Control of marine bacterioplankton populations: measurement and significance of grazing. Hydrobiologia 159: 51–62.

Riemann, B., P. K. Bjornsen, S. I. Newell & R. D. Fallon, 1987. Calculation of bacterioplankton production from measurements of ^{3}H-thymidine incorporation. Limnol. Oceanogr. 32: 471–476.

Servais, P., 1989. Bacterioplanktonic biomass and production in the river Meuse (Belgium). Hydrobiologia. 174: 99–110.

Servais, P., G. Billen & J. Vives-Rego, 1985. Rate of bacterial mortality in aquatic environments. Appl. envir. Microbiol. 49: 1448–1454.

Servais, P., G. Billen, J. Martinez & J. Vives-Rego, 1989. Estimating bacterial mortality by the disappearance of ^{3}H-labeled intracellular DNA. FEMS Microb. Ecol. 62: 119–126.

Servais, P., G. Billen, C. Ventresque, M. Benezet. Bacterial regrowth in distribution systems: investigation in the Eastern Parisian suburbs network. Wat. Res., in press.

Sherr, B. F., E. B. Sherr & C. S. Hopkinson, 1988. Trophic interactions within pelagic microbial communities: Indications of feedback regulation of carbon flow. Hydrobiologia 159: 19–26.

Simon, M. & F. Azam, 1989. Protein content and protein synthesis rates of planktonic marine bacteria. Mar. Ecol. Prog. Ser. 51: 201–213.

Received 3 August 1989; in revised form 7 March 1990; accepted 21 April 1990

Hydrobiologia **207**: 43–52, 1990.
D. J. Bonin & H. L. Golterman (eds), Fluxes Between Trophic Levels and Through the Water–Sediment Interface.

Trophic Structure and Productivity of the lagoonal communities of Tikehau atoll (Tuamotu Archipelago, French Polynesia)

Loïc Charpy & Claude J. Charpy-Roubaud
ORSTOM et Centre d'Océanologie de Marseille, Station Marine d'Endoume, Rue de la Batterie des Lions, 13007 Marseille, France

Key words: atoll, suspended matter, detritus, phytoplankton, microphytobenthos, zooplankton

Abstract

Carbon standing stocks and fluxes were studied in the lagoon of Tikehau atoll (Tuamotu archipelago, French Polynesia), from 1983 to 1988.

The average POC concentration (0.7–2000 μm) was 203 mg C m^{-3}. The suspended living carbon (31.6 mg C m^{-3}) was made up of bacteria (53%), phytoplankton <5 μm (14.2%), phytoplankton >5 μm (14.2%), nanozooplankton 5–35 μm (5.7%), microzooplankton 35–200 μm (4.7%) and mesozooplankton 200–2000 μm (7.9%). The microphytobenthos biomass was 480 mg C m^{-2}.

Suspended detritus (84.4% of the total POC) did not originate from the reef flat but from lagoonal primary productions. Their sedimentation exceeded phytobenthos production.

It was estimated that 50% of bacterial biomass was adsorbed on particles. the bacterial biomass dominance was explained by the utilisation of 1) DOC excreted by phytoplankton (44–175 mg C m^{-2} day^{-1}) and zooplankton (50 mg C m^{-2} day^{-1}) 2) organic compounds produced by solar-induced photochemical reactions 3) coral mucus.

50% of the phytoplankton biomass belongs to the <5 μm fraction. This production (440 mg C m^{-2} day^{-1}) exceeded phytobenthos production (250 mg C m^{-2} day^{-1}) when the whole lagoon was considered.

The zooplankton >35 μm ingested 315 mg C m^{-2} day^{-1}, made up of phytoplankton, nanozooplankton and detritus. Its production was 132 mg C m^{-2} day^{-1}.

Introduction

Recent upward revisions in estimates of biomass per bacterium indicate that many earlier biomass values may have to be almost doubled and that bacteria are major components of marine systems (Fuhrman *et al.*, 1989). In addition to the classic type of grazing food web (phytoplankton-zooplankton-fish), a 'microbial loop' (phytoplankton-bacteria-heterotrophic flagellates or protozooplankton) was described by Azam *et al.* (1983). A stratified, oligotrophic water column may give rise to dominance of small phytoplankters, relatively high DOM-production rates and a long 'microbial loop' type of food chain (Kiørboe *et al.*, 1990).

Polynesian atoll lagoons are located in oligotrophic waters and dominance of small phytoplankters (cyanobacteria) was demonstrated in the atoll of Tikehau by Blanchot *et al.* (1989). In such shallow ecosystems, benthic primary production can play an important role. Charpy-Roubaud (1988) observed that microphytobenthic production represented 55% of the phytoplanktonic production of the Tikehau lagoon.

The aim of this paper was to investigate the activities of lagoonal primary producers (phytoplankton and phytobenthos), zooplankton (nano, micro and mesoplankton) and bacteria.

All data used in this study were obtained during the ATOLL program of the Tahiti ORSTOM Center.

Description of site studied

The atoll chosen as a study site was Tikehau, situated in the north west of Tuamotu archipelago (Fig. 1); its characteristics make it a suitable model of a mid-sized open atoll. A preliminary description of this atoll was given by Harmelin-Vivien (1985). Tikehau is almost circular: its widest diameter is nearly 28 km. The surface of the lagoon is 400 km^2, 91% of the lagoon bottom is deeper than 15 m and the average depth is 25 m (Lenhardt, 1987). Oceanic water enters by the east and south east reef-flat spillways, and exits after a mean residence time of 176 days by the passage located to the west (Lenhardt, 1988).

Material and methods

Suspended material (0.5–35 μm)

Standing stocks: eleven surveys were made in the lagoon between 1983 and 1985 (Table 1); sam-

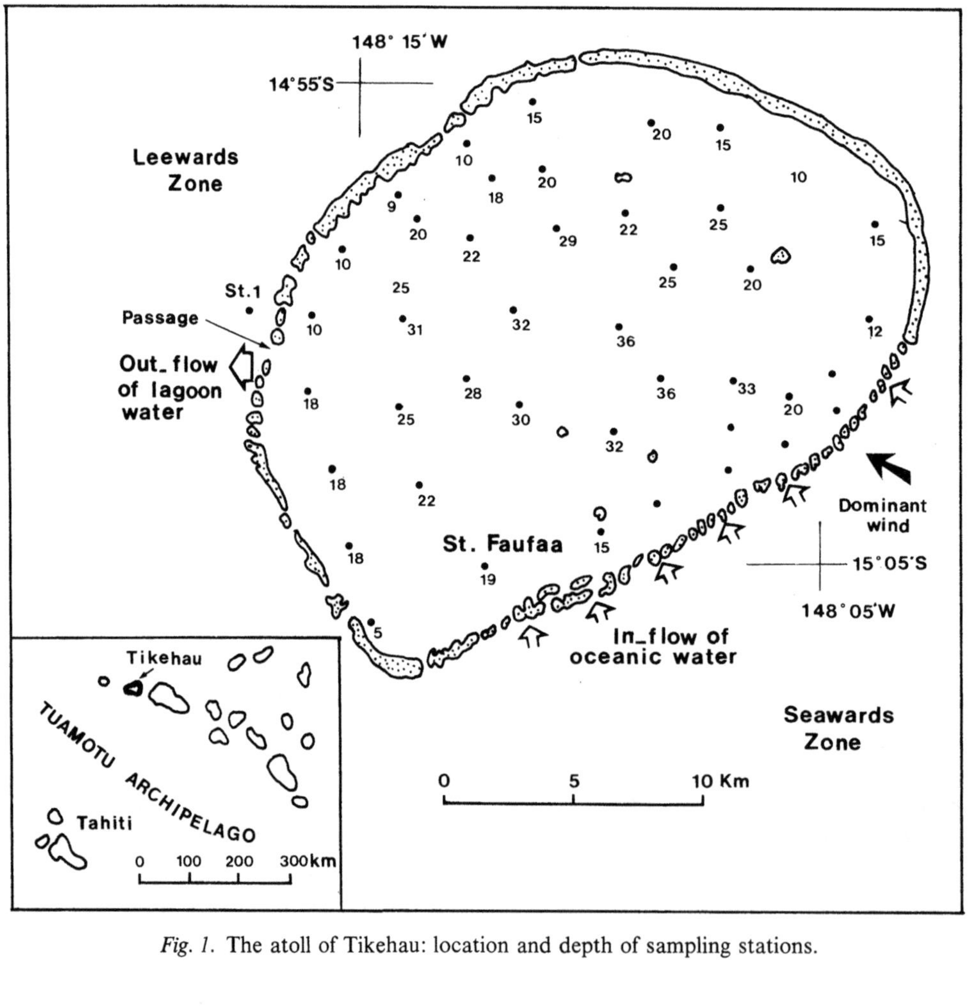

Fig. 1. The atoll of Tikehau: location and depth of sampling stations.

pling was performed at 46 stations, every 10 m depth and 1 m above the bottom using 5 l Niskin bottles. Chlorophyll (chl·*a*) and particulate organic carbon (POC) concentrations were also analysed at the Faufaa station (19 m deep), every week, between July 1985 and February 1987 (see Fig. 1).

Sea water samples were prefiltered through a 35 μm mesh polyamide screen. Then, a) 100 to 300 ml were filtered through a Whatman GF/F filter for chl·*a* analysis; b) 500 to 1000 ml through a GF/F procesed filter (precombusted 4 hours at 450 °C and precleaned with HCl 1 N for POC analysis and c) 250 to 500 ml were filtered through a Millipore 0.45 μm filter for ATP analysis.

Chl·*a* was determined by fluorescence (Yentsch & Menzel, 1963) using a TURNER 111 fluorometer. POC concentrations were determined after rinsing the filter with 20 ml of HCl (0.1 N) using a CHN analyzer 185-B Hewlett-Packard (Gordon & Sutcliffe, 1973); the combustion temperature of 720 °C was chosen to minimize carbonate interference on the filters (Telek & Marshall, 1974). ATP extractions were performed immediately in 5 ml of boiling TRIS (0.02 M, pH = 7.85); ATP extracts were frozen at − 20 °C for later analysis. ATP concentrations were measured according to the method described by Holm-Hansen & Booth (1966), using a LKB Luminometer and Luciferine-Luciferase preparations from SIGMA (ref. FLE 50).

In April 1986, plankton counts were performed using epifluorescence microscopy as described in Blanchot *et al.* (1989). To measure the size structure of suspended particles, water samples were successively filtered through a polyamide net with a 35 μm size mesh, through a 5 μm filter (Millipore) and finally through a GF/F (Whatman) filter.

Conversion factors: The living carbon (liv C) was estimated from ATP concentration and a C/ATP ratio of 250 (Holm-Hansen & Booth, 1966 and Laws *et al.*, 1984). The liv C < 5 μm was estimated from the average percentage of ATP passing through a filter of 5 μm pore size. Phytoplanktonic carbon (phy C) was estimated from chlorophyll concentration and a C/chl·*a* ratio (Riemann *et al.*, 1982; Søndergaard *et al.*, 1988). The C/chl·*a* ratio of the phytoplankton of Tikehau was determined experimentally on three occasions: lagoon water prefiltered onto a 5 μm pore size filter was incubated *in situ* with ^{14}C; the ratio assimilated C/increase in chl·*a* inside the incubation bottle represents the phytoplanktonic C/chl·*a* ratio. The carbon content of heterotrophs smaller than 5 μm was estimated from the difference between liv $C_{(<5\ \mu m)}$ and phy $C_{(<5\ \mu m)}$. The carbon content of heterotrophs between 5 μm and 35 μm was estimated from the difference between the liv $C_{(5-35\ \mu m)}$ and phy $C_{(5-35\ \mu m)}$. The carbon biomasses of ciliates and heteroflagellates were calculated using counts performed in April 1986 and conversion factors given by Blanchot *et al.* (1989).

Fluxes: The carbon assimilation of phytoplankton was estimated by measuring *in situ* ^{14}C uptake between 1983 and 1987 (292 incubations). The method was that described in Charpy-Roubaud *et al.* (1988). Fourteen measurements of sedimentation rate of particulate matter were performed between January 1986 and May 1987, at the Faufaa station. The sediment trap consisted of a PVC plastic jar with a ratio of height to width of 2.5 : 1. Poison was not used, so as to allow measurements of ATP concentration inside the trap. The jar was mounted 15 m below the surface and 4 m above the bottom to prevent resuspension. Material was collected for 6 to 20 hours. The POC export out of the lagoon was calculated by

Table 1. Date and number of stations (N) sampled during the surveys in the lagoon.

Date	03/83	07/83	09/83	12/83	02/84	11/84	01/85	03/85	04/85	07/85	08/85
N	21	14	22	17	9	10	7	7	7	7	7

the equation:

$$\text{Export (mg C m}^{-2}\text{ day}^{-1}) = F \cdot POC_L/S_L$$

F = annual average flow through the passage and the reef flat spillways; it was estimated as $6 \cdot 10^7$ m^3 day^{-1} by Lenhardt (1988).
POC_L = average POC concentration in the lagoon (mg C m^{-3}).
S_L = lagoon area = 4 10^8 m^2.

The zooplankton (35 μm–2000 μm)

The standing stocks and taxonomic composition of the zooplankton were monitored in the lagoon of Tikehau between April 1985 and April 1986. Microzooplankton (35–200 μm) and mesozooplankton (200–2000 μm) were collected by vertical hauls from the bottom to the surface. These data were supplemented by two 10 day studies of the variability, structure and functioning of the pelagic ecosystem. Excretion and respiration rates were determined from incubation experiments. Using the C:N:P ratio method, net growth efficiencies (K_2) were calculated for the zooplankton. Combined with nitrogen and phosphorus excretion rates, K_2 values enabled the assessment of production rates. Ingestion by animals > 35 μm was calculated by means of assimilation efficiencies. These methods are described in Le Borgne *et al.* (1989).

Microphytobenthic carbon at the sediment water interface

Microphytobenthos biomass and production were studied between 1985 and 1987 (Charpy-Roubaud, 1988):

- microphytobenthos biomass was estimated by sediment chl · *a* concentration measurements following the procedure described by Plante-Cuny (1984). Results are in mg chl · *a* m^{-2} for the upper 0.5 cm of sediment. Microphytobenthic carbon was estimated using the same C/chl · *a* ratio as for the phytoplankton.
- phytobenthic production was determined by O_2 budgets, measured within clear and dark plexiglass domes. The production of O_2 (BP_{O_2}) may be converted into the gross production of carbon (BP) by the equation:

$$Bp = (BP_{O_2} \cdot 0.375 \cdot PQ) + (R \cdot 0.375 \cdot RQ)$$

R = Respiration during daytime; PQ and RQ = photosynthetic and respiratory coefficients.

Results

Particulate organic matter

POC concentration values (N = 522) ranged from 82 to 893 mg C m^{-3}. The yearly average was significantly ($P < 0.05$) higher in 1983 than in other years (Fig. 2). The occurrence of two hurricanes in March and May 1983 in the Western Tuamotu Archipelago is believed to be responsible for this high POC level. Monthly averages (all years included), presented wide variations but the station location had no influence on POC concentration. The POC concentrations were 40% higher in samples taken near the bottom than in the water column.

Chl · *a* concentration values (Fig. 2) ranged (N = 782) from 0.02 to 1.01 mg m^{-3}. Yearly averages were not significantly ($P > 0.05$) different but monthly averages differed significantly ($P < 0.05$). Chl · *a* concentrations were 37% higher close to the bottom than in the water column.

ATP concentrations (Fig. 2) were lower in 1985 than in other years. Monthly averages differed significantly ($P < 0.05$); ATP concentration was independent of the proximity to the bottom.

Therefore, in order to estimate the lagoonal POM averages, we consider neither 1983 data nor data collected close to the bottom, and we use the equation:

$$\text{Average } POM = \left[\sum_{i=N_i}^{i=1} POM_i\right] \Big/ N_i$$

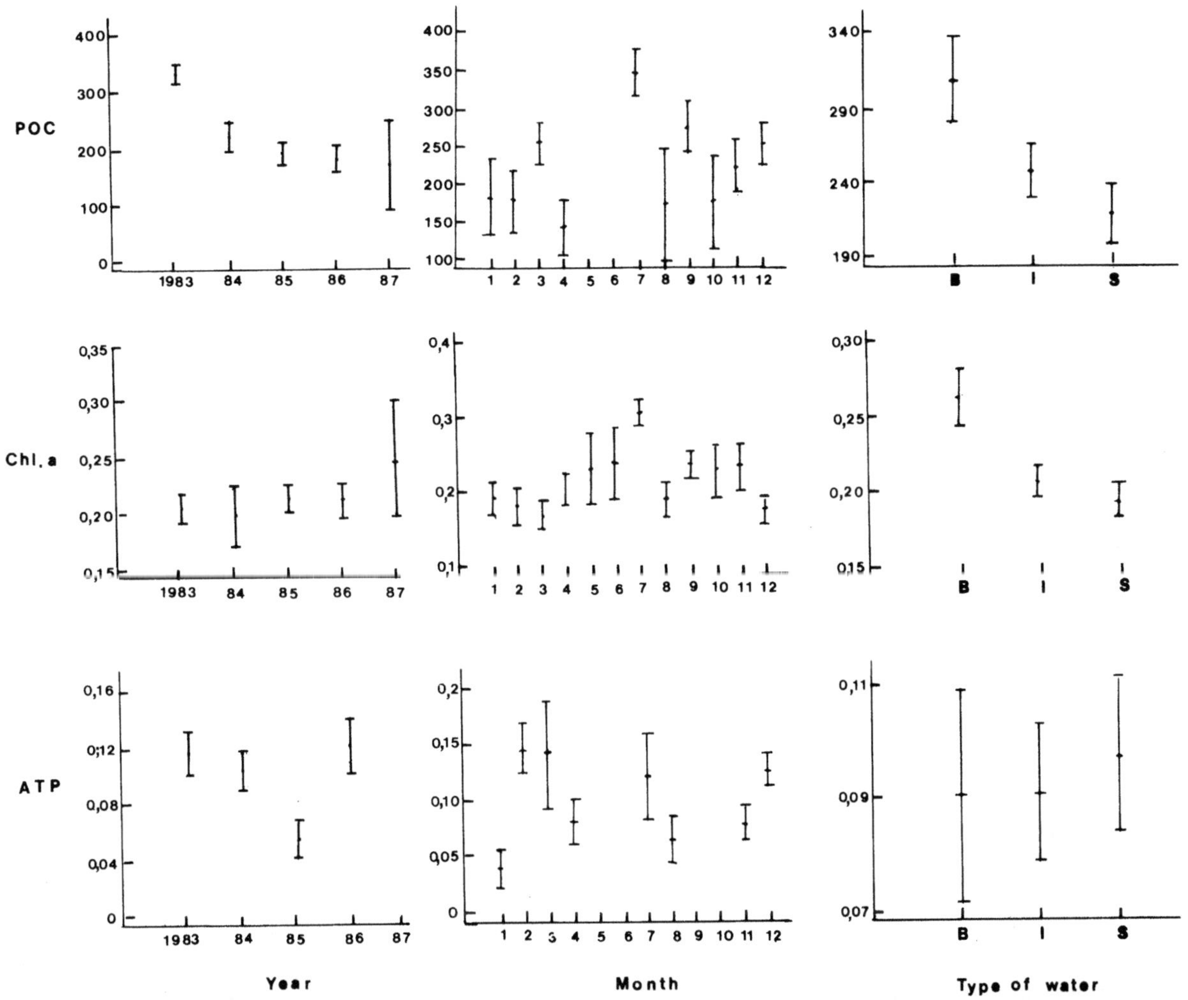

Fig. 2. Means and confidence intervals ($P = 95\%$) for factors 'year', 'month' and 'type of water'; B = samples taken at 1 m above the bottom, I = samples taken between 2 m and the bottom, S = samples taken between 0 m and 2 m.

POM_i = POM average on month i.
N_i = number of months prospected.

The weighted averages and confidence intervals are summarized in Table 2.

The results of seven filtering experiments performed in 1984, 1985 and 1986 with 35 μm mesh size (polyamide net) and then 5 μm pore size (Nuclepore filters) appear in Table 2.

The average sedimentation rates are summarized in Table 2.

The POC export was calculated at 29 mg m^{-2} day^{-1}.

Table 2. Weighted averages and confidence intervals ($P = 95\%$) of chl · *a*, ATP and POC concentration (mg m^{-3}); percentages of particles passing through 5 μm (% < 5 μm) and averages sedimentation rates (SR: mg m^{-2} day^{-1}).

	chl · *a*	ATP	POC
Average	0.18 ± 0.01	0.11 ± 0.01	192 ± 10
N	409	162	290
% < 5 μm	50 ± 11	50 ± 11	46 ± 11
N	27	23	21
SR	0.11 ± 0.06	0	350 ± 218
N	11	11	11

Table 3. Phytoplankton, zooplankton and phytobenthos biomass and fluxes measured in the lagoon of Tikehau. Data from Charpy-Roubaud (1988), Charpy-Roubaud *et al.* (1988), Le Borgne *et al.* (1989), Blanchot *et al.* (1989). B = biomass (mg C m^{-3}); A = assimilation, I = ingestion, P = production (A, I, P = mg C m^{-3} day^{-1}); E = Excretion of organic P (mg P m^{-3} day^{-1}).

Organism	B	A	I	P	E	Ratios
Phytoplankton						
<5 μm	4.5					
>5 μm	4.5					
Total	9.0	17.6		?	?	C/chl · *a* = 50
Zooplankton						
<35 μm	1.8		?	?	?	
Zooplankton						
35– 200 μm	1.5		2.6	1.2	8.4	
200–2000 μm	2.5		10.0	4.1	19.2	
Total	4.0		12.0	5.3	27.6	C/P = 52
Detritus						
35– 200 μm	2.3					
200–2000 μm	4.4					
Total	6.7					
Phytobenthos	B (mg chl · *a* m^{-2})			P (mg C m^{-2} day^{-1})		
	9.6			250		

Plankton and microphytobenthos biomasses and fluxes

These results are presented in Charpy-Roubaud (1988), Charpy-Roubaud *et al.* (1988), Le Borgne *et al.* (1989) and Blanchot *et al.* (1989). They are summarized in Table 3.

Results of the 5 experiments for determining the C/chl · *a* ratio appear in Table 4. This ratio ranged from 34 to 73 with an average of 50.

Table 4. Determination of the phytoplanktonic C/chl · *a* ratio in the lagoon of Tikehau. chl · a_{t0} = concentration (mg m^{-3}) at zero time; chl · a_{te} = concentration at $t0 + \delta t$ time; $\delta C/\delta t$ = carbon assimilation rate (mg C m^{-3} h^{-1}).

Date	δt	chl · a_{t0}	chl · a_{te}	δchl · $a/\delta t$	$\delta C/\delta t$	C/chl · *a*
01/85	2	0.023	0.030	0.0035	0.12	34
	2	0.030	0.033	0.0015	0.11	73
04/85	4	0.040	0.075	0.0088	0.35	40
	4	0.036	0.077	0.0103	0.67	65
12/85	4	0.120	0.190	0.0175	0.69	39

Discussion

Planktonic trophic web

50% of the POC were in particles <5 μm.

The detritus pool <35 μm can be estimated by the difference: POC − liv $C_{(<35\,\mu m)}$ = 192 − (0.11 × 250) = 164 mg C m^{-3}. Such a large proportion of detritic carbon in the POM is generally observed in tropical coastal waters: Gerber & Marshall (1982) found 77 to 84% of detritic carbon in the Enewetok lagoon, and Winn & Karl (1984) found 70 to 90% in the waters close to Hawaii.

We did not measure directly the biomass of bacteria, but we think that we can obtain an order of magnitude by the difference between liv C estimated from ATP and the other biomasses measured or estimated. Therefore, free bacteria biomass (BB) may be calculated with the equation:

$$BB = liv\ C_{(<5\,\mu m)} - phy\ C_{(<5\,\mu m)}$$

with: $liv\ C_{(<5\ \mu m)}$ = percentage of $ATP_{(<5\ \mu m)} \times$ ATP × 250
$phy\ C_{(<5\ \mu m)}$ = percentage of $chl \cdot a_{(<5\ \mu m)} \times$ chl.a × (C/chl · a ratio).

The average value C/chl · a = 50 (from Table 4) lies within the range reported by Takahashi *et al.* (1985) for picoplankton and is very close to the ratio of 46 found by Laws *et al.* (1987) for oligotrophic Pacific waters. Therefore:

$$BB = (0.46 \times 0.11 \times 250) - (0.5 \times 0.18 \times 50) = 8.2\ \text{mg C m}^{-3}$$

The liv $C_{(5-35\ \mu m)}$ is made up of heterotrophs and phytoplankton (4.5 mg C m^{-3}). It can be calculated using $ATP_{(5-35\ \mu m)}$ data: 0.54 × 0.11 × 250 = 14.9 mg C m^{-3}. The carbon content of heterotrophs in the size range from 5 to 35 μm was therefore equal to: 14.9 − 4.5 = 10.4 mg C m^{-3}. The biomass of ciliates and heteroflagellates was equal to 1.8 mg C m^{-3} (Table 3), and the difference (= 8.6 mg C m^{-3}) was certainly due to bacteria adsorbed onto the detritus (Charpy, 1985).

The total biomass of bacteria was therefore equal to: free bacteria + adsorbed bacteria = 16.8 mg C m^{-3}. Such a biomass is commonly observed in waters over reefs; Sorokin (1974) summarizes data for biomasses of bacteria which range from 11 to 170 mg C m^{-3}. More recently, Moriarty *et al.* (1985) reviewed the productivity and trophic role of bacteria on coral reefs. They give biomass values ranging from 19 to 150 mg C m^{-3}. Linley & Koop (1986) observed in the coral reef lagoon of One Tree Island (Great Barrier Reef) a biomass of heterotrophic bacteria ranging from 1.2 to 16.2 mg C m^{-3}, and Hopkinson *et al.* (1987) observed a bacterial biomass of 2 mg C m^{-3} in the water column of Davies Reef (Australia). In Tikehau, in April 1986, the biomass of bacteria was estimated at 17.1 mg C m^{-3} by Blanchot *et al.* (1989). The observed ratio free bacteria/adsorbed bacteria = 2 is consistent with the ratios given by Moriarty (1979) and Moriarty *et al.* (1985) in coral reef areas.

The estimated bacterial biomass was 2 times higher than the phytoplankton C in the Tikehau lagoon. Dominance of bacterial biomass was also observed in the oligotrophic waters of the Sargasso Sea by Fuhrman *et al.* (1989); the interpretation of these authors was that bacteria consume significant amounts of carbon probably released from phytoplankton directly or via herbivores. In Tikehau, the excretion rate of dissolved organic phosphorus (DOP) of the zooplankton > 35 μm was measured in 1986 at 27.6 10^{-3} mg P m^{-3} day^{-1} (Table 3). Using the ratio C/P of zooplankton body constituent: 52 (Table 3), the organic carbon excretion was 1.4 mg C m^{-3} day^{-1}. Assuming a constant excretion per unit of biomass in the zooplankton, the nanozooplankton excretion was estimates as: (1.4/4) × 1.8 = 0.6 mg C m^{-3} day^{-1}. The release of carbon by the phytoplankton was not measured. Values recorded in the literature vary from 10 to 75% but are generally less than 40% of the photoassimilated carbon (Newell *et al.*, 1981). Therefore, the phytoplankton excretion ranged from 0.1 × 17.6 (average phytoplankton assimilation) = 1.7 to 0.4 × 17.6 = 7 mg C m^{-3} day^{-1}. Fuhrman *et al.* (1989) gave another interesting alternative to plankton as a source of DOM: solar-induced photochemical production of labile low molecular weight organic compounds from high molecular weight refractory DOM. This source may be important for the lagoon due to the low depth and high energy level. In the coral reef area, the role of coral mucus as a substrate for microbial growth (Moriarty *et al.*, 1985) is particulary interesting.

Representation of the lagoonal ecosystem

Figure 3 illustrates biomass and fluxes of matter in the plankton and in the benthos measured or estimated. In order to be able to compare benthic and planktonic ecosystems, standing stocks are in mg C m^{-2} and fluxes in mg C m^{-2} day^{-1}. The depth of integration is 25 m (average depth of the lagoon). We assume that the micro and mesozooplankton ingest all the nanoplankton production and that bacteria are ingested by the nanozoo-

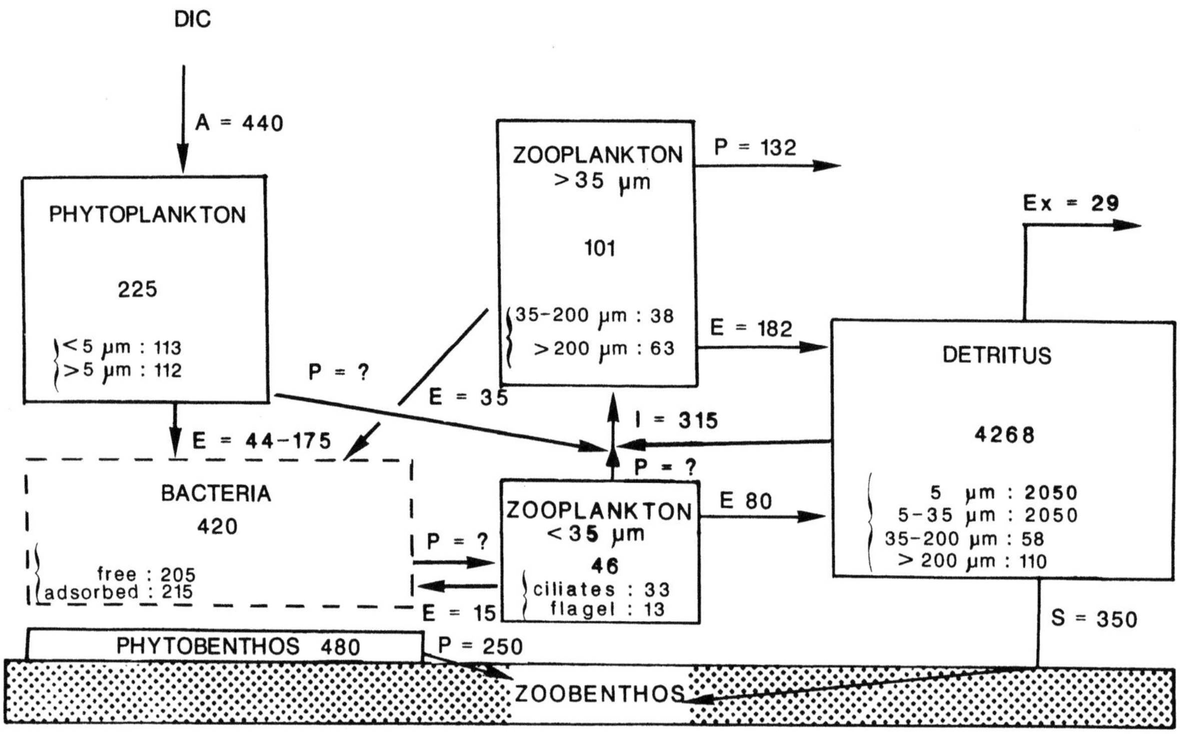

Fig. 3. Trophic structure and productivity of the lagoonal communities. Standing stocks (mg C m^{-2}) are in boxes, and fluxes (mg C m^{-2} day^{-1}) are represented by arrows. *A* = assimilation, *E* = Excretion, *Ex* = Export, *I* = Ingestion, *P* = Production, *S* = Sedimentation, *DIC* = dissolved inorganic carbon.

plankton. Indeed, the major consumers of bacteria in the plankton are flagellate and ciliate protozoans (Landry *et al.*, 1984).

Origin of the detritus: If we consider that the detritus pool is in a steady state, its production is equal to its disappearance from the water column (sedimentation + export out of the lagoon + ingestion). We do not know the ingestion rate of the detritus; however, detritus production is at least 379 mg C m^{-2} day^{-1} (sedimentation + export). The lagoonal suspended detritus can originate from the lagoonal production or from the reef-flat primary production. The reef flat production was not measured, but it may be assumed to be close to the proposed 'standard' reef-flat production: 7 ± 1 g C m^{-2} day^{-1} (Kinsey, 1983). Assuming a reef flat surface of 10 km^2, the total reef production would be close to $7 \cdot 10^{10}$ mg C day^{-1}. Hatcher (1983) has reconsidered export of organics from reef-flats and concluded that many systems may be exporting more than 20% (as much as 60%) of their gross primary production to downstream areas. However most of the organic export from the reef-flat is towards oceanic waters. We can roughly estimate the percentage of oceanic waters coming into the lagoon in relation to the total oceanic flux. The current speed of oceanic waters close to the atoll was measured in March 1984, and values in the upper 50 m were about 0.4 m s^{-1} (Charpy *et al.*, 1985). Therefore, the flux of waters passing outside the atoll can be estimated as the flux of water passing through an area 30 km wide (atoll diameter) and 20 metres deep: $0.24 \cdot 10^6$ m^3 s^{-1}. The average flux of water passing through the lagoon is 700 m^3 s^{-1} (Lenhardt, 1988), i.e. 0.3% of the oceanic flux. Assuming that the reef flat exports between 20% and 60% of its gross production, $42 \cdot 10^6$ to $126 \cdot 10^6$ mg C enter the lagoon each day. These values represent 0.1 to 0.3 mg C day^{-1} per square metre of lagoon. Therefore, the detritus reef export towards the lagoon is insignificant and the detritus pool originates from lagoonal primary productions.

The role of the phytoplankton: In spite of a low biomass (225 mg C m^{-2}), phytoplankton production is relatively high (440 mg C m^{-2} day^{-1}). The annual mean assimilation number calculated by Charpy-Roubaud *et al.* (1988) was 9.8 mg C mg^{-1} chl · *a* h^{-1}. This value is high but commonly observed in tropical coastal waters. Indeed, Takahashi & Bienfang (1983) gave an assimilation number of 8.09 mg C mg^{-1} chl · *a* h^{-1} for the picoplankton of the coastal Hawaiian waters; however, this value remains lower than the maximum potential production per chl · *a* observed by Legendre *et al.* (1988) in waters close to Moorea Island (French Polynesia): 5 to 25 mg C mg^{-1} chl · *a* h^{-1}.

The role of the zooplankton: The mesoplankton (200–2000 μm), microzooplankton (35–200 μm) and nanozooplankton (5–35 μm) biomasses are not very different: 63, 38 and 46 mg C m^{-2}. They represent in total 26% of the heterotrophs. In April 1986, Le Borgne *et al.* (1989) calculated the assimilation efficiency for mixed copepods ($D = 0.9$) and the net growth efficiency of carbon ($K_2 = 0.21 - 0.46$). The gross growth efficiency (K_1) is equal to the product of the assimilation efficiency, D, by the net growth efficiency, K_2 (Le Borgne, 1978). The high values of K_1 (19%–41%) could explain the rapid zooplankton turnover (one day). Zooplankton (35–2000 μm) ingestion is in the same order as the phytoplankton production and probably regulates its biomass. The zooplankton production (> 35 μm): 132 mg C m^{-2} day^{-1} represents the net pelagic production of the lagoon, available for the fish.

The role of the microphytobenthos: Microphyte biomass was higher in the benthos than in the plankton; however, production was lower in the benthos. Charpy-Roubaud *et al.* (1988) have demonstrated that phytobenthos production decreased with depth, while phytoplankton production increased (integrated production), but the sum remained constant. Phytobenthic production appears to be lower than the detritus sedimentation which probably makes a major contribution to the benthic food web. The biomass of the animals living in the sediments is not known, but they probably ingest the phytobenthic production plus a large part of the organic carbon sedimentation. Indeed, the organic C content of the interstitial waters of the sediments in Tikehau is very low 0.5 mg C g^{-1} of dry sediment (Sarazin *et al.*, 1988).

Conclusions

The detritus < 35 μm represent the most important particulate organic carbon pool in the lagoon. They originate from lagoonal primary production, and their sedimentation onto the bottom exceeds benthic primary production. Planktonic bacteria biomass is in the same order as the microphytobenthos, and is equal to twice the phytoplanktonic biomass. We interpret pelagic bacteria dominance by a 'microbial loop', returning energy released as DOM by phytoplankton and zooplankton, but also energy released as mucus from lagoonal coral communities.

Direct measurements of bacterial biomass, bacterial and nanozooplankton productions and nanoplankton and phytoplankton excretions could complete this carbon cycle.

References

Azam, F., T. Frenchel, J. G. Field, J. S. Gray, L. A. Meyer-Reil & F. Thingstad, 1983. The ecological role of water-column microbes in the sea. Mar. Ecol. Prog. Ser. 10: 257–263.

Blanchot, J., L. Charpy & R. Le Borgne, 1989. Size composition of particulate organic matter in the lagoon of Tikehau atoll (Tuamotu archipelago). Mar. Biol. 101: 329–339.

Charpy, L., 1985. Distribution and composition of particulate organic matter in the lagoon of Tikehau (Tuamotu archipelago, French Polynesia). Proc. 5th int. Coral Reef Congress, Tahiti, 3: 353–357.

Charpy, L., J. Marchand, F. Rougerie, J. Teuri, P.-J. Vienney & B. Wauthy, 1985. Résultats de la mission TATI du N.O. CORIOLIS (Tahiti-Tikehau) – Mars 1984 –. Archives d'Océanographie du Centre Orstom de Tahiti, 85-09: 1–57.

Charpy-Roubaud, C. J., 1988. Production primaire des fonds meubles du lagon de Tikehau (Atoll des Tuamotu, Polynésie Française). Oceanol. Acta. 11: 241–248.

Charpy-Roubaud, C. J., L. Charpy & L. Lemasson, 1988. Benthic and Planktonic primary production of an open atoll lagoon (Tikehau, French Polynesia). Proc. 6th int. Coral Reef Symposium, Australia, 2: 551–556.

Fuhrman, J. A., T. D. Sleeter, C. A. Carlson & L. M. Proctor, 1989. Dominance of bacterial biomass in the Sargasso Sea and its ecological implications. Mar. Ecol. Prog. Ser. 57: 207–217.

Gerber, R. P. & N. Marshall, 1982. Characterization of the suspended particulate organic matter and feeding by the lagoon zooplankton at Enewetak atoll. Bull. mar. Sci. 32: 290–300.

Gordon, D. C. Jr & W. H. Sutcliffe Jr, 1973. A new dry combustion method for the simultaneous determination of total organic carbon and nitrogen in sea water. Mar. Chem. 1: 231–244.

Harmelin-Vivien, M., 1985. Atoll de Tikehau, Archipel des Tuamotu. Proc. 5th int. Coral Reef Congress, Tahiti, 1: 211–268.

Hatcher, B. G., 1983. The role of detritus in the metabolism and secondary production of coral reef ecosystems. In J. T. Baker *et al.* (eds), Proc. Great Barrier Reef Conference, Australia: 317–325.

Holm-Hansen, O. & C. R. Booth, 1966. The measurement of adenosine triphosphate in the ocean and its ecological significance. Limnol. Oceanogr. 11: 510–519.

Hopkinson Jr, C. S., B. F. Sherr & H. W. Ducklow, 1987. Microbial regeneration in the water column of Davies Reef, Australia. Mar. Ecol. Prog. Ser. 41: 147–153.

Kinsey, D. W., 1983. Standards of performance in coral reef primary production and carbon turnover. In D. J. Barnes (ed.), Perspectives on coral reefs, Brian Clousten Publisher, Manuka, A.C.T.: 209–220.

Kiørboe, T., H. Kaas, B. Kruse, F. Møhlenberg, P. Tiselius & G. Ertebjerg, 1990. The structure of the pelagic food web in relation to water column structure in the Skagerrak. Mar. Ecol. Prog. Ser. 59: 19–32.

Landry, M. R., L. W. Haas & V. L. Fagerness, 1984. Dynamics of microbial plankton communities: experiments in Kaneohe Bay, Hawaii. Mar. Ecol. Prog. Ser. 16: 127–133.

Laws, E. A., G. R. DiTullio & D. G. Redalje, 1987. High phytoplankton growth and production rates in the North Pacific subtropical gyre. Limnol. Oceanogr. 32: 905–918.

Laws, E. A., D. G. Redalje, L. W. Haas, P. K. Bienfang, R. W. Eppley, W. G. Harrison, D. M. Karl & J. Mara, 1984. High phytoplankton growth and production rates in oligotrophic Hawaiian coastal waters. Limnol. Oceanogr. 29: 1161–1169.

Le Borgne, R., 1978. Evaluation de la production secondaire planctonique en milieu océanique par la methode des rapports C/N/P. Oceanol. Acta 1: 107–118.

Le Borgne, R. P., J. Blanchot & L. Charpy, 1989. Zooplankton of the atoll of Tikehau (Tuamotu Archipelago) and its relationship to particulate matter. Mar. Biol. 102: 341–353.

Legendre, L., S. Demers, B. Delesalle & C. Harnois, 1988. Biomass and photosynthetic activity of phototrophic picoplankton in coral reef waters (Moorea Island, French Polynesia). Mar. Ecol. Prog. Ser. 47: 153–160.

Lenhardt, X., 1987. Etude bathymetrique du lagon de l'atoll de Tikehau, ORSTOM Tahiti, Notes et Doc. ORSTOM Tahiti Ser. Oceanogr. 35: 53–70.

Lenhardt, X., 1988. Hydrodynamique des lagons d'atoll et d'île haute en Polynésie Française. Thèse du Museum National d'Histoire Naturelle, Paris, 156 pp.

Linley, E. A. S. & K. Koop, 1986. Significance of pelagic bacteria in a coral reef lagoon, One Tree Island, Great Barrier Reef. Mar. Biol. 92: 457–464.

Moriarty, D. J. W., 1979. Biomass of suspended bacteria over coral reefs. Mar. Biol. 53: 193–200.

Moriarty, D. J. W., P. C. Pollard, D. M. Alongi, C. R. Wilkinson & J. S. Gray, 1985. Bacterial productivity and trophic relationships with consumers on a coral reef (MECOR I). Proc. 5th int. Coral Reef Congress, Tahiti, 3: 457–462.

Newell, R. C., M. I. Lucas & E. A. S. Linley, 1981. Rate of Degradation and Efficiency of Conversion of Phytoplankton Debris by Marine Micro-Organisms. Mar. Ecol. Prog. Ser. 6: 123–136.

Plante-Cuny, M.-R., 1984. Le microphytobenthos et son rôle à l'échelon primaire dans le milieu marin. Oceanis 10: 417–427.

Riemann, B., M. Søndergaard, H. H. Schierup, S. Bosselmann, G. Christensen, J. Hansen & B. Nielsen, 1982. Carbon metabolism during a spring diatom bloom in eutrophic Lake Mossø. Int. Revue ges. Hydrobiol. 67: 145–185.

Sarazin, G., C. Charpy-Roubaud & L. Charpy, 1988. Early diagenesis of organic matter in the sediments of the central bassin of Tikehau lagoon-reef (Tuamotu Archipelago – French Polynesia). Proc. 6th int. Coral Reef Symposium, Australia, 3: 373–378.

Søndergaard, M., B. Riemann, L. M. Jensen, N. O. G. Jørgensen, P. K. Bjørnsen, M. Olesen, J. B. Larsen, O. Geertz-Hansen, J. Hansen, K. Christoffersen, A. M. Jespersen, F. Andersen & S. Bosselmann, 1988. Pelagic food web processes in an oligotrophic lake. Hydrobiologia 164: 271–286.

Sorokin, Y. I., 1974. Bacteria as a component of the coral reef community. Proc. 2nd int. Coral Reef Symposium, Manille, 1: 3–10.

Takahashi, M. & P. K. Bienfang, 1983. Size structure of phytoplankton biomass and photosynthesis in subtropical Hawaiian waters, Mar. Biol. 76: 203–211.

Takahashi, M., K. Kikuchi & Y. Hara, 1985. Importance of picocyanobacteria biomass (unicellular, blue-green algae) in the phytoplankton population of the coastal waters off Japan. Mar. Biol. 89: 63–69.

Telek, G. & N. Marshall, 1974. Using a CHN analyser to reduce carbonate interference in particulate organic carbon analysis. Mar. Biol. 24: 219–221.

Winn, C. & D. Karl, 1984. Microbial productivity and community growth rate estimates in the tropical north Pacific Ocean. Biol. Oceanogr. 3: 123–145.

Yentsch, C. S. & D. W. Menzel, 1963. A method for the determination of phytoplankton chlorophyll and phaeophytin by fluorescence. Deep-Sea Res. 10: 221–231.

Received 20 July 1989; in revised form 1 March 1990; accepted 4 May 1990

Hydrobiologia **207**: 53–60, 1990.
D. J. Bonin & H. L. Golterman (eds), Fluxes Between Trophic Levels and Through the Water–Sediment Interface.

Relationships between trophic organization of benthic communities and organic matter content in Tyrrhenian Sea sediments

Silvia Cocito [1], Sandra Fanucci [1], Ilaria Niccolai [1], Carla Morri [2] & Carlo Nike Bianchi [1]
[1] *ENEA-CREA S. Teresa, CP 316, I-19100 La Spezia, Italia;* [2] *Istituto Zoologia, Univ., via Balbi 5, I-16126 Genova, Italy*

Key words: soft-bottom communities, benthic biomass, sediment organic matter, C/N ratio, trophic guilds

Abstract

The organic content of sediments, often considered as an important trophic source for the benthos, can be used as a measurement of the trophic status of marine ecosystems. Soft-bottom communities were studied between Leghorn and Elba Island, from 5 to 500 m depth. Total biomass and trophic structure (percent dominance of trophic guilds) were related to organic carbon content and C/N ratio of sediments. Sediment organic matter, depending on its quantity and quality, could act as a food source or a stress source for the community.

Introduction

One of the major synecological questions in studying benthic communities is how far structural aspects are consistent with functional ones: in particular, to what extent the trophic adaptations of organisms can be a response to the variation of different environmental conditions. Sanders (1958) was one of the first to explore the relationships between sediment characteristics and trophic-functional adaptations of soft bottom communities. He found that filter-feeding organisms are typical of relatively coarse sediments (high energy environments), whereas deposit-feeders are more abundant in sediments with high percentage of fine particles (low energy environments). His observations have been later confirmed by other studies (see Gray, 1981, for references). Bachelet (1981) described the trophic organisation of soft-bottom benthic communities using the relative dominance of different feeding guilds. Trophic organisation has usually been related to the nature of the substratum and the organic matter content in sediments (Fresi *et al.*, 1983; Eleftheriou & Basford, 1989).

Organic matter in sediments is often considered as an important food resource for the sediment community (Mirza & Gray, 1981; Grebmeier *et al.*, 1989); however, organic matter composition is known to change according to its concentration (Bordovskiy, 1965), with consequences in degradative and biogeochemical processes that lead to its use by benthic organisms (see also Albaigés *et al.*, 1985).

The aim of this study is to report biomass and trophic structure of soft-bottom communities in the northern Tuscan Archipelago, and to evaluate whether the organic matter accumulated in the sediments represents a trophic input to the benthic food chain.

Description of site studied

The study area is located in the northern part of the Tyrrhenian sea (western Mediterranean), between the town of Leghorn and the island of Elba. The morphology of the area is characterized by the presence of the Elba walloon – due to submarine accumulation and erosion phenomena – to the west, and of sedimentation basins – orientated NNW-SSE – in the centre-east. The edge of the shelf drops sharply to the west, where it is situated at depths ranging between 120 and 140 m, and is rather blunt to the east, where it could be assumed to lie at a depth of about 110–115 m (Bartolini *et al.*, 1979). The continental hydrography of the region has limited importance for the marine environment: only the Cecina river has a significant flow and suspended load, thus exercising an influence on the shelf sedimentation. The circulation of the water masses is greatly affected by morphology, which leads to the formation of vorticose structures, limiting water mass exchanges in coastal areas. On the shelf, a main stream is directed northward, and probably depends on the large scale circulation between Tyrrhenian and Ligurian basins (Elliott, 1979).

Materials and methods

Benthos was sampled with a modified Van Veen grab, covering a surface of 1500 cm^2, in 55 stations between 5 and 500 m deep; for each station, 2 replicates were collected. Sediments were sampled with a Shipeck grab for the determination of granulometry (standard sieving and sedigraph) and organic matter content.

C and N in the sediments were analyzed with CHN analyzer standardized with cystine (Reference Material 143, NBS). The organic component was determined as the difference between the total content and the fraction remaining after combustion of the samples at 480 °C for 5 hours. The accuracy of the analytical measures was estimated by using standard sediment MESS1 (Nat. Res. Counc. Canada, Ottawa) certified for C content.

Biomass was calculated by fresh weight ($g \cdot m^{-2}$), in order not to destroy organisms. Values were then transformed to dry weight equivalents, using dry weight/wet ratios from literature.

Homogeneous groups of stations were singled out by cluster analysis applied to a similarity matrix (Van der Maarel's coefficient); clusters were then considered as representative of different communities, and named after the most characteristic and/or dominant organisms.

Trophic structure of the communities was described by the quantitative dominance, computed as percent biomass, of different feeding guilds. Such guilds were defined on the basis of a three level criterion, partially inspired by Fredj (1972) and Word (1980), which takes into account:

Table 1. Depth ranges, granulometries and organic matter contents of the sediments, and total biomass values in the 11 considered communities.

Communities	Depth range (m)	Gravel (%)	Sand (%)	Silt (%)	Clay (%)	C org (%)	C/N	Biomass ($g \cdot m^{-2}$)
Owenia fusiformis	10– 25	0.2	89.7	5.7	4.4	0.25	14.17	3.29
Cymodocea nodosa	5– 11	0.1	86.4	6.5	7.0	0.28	4.00	1.82
Bryozoan	35– 55	14.4	79.4	3.2	3.0	0.32	8.03	6.51
Maërl	70–117	3.0	88.8	5.4	2.8	0.36	10.80	1.44
Amphicteis gunneri	68–115	0.9	35.5	20.4	43.2	1.18	14.90	1.54
Lumbrineris latreilli	20– 50	1.0	20.3	36.1	42.6	1.49	20.27	4.19
Drilonereis filum	45– 80	0.0	2.0	40.7	57.3	1.92	17.08	2.38
Labidoplax digitata	48– 70	0.0	0.0	45.5	54.5	1.68	18.32	1.95
Onchnesoma steenstrupi	99–100	0.0	0.0	32.0	68.0	1.62	16.03	3.80
Apseudes grossimanus	100–185	0.2	13.4	23.0	63.4	1.35	15.84	3.52
Auchenoplax crinita	122–490	0.1	22.8	21.0	56.1	1.36	16.99	0.65

1) food size; 2) food composition; 3) food-gathering mechanism. For each of these levels, different modalities were identified. At the first level it was possible to distinguish: macrophages, mixtophages, microphages, fluidiphages; at the second level: autotrophs, carnivores, herbivores, omnivores; at the third level: predators, scrapers, scavengers, suspension-feeders, detritus-feeders, limivores, suckers.

Biomass and trophic organization of the different communities were related to the organic matter content of sediments, expressed as percentage of organic carbon; the organic carbon to nitrogen ratio was used to estimate potential food quality (Parsons *et al.*, 1984).

Results and discussion

Cluster analysis was performed on a matrix of 55 stations × 420 species, and evidenced 11 major infaunal communities (Table 1, Fig. 1). They were identified as follows: *Cymodocea nodosa* (Ucria) Aschers, and *Owenia fusiformis* Delle Chiaje communities on fine sands; bryozoan detritic and 'maërl' communities, on gravelly sands with high biodetritic component; *Amphicteis gunneri* (Sars), and *Lumbrineris latreilli* Audouin & Milne-Edwards communities, on more or less sandy muds; *Labidoplax digitata* (Montagu), and *Drilonereis filum* (Claparède) communities on silty clays; *Onchnesoma steenstrupi* Koren & Danielssen, *Apseudes grossimanus* Norman, and *Auchenoplax crinita* Ehlers communities on deeper clayish muds with different amounts of sand. Two other communities were recognized (*i.e.*, *Posidonia oceanica* and *Gryphus vitreus*), but they were poorly represented in our samples and will not be considered in the present paper.

Total biomass values, computed for each community as dry weight equivalents in $g \cdot m^{-2}$ (Table 1), are in agreement with those from similar communities in the Mediterranean Sea, as far as sandy- and detric-bottom communities are concerned (Bellan-Santini, 1985); for muddy-bottom communities, we found values less than those reported by Albertelli & Cattaneo (1985) in the case of coastal communities (*L. latreilli*, *D. filum*, and *L. digitata*), and greater in the case of two of the deep communities (*O. steenstrupi* and *A. grossimanus*), the third (*A. crinita*) being in the previously reported range.

A total of 18 trophic guilds were identified in the 11 communities (Fig. 2), but only 13 of them were important in terms of biomass percent dominance. Coastal muddy-bottom communities (*i.e.*, *A. gunneri*, *L. latreilli*, *D. filum*, *L. digitata*) were all dominated by the MIC-OMN-LIM (microphages-omnivores-limivores) guild, largely corresponding to sub-surface deposit feeders. Suspension-feeders or MIC-OMN-SUS (microphages-omnivores-suspension feeders) and benthic macroalgae or MIC-AUT-SUS (microphages-autotrophs-suspension feeders), dominated in biodetric communities (*i.e.*, bryozoan and maërl). Sandy-bottom communities exhibited two different trophic organisations. *O. fusiformis* has a fairly even dominance pattern, with MIC-OMN-LIM and MAC-CAR-PRE (macrophages-carnivores-predators), followed by MIC-OMN-DET (microphages-omnivores-detritus feeders) and MIC-OMN-SUS, as major guilds; *C. nodosa* was characterized by the great biomass values of this phanerogam, which constituted the guild of MIC-AUT-SUC (microphages-autotrophs-suckers), with MAC-CAR-PRE as second dominants. Deep muddy-bottom communities were to be found in three different situations: in the *O. steenstrupi* community, MAC-CAR-PRE dominated heavily; *A. grossimanus* presented high biomass values of MIC-(microphages) and MIX-CAR-SUS (mixtophages-carnivores-suspension feeders), both represented by large benthic cnidarians; finally *A. crinita* was too poor to evidence a clear dominance pattern, many guilds being equally scarce. Organic carbon contents in sediments vary from a minimum value of 0.08% (*C. nodosa* community) to a maximum of 1.92% (*D. filum* community) (Table 1); all such values fall within the range reported by Cosma *et al.* (1983) for Ligurian sea sediments, and in general agree with data concerning the whole of the western Mediterranean (Albaigés *et al.*, 1985; Fredj & Laubier, 1985).

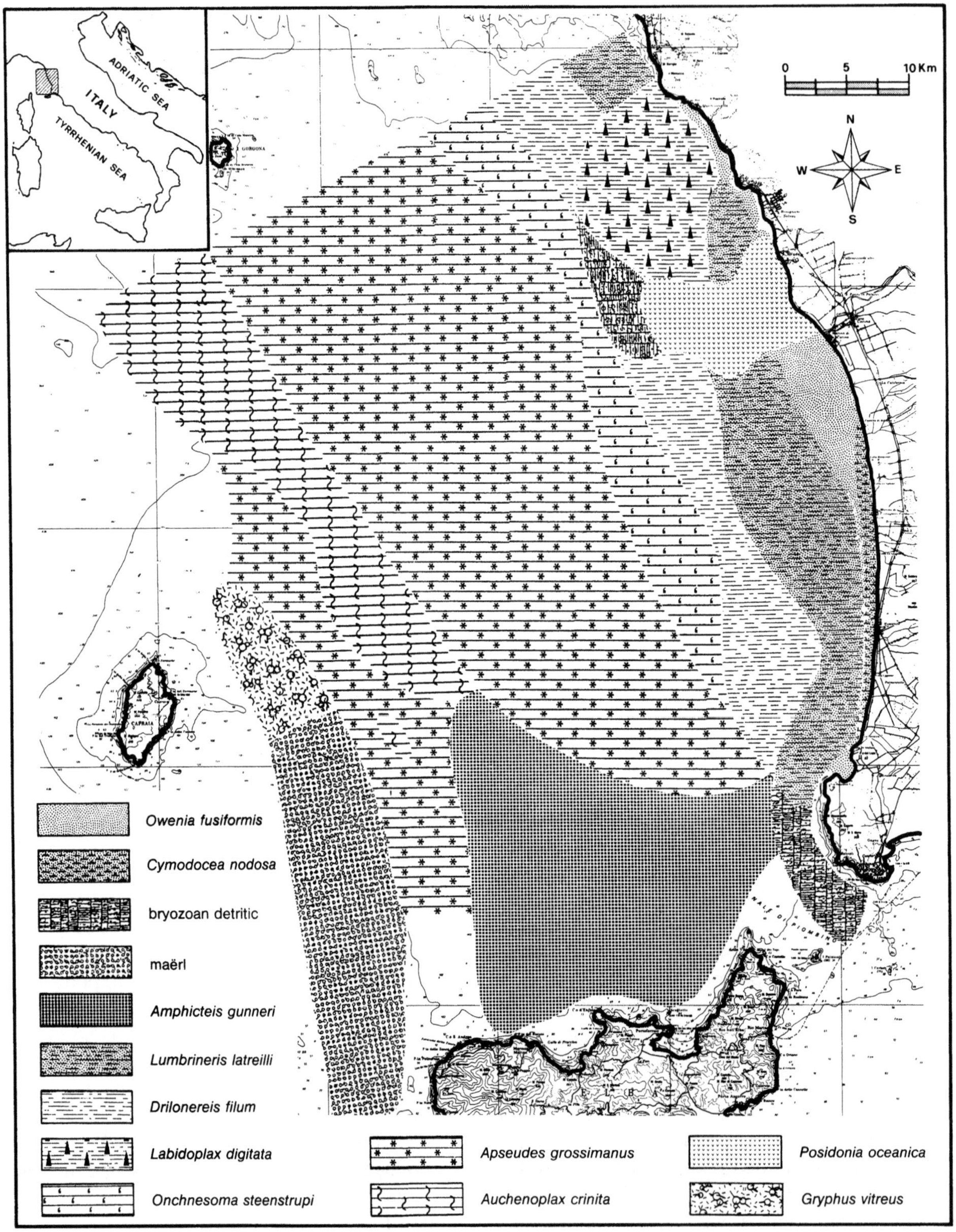

Fig. 1. Map of the major marine benthic communities between Leghorn and Elba Island: *Posidonia oceanica* (L.) Delile, and *Gryphus vitreus* (Born) communities are not considered in the present paper (see text).

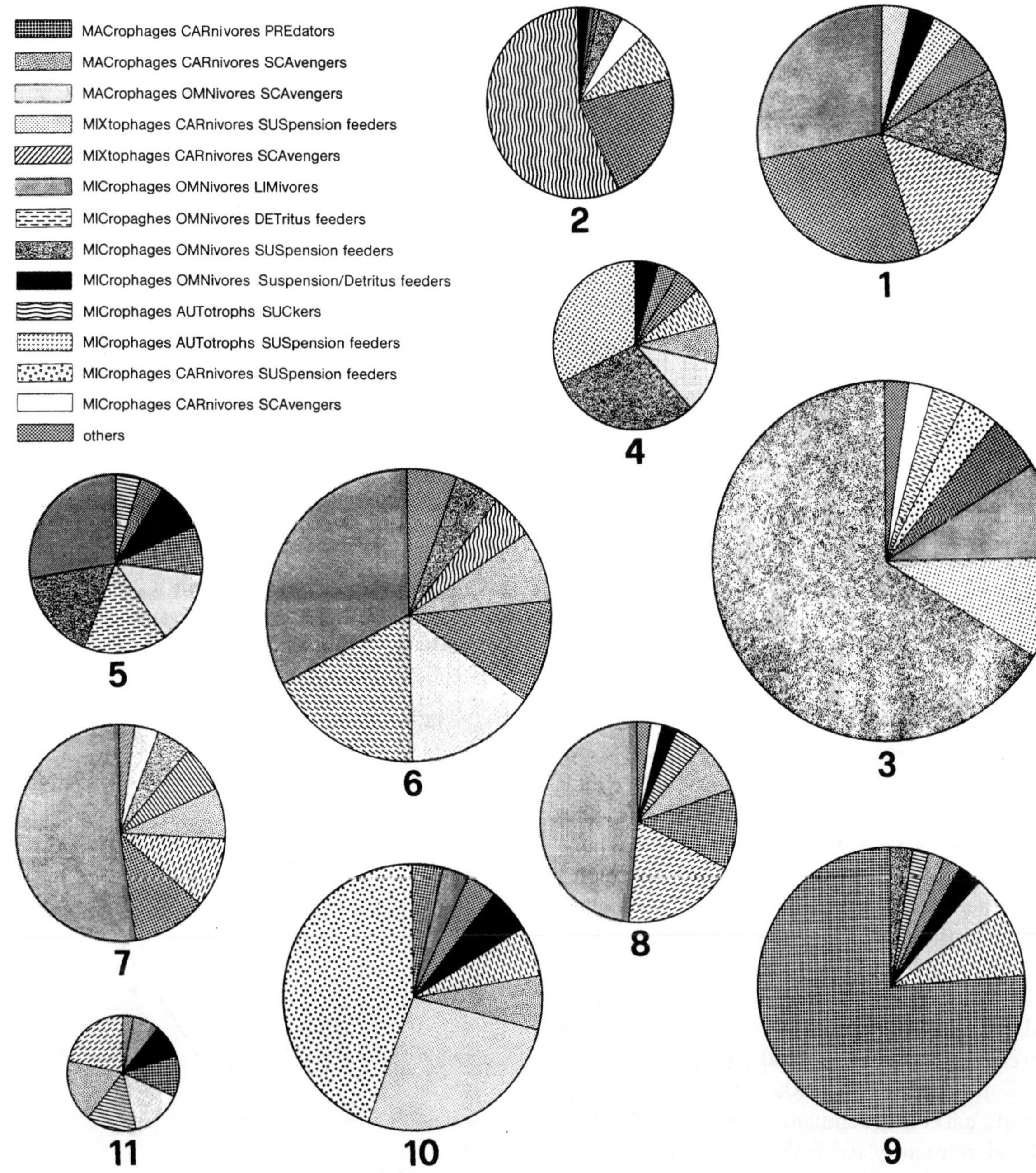

Fig. 2. Trophic structure (percent dominance of feeding guilds) of the 11 benthic communities considered (diameters of the pie diagrams are proportional to total biomass values). 1: *O. fusiformis*; 2: *C. nodosa*; 3: bryozoan detritic; 4: maërl; 5: *A. gunneri*; 6: *L. latreilli*; 7: *D. filum*; 8: *L. digitata*; 9: *O. steenstrupi*; 10: *A. grossimanus*; 11: *A. crinita.*

No correlation exists between total benthic biomass and organic carbon content of sediments (Fig. 3a): the same result was obtained by Sanders & Hessler (1969). Nevertheless, a positive correlation, although low ($r = 0.63$, $0.05 < p < 0.1$), can be found between organic carbon and MIC-OMN-DET biomass (Fig. 3b), when two communities (*L. latreilli* and *O. fusiformis*) are excluded from the analysis. It is commonly supposed that deposit-feeders, like MIC-OMN-

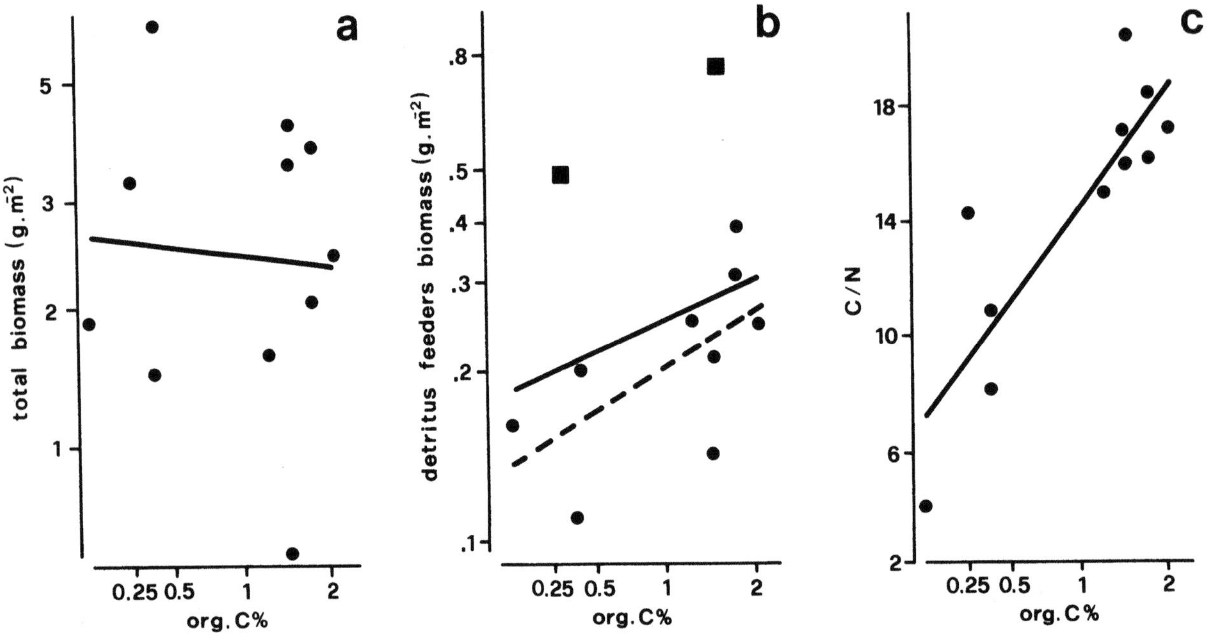

Fig. 3. Regressions of total benthic biomass (**a**: $r = -0.14$, n.s), detritus-feeders biomass (**b**: $r = 0.63$, $0.05 < p < 0.1$), and C/N ratio (**c**: $r = 0.86$, $p < 0.001$), on the percent organic carbon content of sediments. (Solid line = all communities considered; dashed line = 2 communities – *i.e.*, quadrats – excluded).

DET, could utilize sediment organic matter as a food source, especially via the microbial loop (Cammen, 1980). Thus, such a correlation could express a trophic relationship. However, the values of C/N ratio we found in many communities were too high to consider the sediment organic matter as a potential food source: the higher C/N ratio, the lower the quality of food and consequently the assimilation efficiency (Valiela, 1984).

The total organic carbon content of sediments and the C/N ratio turned out to be positively correlated ($r = 0.86$, $p < 0.001$) (Fig. 3c). Such a correlation implies that when high quantities of organic carbon accumulate, they are to be considered 'refractory' to benthic consumers for their major part, being poor in nitrogen. C/N ratios vary from a minimum of 4.00 (*C. nodosa* community) to a maximum of 20.27 (*L. latreilli* community) (Table 1): a ratio not exceeding 17/1 is required for animal consumption (Russel-Hunter, 1970).

The communities with the highest C/N ratio in the sediments also showed an increased dominance of species indicating organic matter accu-

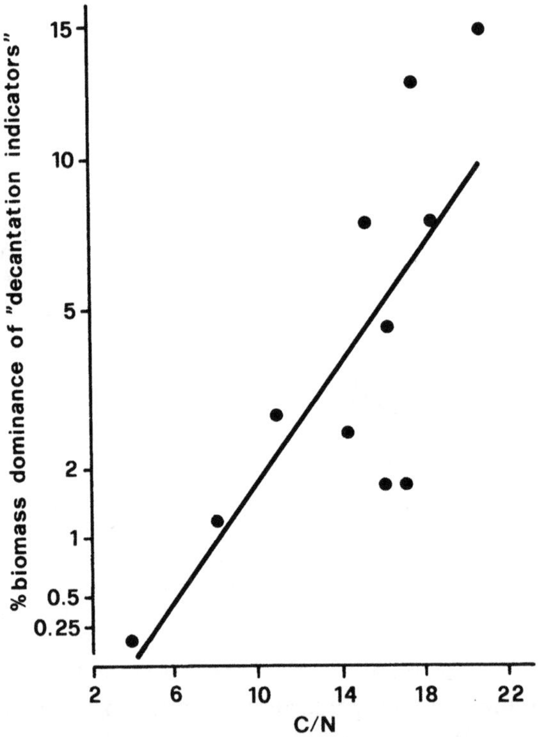

Fig. 4. Correlation between percent biomass dominance of 'decantation indicators' (see text), and C/N ratio of sediments ($r = 0.76$, $p < 0.01$).

mulation. These were represented by *Lumbrineris latreilli* Audouin & Milne-Edwards, *Corbula gibba* (Olivi), *Paralacydonia paradoxa* Fauvel, *Myrthea spinifera* (Montagu), and other, which are known to be very tolerant with respect to high organic sedimentation rates (Nodot *et al.*, 1984). Salen-Picard (1985) defined such species as 'indicators of decantation'; their proliferation is thought to indicate a perturbation or even a degradation of the community (Bellan *et al.*, 1985). We found a positive correlation ($r = 0.76$, $p < 0.01$) between their biomass percent dominance and the C/N ratio of sediments (Fig. 4).

Conclusion

The positive correlation between organic carbon and C/N ratio on sediments gave evidence that sediment organic matter accumulates mainly as refractory matter, and thus cannot be considered as an important food source for infauna (Degens & Mopper, 1976). Detrivores, however, turned out to be slightly correlated with sediment organic carbon, thus suggesting that they may be able to utilize it. Valiela (1984) hypothesized that marine consumers feeding on detritus are limited by the quantity and quality of their food supply. Although the quantities we found were usually abundant, quality was generally poor, as suggested by the high C/N ratio. Even where C/N ratios of organic matter are low, detritus-feeders may not get enough nitrogen, due to unavailability (Tenore *et al.*, 1979). This could explain the weak correlation with detritus-feeders dominance, and also the fact that the role they played in the trophic organization of our communities was not always as great as we expected, on the basis of sediment nature and depth (Gray, 1981). We found a sharp and absolute dominance of (sub-surface) detritus-feeders (MIC-OMN-LIM) only in 4 (*L. latreilli*, *L. digitata*, *D. filum*, *A. gunneri*) of the 11 communities we had taken into account, whereas other Mediterranean workers reported the detrivores as the usual dominant feeding guild in soft-bottom communities (Gambi *et al.*, 1986; Tunesi & Peirano, 1986).

Thus, we raise doubts that organic matter in sediment could really represent a food resource for the infaunal communities we studied. On the contrary, we hypothesize that the positive correlation between C/N ratio and the dominance of 'decantation indicators' was a clue that organic matter in sediment could rather be a source of stress (Pearson & Rosenberg, 1978; Bellan *et al.*, 1980). This might be a site-specific effect; other authors, however, have reported a significant negative correlation between sediment organic content and total biomass of the infaunal community (Grebmeier *et al.*, 1988). Silt content, which was high in the areas where organic matter accumulated, is to be considered a component of the stress factor, since silt may be an inhibitor of certain animal activities.

Of course we are well aware of the difficulty in making a general rule from our own data and those found in literature. Data from Bianchi *et al.* (1986) indicate that there is not a univocal relationship between sediment organic matter, microbial activity, and detritus-feeder biomass and production. Our main results are apparently in contradiction with each other; sediment organic carbon had a positive relation with both detritus-feeders and 'stress-indicating' species, but had no relation with total biomass. Our interpretation is that relationships between benthic consumers and organic matter do not follow a single pattern: a mosaic of situations exists, in which the energetic input of sediment organic matter could in turn act as a feeding or stress source, depending on its quantity and quality, and on a number of abiotic and biotic factors we are not yet able to distinguish completely.

Acknowledgements

We wish to thank Dr. T. H. Blackburn (Department of Ecology and Genetics, Aarhus University, Denmark) for his critical reading of the manuscript and for his suggestions.

References

Albaigés, J., M. Aubert, J. Aubert, 1985. The footprints of life and of man. In R. Margalef (ed.), Western Mediterranean. Pergamon Press., Oxford: 317–351.

Albertelli, G. & M. Cattaneo, 1985. Macrobenthos dei fondi molli del Mar Ligure. In F. Cinelli & M. Fabiano (eds), Atti 6° Congr. Ass. It. Oceanol. Limnol., Livorno: 87–98.

Bachelet, G., 1981. Données préliminaires sur l'organization trophique d'un peuplement benthique marin. Vie Milieu 31: 205–231.

Bartolini, C., F. Fanucci, G. Fabbani, S. Rossi, G. Valleri & R. Lenaz, 1979. Studio della piattaforma continentale medio-tirrenica per la ricerca di sabbie metallifere: 2) dall'isola d'Elba a Livorno. Boll. Soc. Geol. It. 98: 327–352.

Bellan, G., D. Bellan-Santini, J. Picard, 1980. Mise en évidence de modèles éco-biologiques dans des zones soumises à pertubations par matière organiques. Oceanol. Acta. 3: 383–390.

Bellan, G., M. Bourcier, J. Picard, S. Salen-Picard & G. Stora, 1985. Conséquences structurelles dues aux perturbations affectant les biocénoses benthiques méditerranéennes de substrat meuble. Rapp. Comm. int. Mer Médit. 29: 215–221.

Bellan-Santini, D., 1985. The Mediterranean benthos: reflection and problems raised by a classification of the benthic assemblages. In M. Moraitou-Apostolopoulou & V. Kiortsis (eds), Mediterranean ecosystems. Plenum Press, New York: 19–49.

Bianchi, C. N., L. Biondi, R. Cattaneo Vietti, C. Morri, I. Niccolai, C. Peroni, G. Rossi, R. Ruggiero & S. Tucci, 1986. Ecologie trophique de quelques biocénoses de fonds meubles en mer Tyrrhénienne. Rapp. Comm. int. Mer Médit. 30: 253.

Bordovskiy, O. Y., 1965. Accumulation and transformation of organic substance in marine sediments. Parts 1–3. Mar. Geol. 3: 3–144.

Cammen, L. M., 1980. The significance of microbial carbon in the nutrition of the deposit feeding polychaete *Nereis succinea*. Mar. Biol. 61: 9–20.

Cosma, B., V. Contardi, G. Zanicchi & R. Capelli, 1983. Heavy metals in superficial sediments from the Ligurian Sea, Italy. Chemistry in Ecology 1: 331–344.

Degens, E. & K. Mopper, 1976. Factors controlling the distribution and early diagenesis of organic material in marine sediments. In J. P. Riley & R. Chester (eds), Chemical Oceanography, vol. 6, 2nd Ed., Academic Press, S. Diego: 59–113.

Eleftheriou, A. & D. J. Basford, 1989. The macrobenthic infauna of the offshore Northern North Sea. J. mar biol. Ass. U.K. 69: 123–143.

Elliott, A. J., 1979. Low-frequency sea level and current fluctuation along the coast of Northwest Italy. J. geophys. Res. 84: 3752–3760.

Fredj, G., 1972. Stockage et exploitation des données en écologie marine. A. Un fichier sur ordinateur des Invertébre macrobentiques. Mém. Inst. océanogr. Monaco 4: 1–61.

Fredj, G. & L. Laubier, 1985. The deep Mediterranean benthos. In M. Moraitou-Apostolopoulou & V. Kiortsis (eds), Mediterranean marine ecosystems. Plenum Press, New York: 109–145.

Fresi, E., M. C. Gambi, S. Focardi, R. Bargagli, F. Baldi & L. Falciai, 1983. Benthic communities and sediment types: a structural analysis. P.S.Z.N. I. Mar. Ecol. 4: 101–121.

Gambi, M. C., M. Lorenti, G. F. Russo & M. B. Scipione, 1986. Feeding-group distribution in soft-bottom macrobenthos: an example in the Gulf of Salerno (Tyrrhenian Sea, Italy). Rapp. Comm. int. Mer Médit. 30: 253.

Gray, J. S., 1981. The ecology of marine sediments. Cambridge Univ. Press, Cambridge, 185pp.

Grebmeier, J. M., C. P. McRoy & H. M. Feder, 1988. Pelagic-benthic coupling on the shelf of the northern Bering and Chukchi Seas. I. Food supply source and benthic biomass. Mar. Ecol. Prog. Ser. 48: 57–67.

Grebmeier, J. M., H. M. Feder & C. P. McRoy, 1989. Pelagic-benthic coupling on the shelf of the northern Bering and Chukchi Seas. II. Benthic community structure. Mar. Ecol. Prog. Ser. 51: 253–268.

Mirza, F. B. & J. S. Gray, 1981. The fauna of benthic sediments from the organically enriched Oslofjord, Norway. J. exp. mar. Biol. Ecol. 54: 181–207.

Nodot, C., M. Bourcier, A. Jeudy de Grissac, S. Heusner, J. Régis & J. Tiné, 1984. Répartition des biocoenoses benthiques en fonction des substrats sédimentaires de la rade de Toulon (France). 2. La Grande Rade. Téthys 11: 141–153.

Parsons, T. R., M. Takahashi & B. Hargrave, 1984. Biological oceanographic processes. Pergamon Press, Oxford, 3rd Ed., 330 pp.

Pearson, T. H. & R. Rosenberg, 1978. Macrobenthic succession in relation to organic enrichment and pollution of the marine environment. Oceanogr. mar. Biol. annu. Rev. 16: 229–311.

Russel-Hunter, W. D., 1970. Aquatic productivity. McMillan, 187 pp.

Salen-Picard, C., 1985. Indicateurs biologiques et sedimentation en milieu circalittoral Mediterraneen. Rapp. Comm. int. Mer Médit. 29: 211–212.

Sanders, H. L., 1958. Benthic studies in Buzzards Bay. I. Animal-sediment relationships. Limnol. Oceanogr. 3: 245–258.

Sanders, H. L. & R. R. Hessler, 1969. Ecology of the deep-sea benthos. Science 163: 1419–1424.

Tenore, K. R., R. B. Hanson, B. E. Donseif & C. N. Wiederhold, 1979. The effect of organic nitrogen supplement on the utilization of different sources of detritus. Limnol. Oceanogr. 84: 350–355.

Tunesi, L. & A. Peirano, 1986. Organization trophique d'un peuplement des vases terrigènes côtières dans le golfe Marconi (Mer Ligure). Rapp. Comm. int. Mer Médit. 30: 254.

Valiela, I., 1984. Marine Ecological Processes. Springer Verlag, New York, 546 pp.

Word, J. Q., 1980. Classification of benthic invertebrates into infaunal trophic index feeding groups. Coastal Water Research Project, Annual Report 1979, El Segundo, California: 103–121.

Received 9 August 1989; in revised form 18 January 1990; accepted 28 February 1990

Hydrobiologia **207**: 61–69, 1990.
D. J. Bonin & H. L. Golterman (eds), Fluxes Between Trophic Levels and Through the Water–Sediment Interface.

Decrease in allochthonous organic inputs in dark submarine caves, connection with lowering in benthic community richness

Renaud Fichez
Centre d'Océanologie de Marseille, Station Marine d'Endoume, rue de la Batterie des Lions, 13007 Marseille, France; Present address: School of Environmental Sciences, University of East Anglia, NR4 7TJ, Norwich, England

Key words: vertical flux, sediment trap, organic compounds, oligotrophy, submarine cave, deep-sea

Abstract

Ten sediment trap arrays were deployed over two years for periods of 3 to 40 days in three different sampling points along a 50 m long Mediterranean submarine cave. Mean total particulate matter flux decreased strongly from the semi-dark area (3.3 g m^{-2} d^{-1}) to the dark area (0.8 and 0.6 g m^{-2} d^{-1}). Carbon represented 3.3% to 3.5% and nitrogen 0.34% to 0.38% of settling dry matter. The decrease in organic input from the entrance to the terminal part of the cave results in increasingly oligotrophic conditions with distance from the cave entrance. Horizontal resource limitation can be connected with a strong zonal decrease in fauna richness. Biomass declines both in hard substrate and soft bottom communities. Despite major differences, some similarities are noticed between oligotrophic conditions that may occur in the dark cave and those in around 1000 m depth ecosystems. Dark oligotrophic submarine caves can be considered to be good scale models for the study of some aspects of general trophic pathways.

Introduction

Decrease of fauna richness has been noticed in early studies of Mediterranean cave communities, both in hard substrata communities (Pérès & Picard, 1949; Laborel & Vacelet, 1958; 1959) and in soft bottom communities (Monteiro-Marques, 1981) when compared to communities outside the cave. More recently a biomass study gave new evidence of a strong decrease in benthic populations in a Mediterranean dark cave (Gili *et al.*, 1986) which is presumed to be related to a decrease in trophic resources inside the cave (Harmelin *et al.*, 1985).

In an attempt to examine the validity of this hypothesis, vertical flux studies were performed for two years (10 sampling operations) in a Mediterranean sea cave close to Marseille (France). Vertical flux provides information on particulate flows to the benthic boundary layer. Since under aphotic conditions in submarine caves photosynthetic production does not occur, the supply of dissolved and particulate organic material to the cave sediment emanates from outside sources. It enters the cave through water exchange and subsequent sedimentation of particles to the water-sediment interface.

Direct measurement of the particulate flux to sediments is commonly carried out using sediment traps (Hargrave & Phillips, 1986; Noriki & Tsunogai, 1986). Comparative studies between sediment trapping and other methods of measuring sedimentation rates have been made (Spencer *et al.*, 1978; Lorenzen *et al.*, 1981). Other studies focus on the effect of trap geometry and current velocity on the amount of collected particles (Hargrave & Burns, 1979; Gardner, 1980a; 1980b; Blomqvist & Kofoed, 1981; Butman, 1986). The main results from these studies have demonstrated that cylinders should be used as sediment traps (Hargrave & Burns, 1979; Gardner, 1980b) both in turbulent and calm hydrodynamic conditions (Blomqvist & Hakanson, 1981). The aspect ratio, h/d (h = trap height, d = trap diameter), is considered to be the main geometrical parameter for traps. Its minimum values is estimated at 2.3 (Gardner, 1980a), 3 (Butman, 1986) or more than 3 (Blomqvist & Kofoed, 1981). Results obtained with sediment traps provide valuable information about vertical fluxes; nevertheless, they should be treated with caution (Smetacek, 1984).

Description of site studied

The submarine cave studies is located on the calcareous coastline close to Marseille (France) (Fig. 1). Abundant karst formations in this area have been transformed into submarine caves by the last marine transgression.

Trémies cave is 60 long, 35 m wide and up to 9 m high. A 8 to 10 m high step 30 m from the entrance separates two topographic areas. The first area extending from the entrance to the step, is a wide chamber (6600 m^3) where the light level quickly decreases (semi-dark area). About 15 m from the entrance close to a *Corallium rubrum* population, light is expected to range from 0.04% to 0.06% of surface illumination (Marinopoulos, 1989). The second area is smaller (2800 m^3) and sinuous and includes two distinctive terminal parts. The very poor benthic fauna covering the hard substrata in full darkness is termed the dark cave community, the corresponding area being the dark area. It should be noted that, in Trémies cave, the dark cave community is more extensive than in any of the surrounding caves (Harmelin, 1969).

Sampling stations were numbered TR1, TR2,

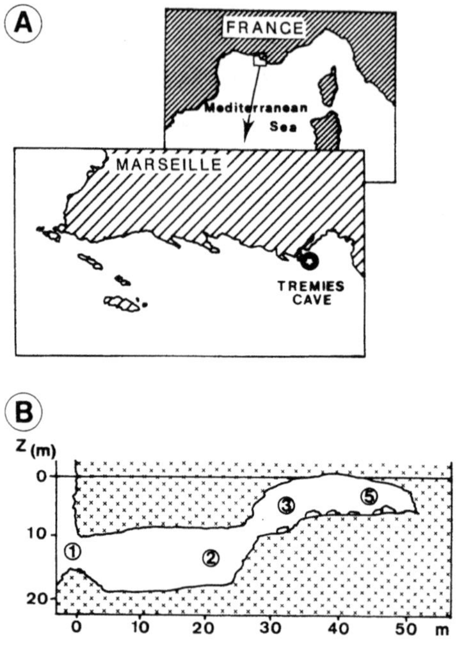

Fig. 1. Trémies submarine cave location (A) and cross section (B). Numbers indicate corresponding sampling points. Point 1 (TR1) at the entrance, point 2 (TR2) in the semi-dark area, points 3 (TR3) and 5 (TR5) in the dark area.

TR3, TR5. TR1 was located at the entrance where the bottom is covered with fallen rocks. No vertical flux measurements were performed at this point exposed to various perturbations (slumping, current effects, human activities). TR2 was located 22 m from the entrance in the semi-dark area over a muddy grey sediment layer. TR3 was located at the beginning of the dark area 35 m from the entrance over a muddy brown sediment layer close to the rocky vertical wall. TR5 was in the terminal part of the cave 45 m from the entrance over a muddy brown sediment layer.

Materials and methods

Sediment trapping

Each sediment trap (Fig. 2) consisted of three to four cylindrical glass collectors set on a central axis holder terminated by a cylindrical guiding

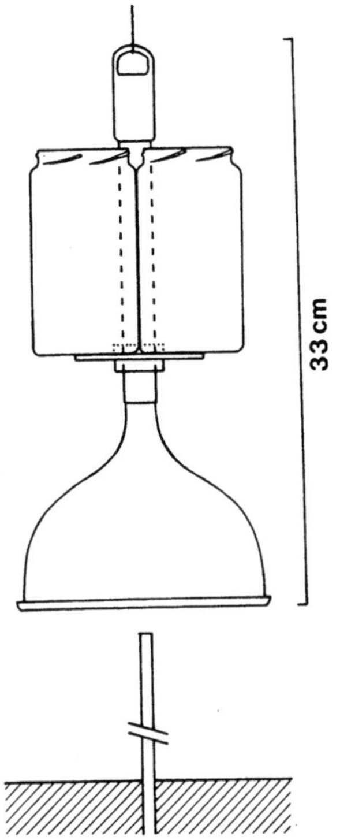

Fig. 2. Sediment trap design used in this study. Trap carrier stick permanently anchored in the sediment (1), guiding cone (2), collector holder (3), collectors (4) (three to four for each sediment trap), handling rope and float (5).

cone. Each collector was 5.5 cm in diameter (24 cm^2 collecting area) and 12.5 cm high (h/d = 2.3). No poison was added to these collectors.

Sediment trap handling was performed using SCUBA diving. The traps were brought underwater upside down to the sampling point. The collecting apparatus was released close to the sampling point and held by a 2.5 m rope equipped with a terminal float. With the cone acting as a guide, it was then carefully lowered and placed onto a stick permanently anchored in the sediment. Sediment traps were then held 30 cm above the bottom sediment. Collectors remained in place for 3 to 40 days. During retrieval, each collector was directly closed with a 'twist off' cover and the sediment trap was carried on board and brought back to the laboratory in the dark.

Treatment

After sedimentation under calm conditions, the supernatant was poured into a vessel and the residue was rinsed twice using an isotonic ammonium formiate (NH_4 COOH) solution. Ammonium formiate was proved to be a volatile rinsing solution well suited to the necessary low temperature drying. Rinsed material was then centrifuged at 5000 rev./min for 20 minutes and the new supernatant was added to the previous supernatant. The residue was then dried at 40 °C for 40 hours. The whole supernatant collected was filtered through a preweighed Whatman GF/C filter. Residue and filter retained dry weights were added and were then correlated with collection time and surface and expressed in g m^{-2} d^{-1}.

Analysis

Carbon and nitrogen were measured using a CHN autoanalyzer (Perkin Elmer 240). To remove carbonate, the filters were treated with concentrated HCl vapour for 2 hours and then redried at 40 °C for 12 hours. This treatment could not be applied to sediment which had undergone a direct acid attack, as the autoanalyzer was not equipped with an HCl vapour trap and would have been severely damaged. Therefore the difference on ignition (DOI) method (Hirota & Szyper, 1975; Byers *et al.*, 1978) was used to measure organic carbon (OC) and organic nitrogen (ON) in the sediment. Comparison of these two methods showed that the DOI method is recommended for routine measurements of OC in sediments (Kristensen & Andersen, 1987).

Sediment from sediment traps was homogenized in a mortar and three subsamples were analyzed to check reproducibility. OC and ON concentrations expressed as percentage of dry matter were then associated with the correponding vertical particle fluxes and converted into sedimentation rate units (mgC or mgN m^{-2} d^{-1}). Despite the great interest of CHN analysis, it should be pointed out that the detection limits of

Table 1. Sedimentation rate measurements for each sediment trap array and each sampling point. Values are expressed in mg m^{-2} d^{-1} of dry matter. Mean and standard deviation (SD) of point/array couple values are listed in the table. Mean and standard deviation over the two years survey for each sampling point appear in the last part of the table.

Start	End	Sampling point TR2		Sampling point TR3		Sampling point TR5	
		Flux/collector	Mean (SD)	Flux/collector	Mean (SD)	Flux/collector	Mean (SD)
28/06/85	08/07/85	3583.5		1122.3		1070.5	
		3032.1	3303.2	1238.7	1128.0	1083.8	1088.4
		3447.3	(255.8)	1009.7	(93.9)	1051.3	(41.8)
		3149.8		1141.2		1147.8	
12/09/85	24/09/85	2509.2		571.3		483.1	
		2425.1	2382.8	537.2	569.2	510.4	505.2
		2214.0	(152.1)	599.0	(31.0)	522.0	(20.0)
24/09/85	15/10/85	2555.2		987.3		987.2	
		2318.5	2648.8	1122.6	1096.0	883.5	923.6
		2729.0	(284.2)	1145.3	(73.1)	930.3	(46.9)
		2992.4		1128.8		893.5	
15/10/85	13/11/85	6840.5		1941.0		939.7	
		6703.2	6720.0	1635.8	1600.8	1090.5	1463.6
		6639.7	(85.2)	1455.3	(252.3)	1854.0	(523.7)
		6696.6		1371.0		1970.2	
29/11/85	08/01/86	834.3		121.2		103.9	
		1512.0	1296.8	137.3	130.0	112.4	116.8
		1122.5	(395.1)	142.3	(11.5)	121.5	(11.1)
		1718.3		119.2		129.4	
04/03/86	01/04/86	6922.9		1005.6		831.7	
		7220.4	7227.2	1533.4	1307.2	1332.4	1015.2
		7342.7	(219.2)	1312.1	(221.5)	1188.4	(226.3)
		7422.6		1377.7		808.2	
06/06/86	09/06/86	3859.7		875.2		482.3	
		3677.2	4320.0	930.3	920.0	567.0	543.2
		4803.5	(643.6)	938.1	(30.1)	550.6	(41.7)
		4939.6		936.4		572.8	
30/09/86	07/10/86	2540.5		493.2		559.4	
		2374.3	2580.4	685.7	648.0	511.3	519.2
		2896.0	(222.5)	763.9	(113.8)	497.2	(27.5)
		2510.8		649.2		508.9	
05/02/87	17/02/87	1283.2		259.4	99.5		
		1307.6	1361.3	310.2	250.7	110.7	106.4
		1399.4	(80.0)	248.6	(51.6)	103.3	(6.0)
		1455,0		184.6		112.1	
19/05/87	25/05/87	925.6		155.6		70.2	
		946.8	912.4	250.4	234.0	54.8	68.0
		847.8	(44.0)	239.2	(56.9)	90.0	(16.2)
		929.2		291.2		56.8	
Mean			3275.3		788.4		635.0
Standard deviation			2196.2		500.5		474.7

Table 2. Vertical organic carbon and nitrogen flux values for the two years survey in Trémies cave (mg m^{-2} d^{-1}). Sampling points are numbered TR 2, TR 3, TR 5. Mean and standard deviation of survey (n = 10) appear at the end of the table.

Start	End	Organic carbon			Organic nitrogen		
		TR 2	TR 3	TR 5	TR 2	TR 3	TR 5
28/06/85	08/07/85	102.72	35.20	29.00	11.58	4.22	2.67
12/09/85	24/09/85	122.96	42.80	22.64	10.18	3.65	2.14
24/09/85	15/10/85	61.98	44.83	41.10	13.69	4.99	5.79
15/10/85	13/11/85	176.06	37.14	52.40	13.89	3.16	3.95
29/11/85	08/01/86	43.36	5.00	3.96	6.08	0.84	0.62
04/03/86	01/04/86	136.35	28.50	19.39	9.40	1.70	0.91
06/06/86	09/06/86	160.52	20.32	18.36	16.54	2.86	1.80
30/09/86	07/10/86	150.56	32.52	23.20	13.38	5.45	2.85
05/02/87	17/02/87	72.21	10.80	6.72	9.31	0.98	0.66
19/05/87	25/05/87	49.12	11.24	4.68	8.46	2.04	0.45
Mean		107.58	26.84	22.15	11.25	2.99	2.19
Standard deviation		108.12	25.91	21.38	11.21	2.85	2.13

available industrial autoanalysers are poorly suited to sediments with low contents of organic matter.

Results

Sedimentation rate evolution over time and in space and the corresponding mean values over 2

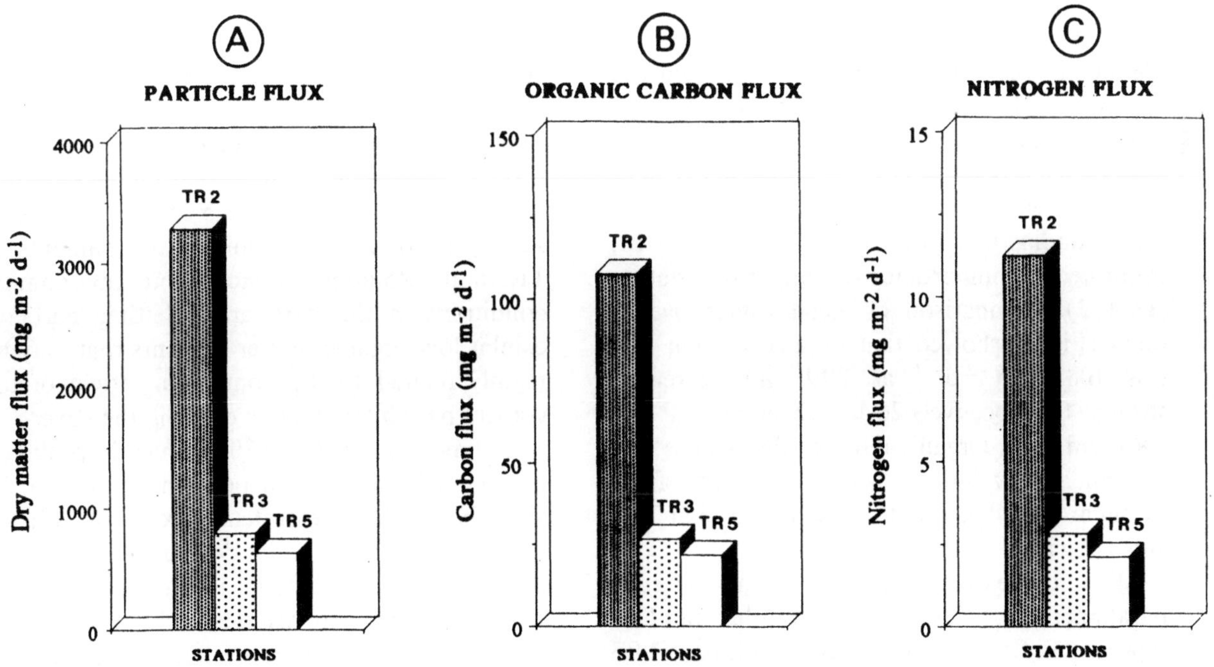

Fig. 3. Mean vertical fluxes (n = 10) over two years at sampling stations TR2, TR3 and TR5 in Trémies cave. Total particle flux (A), organic carbon flux (B), organic nitrogen flux (C).

Table 3. Comparison of the organic carbon (OC) and organic nitrogen (ON) content of suspended and sedimented particles. Suspended particle carbon and nitrogen content values are means of a two years survey (n = 27 for each sampling station) from Fichez (1989). Theoretical fluxes are calculated by multiplying total dry particulate flux with suspended particulate carbon or nitrogen content and are compared with calculated fluxes.

		TR2	TR3	TR5
Measured flux	($g\,m^{-2}\,y^{-1}$)	1195	288	232
Suspended particle OC	(% dry matter)	4.07	3.68	3.45
Sedimented particle ON	(% dry matter)	3.30	3.40	3.49
Suspended particle ON	(% dry matter)	0.46	0.37	0.32
Sedimented particle ON	(% dry matter)	0.34	0.38	0.34
OC theoretical flux	($g\,m^{-2}\,y^{-1}$)	48.6	10.6	8.0
OC measured flux	($g\,m^{-2}\,y^{-1}$)	39.3	9.8	8.1
ON theoretical flux	($g\,m^{-2}\,y^{-1}$)	5.5	1.1	0.7
ON measured flux	($g\,m^{-2}\,y^{-1}$)	4.1	1.1	0.8

years (n = 10) are listed in Table 1. At each sampling point spatial variation of fluxes is moderate; however temporal variation is high with strong variations in particulate input into the sediment. The mean values over two years was 3275 mg $m^{-2}\,d^{-1}$ at station TR2 in the proximal part of the cave 25 m from the entrance. In the distal part, mean values strongly decreased to 788 mg $m^{-2}\,d^{-1}$ at station TR3 35 m from the entrance and 635 mg $m^{-2}\,d^{-1}$ at station TR5 55 m from the entrance.

Organic carbon and organic nitrogen sedimentation values (mean for 3 or 4 collectors) and corresponding mean values over 2 years (n = 10) confirmed the importance of temporal variation (Table 2). Comparison of mean values over 2 years (Fig. 3) showed that organic carbon flux was 108 mg $m^{-2}\,d^{-1}$ at TR2 and decreased strongly to respectively 26 and 22 mg $m^{-2}\,d^{-1}$ at TR3 and TR5. Organic nitrogen flux values had a similar evolution and decreased from 11 mg $m^{-2}\,d^{-1}$ at TR2 to 3 mg $m^{-2}\,d^{-1}$ at TR3 and 2 mg $m^{-2}\,d^{-1}$ at TR5.

Mean decrease remained constant for each parameter (particles, OC, ON). Flux at TR3 represented 25% of flux at TR2 and flux at TR5 represented 20% of flux at TR2. This constancy between different component fluxes demonstrates a stability in particulate material composition. Organic carbon content ranges from 3.3% to 3.5% of dry particulate matter with an average C/N values close to 10. Organic carbon, measured using ignition loss technic, accounted for 13% to 24% of total dry matter.

Discussion

The results demonstrate the general decrease in vertical particulate matter flux from the entrance to the inner dark parts of the cave. Similarity in settling particle and organic matter amounts at TR3 and TR5 attests homogeneous environment conditions in the dark area. Settling particles exhibit low organic matter contents that may be mainly related to the continental origin of the coastal particulate matter entering the cave.

Sediment traps placed 30 cm over the sediment may be affected by resuspension but, due to its size, the whole cave system is certainly affected too. On the other hand degradative processes occur on organic material in unpoisoned traps. Both phenomena are responsible for a decrease in the organic content of trapped particles, thus their importance could be estimated by the difference between the suspended and the sedimented particle organic matter composition (Table 3). Few differences were observed in the dark parts of the cave (TR3 and TR5) indicating insignificant resuspension and degradative processes. It further demonstrates that the suspended organic matter is resistant to decomposition processes. In contrast, at TR2, sedimented particles showed lower organic carbon and nitrogen content than suspended particles and the measured flow is less than the theoretical flow. This indicates the influence of both resuspension and degradation, but these processes cannot be separated in this study.

There is no apparent seasonal cycle in vertical flows, because of the irregular sampling and collecting times and the variability of vertical fluxes. Temporal variability can be explained by temporal variations in both suspended matter and hydrodynamic conditions. The scale of water exchanges between the inner water mass and the open sea directly influences the amount of suspended matter reaching the water sediment interface. High vertical fluxes at TR2 show the importance of water exchange between the productive open sea and the anterior part of the cave. This area is characterized by a dense hard substrate community mainly composed of active filter feeders; passive filter feeders (Gorgonians) are present only near the entrance (Harmelin *et al.*, 1985). The hard substrata biomass is about 250 g m^{-2} and soft bottom community biomass is about 3.4 g m^{-2} 22% of which are meiofauna and 78% are macrofauna (>250 μg) (Fichez, 1989). In the dark area (TR3 and TR5) considerably lower vertical fluxes were observed with very similar sedimentation values at both sampling points. Benthic communities are known to be poor in dark areas (Monteiro-Marques, 1981; Harmelin *et al.*, 1985; Gili *et al.*, 1986) the cover being always much less than 100% of rocky substrata. Recorded biomass for sessile fauna was 36 g m^{-2} at TR3 and 25 g m^{-2} at TR5 and sediment fauna was 0.3 g m^{-2} at each sampling point. Repartition ratio between meiofauna and macrofauna was the same as in TR2 (Fichez, 1989).

Along with these results a strong decrease in suspended chloropigment amount was observed in water between TR2 and TR3, with similar concentrations in TR3 and TR5 (Fichez, 1990). On the other hand, the sediments from the dark area (TR3 and TR5) were characterized by the absence of any redox potential discontinuity layer. This suggests a very low flow of organic material to the sediment, implying low degradative processes and thus low oxygen consumption (Fichez, 1989).

The cave studied can therefore be separated into two subunits with different hydrodynamic features. The proximal part (semi-dark area) has a rich benthic population related to efficient water exchange and high organic inputs. Nevertheless, a general hydrodynamic slowdown can already be seen in the disappearance of passive filter feeders a few meters away from the entrance. In the distal part (dark area) the overall decrease in benthic population density and biomass can be related to an oligotrophic water mass with low organic inputs. Cave topography (the dark area being separated from the semi dark area by a 9 m high step) made such a specific water mass system possible. This structure allowed for heat stratification and isolation of a slowed hydrodynamic system with a long renewal time (several days). This general feature of water mass confinement may be broken by high hydrodynamic activity or by heat counterbalancing currents (Fichez, 1989). In fact, storms, mainly occurring during winter, may affect the whole cave and cause a rapid renewal of the inner waters. On the other hand, spring warming of the surface layer outside the cave results in a density differences between open sea and inside trapped waters. Because of the topography and because water salinity always stays close to 38‰, colder inner waters will be replaced by warmer waters. This second feature, occurring during the spring bloom period when particulate organic matter reaches high values, increases water exchange and influences benthic metabolism (Fichez, 1989).

The general oligotrophic conditions suggest that dark cave heterotrophs are drastically food-limited. On the basis of this feature, submarine cave ecosystems can be compared with ecosystems in the aphotic deep sea. Some values of organic carbon and nitrogen fluxes in the deep ocean, near the benthic boundary, come close to those obtained in the present study, particularly at depths ranging from 1000 to 2000 m all over the world (Hinga *et al.*, 1979; Knauer *et al.*, 1979; Lorenzen *et al.*, 1983; Karl & Knauer, 1984; Tsunogai & Noriki, 1987) or in the north-western Mediterranean (Cauwet, 1985; Buscail *et al.*, 1987). Of course, environmental conditions are extremely diverse in the deep sea and divergent results can also be found in numerous works. Comparing different oceanic regions is difficult and no general vertical flux value may correspond with a precise depth in the whole ocean. On the other hand large differences, e.g. of temperature, pressure, residence time of particles or transport processes characterize deep sea and littoral caves. Accordingly, these two ecosystems should be compared with great caution in view of a simple parallel between two oligotrophic systems having some comparable features.

However, the decrease in vertical flux from the entrance to the dark terminal area in Trémies submarine cave indicates an input lowering. Trophic resource limitation has produced an oligotrophic ecosystem where degradative processes are extremely low. Submarine caves thus appear to be very interesting environments for trophic pathway studies. The energetic flow is clearly channelled from the entrance to the innermost part of the cave with a strong decrease in resources. This simplified oligotrophic ecosystem may constitute a good model for the general study of benthic trophic transfers and particularly for organic carbon fluxes.

References

Blomqvist, S. & L. Hakanson, 1981. A review on sediment traps in aquatic environments. Arch. Hydrobiol. 91: 101–132.

Blomqvist, S. & C. Kofoed, 1981. Sediment trapping a subaquatic *in situ* experiment. Limnol. Oceanogr. 26: 585–590.

Buscail, R., L. Guidi & F. de Bovée, 1987. Flux de matière organique, variations des paramètres biochimiques et dynamique biologique dans le canyon Lacaze-Duthiers (golfe du Lion). Coll. int. Oceanol., Perpignan, Comm. int. Mer Médit. (CIESM), Monaco: 75.

Butman, C. A., 1986. Sediment trap biases in turbulent flows: result from a laboratory flume study. J. mar. Res. 44: 645–693.

Byers, S. C., E. L. Mills & P. L. Stewart, 1978. A comparison of methods of determining organic carbon in marine sediments with sugestion for a standard method. Hydrobiologia 58: 43–47.

Cauwet, G., 1985. Dynamique de la matière organique dans les milieux marins et polyhalins: son role dans les processus géochimiques aux interfaces. Thèse Doct. ès Sci. Univ. Perpignan, 389 p.

Fichez, R., 1989. Phénomènes d'oligotrophie en milieu aphotique, étude des grottes sous-marines, comparaison avec les milieux profonds et bilans énergétiques. Thèse Doct. Univ. d'Aix-Marseille II, 251 p.

Fichez, R., 1990. Les pigments chlorophylliens: indices d'oligotrophie dans les grottes sous-marines. C. r. Acad. Sci. Paris (ser. III). 310: 155–161.

Gardner, W. D., 1980a. Sediment trap dynamics and calibration: a laboratory evaluation. J. mar. Res. 38: 17–39.

Gardner, W. D., 1980b. Field assessment of sediment traps. J. mar. Res. 38: 41–52.

Gili, J. M., T. Riera & M. Zabala, 1986. Physical and biological gradients in a submarine cave on the Western Mediterranean coast, (North-East Spain). Mar. Biol. 90: 291–297.

Hargrave, B. T. & N. M. Burns, 1979. Assessment of sediment trap collection efficiency. Limnol. Oceanogr. 24: 1124–1135.

Hargrave, B. T. & G. A. Phillips, 1986. Dynamics of the benthic foods web; St. Georges Bay, southern Gulf of St. Laurence. Mar. Ecol. Prog. Ser. 31: 277–294.

Harmelin, J. G., 1969. Bryozoaires des grottes sous-marines obscures de la région marseillaise, faunistique et écologie. Téthys 1: 793–806.

Harmelin, J. G., J. Vacelet & P. Vasseur, 1985. Les grottes sous-marines obscures: un milieu extrème et un remarquable biotope refuge. Téthys 11: 214–229.

Hinga, K. R., J. Mc N. Sieburth & G. R. Heath, 1979. The supply and use of organic material at the deep-sea floor. J. mar. Res. 37: 557–579.

Hirota, J. & J. P. Szyper, 1975. Separation of total carbon into inorganic and organic components. Limnol. Oceanogr. 20: 896–900.

Karl, D. M. & G. A. Knauer, 1984. Vertical distribution, transport and exchange of carbon in the northeast Pacific Ocean: evidence for multiple zones of biological activity. Deep-Sea Res. 31: 221–243.

Knauer, G. A., J. H. Martin & K. W. Bruland, 1979. Fluxes of particulate carbon, nitrogen and phosphorus in the upper water column of the northeast Pacific. Deep-Sea Res. 26: 97–108.

Kristensen, E. & F. O. Andersen, 1987. Determination of organic carbon in marine sediments: a comparison of two CHN-analyzer methods. J. exp. mar. Biol. Ecol. 109: 15–23.

Laborel, J. & J. Vacelet, 1958. Etude des peuplements d'une

grotte sous-marine du golfe de Marseille. Bull. Inst. Océanogr., Monaco 55 (1120): 1–20.
Laborel, J. & J. Vacelet, 1959. Les grottes sous-marines obscures en Méditerranée. Cah. océanogr. 13: 73–107.
Lorenzen, C. J., F. R. Shuman, J. T. Bennett, 1981. *In situ* calibration of a sediment trap. Limnol. Oceanogr. 26: 580–585.
Lorenzen, C. J., N. A. Welschmeyer & A. E. Copping, 1983. Particulate organic carbon flux in the subartic Pacific. Deep-Sea Res. 30: 639–643.
Marinopoulos, J., 1989. Etude des peuplements infralittoraux de substrats rocheux de la région de Marseille et des facteurs abiotiques (lumière, hydrodynamique) les influençant. Thèse Doct. ès Sci., Univ. d'Aix-Marseille II, 646 p.
Monteiro-Marques, V., 1981. Peuplements des planchers envasés de trois grottes sous-marines de la région de Marseille. Etude préliminaire. Téthys 10: 89–96.
Noriki, S. & S. Tsunogai, 1986. Particulate fluxes and major components of settling particles from sediment trap experiments in the Pacific Ocean. Deep-Sea Res. 33: 903–917.
Pérès, J. M. & J. Picard, 1949. Notes sommaires sur le peuplement des grottes sous-marines de la région de Marseille. C. R. Soc. Biogéogr. 227: 42–45.
Smetacek, V., 1984. The supply of food to the benthos. In M. J. R. Fasham (ed.), Flows of energy and materials in marine ecosystems: theory and practice. Plenum Press, New-York: 517–547.
Spencer, D. W., P. G. Brewer, A. Fleer, S. Honjo, S. Krisnaswami & Y. Nozaki, 1978. Chemical fluxes from a sediment trap experiment in the deep Sargasso sea. J. mar. Res. 36: 493–523.
Tsunogai, S. & S. Noriki, 1987. Organic matter fluxes and the sites of oxygen consumption in deep water. Deep-Sea Res. 34: 755–767.

Received 28 July 1989; in revised form 22 January 1990; accepted 27 March 1990

Hydrobiologia **207**: 71–77, 1990.
D. J. Bonin & H. L. Golterman (eds), Fluxes Between Trophic Levels and Through the Water–Sediment Interface.

Seasonal coupling between phyto- and bacterioplankton in a sand pit lake (Créteil lake, France)

Josette Garnier & Danielle Benest
Ecole Normale Supérieure, Laboratoire d'Ecologie, 46, rue d'Ulm, 75005 Paris, France

Key words: phytoplankton, bacterioplankton, biomass, activity, seasonal coupling, shallow lake

Abstract

Phyto- and bacterioplankton biomass and activity were simultaneously measured during the course of one year in the shallow Créteil Lake (France).

Phytoplankton was dominated, during the whole year, by small-sized organisms (10 to 25 μm). Bacteria were in a majority small coccoids (<0.3 μm). Phyto- and bacterioplankton abundances averaged respectively 3.3×10^6 cells l^{-1} and 6×10^9 cells l^{-1}.

The phasing of the activity and biomass periods suggest a close coupling between phyto- and bacterioplankton. There were two distinct periods of high activity and biomass. Maximal values were observed in summer but an early increase occurred also in winter. Low or undetectable phytoplankton excretion rates, when heterotrophic activity was maximum, indicated a bacterial uptake of up to 100% of the released algal products during the incubation period. Heterotrophic uptake measurements with both glucose and amino acids revealed a seasonal change of the substrates in the lake, glucose uptake being associated more with the maximum activity of the algae, while the amino acids uptake was relatively higher during their decline.

The maximal photosynthetic rate averaged 21.5 mgC m^{-3} h^{-1} and mean Vmax values were 0.056 and 0.050 mgC m^{-3} h^{-1} respectively for glucose and amino acids uptake.

Introduction

While particulate organic matter produced by phytoplankton is partly grazed by zooplankton, dissolved organic matter provided either actively through phytoplankton excretion or indirectly by algal decay and sloppy feeding by zooplankton constitutes a source of food for planktonic bacteria. Among these release processes, primary attention has been paid to phytoplankton excretion in the last few years, and results have shown that excreted molecules could sustain the major part – from 32 to 99% – of bacterial activity (Coveney, 1982; Riemann *et al.*, 1982; Søndergaard *et al.*, 1985; Feuillade *et al.*, 1988). A close coupling between phytoplankton and bacteria has generally been found during the period of an algal bloom, the peaks of bacterial activity and/or biomass following those of phytoplankton with a delay of only a few days (Riemann *et al.*, 1982; Bell & Kuparinen, 1984; Lancelot & Billen, 1984; Billen & Fontigny, 1987). Such close interactions did not however appear so clearly during the course of a seasonal cycle, probably because sources of dissolved organic matter are multiple and cannot be distinguished from each other (Pedrós-Alió & Brock, 1982; Simon & Tilzer, 1987; Robarts, 1988).

An indirect method can be however developed on the basis of assumptions concerning the nature of molecules originated from the different processes producing dissolved organic matter.

Excreted molecules are indeed mostly composed of carbohydrates (Hanson & Snyder, 1980; Chróst & Faust, 1983; Lancelot, 1984; Hama & Handa, 1987) whereas amino acids seem to originate mostly from cellular lysis or algal decay (Coveney *et al.*, 1977; Fuhrman *et al.*, 1980; Søndergaard, 1981; Jørgensen *et al.*, 1981) and sloppy feeding by zooplankton (Lampert, 1978; Fuhrman *et al.*, 1980; Riemann *et al.*, 1986). Heterotrophic bacterial activity measured with both glucose and amino acids should therefore reveal a qualitative variation in the substrates available in the aquatic environment if one considers that the activity of bacteria observed on a substrate during incubation could reflect their *in situ* behaviour.

This methodology was applied in the mesotrophic Lake Créteil (France) where phyto- and bacterioplankton biomass and activity were measured during the course of a year.

Description of the site studied

Créteil Lake originates from an excavation of alluvial sediments near the confluence of the Rivers Seine and Marne (S-E Paris). It varies in depth from 4 m to 4.5 m for most of its surface area (0.42 km^2). The lake is mainly supplied with anoxic phreatic waters circulating through alluvial deposits and diverse filling materials. It is also supplied by a rainwater sewer.

Temperature ranged from 3 °C in winter to 22.9 °C in summer. Well exposed to wind, the lake stratified only weakly and intermittently from May to October. Differences in temperature never exceeded 3 °C between the surface and the bottom. Thus, the water column is homogeneous and well oxygenated; percentage of saturation never fell below 60% at the bottom. Water transparency varied from 1.1 to 4.2 m (2.5 m as an annual average). pH and ΣCO_2 averaged 8 and 24 $mgC\,l^{-1}$ respectively. For the whole water column, concentrations in total phosphorus were high, varying from 25 to 72 $\mu g\,l^{-1}$ (43 $\mu g\,l^{-1}$ as an annual mean) while concentrations in $P\text{-}PO_4$ and $N\text{-}NO_3$ were relatively low ($P\text{-}PO_4$: <1–13 $\mu g\,l^{-1}$, annual mean: 4.9 $\mu g\,l^{-1}$; $N\text{-}NO_3$: 0.02–0.22 $mg\,l^{-1}$, annual mean: 0.11 $mg\,l^{-1}$).

Materials and methods

Samples were collected in the central area of the lake from April 1985 to April 1986, at six depths 0– 0.5– 1.0– 1.5–2.5 and 4 m at intervals ranging from two weeks in summer to three weeks in winter.

Phytoplankton

Algal enumeration and chlorophyll *a* analysis were performed on a mixed sample from the six depths. A sample for phytoplankton enumeration was fixed with formol at a final concentration of 2%. Phytoplankton were counted with an inverted microscope after the sample had been in a sedimentation chamber for 24 h (Utermöhl, 1958). For chlorophyll determination, 1 to 3 litres of lake water were filtered with 2×10^4 Pa of vacuum through a GF/C glass fiber filter. Duplicate samples were extracted in 90% acetone and analysed spectrophotometrically according to Lorenzen (1967).

Particulate primary production was measured at the six depths by means of the radiocarbon method of Steemann-Nielsen (1952). 150 ml of water were inoculated with 1 μCi Na_2CO_3 (5 mCi $mmole^{-1}$) and incubated for 1/3 day around midday. After incubation, 100 ml were filtered through 0.45 μm pore size membrane filters (Millipore HA). Filters were rinced with distilled water and HCl-fumed. Dissolved primary production was measured at 1 and 4 m, representing the depths of the maximum and the minimum of photosynthetic rate. The procedure used to determine the excretion rate was adapted from Blaauboer *et al.* (1982). The remaining 50 ml of the bottles were filtered through a 0.22 μm membrane filter. Filtrates were acidified with HCl 1N to pH 2.5, bubbled for 1 h and 40 ml of the filtrate were lyophilised. The lyophilisat was dissolved in 5 ml of distilled water and 10 ml of Instagel were added. Two controls were obtained with inoculated lake water which was immediately filtered, the filtrate being analysed as above. Low vacuum was applied ($< 1 \times 10^4$ Pa). Filters and filtrates were counted by the liquid scintillation technique. Quenching was corrected by the external standard method. All measurements were duplicated.

Bacteria

Bacterioplankton was enumerated by acridine orange staining and epifluorescence microscopy on black Nucleopore 0.2 μm pore size membrane filters (Hobbie *et al.*, 1977). Enumerations were made at 1 and 4 m. Duplicate samples were analysed and about 170 bacteria were counted on 12 fields (i.e. 340 bacteria for each depth). The bacteria were classified according to shape and size, and for each size and shape fraction an average volume was estimated. The total cell volume was converted to carbon units, assuming a conversion factor of $3.77\ 10^{-13}$ gC μm^{-3} corresponding to the average cell volume (0.03 μm^3) found during the study (Simon & Azam, 1989). The uptake kinetics were also determined at 1 m and 4 m (Wright & Hobbie, 1965; 1966). Five concentrations on 20 ml triplicates were tested. Glucose (U-^{14}C, 240 mCi mmole^{-1}, CEA) and amino acids (U-^{14}C, 45 mCi matC^{-1}, CEA) were added to give a final concentration in the range of 0.05–1 μgC l^{-1} and 0.08–1.6 μgC l^{-1} respectively. A control for substrate absorption were determined in triplicate at the low and high concentrations only. Vmax were determined using Michaelis-Menten kinetics and the Lineweaver Burke transformation (Wright & Hobbie, 1965). The radioactivity of the filters was counted as above.

As the values obtained at 1 m and 4 m did not differ significantly (paired samples t-test, $P > 0.05$), results are given in average of the two depths.

Results

General features of phyto- and bacterioplankton

Throughout the year, small cells were predominant in both phyto- and bacterioplankton.

The greatest dimension of algae ranged from 10 to 25 μm. Cryptophyceae (*Chroomonas sp.*, *Cryptomonas sp.*) formed a major part of the phytoplankton abundance (50%). Chlorophyceae (*Oocystis parva* mainly, *Chlamydomonas spp.*) and small centric diatoms constituted 28% and 21.2% of the phytoplankton abundance. Cyanophyceae appeared only occasionally in summer. The total abundance of phytoplankton ranged from 0.2 to 11.9×10^6 cells l^{-1} (Fig. 1). Chlorophyll *a* concentrations varied from 0.8 to 12.4 μg l^{-1}. Using a C/Chl ratio of 80 (Garnier *et al.* 1989), phytoplankton biomass varied from 64 to 992 μgC l^{-1} (Fig. 1).

The bacterial population was dominated by small sized coccoids of less than 0.3 μm in diame-

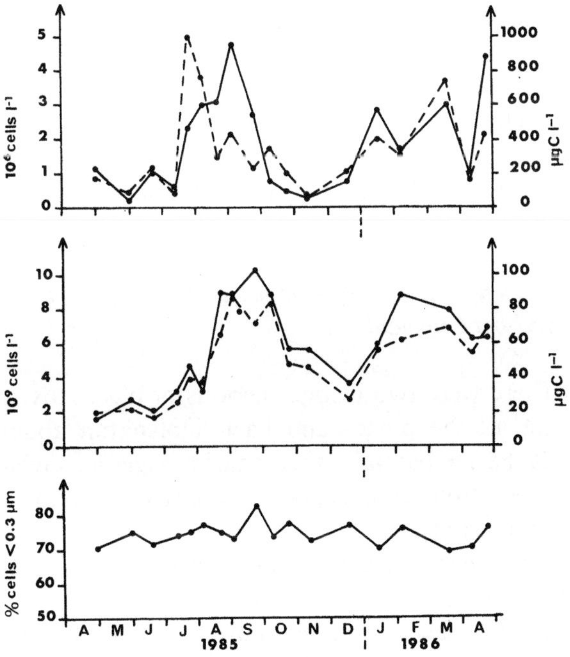

Fig. 1. Seasonal variations in a) phytoplankton abundance (●——●) and biomass (●---●), b) bacterioplankton abundance (●——●) and biomass (●---●) and c) the proportion of small bacteria.

ter which comprised 74% of the bacterial abundance. Average cell volume was 0.03 μm^3. Most of the bacteria were free living. Attached bacteria represented only 5% (as an annual average). Abundance ranged from 1.6 to 10.3 x 10^9 cells l^{-1} and biomass from 16.6 to 87.6 $\mu g\ C\ l^{-1}$ (Fig. 1).

The maximal photosynthetic rate (Am) varied in the same way as the algal biomass, except at the end of June when a low phytoplankton biomass coincided with a high photosynthetic activity (Fig. 2). Highest values were about 40–50 mgC $m^{-3}\ h^{-1}$ in summer, while minimum values equalling 5 mgC $m^{-3}\ h^{-1}$ were observed in autumn. Integral daily primary production (ΣP) showed the same trend with extreme values equalling 1720 mgC $m^{-2}\ d^{-1}$ in July and about 60 mgC $m^{-2}\ d^{-1}$ in November and December (Fig. 2). Regarding the annual mean, ΣP was 530 mgC $m^{-2}\ d^{-1}$ for the period studied.

As an average of the two depths, excretion rates varied from 0 to 8 mgC $m^{-3}\ h^{-1}$. As an annual mean, the percent excretion rate (PER) represented 15% of the total assimilation (Fig. 2). The values were underestimated as they did not take into account the bacterial assimilation of excreted molecules.

The rate of substrate uptake (Vmax) varied from 0.012 to 0.167 $\mu gC\ l^{-1}$ for glucose and from 0.011 to 0.153 $\mu gC\ l^{-1}$ for amino acids (Fig. 2).

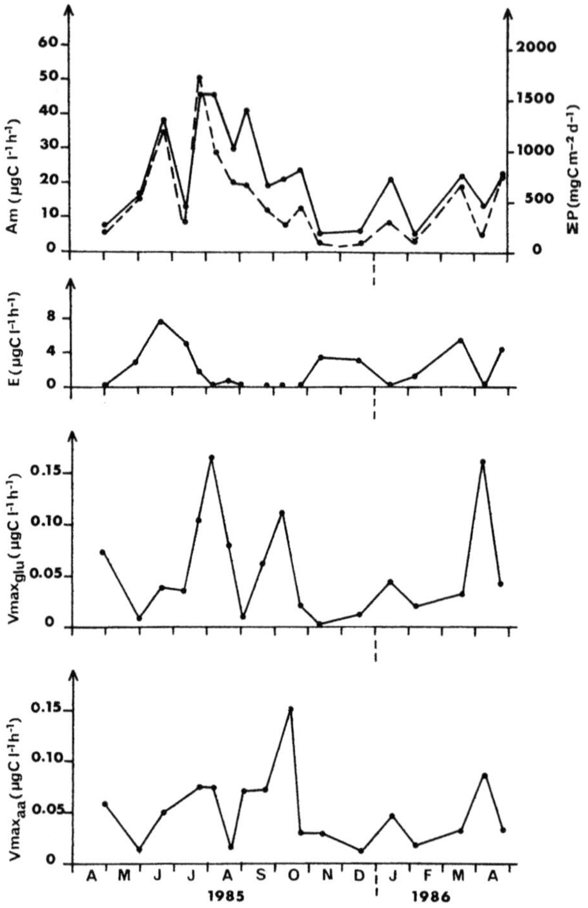

Fig. 2. Seasonal changes in a) maximal assimilation rate of phytoplankton (Am ●——●) and integral daily primary production (ΣP ●---●, b) phytoplankton excretion rate (E), c) maximum rate of glucose uptake ($Vmax_{glu}$) and d) maximum rate of amino acids uptake ($Vmax_{aa}$).

Seasonal coupling of biomass and activity of phyto- and bacterioplankton

There were two distinct periods of high activity and of the phyto- and bacterioplankton abundance and biomass; their phasing suggests a relatively strong coupling between the two planktonic compartments.

During the spring period, the bacterial and algal abundances and biomasses were relatively low and increased from July onwards (Fig. 1). High algal and bacterial abundances and biomasses were found during the summer period. At the end of the year, the values for bacteria remained relatively high after the decline of the phytoplankton. The winter development of phytoplankton coincided with an increase of the bacterial populations.

The peak of photosynthetic activity observed at the end of June was not accompanied by a peak of bacterial activity, although an increase was observed (Fig. 2). It is however not excluded that a delayed bacterial response occurred, but was not observed, because of the too low frequency of the sampling. Maximal values of both photosynthetic and heterotrophic activity were observed during the summer period from July to October (Fig. 2). In November and December, minimum values of Am and ΣP were associated with low values of bacterial activity. The early increase of algal activity in winter was followed by an enhanced bacterial activity. At the beginning of April Vmax values were as high as summer values. The peak of heterotrophic activity observed at the beginning of April occurred when photosynthetic activity had already declined.

When bacterial activity was low, algal excretion varied in the same way than assimilation. This was the case in spring 1985 and winter 1986 (Fig. 2). On the other hand, when bacterial activity was maximum, the excretion rates were low or undetectable, suggesting that bacterial activity took place at the expense of labile substrates produced during photosynthesis, but the degree to which bacteria rely on phytoplankton exudates cannot be determined.

A seasonal variation of the $Vmax_{aa}$ to $Vmax_{glu}$ ratio occurred in the course of the year (Fig. 3). The higher activity on amino acids at the beginning of July and then from September to November corresponded to the decline of phytoplankton. The enhanced assimilation on glucose in late July and early August corresponded to the

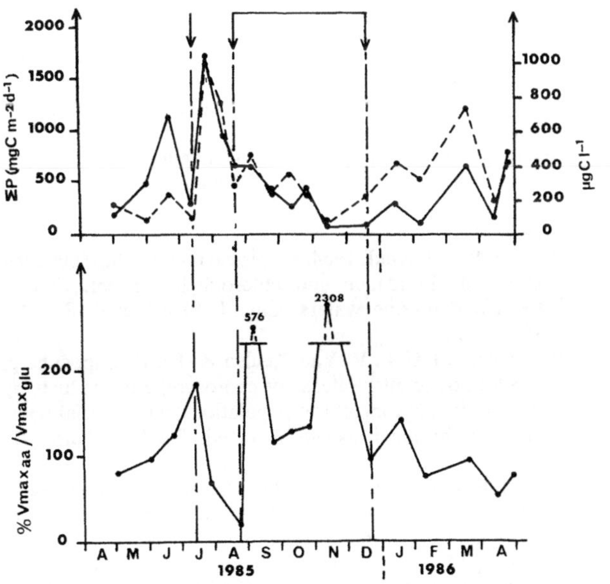

Fig. 3. Comparative evolution a) of the integral daily primary production (ΣP ●——●) and the biomass of phytoplankton (●---●) and b) of the $Vmax_{aa}$ – to – $Vmax_{glu}$ ratio. Arrows indicate the summer periods of phytoplankton decline.

maximum photosynthetic activity of phytoplankton. In late winter and early spring, at the onset of the algal increase, heterotrophic activity was again predominant on glucose.

Discussion

The phyto- and bacterioplankton activity and biomass indicated that Créteil Lake is mesotrophic. However, the dominant algal species and bacteria were small sized organisms generally found in oligotrophic system. According to Reynolds *et al.* (1983) and Reynolds (1984), the small size of the algae has been interpreted in function of both intermittent perturbations of the water column under wind effect and the continuous replenishment of the nutrients (Garnier, 1989: Lacroix *et al.*, 1989). Pedrós-Alió & Brock (1982) suggested that bacteria were bigger in eutrophic freshwater lakes than in oligotrophic lakes or the marine system. However, Krambeck *et al.* (1981), Bird & Kalff (1984) and Robarts (1988) also reported a trend toward smaller bacteria in productive systems. In Créteil Lake, as with phytoplankton, it is tempting to relate the small size of bacteria to the instability of the environment but predators are also known to influencc thc size of both algac (Pccr 1986) and bacteria (Ammerman *et al.*, 1984).

Phytoplankton excretion has been underestimated, as a subsequent assimilation of excreted molecules by bacteria was not taken into account. Nevertheless, the process appeared to be quantitatively important in Créteil Lake as the apparent percent excretion rate reached 15% of the total fixed carbon. Higher PER values are generally reported for scenesent algal cells (Mague *et al.*, 1980; Riemann *et al.*, 1982; Søndergaard & Shierup, 1982; Søndergaard *et al.*, 1985) or for oligotrophic systems where nutrients are limited (Fogg, 1971; Thomas, 1971). In Créteil Lake, the high values could be related to the dominance of nanoplankton, which is known to be metabolically more active.

The seasonal change of the $Vmax_{aa}$ to $Vmax_{glu}$ ratio observed, whereas bacterial biomass did not show such variations, can be interpreted as an adaptation to a qualitative change in the available substrates in the lake. According to Bell & Albright (1982), a maximum activity on amino acids would correspond to an aptitude to break protein chains or other complex substrates, while a maximum activity on glucose would express an adaptation to utilize smaller molecules.

Input to the DOC pool through phytoplankton excretion must have been higher at the beginning of the summer when phytoplankton activity was also higher. A parallel has indeed been shown between the assimilation and excretion rates of algae growing in an axenic culture (Nalewajko *et al.*, 1980). Similar results are obtained on natural populations when assimilation of excreted molecules is measured (Kato, 1984). Considering that algal excretion produces more carbohydrates (Hanson & Snyder, 1980; Chróst & Faust, 1983; Lancelot, 1984; Hama & Handa, 1987), the relatively higher heterotrophic activity on glucose tends to show that bacteria were, at that time, more adapted to the labile fraction of dissolved organic carbon originating from excretion. Excretion appeared also to be the predominant source of DOC for bacteria in late winter and early spring when phytoplankton activity increased.

At the end of summer, algal activity and biomass declined but bacterial activity and biomass remained high. The DOC pool originated from algal decay, lysis or autolysis, sloppy feeding by zooplankton might have been a significant complementary source of carbon for bacteria. This is supported by the relatively higher activity on amino acids. These release processes are indeed known to produce mainly amino acids (Coveney *et al.*, 1977; Lampert, 1978; Fuhrman *et al.*, 1980; Riemann *et al.*, 1986). The DOC pool might also originate from a lateral transport of labile DOC from macrophytes which developed at that time. However, the small differences between activity on glucose and amino acids indicate that throughout the seasonal cycle all the release processes occurred simultaneously and we do not know the contribution of each of these processes to the total release. Moreover, each process produces a great variety of molecules.

The high Vmax values obtained both in summer and winter suggest that the organic substrate supply is probably important in regulating bacterial activity and the biomass. Temperature could influence the delay of the response of bacteria to phytoplankton.

Acknowledgements

This research has been supported by the PIREN in the framework of the GRECO-lacs programme.

We are grateful to Dr. Roger Pourriot for his valuable comments on the manuscript. Thanks are also due to Ms Mary Delahaye for improving the english text.

References

Ammerman, J. W., J. A. Fuhrman, Å. Hagström & F. Azam, 1984. Bacterioplankton growth in seawater: I. Growth kinetics and cellular characteristics in seawater cultures. Mar. Ecol. Prog. Ser. 18: 31–39.

Bell, C. R. & L. J. Albright, 1982. Attached and free-floating bacteria in a diverse selection of water bodies. Appl. environ. Microbiol. 43: 1227–1237.

Bell, R. T. & J. Kuparinen, 1984. Assessing phytoplankton and bacterioplankton production during early spring in Lake Erken, Sweden. Appl. envir. Microbiol. 48: 1221–1230.

Billen, G. & A. Fontigny, 1987. Dynamics of *Phaeocystis*-dominated spring bloom in Belgian coastal waters. 2. Bacterioplankton dynamics. Mar. Ecol. Prog. Ser. 37: 249–257.

Bird, D. F. & J. Kalff, 1984. Empirical relationships between bacterial abundance and chlorophyll concentration in fresh and marine waters. Can. J. Fish. aquat. Sci. 41: 1015–1023.

Blaauboer, M. C. I., R. Van Keulen & Th. E. Cappenberg, 1982. Extracellular release of photosynthetic products by freshwater phytoplankton populations, with special reference to the algal species involved. Freshwat. Biol. 12: 559–572.

Chróst, R. J. & M. A. Faust, 1983. Organic carbon release by phytoplankton: its composition and utilization by bacterioplankton. J. Plankton Res. 5: 477–493.

Coveney, M. F., 1982. Bacterial uptake of photosynthetic carbon from freshwater phytoplankton. Oikos 38: 8–20.

Coveney, M. F., G. Cronberg, M. Enell, K. Larsson & L. Olofsson, 1977. Phytoplankton, zooplankton and bacteria-standing crop and production relationships in a eutrophic lake. Oikos 29: 5–21.

Feuillade, M., Ph. Dufour & J. Feuillade, 1988. Organic

carbon release by phytoplankton and bacterial assimilation. Schweiz. Z. Hydrol. 50: 115–13.

Fogg, G. E., 1971. Extracellular products of phytoplankton photosynthesis. Arch. Hydrobiol. 5: 1–25.

Fuhrman, J. A., J. W. Ammerman & F. Azam, 1980. Bacterioplankton in the coastal euphotic zone: distribution, activity and possible relationships with phytoplankton. Mar. Biol. 60: 201–207.

Garnier, J., 1989. Peuplement phytoplanctonique et bactéries hétérotrophes d'un lac peu profond (lac de Créteil, région parisienne). Production, fonctionnement, évolution. Thèse Doct. ès Sci. (Ecologie). Univ. Paris. VI., 331 pp.

Garnier, J., P. Blanc & D. Benest, 1989. Estimating a carbon/chlorophyll ratio in nannoplankton (Créteil Lake, S-E Paris, France). Wat. Res. Bull. 25: 751–754.

Hama, T. & N. Handa, 1987. Pattern of organic matter production by natural phytoplankton populations in a eutrophic lake. 2. Extracellular products. Arch. Hydrobiol. 109: 227–243.

Hanson, R. B. & J. Snyder, 1980. Glucose exchanges in a salt marsh estuary: biological activity and chemical measurements. Limnol. Oceanogr. 25: 633–642.

Hobbie, J. E., R. J. Daley & S. Jasper, 1977. Use of nucleopore filters for counting bacteria by fluorescence microscopy. Appl. envir. Microbiol. 33: 1225–1228.

Jørgensen, N. O. G., T. H. Blackburn, K. Henriksen & D. Bay, 1981. The importance of *Posidonia oceanica* and *Cymodocea nodosa* as contributors of free amino acids in water and sediment of seagrass beds. Mar. Ecol. 2: 97–112.

Kato, K., 1984. Functional difference in the utilization of phytoplankton cell materials and dissolved organic compounds in free-living and particulate-associated bacteria in an aquatic system. Arch. Hydrobiol. 100: 521–532.

Krambeck, C., H.-J. Krambeck & J. Overbeck, 1981. Microcomputer assisted biomass determinations of plankton bacteria on scanning electron microscopy. Appl. envir. Microbiol. 42: 142–149.

Lacroix, G., Ph. Boët, J. Garnier, F. Lescher-Moutoué, R. Pourriot & P. Testard, 1989. Controlling factors of the planktonic community in the shallow lake of Créteil. Int. Revue ges. Hydrobiol. 74: 353–370.

Lampert, W., 1978. Release of dissolved organic carbon by grazing zooplankton. Limnol. Oceanogr. 23: 831–834.

Lancelot, C., 1984. Extracellular release of small and large molecules by phytoplankton in the southern bight of the North Sea. Estuar. Coast. Shelf Sci. 18: 65–77.

Lancelot, C. & G. Billen, 1984. Activity of heterotrophic bacteria and its coupling to primary production during the spring phytoplankton bloom in the southern bight of the North Sea. Limnol. Oceanogr. 29: 721–730.

Lorenzen, C. J., 1967. Determination of chlorophyll and pheopigments: spectrophotometric equations. Limnol. Oceanogr. 12: 343–346.

Mague, T. H., E. Friberg, D. J. Hugues & I. Morris, 1980. Extracellular release of carbon by marine phytoplankton; a physiological approach. Limnol. Oceanogr. 25: 262–279.

Nalewajko, C., K. Lee & P. Fay, 1980. Significance of algal extracellular products to bacteria in lakes and in cultures. Microb. Ecol. 6: 199–207.

Pedrós-Alió, C. & T. D. Brock, 1982. Assessing biomass and production of bacteria in eutrophic Lake Mendota, Wisconsin. Appl. envir. Microbiol. 44: 203–218.

Peer, R. L., 1986. The effect of microcrustaceans on succession and diversity of an algal microcosm community. Oecologia 68: 308–314.

Reynolds, C. S., 1984. Phytoplankton periodicity: the interactions of form, function and environmental variability. Freshwat. Biol. 14: 111–142.

Reynolds, C. S., S. W. Wiseman, B. M. Godfrey & C. Butterwick, 1983. Some effects of artificial mixing on the dynamics of phytoplankton populations in large limnetic enclosures. J. Plankton Res. 5: 203–234.

Riemann, B., N. O. G. Jørgensen, W. Lampert & J. A. Fuhrman, 1986. Zooplankton-induced changes in dissolved free amino acids and in production rates of freshwater bacteria. Microb. Ecol. 12: 247–258.

Riemann, B., M. Søndergaard, H. H. Schierup, S. Bosselman, G. Christensen, J. Hansen & B. Nielsen, 1982. Carbon metabolism during a spring bloom Diatom bloom in the eutrophic Lake Mossø. Int. Revue ges. Hydrobiol. 67: 145–185.

Robarts, R. D., 1988. Heterotrophic bacterial activity and primary production in a hypertrophic African lake. Hydrobiologia 162: 97–107.

Simon, M. & F. Azam, 1989. Protein content and protein synthesis rates of planktonic marine bacteria. Mar. Ecol. Prog. Ser. 51: 201–213.

Simon, M. & M. M. Tilzer, 1987. Bacterial response to seasonal changes in primary production and phytoplankton biomass in Lake Constance. J. Plankton Res. 9: 535–552.

Søndergaard, M., 1981. Kinetics of extracellular release of ^{14}C-labelled organic carbon by submerged macrophytes. Oikos 36: 331–347.

Søndergaard, M. & H. H. Schierup, 1982. Release of extracellular organic carbon during a diatom bloom in Lake Mossø: molecular weight fractionation. Freshwat. Biol. 12: 313–320.

Søndergaard, M., B. Riemann & N. O. G. Jørgensen, 1985. Extracellular organic carbon (EOC) released by phytoplankton and bacterial production. Oikos 45: 323–332.

Steemann-Nielsen, E., 1952. The use of radioactive carbon (C^{14}) for measuring organic production in the sea. J. Cons. perm. int. Explor. Mer. 18: 117–140.

Thomas, J. P., 1971. Release of dissolved organic matter from natural populations of marine phytoplankton. Mar. Biol. 11: 311–323.

Utermöhl, H., 1958. Zür Vervollkommung der quantitativen Phytoplankton Methodik. Mitt. int. Ver. Limnol. 9: 1–38.

Wright, R. T. & J. E. Hobbie, 1965. The uptake of organic solute in lake water. Limnol. Oceanogr. 10: 22–28.

Wright, R. T. & J. E. Hobbie, 1966. Use of glucose and acetate by bacteria and algae in aquatic ecosystems. Ecology 47: 447–464.

Received 4 August 1989; in revised form 18 January 1990; accepted 19 March 1990

Hydrobiologia **207**: 79–86, 1990.
D. J. Bonin & H. L. Golterman (eds), Fluxes Between Trophic Levels and Through the Water–Sediment Interface.

Dynamics of bacterioplankton in a mesotrophic French reservoir (Pareloup)

Pierre Lavandier
Laboratoire d'Hydrobiologie, Université Paul Sabatier, 118 route de Narbonne, 31062 Toulouse Cedex, France

Key words: Bacterioplankton, heterotrophic production, thymidine uptake

Abstract

Bacterioplankton abundance, biomass and production were studied at a central station (35 m depth) from April 1987 to September 1988 in a mesotrophic reservoir. Bacterial production was calculated by the (^{3}H) thymidine method.

For the water column, integrated estimates of bacterioplankton abundance ranged from 2.3 10^9 to 4.6 10^9 cells l^{-1}, and carbon biomass from 0.037 to 0.068 mg C l^{-1}; the thymidine incorporation rates ranged from 0.8 to 17.2 picomoles l^{-1} h^{-1}, leading to net bacterial production estimates of less than 0.7 μg C l^{-1} d^{-1} in winter to 18 μg C l^{-1} d^{-1} in summer. About 55% of the production occurred in the euphotic layers.

Over the year, the bacterial carbon requirement represented 90% of the autotrophic production for the whole lake. It was five times lower than autotrophic production in spring, but twice as high in summer. This important temporal lack of balance suggests that not all the spring primary production products are consumed immediately and/or that other carbon sources probably support bacterial growth in summer.

Introduction

Since recent methods for estimating bacterial biomass and production were proposed (Hobbie *et al.*, 1977; Hagström *et al.*, 1979; Fuhrman & Azam, 1980), a lot of studies have emphasized the quantitative importance of the heterotrophic bacterioplankton in the carbon cycle of aquatic ecosystems. Carbon bacterial production budgets have been established over one year and over longer periods in several lakes of various trophic states (Jordan & Likens, 1980; Pedrós-Alió & Brock, 1982; Riemann, 1983; Güde *et al.*, 1985; Lovell & Konopka, 1985a; Scavia & Laird, 1987; Simon & Tilzer, 1987; Dufour *et al.*, 1989). They indicate that an important part of the carbon flow is channelled through the bacteria, but the quantitative relationships of bacterial production to primary production are still largely unknown.

In this study, we present the vertical and seasonal patterns of bacterioplankton density and production over one and a half years in a mesotrophic lake where phytoplanktonic parameters are already known (Capblancq *et al.*, 1988). The results are compared to simultaneous autotrophic production estimates.

Material and methods

Study area and sampling

The reservoir of Pareloup (area = 13.5 km^2, mean depth 12.5 m, maximum depth 35 m) is a mesotrophic lake located to the south-west of the Massif Central (France) at an altitude of 850 m.

Twenty eight sets of samples were collected at a central station from April 1987 to September 1988. Samples were taken monthly (in winter) or weekly (in summer) at five to twelve different depths with a sterilized bottle, an acid-washed Van Dorn bottle or a peristaltic pump with an autoclaved plastic tube.

Bacterioplankton numbers and biomass

Cell counts were made using the acridine orange procedure (Hobbie *et al.*, 1977) from fixed samples (formalin 2% final concentration). Cell volume was estimated from photographs of acridine orange-stained cells and was converted to carbon biomass using a factor ranging from 14 to 21 fg C according to the cell size (Simon & Azam, 1989). Most counts were done in duplicate; at least 1000 cells were counted and 150 cells were measured for each subsample. The mean difference between duplicate bacteria counts was 6,5%.

Bacterial production

Rates of bacterial production were calculated from the rates of incorporation of methyl-^3H thymidine into the cold-TCA insoluble precipitate (Fuhrman & Azam, 1980; 1982). Duplicate or triplicate 8 ml samples and duplicate formalin-killed blanks (2% final concentration) were incubated *in situ*, at the depth from which the sample was taken, with 20 nM ^3H thymidine (40–60 Ci $mmol^{-1}$, CEA France). To prevent aeration of the anoxic deep water, the samples were prepared as follows: deep water was injected into the bottom of 12 ml autoclaved tubes with a syringe needle fixed at the end of the outlet tube of the peristaltic pump. After about 40 ml had overflowed, the tubes were capped with a rubber stopper into which a syringe needle was inserted to allow excess water to be evacuated; ^3H thymidine and formol were injected into the samples through the syringe needle. Incubation times varied from 20 to 40 minutes, depending on the temperature and bacterial activity. ^3H thymidine incorporation was stopped by adding 2% formalin. It was occasionally verified, by adding increasing amounts of radiolabelled thymidine, that the concentration of ^3H thymidine used during the study was a saturating concentration.

Four diel cycles (in April, June, July and August) showed that the hourly thymidine incorporation rates, always estimated between 12.00 a.m. and 3 p.m. represented 3 to 5% daily incorporation, with a mean of 3.7%; therefore, they were multiplied by 27 to be converted into daily results.

Phytoplanktonic parameters

Photosynthetic pigments were determined spectrophotometrically from boiling ethanol extracts of algae filtered onto glass fibre filters (Marker *et al.*, 1980). Autotrophic production was measured fourteen times during the study, from rates of uptake of $NaH^{14}CO_3$ over 2–4 hours (personal measurements and Capblancq *et al.*, 1988). Daily autotrophic production was estimated by integrating curves of production versus irradiance over time and depth.

Presentation of the results

Since photosynthetic products are an important carbon source for bacteria, we made a distinction between the euphotic zone, 'e.z.' (from the surface to the depth at which the light intensity is 1% of the subsurface light intensity), and the aphotic zone 'a.z.'. The patterns of seasonal changes of the mean values of the variables in the euphotic and aphotic zones (see Fig. 1) were established from the 28 profiles determined during the study; three examples of these profiles are shown in Fig. 2. Data are presented as mean ± standard error in the text. The monthly means of the estimates were used to compare heterotrophic and eutrophic productions.

Results

Environmental data

Lake Pareloup is isothermal from October to May with temperatures ranging from 16.5 °C after destratification (October) to 4.5 °C (February). It is thermally stratified from June to September with highest temperatures of about 22 °C. Depending on the year, the metalimnion is more or less accentuated and temperatures in the deep layers increase progressively up to 14 °C in September. Maximum chlorophyll concentrations occurred in spring (24 $\mu g\,l^{-1}$) and October (12 $\mu g\,l^{-1}$) during the phytoplanktonic blooms dominated by diatoms (*Asterionella formosa*, *Melosira ambigua*). Minimum concentrations are observed in winter (<3 $\mu g\,l^{-1}$), and in June (<5 $\mu g\,l^{-1}$). The euphotic zone ranged from 5.5 to 12.5 m depth, according to the phytoplankton biomass; it may be stratified in summer, with high densities of *Gonyostomum semen* (Raphidophyceae) building up in the lower part and bringing about high pigment concentrations at this depth (LeCohu *et al.*, 1989).

Bacterial density and biomass

Over the year, the monthly mean densities of bacterial populations fluctuated more in the euphotic zone (N.e.z. = 3.88 ± 0.33 10^9 cells l^{-1}) than in the aphotic zone of the lake (N.a.z. = 3.18 ± 0.22 10^9 cells l^{-1}) (Fig. 1). The lowest densities were during the clearwater phase (in June – N.e.z. = 2.5 10^9 cells l^{-1}, N.a.z. = 2.2 10^9 cells l^{-1}); the highest were during the algal bloom (February-April 1988 – 4.5–6.3 10^9 cells l^{-1}),

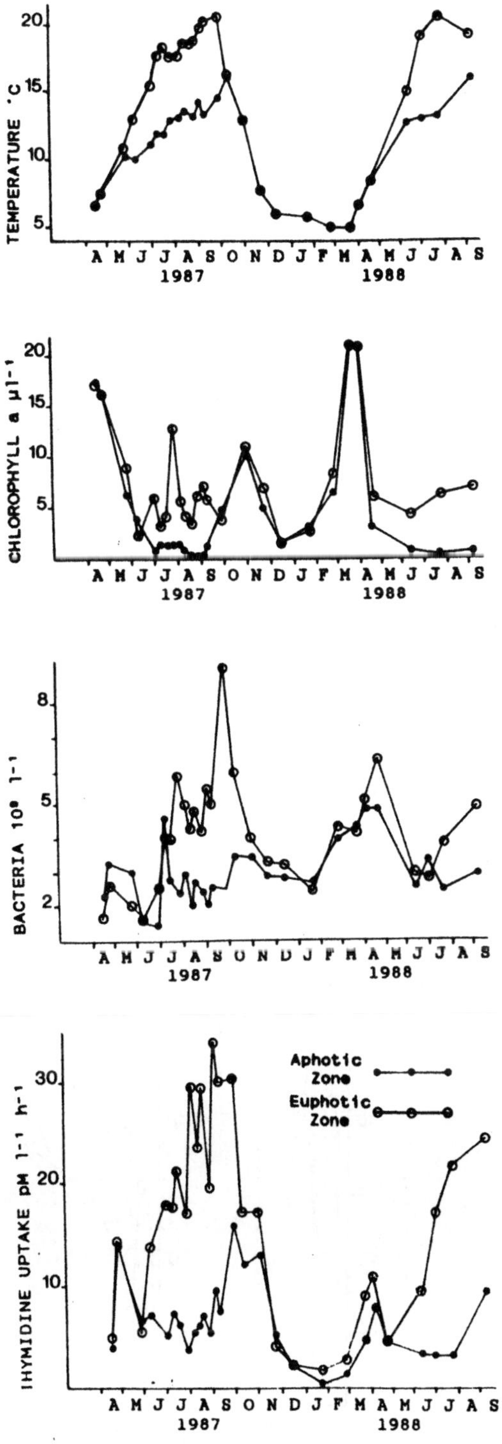

Fig. 1. Seasonal changes in ^3H thymidine incorporation rates, bacterial densities, chlorophyll *a* concentrations and temperatures in the euphotic and aphotic zones of a water column of Lake Pareloup from April 1987 to September 1988.

and at the end of the stratification period (September, N.e.z. = 5.9 cells l^{-1}, October, N.a.z. = 3,5 cells l^{-1}). Bacterial populations were relatively homogeneously distributed in the whole water column when the lake was isothermal (Fig. 2); during the period of stratification, the highest densities of bacteria occurred in the euphotic zone, usually in the deep epilimnic water layers.

Cell size ranged from 0.045 μm^3 in winter to 0.08 μm^3 in autumn and bacterioplankton biomass, resulting from both cell numbers and cell biomass, ranged from 35 $\mu g\, l^{-1}$ (November, December) to 55 $\mu g\, l^{-1}$ (October) at depth and from 30–35 $\mu g\, l^{-1}$ (January, May) to 100 $\mu g\, l^{-1}$ (September) in the euphotic zone. Integrated in time and depth for the whole water column, bacterial number and biomass respectively average 3.4 ± 0.21 10^9 cells l^{-1} (2.3 to 4.6 10^9) and 51 ± 3.46 μg C l^{-1} (37 to 68) during the period of study.

Thymidine incorporation

Thymidine incorporation rates (Fig. 1) fluctuated much more than cell density did. They are higher and more variable in the euphotic zone than in the aphotic zone (monthly mean estimates of ^3H uptake = 12.1 ± 2.88 and 5.95 ± 1.1 picomoles $l^{-1} h^{-1}$, respectively); these differences result from variations in both cell numbers and specific incorporation rates, which are higher in the upper-water column than in deep layers.

Assimilation rates are lowest in winter (less than 1 picomole $l^{-1} h^{-1}$ in January). There was a small peak in early spring followed by a fall, roughly corresponding to the clear water phase. Thymidine incorporation increased in late spring and the highest incorporation rates occurred in August and September (up to 30 picomoles $l^{-1} h^{-1}$ in the euphotic zone). The destratification of the lake induces a fall of production corresponding to the cooling down of the surface layers. In September, the profile was characterized by two peaks: one peak is in the low layers of the epilimnion, another, less pronounced, is near the bottom in the anoxic zone (Fig. 2). When the oxygen concentrations were very low, we aerated deep water 'anaerobic samples' to compare ^3H thymidine incorporation rates in aerobic and anaerobic conditions. For five experiments, the estimates were very close. On the other hand, twice in September, aeration of the anaerobic

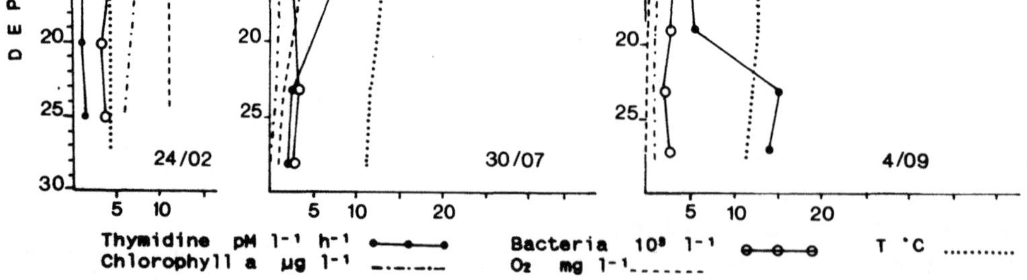

Fig. 2. Examples of profiles of ^3H thymidine incorporation rates, bacterial densities, concentration of chlorophyll *a*, oxygen and temperature obtained for a central water column of Lake Pareloup. The vertical arrows indicate the depth of the euphotic zone.

samples inhibited thymidine incorporation by 50%. These results suggest that the heterotrophic population in the anoxic hypolimnion, is mainly composed of non-obligate anaerobic bacteria, though strict anaerobic bacteria could develop during the anoxic period.

Using a conversion factor of 2.1 10^9 bacteria per nM ^{3}H thymidine incorporated, 457 10^9 new cells were produced per m^3 and per day during the study for the water column. Taking account of cell size and carbon cell content, net bacterial production averaged 7.5 mg C m^{-3} per day (72 g m^{-2} per year at the sampling station where the lake is deepest). Two-thirds of the production occurred during the summer months and more than 55% in the euphotic zone, although it constitutes only 1/3 of the water column.

Discussion

In the reservoir of Pareloup, mean bacterial density and biomass is in the range of other lakes of similar trophic status (Güde *et al.*, 1985; Lovell & Konopka, 1985a; Dufour *et al.*, 1989). As for the pattern of bacterial distribution in time and space, which is characterized by a rather homogeneous density when the lake is isothermal, a spring peak and higher cell numbers in the epilimnion during the stratification period are often observed in lakes (Riemann, 1983; Pedrós-Alió & Brock, 1982; Güde *et al.*,1985; Lovell & Konopka, 1985a; Scavia & Laird, 1987).

^{3}H thymidine incorporation can be used to estimate bacterial production in both aerobic and anaerobic layers (Pollard & Moriarty, 1984). However, as in previous estimates (Mc Donough *et al.*, 1986), our results showed that the method must be applied carefully; in particular, oxygenation of anaerobic samples during incubation may greatly modify the assimilation rates. In Lake Pareloup, we used the same conversion factors for both aerobic and anaerobic layers. This is probably inaccurate. However, this approximation has only little repercussion on the annual budget of production, because the anoxic period was less than 2 months and the anoxic water volume represented only about 5% of the total volume of the lake.

Several studies have shown that planktonic algae are a source of substrates for bacteria, from both their extracellular organic carbon release and their cell biomass. The increase in biomass and activity of bacteria during the spring bloom of phytoplankton in Lake Pareloup agrees with these conclusions. Afterwards, on the other hand, bacterial densities and thymidine incorporation rates presented no significant relationship either with the primary production rates ($r = 0.325$ and 0.185 respectively in the euphotic zone, $n = 11$), or with the mean chlorophyll *a* concentrations ($r = 0.19$ and -0.13 respectively in the whole water column, $n = 21$). This apparent lack of coupling already observed in several lakes (Brock & Clyne, 1984; Scavia & Laird, 1987) merely expresses the dependence of bacterial production on a set of factors acting together, including not only phytoplankton exudates and biomass, but also plankton autolysis, grazing, allochtonous organic matter inputs and temperatures. In Lake Pareloup bacterial activity appears clearly related to temperature if we compare seasonal patterns of thymidine uptake and temperature (Fig. 1). Over the whole year, there was a good correlation ($p < 0.01$) between Log.(^{3}H thymidine incorporation rates) and temperature in both aphotic ($n = 28$, $r = 0.54$) and euphotic zones ($r = 0.86$). Correlations are better if the spring bloom period, during which the bacterial growth is stimulated by the new input of organic matter resulting from phytoplanktonic activity, is not taken into account ($n = 23$, $r = 0.767$ and 0.962 respectively, when data for March and April were not used). Several previous studies also reported a close relationship between natural bacterioplankton growth and temperature (Hobbie & Cole, 1984; Lovell & Konopka, 1985b; Pomeroy & Deibel, 1986; Scavia & Laird, 1987).

Taking into account the morphometry and level fluctuations of the lake over the sampling period, net bacterial production averages 39.2 g m^{-2} per year. If we assume that 50% of the organic matter consumed is converted into bacterial biomass (Calow, 1977), the bacterial carbon requirement is

78.5 g m^{-2} per year, about 90% of the gross primary production as estimated during the present study. This high proportion is quite close to the results obtained for several mesotrophic (Güde *et al.*, 1985; Lovell & Konopka, 1985a; Scavia & Laird, 1987; Simon & Tilzer, 1987; Dufour *et al.*, 1989) and eutrophic lakes (Pedrós-Alió & Brock, 1982; Riemann, 1983), if we take into account the different values of the conversion factors used in the various studies.

The ratio of bacterial production to primary production was about 20% in spring and higher than 200% in late summer. Similar values can be deduced from the results of Riemann (1983) from Lake Tystrup, where bacterial production estimates, based on the FDC method, with a growth yield of 60% and a carbon cell content about twice as low as that used in Lake Pareloup, represented less than 10% of the carbon fixed by the algae in early spring and more than 120% in August. A similar seasonal pattern was also reported by Scavia & Laird (1987) from Lake Michigan.

To estimate bacterial production from ^{3}H thymidine incorporation rates, we used the conservative conversion factor of 2.10^{9} cells per nM thymidine incorporated into TCA insoluble precipitate (Fuhrman & Azam, 1982; Bell, 1988; Smits & Riemann, 1988), rather than the conversion factors calculated during the study from diluted water cultures in the exponential phase of growth (Lavandier & Faure, 1989). Indeed, if we consider that recent and converging estimates of cell carbon content (Bjornsen, 1986; Lee & Fuhrman, 1987; Simon & Azam, 1989) are valid, these empirical conversion factors (which varied through the year from 3.5 to 10.7 10^{9} cells per nM thymidine incorporated) gave bacterial production estimates which represented about 60% of the primary production in spring and up to 1200% in summer! Over the whole year, the heterotrophic production could be 3.2 fold higher than autotrophic production, which is ecologically unrealistic in Lake Pareloup, even if gross primary production estimates based on two-four hours incubations were probably slightly underestimated (Harris, 1978), and even if exogenous organic sources may enrich the middle (see below).

Whatever the real magnitude of the estimates, our results point out two main facts (Fig. 3): (1) bacterial production is important compared with autotrophic production, and bacteria provide a major pathway for the carbon flux in the lake, (2) the ecosystem is characterized by a significant lack of balance in time between primary and bacterial productions. Primary production is higher than bacterial production in spring; bacterial production is higher than autotrophic production in late summer. A pattern of algal production utilization by bacteria, similar to that observed in Lake Michigan (Scavia & Laird, 1987), can be assumed: during the phytoplankton spring bloom, algal exudates provide a good source of nutrients for the bacteria, but the utilization of the extracellular release is probably limited by the low temperatures and by a strong grazing pressure, particularly during the clear water phase; consequently, the amount of organic carbon in the lake rises. This reserve would be consumed later, when the temperature increases, together with the algal lysis products, the extracellular release of growing algae and the excretion products resulting from grazing. Note that this pattern of organic matter use, applies not only to the algal products, but also to other carbon

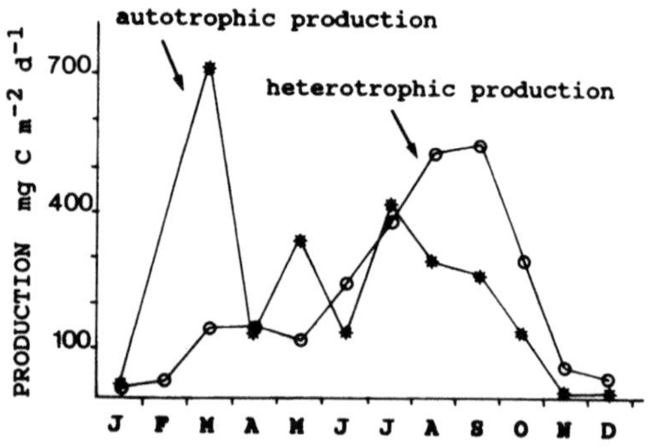

Fig. 3. Monthly mean estimates of autotrophic and heterotrophic gross productions. Bacterial production estimates were calculated from the ^{3}H thymidine incorporation rates assuming a conversion factor of 2.1 10^{9} bacteria produced per nanomole thymidine incorporated, a bacterial growth yield of 50% and the carbon-to-volume ratios from Simon & Azam (1989).

sources in the lake. Although we have rarely observed great amounts of suspended matter at the station, the resuspension of bottom material during the isothermal period and in the shallow zones of the lake, when the wind blows, could provide additional organic substrate for bacteria. Besides, in Lake Pareloup, the high bacterial production estimates in summer are probably enhanced by carbon nutrient sources other than autotrophic production, considering the catchment-area, where more than 20000 people are estimated to spend their holidays in July and August.

Acknowledgements

I thank J. Capblancq, C. Gossart, L. Labroue, C. Marchal, R. LeCohu and J.N. Tourenq for field assistance.

References

Bell, R. T., 1988. Thymidine incorporation and estimates of bacterioplankton production: Are the conversion factors valid? Ergebn. Limnol. 31: 163–171.

Bjornsen, P. K., 1986. Bacterioplankton growth yield in continuous seawater cultures. Mar. Ecol. Prog. Ser. 30: 191–196.

Brock, T. H. & J. Clyne, 1984. Significance of algal excretory products for growth of epilimnic bacteria. Appl. envir. Microbiol. 47: 731–734.

Calow, P. 1977. Conversion efficiencies in heterotrophic organisms. Biol. Rev. 52: 385–400.

Capblancq, J., A. Jrad & J. M. Thébault, 1988. Relation entre la lumière et la photosynthèse du phytoplancton dans un réservoir mésotrophe (Pareloup): variations saisonnières des paramètres. Ann. Limnol. 24: 39–48.

Dufour, P., S. Stroffek & M. Colin, 1989. Le bacterioplancton du Léman. Rapport commission internationale pour la protection des eaux du Léman contre la pollution., campagne 1988. C.I.P.E.L. Lausanne (CH): 71–77.

Fuhrman, J. A. & F. Azam, 1980. Bacterioplankton secondary production estimates for coastal waters of British columbia, Antarctica and California. Appl. envir. Microbiol. 39: 1085–1095.

Fuhrman, J. A. & F. Azam, 1982. Thymidine incorporation as a measure of heterotrophic bacterioplankton production in marine surface waters: evaluation and field results. Mar.Biol. 66: 109–120.

Güde, H., B. Haibel & H. Muller, 1985. Development of bacterial planktonic population in a water column of Lake Constance (Bodensee-Obersee). Arch. Hydrobiol. 105: 59–77.

Hagström, A., U. Larssen, P. Horstedt & S. Normark, 1979. Frequency of dividing cells, a new approach to determination of the bacterial growth rates in aquatic environments. Appl. envir. Microbiol. 37: 805–812.

Harris, G. P., 1978. Photosynthesis productivity and growth. The physiological ecology of phytoplankton. Ergebn. Limnol. 10: 1–171.

Hobbie, J. E. & J. J. Cole, 1984. Response of a detrital foodweb to eutrophication. Bull. mar. Sci. 35: 357–363.

Hobbie, J. E., R. J. Daley & S. Jasper, 1977. Use of nuclepore filters for counting bacteria by fluorescence microscopy. Appl. envir. Microbiol. 44: 1296–1307.

Jordan, M. J. & G. E. Likens, 1980. Measurements of planktonic bacterial production in an oligotrophic lake. Limnol. Oceanogr. 25: 719–731.

Lavandier, P. & F. Faure, 1989. Validity of the empirical conversion factors for assessing bacterial production from ^{3}H thymidine incorporation rates. Ann. Limnol. 25: 93–100.

LeCohu, R., J. Guitard, N. Comoy & J. Brabet, 1989. *Gonyostomum semen* (Raphidophycées), nuisance potentielle des grands réservoirs français? l'exemple du lac de Pareloup. Arch. Hydrobiol. 117: 225–236.

Lee, S. & J. A. Fuhrman, 1987. Relationship between biovolume and biomass of naturally derived marine bacterioplankton. Appl. envir. Microbiol. 53: 1298–1303.

Lovell, C. R. & A. Konopka, 1985a. Seasonal bacterial production in a dimictic lake as measured by increases in cell number and thymidine incorporation. Appl. envir. Microbiol. 49: 492–500.

Lovell, C. R. & A. Konopka, 1985b. The effects of temperature on bacterial production in a dimictic eutrophic lake. Microbiol. Ecol. 31: 135–140.

Mc Donough, R. J., R. W. Sanders, K. G. Porter & D. L. Kirchman, 1986. Depth distribution of bacterial production in a stratified lake with an anoxic hypolimnion. Appl. envir. Microbiol. 52: 992–1000.

Marker, A. F. F., E. A. Nuush & B. Riemann, 1980. The measurements of photosynthetic pigments in freshwater and standardization of methods: conclusions and considerations. Ergebn. Limnol. 14: 91–106.

Pedrós-Alió, C. & T. D. Brock, 1982. Assessing biomass and production of bacteria in eutrophic Lake Mendota, Wisconsin. Appl. envir. Microbiol. 44: 203–218.

Pollard, P. C. & D. J. W. Moriarty, 1984. Validity of the tritiated thymidine method for estimating bacterial growth rates: measurements of isotope dilution during DNA synthesis. Appl. envir. Microbiol. 48: 1076–1083.

Pomeroy, L. R. & D. Deibel, 1986. Temperature regulation of bacterial activity during the spring bloom in Newfoundland and coastal waters. Science. 233: 359–361.

Riemann, B., 1983. Biomass and production of phyto and

bacterioplankton in eutrophic Lake Tystrup, Denmark. Freshwat. Biol. 13: 389–398.
Scavia, D. & G. A. Laird, 1987. Bacterioplankton in Lake Michigan: dynamics, controls, and significance to carbon flux. Limnol. Oceanogr. 32: 1017–1033.
Simon, M. & F. Azam, 1989. Protein content and protein synthesis rates of planktonic marine bacteria. Mar. Ecol. Prog. Ser. 51: 201–213.
Simon, M. & M. M. Tilzer, 1987. Bacterial response to seasonal changes in primary production and phytoplankton biomass in Lake Constance. J. Plankton Res. 9: 535–552.
Smits, J. D. & B. Riemann, 1988. Calculation of cell production from (^{3}H) thymidine incorporation with freshwater bacteria. Appl. envir. Microbiol. 54: 2213–2219.

Received 22 July 1989; in revised form 8 February 1990; accepted 27 March 1990

Hydrobiologia **207**: 87–93, 1990.
D. J. Bonin & H. L. Golterman (eds), Fluxes Between Trophic Levels and Through the Water–Sediment Interface.

Effects of a plume front on the distribution of inorganic and organic matter off the Rhône River

Michel Leveau, François Lochet, Madeleine Goutx & François Blanc
Centre d'Océanologie de Marseille, Faculté des Sciences de Luminy, Case 901, 13288 Marseille Cedex 9, France

Key words: Rhône River, plume front, organic matter accumulation

Abstract

Hydrological and chemical structures off the Rhône River estuary resulting from the introduction of the river flow into the Mediterranean Sea are described. The effect of the fresh-water/sea-water interface on the distribution of inorganic and organic matter off the Rhône river is investigated. Strong vertical gradients of inorganic and dissolved organic matter such as lipids characterized the first few meters in this area (from 83.7 to 0.6 $N-NO_3$ μgat l^{-1}, from 6.39 to 0.92 $N-NH_4$ μgat l^{-1} and from 299 to 73 $\mu g\,l^{-1}$ of total dissolved lipids). At the interface, substantial increases of particulate organic (PON: from 45 $\mu g\,l^{-1}$ at surface to 118 $\mu g\,l^{-1}$ at the interface, POC: from 462 to 876 $\mu g\,l^{-1}$, total particulate lipids: from 33 to 648 $\mu g\,l^{-1}$) and suspended matter in general (from 18 to 22.2 $mg\,l^{-1}$) were observed. High phytoplanktonic production may account for some of this enrichment, although passive accumulation might also be involved.

Introduction

The Rhône River, with its large tributary catchment and its high flow rate, may be considered as the most eutrophicated telluric supply to the North Western Mediterranean Sea (Blanc *et al.*, 1969; Coste, 1974; Arfi, 1987; Leveau & Coste, 1987).

Previous investigations in this area have been mainly concerned with nutrients and their effect on the primary and secondary productions in the Rhône River plume and neighbouring areas (Blanc *et al.*, 1969). The distribution of organic matter within the whole Gulf of Lions or in the deep nepheloid layer located at the outlet of the Rhône River has been studied by Aloisi *et al.* (1982) and Buscail (1988). So far, no work has been devoted to on processes occurring at the interface between plume waters and sea waters off the Rhône River. The present study was carried out within the framework of the 'Programme National Frontal, Front de Panache du Rhône'.

The aim of our research program is to answer the following questions:

1. Is the plume front characterized by an increase in particulate or dissolved organic matter?
2. If such an increase is detected, is it due to a passive accumulation of organic matter or to an enhancement of biological production?
3. By which processes and to what extent is the organic matter used inside the frontal envelope or transported towards the surrounding ecosystem?

In the present paper, we describe the hydrological and chemical structures at the fresh-water/seawater interface at greater and lesser distances from the river mouth. Changes in the organic matter content and composition at the interface are discussed. Biological processes, which account for a new food web model (bacteria-microzooplankton-mesoplankton) located at the interface, were studied during the same cruise, but are not included in this paper.

Description of the site

A basin of approximately 95 000 km^2, rich in industrial and agricultural activities is drained by the Rhône River. The amounts of nutrients produced by the basin have been estimated by Coste (1974) as follows: 4.8 10^3 metric tons of $P-PO_4$ and 5.37 10^4 metric tons of inorganic nitrogen. Seasonal variations of the flow rate of the river range between 500 and 5 000 m^3 s^{-1}. The highly eutrophic and turbid waters of the Rhône flow into the transparent oligotrophic waters of the Liguro-Provençal current of the northern Gulf of Lions running westwards (Fig. 1).

At the mouth of the river, the water flows NW-SE at a velocity of 3 knots (Brun, 1967). When it reaches the Gulf of Lions, its direction changes to SW (Tournier, 1969). The size of the dilution zone varies with the flow rate of the river and the difference of temperature between fresh water and sea water, but depends mainly on the strength and the direction of the winds (Tournier, 1969).

The depth of the dilution zone decreased progressively from approximately 2 m near the mouth to a few cm offshore.

On the eastern side of the dilution zone, the plume front is characterized by a marked change in the colour of the water, a line of foam and a significant decrease in the surface velocity of the currents on both sides of that interface, indicating a frontal convergence (Le Fèvre, 1987). At this level, the Liguro-Provençal current dives under the dilution zone. The frontal envelope is the limit between the two water masses.

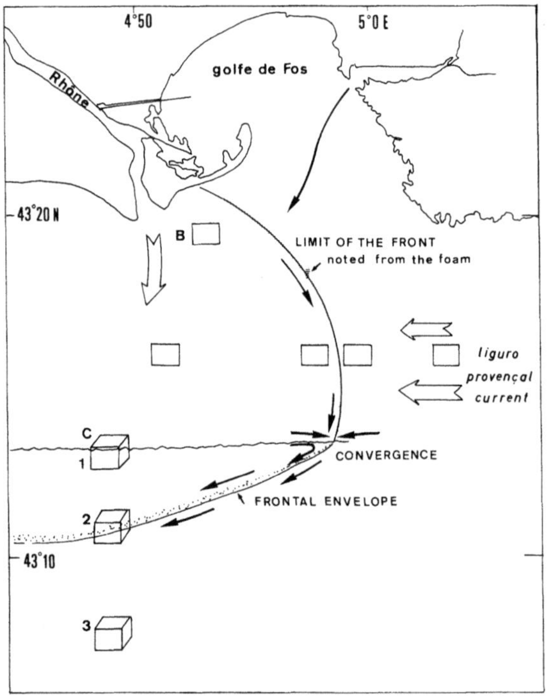

Fig. 1. Location of stations sampled in May 1988, off the Rhône River.

Material and methods

During a spring cruise (May, 1988), samples were taken in an area extending from the mouth of the Rhône River towards the open sea and on both sides of the frontal plume. This paper deals only with stations B (27 m depth) and C (95 m depth) (Fig. 1), which are representative of two hydrological structures of this area. Station B was representative of hydrological conditions near the Rhône River mouth, *i.e.* high surface currents (Rhône jet), high turbidity due to the Rhône river flow, whereas station C was representative of hydrological conditions further from the river mouth, characterized by lower turbidity and lower jet currents, and thus supposed to be more favourable to primary productivity.

Characteristics of water masses and localization of the interface were obtained by *in situ* measurements of temperature, conductivity and depth using a CTD Interocean 513E probe at each station. The submersible unit contains a centrifugal pump and a structure sustaining the CTD probe, connected to the ship by a cable and a 2.5 cm diameter Tygon tubing, which brings the water to the non-submersible instruments (Lochet, 1987).

Shipboard analysis of turbidity, dissolved oxygen, and *in vivo* fluorescence of chlorophyll *a* were performed on a Turner Design nephelometer, an Orbisphère probe and a Turner Design fluorometer (excitation 400–500 nm) respectively.

At each station, sampling was then performed between zero and forty meters depth in the plume waters, at the interface, and in marine waters for measurements of other parameters. Sea water was collected with a conventional Niskin bottle (30 l), for plume and marine water samples, or by direct pumping for the interface. Salinity was obtained with a RS-9B Beckman salinometer, dissolved oxygen measured according to Carpenter's method (1965). Nitrate, phosphate and silicate concentrations were measured with a Technicon Autoanalyzer following the procedure described by Tréguer & Le Corre (1975), ammonium following Koroleff's manual method (1969), and seston (particulate suspended matter, P.S.M.) following Krey's method (1964). Particulate carbon and nitrogen were measured according to Kerambrun & Szekielda (1969) and primary production by the ^{14}C standard method of Steemann-Nielsen (1952).

Table 1. Vertical gradients of phosphates ($P-PO_4$), nitrates ($N-NO_3$), ammonium ($N-NH_4$) and silicates ($Si-SiO_2$), expressed in μatg l^{-1} m^{-1} in surface waters (first two meters) at stations B and C.

Nutrient	St. B	St. C
$P-PO_4$	1	0.6
$N-NO_3$	24	41.5
$N-NH_4$	2.73	1.23
$Si-SiO_2$	17.5	12.5

Dissolved lipids were liquid-liquid extracted, and particulate lipids were extracted according to the method of Bligh & Dyer (1959). Extracts were separated by thin-layer chromatography coupled with flame ionization detection (TLC/FID) (Goutx *et al.*, 1987).

Results

Salinity

At stations B and C, the strong vertical gradients found in the first two meters were respectively 11 and 7‰ per meter. Below 5 m, salinity increased slightly, reaching approximately 38 at 20 m. On the two salinity profiles (Fig. 2A), the interface corresponding to the pycnohalocline was easily

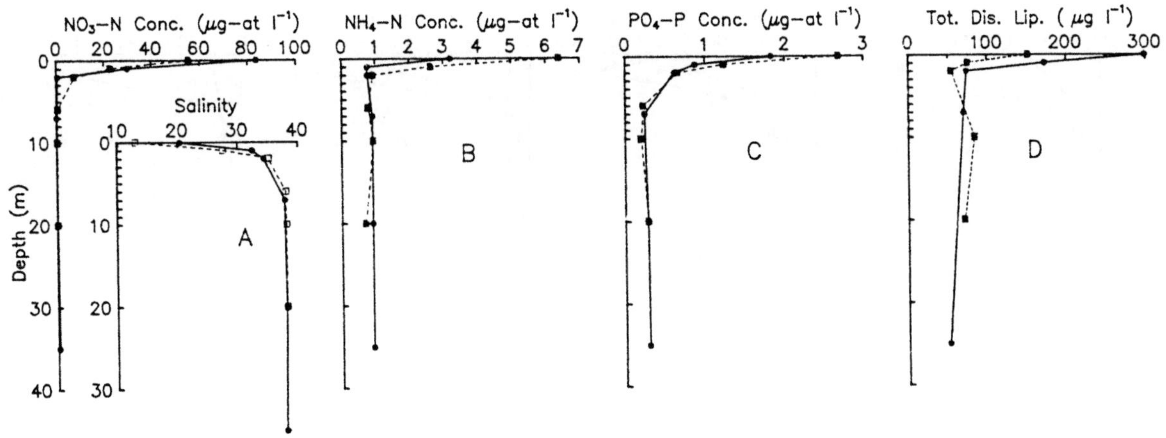

Fig. 2. Salinity, nutrients and dissolved lipids profiles at stations B (– – –) and C (——): salinity and nitrate (A), ammonium (B), phosphate (C) and dissolved lipids (D).

Table 2. Horizontal gradients of phosphates ($P-PO_4$), nitrates ($N-NO_3$) ammonium ($N-NH_4$) and silicates ($Si-SiO_2$) expressed in μatg l^{-1} m^{-1} in plume waters and sea water at stations B and C.

Nutrient	O m	2 m	20 m
$P-PO_4$	$0.1\ 10^{-3}$	$0.6\ 10^{-5}$	0.0
$N-NO_3$	$4.5\ 10^{-3}$	$1.1\ 10^{-3}$	$0.8\ 10^{-5}$
$N-NH_4$	$0.5\ 10^{-3}$	$0.2\ 10^{-4}$	$0.4\ 10^{-4}$
$Si-SiO_2$	$2.4\ 10^{-3}$	$0.7\ 10^{-3}$	$0.1\ 10^{-4}$

detected. At station B, the values of σt increased from 8.85 at the surface to 25.68 at 2 m.

On the other hand, horizontal gradients were very small, since for salinity, between stations B and C (6.2 km), there was at the surface an increase of 22.09‰ and at 2 m depth only 13.99.

Nutrients

Vertical profiles of nutrient concentrations (Fig. 2A, B, C) show drastic decrease in the upper layers; in the plume waters (first two meters) the vertical gradients were very high in stations B and C (Table 1). However, the horizontal gradients between the same stations were very small (Table 2) and the vertical gradients in the first two meters were 3 to 5 orders of magnitude higher than the horizontal gradients respectively for surface and 2 meters depth.

Dissolved lipids

At the interface, the concentrations of dissolved lipids did not increase at all (Fig. 2D), as found for nutrient concentrations.

Particulate suspended matter

In contrast with the observation made for dissolved matter, there was at the interface a substantial increase in total particulate suspended matter (Fig. 3A), which was also reflected in parameters representing the subcomponents of the organic compounds of particulate matter, such as particulate carbon (Fig. 3B), particulate nitrogen (Fig. 3B), chlorophyll (Fig. 3C), and lipids (Fig. 3D); for phospholipids at station B see Table 3. Moreover, the suspended mixture of organic and inorganic material (P.S.M.) did not follow exactly the same vertical distribution as that of the other components in B (Fig. 3A). This could be due to an extensive load of detrital material due to river inputs in B.

For particulate carbon, nitrogen and lipids, higher values were observed at the interface of station C, which may be attributed to higher biomass of phytoplankton and its accumulation at the interface. Higher primary productivity was also noted at station C (Fig. 4A), and reflected different phytoplankton contributions in terms of

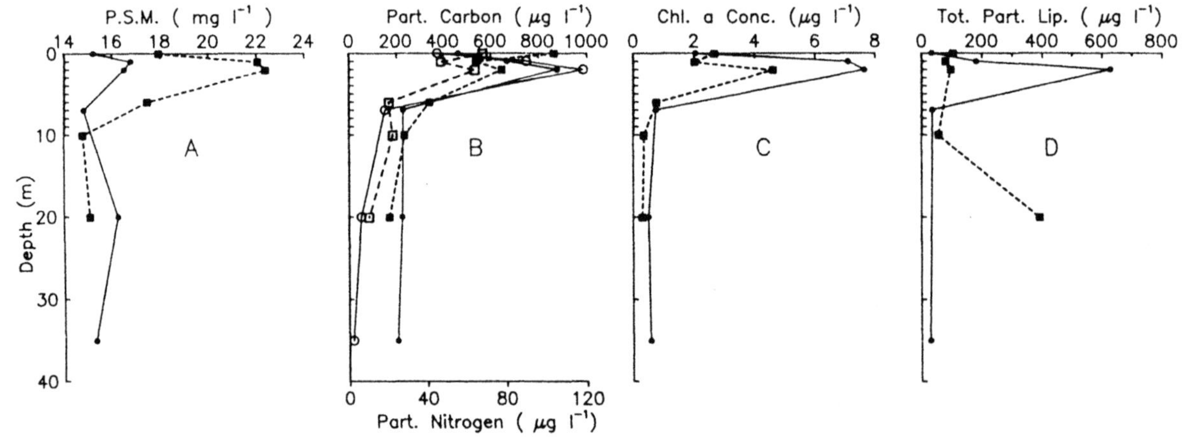

Fig. 3. Characteristics of particulate matter at stations B (- - -) and C (——): concentrations of suspended matter (A), particulate carbon (●, ■) and particulate nitrogen (○, □) (B), chlorophyll *a* (C) and lipids (D).

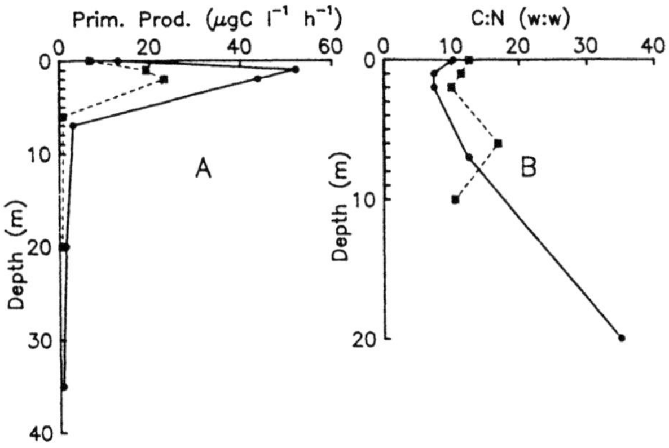

Fig. 4. Primary productions (A) and C : N (w : w) ratios (B) at stations B and C.

species and/or better growth activity at this site compared to station B. This pattern may be explained by lower turbidity at surface and possibly other ecological parameters such as salinity, nutrients, etc.

Particulate lipids were also more concentrated at the level of the interface at station C, but less at station B (Table 3). This accumulation resulted from anthropogenic supplies in station B, close to the Rhône, characterized by a very high hydrocarbon to total lipid ratio, which was not found at station C. Phytoplanktonic supplies in station C, were evidenced by simultaneous high values of neutral lipids (free fatty acids, triglycerides), acetone mobile polar lipids, also called chloroplast lipids in case of phytoplankton sample (Morris *et al.*, 1983; Parrish, 1987), and phospholipids. High levels of phospholipids and triglycerides were also observed as a general pattern at pycnoclines by Parrish *et al.* (1988), over the Scotian slope and in the Bedford Basin.

Discussion and Conclusions

The physical structure of the Rhône River dilution plume could be well described by hydrological parameters. The fresh-water/sea-water interface, quite visible on the surface, dives and forms a frontal envelope corresponding to the vertical limit of the dilution zone. The interface has often been described as a barrier zone, in which the processes of diffusion, flocculation, sorption and complexation occur (Halim & Morcos, 1966; Mantoura, 1987; Zutic & Legovic, 1987). In the present study, the concentration of dissolved inorganic matter and dissolved lipids did not increase at the interface. In contrast, there was a strong accumulation in the mineral and organic, living and non-living, particulate matter. It was difficult to distinguish between these different components in terms of P.S.M., particulate carbon and nitrogen. However, an obvious increase in chlorophyll content and in some particulate lipids enabled us to detect the accumulation of a

Table 3. Particulate lipids (average concentrations of two samples, in $\mu g\,l^{-1}$) at stations B and C for samples taken in the mixed layer (1), at the interface (2), or below the halocline (3).

Lipid class	Stations					
	B			C		
	1	2	3	1	2	3
Aliphatic hydrocarbon	134.68	187.44	134.59	10.12	22.10	13.81
Wax ester and sterol ester	8.32	1.43	1.31	0.05	3.74	0.88
Free fatty acid	18.26	13.58	11.82	14.56	241.40	9.77
Triglycerides	12.84	10.28	2.68	4.95	33.95	2.69
Free sterol	4.77	4.44	0.03	3.88	6.59	1.41
Chloroplast lipids	22.28	29.50	31.80	142.50	306.70	11.40
Phospholipids	11.43	35.96	7.48	13.34	34.18	7.59
Total lipids	212.58	282.63	189.71	189.40	648.46	47.55

phytoplanktonic biomass at the interface. The net increase of particulate triglycerides, which were more abundant at this level than above it, could indicate a stimulation of the phytoplankton production (Parrish, 1987), along with passive accumulation of biomass from water layers above the interface. This was confirmed by high primary production rates, as well as by a strong decrease in the C : N (w : w) ratios with values, at the interface, close to Redfield's ratio of 5.6 for living matter (Fig. 4B).

At station C, the unusually high concentrations of most lipids at the interface, which are known for having surface active properties (Rosenberg, 1986), are consistent with the high surfactant activity estimated at the fresh-water/sea-water interface of the River Krka estuary (Adriatic Sea) by Zutic & Legovic (1987). This suggests the existence of processes, previously demonstrated by Zutic & Legovic (1987), leading to the formation of stable biogenic surface active films at strong haloclines in estuarine environments. The accumulation of particulate matter at the front is probably due to the simultaneous action of various processes, such as, firstly, slow sedimentation rates of particles, due to a change in density at the halocline, allowing low density particles to be trapped, secondly, enrichment due to convergence, and lastly, increase in the production of some phytoplanktonic species, which proliferate at the interface because of increased nutrient and light availability.

Our results show that production and accumulation of particulate matter increase at the fresh-water/sea-water interface of the River Rhône dilution plume. Utilization of this material by higher trophic levels may cause a significant rate of organic production and should thus be considered when studying the large scale effect (Gulf of Lions) of Rhône River water inputs on nutrient fluxes and the functioning of food webs in general.

Acknowledgements

The authors wish to thank Dr. Alain Sournia for scientific coordination of the 'Programme National Frontal' and the unknown referees for their very helpful comments.

References

Aloisi, J-C., J-P. Cambon, J. Carbone, G. Cauwet, C. Millot, A. Monaco & H. Pauc, 1982. Structures et propriétés de la matière en suspension dans les embouchures stratifiées du golfe du Lion. Origine et rôle du néphéloide profond dans le transfert des particules au milieu marin. Oceanol. Acta 5: 481–492.

Arfi, R., 1987. Hydrologie et charge nutritive à l'embouchure du Rhône (saison froide). Bull. Ecol. 18: 123–130.

Blanc, F., M. Leveau & K. H. Szekielda, 1969. Effets eutrophiques au débouché d'un grand fleuve (Grand Rhône). Mar. Biol. 3: 233–242.

Bligh, E. & W. J. Dyer, 1959. A rapid method of total lipid extraction and purification. Can. J. Biochem. Physiol. 37: 911–917.

Brun, G., 1967. Etude écologique de l'estuaire du Grand Rhône. Bull. Inst. océanogr. Monaco 66: 46 pp.

Buscail, R., 1988. Bilan des transferts de matière organique à l'interface eau-sédiment sur le marge de Méditerranée Occidentale. Rapp. Comm. int. Mer Médit. 31: 294.

Carpenter, J. H., 1965. The accuracy of the Winkler method for dissolved oxygen analysis. Limnol. Oceanogr. 10: 135–140.

Coste, B., 1974. Role des apports nutritifs minéraux rhodaniens sur la production organique des eaux du golfe du Lion. Téthys 6: 727–740.

Goutx, M., S. Mutaftshiev & J. C. Bertrand, 1987. Lipid and exopolysaccharide production during hydrocarbon growth of a marine bacterium from the sea surface. Mar. Ecol. Prog. Ser. 40: 259–267.

Halim, Y. & S. A. Morcos, 1966. Le rôle des particules en suspension dans l'eau du Nil en crue dans la répartition des sels nutritifs au large de ses embouchures. Rapp. Comm. Mer Médit. 18: 733–736.

Kerambrun, P. & K. H. Szekielda, 1969. Dosage simultané du carbone, de l'hydrogène et de l'azote dans le matériel en suspension dans l'eau de mer. Note technique. Téthys 1: 581–584.

Koroleff, F., 1969. Direct determination of ammonia in natural waters as indophenol blue. Int. Counc. Explor. Sea, C.M., C: 9: 1–6.

Krey, J., 1964. Die mittlere Tiefenverteilung von Seston, Microbiomasse und Detritus im nordlichen Nordatlantik. Kieler Meeresforsch. 20: 18–29.

Le Fèvre, J., 1987. Aspects of the biology of frontal systems. Adv. mar. Biol. 23: 163–299.

Leveau, M. & B. Coste, 1987. Impact des apports rhodaniens sur le milieu pélagique du golfe du Lion. Bull. Ecol. 18: 119–122.

Lochet, F., 1987. Etude de la structure frontale du Rhône

grâce à la mesure en continu de descripteurs hydrobiologiques. In: Coll. int. océanol. Ecosystèmes des marges continentales. Comm. int. Mer Médit (CIESM), Monaco: 27.

Mantoura, R. F. C., 1987. Organic films at the halocline. Nature 328: 579–580.

Morris, R. J., M. J. McCartney & G. A. Robinson, 1983. Studies of a spring phytoplankton bloom in an enclosed experimental ecosystem. I. Biochemical changes in relation to the nutrient chemistry of water. J. exp. mar. Biol. Ecol. 70: 249–262.

Parrish, C. C., 1987. Time series of particulate and dissolved lipid classes during spring phytoplankton blooms in Bedford Basin, a marine inlet. Mar. Ecol. Prog. Ser. 35: 129–139.

Parrish, C. C., P. J. Wangersky, R. P. Delmas & R. G. Ackman, 1988. Iatroscan – measured profiles of dissolved and particulate marine lipid classes over the Scotian slope and in Bedford Basin. Mar. Chem. 23: 1–15.

Rosenberg, E., 1986. Microbial surfactants. CRC Crit. Rev. Biotechnol. 3: 109–132.

Steemann-Nielsen, E., 1952. The use of radioactive carbon (^{14}C) for measuring organic production in the sea. J. int. Counc. Explor. Sea 18: 117–140.

Tournier, H., 1969. Hydrologie saisonnière du golfe du Lion. (Travaux de l'Ichtys 1966–1967). Rev. Trav. Inst. Pêches marit. 33: 265–300.

Tréguer, P. & P. Le Corre, 1975. Manuel d'analyse des sels nutritifs dans l'eau de mer (utilisation de l'Autoanalyseur II Technicon R.). Laboratoire d'Océanographie chimique, Université de Bretagne Occidentale, Brest, 2nd edit.: 110 pp.

Zutic, V. & T. Legovic, 1987. A film of organic matter at the fresh-water/sea-water interface. Nature 328: 612–614.

Received 3 August 1989; in revised form 19 December 1989; accepted 27 March 1990

Hydrobiologia **207**: 95–103, 1990.
D. J. Bonin & H. L. Golterman (eds), Fluxes Between Trophic Levels and Through the Water–Sediment Interface.

Transfers between a eutrophic ecosystem, the river Rhône, and an oligotrophic ecosystem, the north-western Mediterranean Sea

François Lochet & Michel Leveau
Centre d'Océanologie de Marseille, Faculté des Sciences de Luminy, case 901, 13288 Marseille cedex 9, France

Key words: Nutrients, chlorophyll, suspended particulate matter, dilution, Rhône, NW Mediterranean Sea

Abstract

Transfer processes from a eutrophic ecosystem, the river Rhône, to an oligotrophic environment, the north-west Mediterranean Sea, were studied during the winter period, when inputs of dissolved and particulate matter are at their highest.

During a cruise carried out in February 1987, salinity, temperature, turbidity, nutrients, pigments, suspended particulate matter and particulate carbon and nitrogen were measured in surface and subsurface waters near the Rhône outlet.

For most of the parameters the results show that during this period a simple dilution process occurs in the river plume of the Rhône without adsorption, desorption, precipitation or biological uptake, contrary to recordings for some other rivers. All the nutrients discharged into the sea by the Rhône are available for utilization by marine phytoplankters.

Introduction

The Mediterranean sea is generally considered to be oligotrophic. Yet the coastal areas that receive a large direct supply of nutrients from atmospheric precipitation (Correll & Ford, 1982) and river discharge are exceptions and can be a major site of primary production.

In the north-western Mediterranean, the Rhône is one of the main rivers with a sufficiently high flow rate and nutrient input to induce eutrophication. Investigations of this eutrophication process (Blanc *et al.*, 1969; Blanc *et al.*, 1975) have shown a significant increase in primary production in an area covering about 50% of the Gulf of Lions (Coste, 1971; Coste, 1974).

In order to determine which proportion of the nutrient inputs from the Rhône may be used for primary production in marine waters, transfer processes from fresh to marine water were studied. Variations in various hydrobiological parameters (nutrients, suspended particulate matter, chlorophyll *a*, *etc*...) with salinity were monitored in the vicinity of the river Rhône outlet. This investigation was carried out in late winter, since the input levels of dissolved and particulate matter are particularly high (river in spate) at this time of year.

Description of the site studied

The study area is located at the mouth of the Rhône (Fig. 1), a river with a catchment area of 95 000 km^2, and an annual mean flow rate of around 1 700 $m^3\ s^{-1}$ (measurements recorded at Beaucaire, 70 km from the river mouth, by the Compagnie du Rhône). Low waters occur in August (minimum 855 $m^3\ s^{-1}$) and spate in February (maximum 5 000 $m^3\ s^{-1}$). Terrigenous input and turbidity are very high (Secchi disk depth about 0.10 m). Large stretches of the Rhône are affected by industrial, agricultural and sewage discharge, which results in high nutrient concentrations.

The Gulf of Lions, is one of the few Mediterranean areas with a very extensive continental shelf (average depth 90 m and maximum depth < 150 m). The liguro-provençal current hugs the continental slope to seaward. Its flow rate is about $10^6\ m^3\ s^{-1}$, and it diverts the waters of the Rhône from their original SE direction towards the south west, 20 km offshore. Nutrient concentrations in the liguro-provençal current are low even in winter, when these waters are enriched by vertical mixing (0.5 μmol NO_3-N l^{-1}, 0.04 μmol NO_2-N l^{-1}, 0.05 μmol PO_4-P l^{-1} and 1 μmol SiO_4-Si l^{-1}; Coste & Minas, 1967).

The position of the river plume is influenced by meteorological factors, the flow of the river and the liguro-provençal current. The wind regime in the area shows seasonal variations. The prevailing wind blowing most consistently in late winter from the north-west represents 45% of total winds.

During the Dypol III cruise in February 1987 the flow rate of the Rhône dropped from 4 900 to 2 489 $m^3\ s^{-1}$ and the wind was moderate (NW, 6 $m\ s^{-1}$). The maximum depth of sampling locations ranged from 23 to 90 m (except in the river).

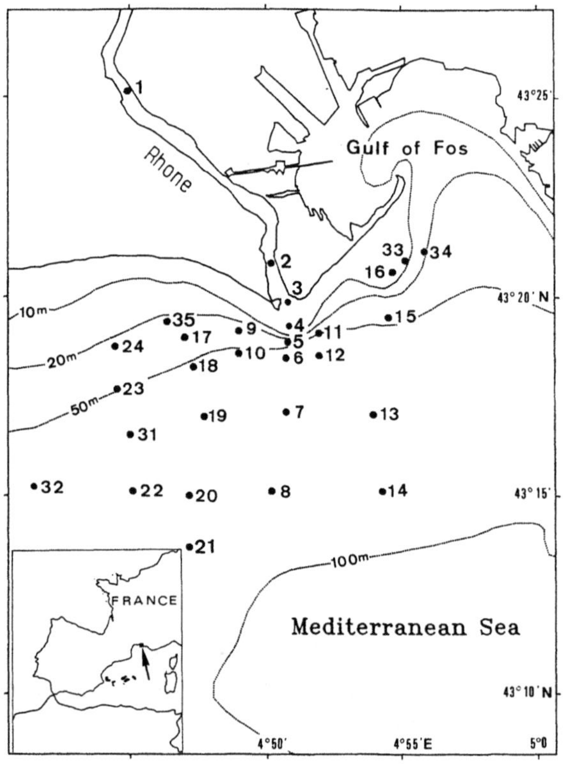

Fig. 1. Station locations of the DYPOL III cruise.

Material and methods

At each station, water was sampled at the surface, at the freshwater/sea water interface and near the bottom. The environment is highly variable (Lochet, 1987), and in order to determine the depth of the freshwater/sea water interface at each station, the hydrological characteristics of the water column (depth, temperature, conductivity, oxygen concentration, pH) were obtained with a multiparameter sensor Hydropolyster HPTC 84 equipped with an integrating diffusiometer (Bapst & Beck, 1987).

Seawater was collected with 10 l Niskin bottles. Salinity was determined with an inductive salinometer. Nitrate, nitrite, silicate and phosphate concentrations were analysed, after thawing of frozen (− 18 °C) samples, with a Technicon Autoanalyzer (Tréguer & Le Corre, 1975). Ammonia was analysed on fresh samples within 24 h of collection with a spectrophotometer according to Koroleff's manual method (1969). Chlorophyll *a* and phaeopigments were measured with a Turner Design Fluorometer after filtration on precombusted Whatman GF/F glass fiber filters (25 mm), according to Holm-Hansen *et al.*

(1965). Suspended particulate matter (SPM) was determined according to Krey (1964). Particulate organic carbon (POC) was measured after filtration on precombusted Whatman GF/F glass-fiber filters and analysed/by Cawet's method (1984). Suspended matter size spectrum was obtained with a TA II model Coulter Counter, equipped with an automatic data acquisition system (Loeillet, 1984). Particles were measured using 70 μm and 280 μm orifices. Data on the flow rate of the Rhône were provided by the Compagnie du Rhône in Beaucaire.

Data analysis

The relationship between salinity and nutrients was studied to determine whether distribution follows a dilution gradient. Since the variables are considered as interdependent, the model II regression method (Laws & Archie, 1981) was used to estimate the slope of the regression lines for each nutrient. The slopes of the theoretical dilution lines were calculated from the equation: $S_{th} = ([N]_{S\ max} - [N]_{FW})/(S_{max})$, where S_{th} is the theoretical slope, $[N]_{S\ max}$ is the nutrient concentration for $S_{max} = 38.11$ (maximum salinity during DYPOL III) and $[N]_{FW}$ is the maximum nutrient concentration for $S = 0$ (freshwater).

The hypothesis of equality between theoretical and estimated slopes was tested with Student's *t* test.

Results

Nutrients

During this survey, we investigated the distribution of nutrients according to salinity. The most outstanding feature of the graphs showing nutrients against salinity is that nutrient distribution is very conservative during mixing in the river plume (Fig. 2 to Fig. 4). This is confirmed by the fact that the model II regression slopes are not significantly different from those forecast by the simple mixing of the river freshwaters (Table 1).

For nitrates (Fig. 2A), we noted conservative behaviour both in the river plume and at salinities higher than 37.5 (sea waters), and no influence of nutrient consumption by freshwater phytoplankton was noted.

Table 1. Observed and theoretical slopes of the principal axes for nutrients and DIN/PO_4 ratio, versus salinity, calculated from DYPOL III data, with values of Student's *t* test (NS: non significant difference).

Nutrient	Equations with obs. data	Theor. slopes	N	$t =$
NO_3-N	-2.968 S + 111.64	-2.954	33	0.13 NS
NO_2-N	-0.022 S + 1.08	-0.023	33	1.61 NS
NH_4-N	-0.318 S + 12.48	-0.301	26	1.66 NS
DIN/PO_4	-1.000 S + 50.20	-0.746	24	2.04 NS
SiO_4-Si	-2.001 S + 76.24	-2.007	33	0.08 NS
PO_4-P	-0.072 S + 3.09	-0.070	33	1.22 NS

During the sampling period (river in spate) the nitrate concentration was about 120 $\mu mol\ l^{-1}$ in the Rhône and 1.5 in the sea. Nitrate was the most abundant form of Dissolved Inorganic Nitrogen (sum of NO_3-N + NO_2-N + NH_4-N) with a percentage of 90%. For nitrites (Fig. 2B) a slight increase in NO_2-N was noticeable in sea waters and at the freshwater/sea water interface. Nitrite concentrations were around 1.20 $\mu mol\ l^{-1}$ in the river and 0.15 in marine waters. Nitrite was a minor component (around 1%) of DIN (Dissolved Inorganic Nitrogen) in the river and increased in the sea (8.7%). The concentration of ammonium decreased from 15.5 $\mu mol\ l^{-1}$ in the river to 0.34 $\mu mol\ l^{-1}$ in the sea (Fig. 2C). Ammonium was about 10% of DIN in the riverine inputs and decreased to 5.8% seaward of the plume. Concentration of phosphorus was fairly high (around 3 $\mu mol\ l^{-1}$) in the freshwater inputs and about 0.08 $\mu mol\ l^{-1}$ in sea waters (Fig. 3A). The silicon concentration in the river was around 77 $\mu mol\ l^{-1}$ and near 1.3 $\mu mol\ l^{-1}$ in the sea (Fig. 3B).

A study of variations, according to salinity, in the ratio of DIN/PO_4, shows that for salinity levels between 0 and 35 this ratio follows a law of simple dilution (Fig. 4A). In the river, the DIN/PO_4 ratio is close to 50, while in the sea it is close to 10. Nevertheless, for stations situated at the perimeter of the river plume (15, 18, 23...), a variance in the ratio DIN/PO_4 may be noted,

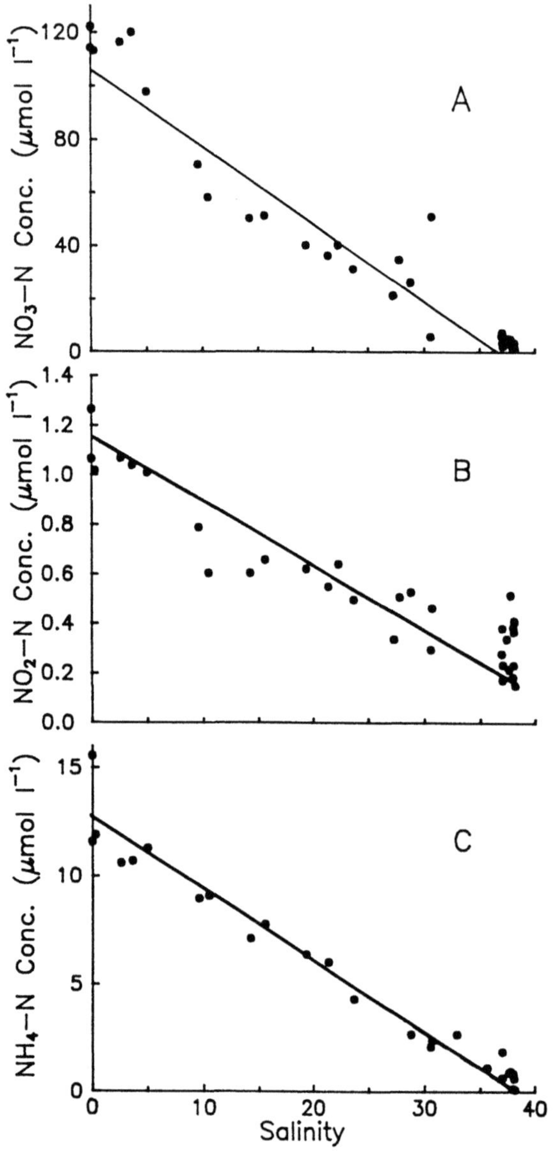

Fig. 2. NO_3-N (a), NO_3-N (B), NH_4-N (C) *vs* Salinity diagrams for DYPOL III showing their principal axes.

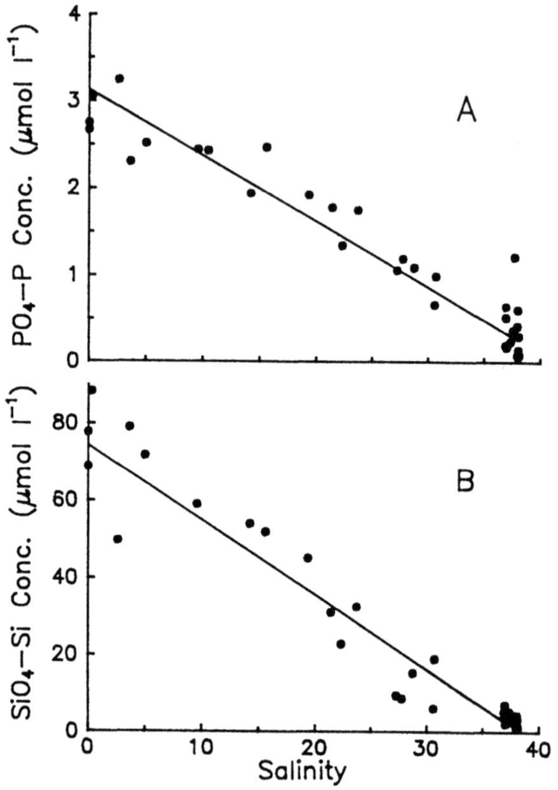

Fig. 3. PO_4-P (A), SiO_4-Si (B) *vs* Salinity diagrams for DYPOL III showing their principal axes.

resulting from the very high variability of the PO_4 values. This variance can be seen in Fig. 4B, which shows that for low NO_3 values (4 μmol l^{-1}) the amount of PO_4 varies from 1 to 0.2 μmol l^{-1} (at these concentrations, desorption phenomena might be significant). On the other hand, in the river plume (values higher than 1 μmol l^{-1}), the relation between the various nutrients shows that they behave in the same way as conservative variables.

Suspended particulate matter

The relation between SPM and salinity is not linear (Fig. 5A). A study of the distribution of various size classes of SPM according to salinity shows that the small particles (Size 1.41 μm) show linear variation according to salinity (Fig. 5B), which is not the case for large particles (Size 28.5 μm) which sediment more rapidly (Fig. 5C).

The distribution of POC according to salinity is identical to that of SPM (Fig. 6A).

For chlorophyll, the situation seems to be more complex (Fig. 6B). The decrease with salinity is higher than it would be in the case of simple dilution. The decrease in chlorophyll concentration levels off at a salinity of around 20, and a very slight increase is apparent in the marine environment ($S > 37.5$), except for bottom samples

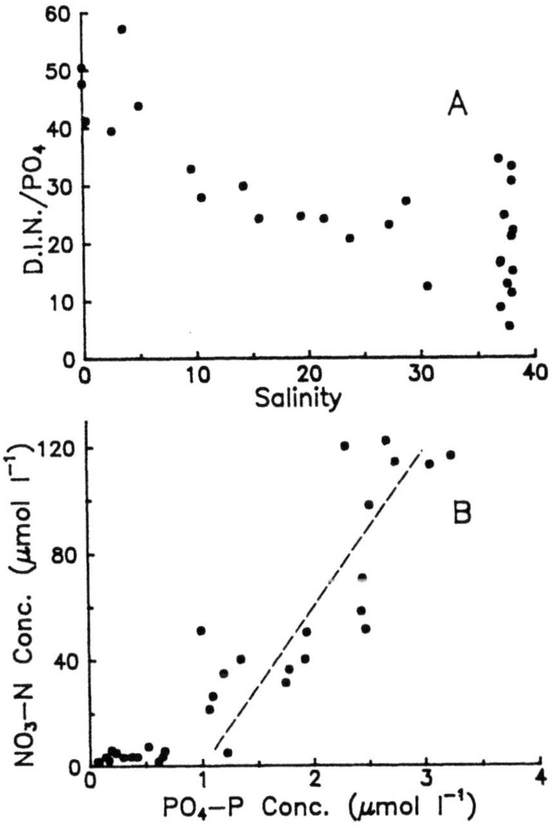

Fig. 4. DIN/PO_4 *vs* Salinity diagram (A); NO_3-N (B) as a function of PO_4-P for DYPOL III.

(St. 8 at 90 m, St. 13 at 75 m), where the values are very low (0.11 $\mu g\,l^{-1}$).

The study of POC fraction in the SPM against salinity (Fig. 7A) shows that this fraction was low and constant in the river plume (3.5%) and increased sharply in the marine environment. The same is true for the chlorophyll fraction in the SPM (4 10^{-5}, Fig. 7B). The percentage of phaeopigments in relation to the sum chlorophyll + phaeopigments is high (57%) in brackish waters (S < 37.5), which shows that the phytoplanktonic matter is severely degraded (Fig. 7C).

Discussion

For nitrates (Fig. 2A), we noted, as did Morris *et al.* (1981), conservative behaviour both in the river plume and at salinities higher than 37.5 (sea

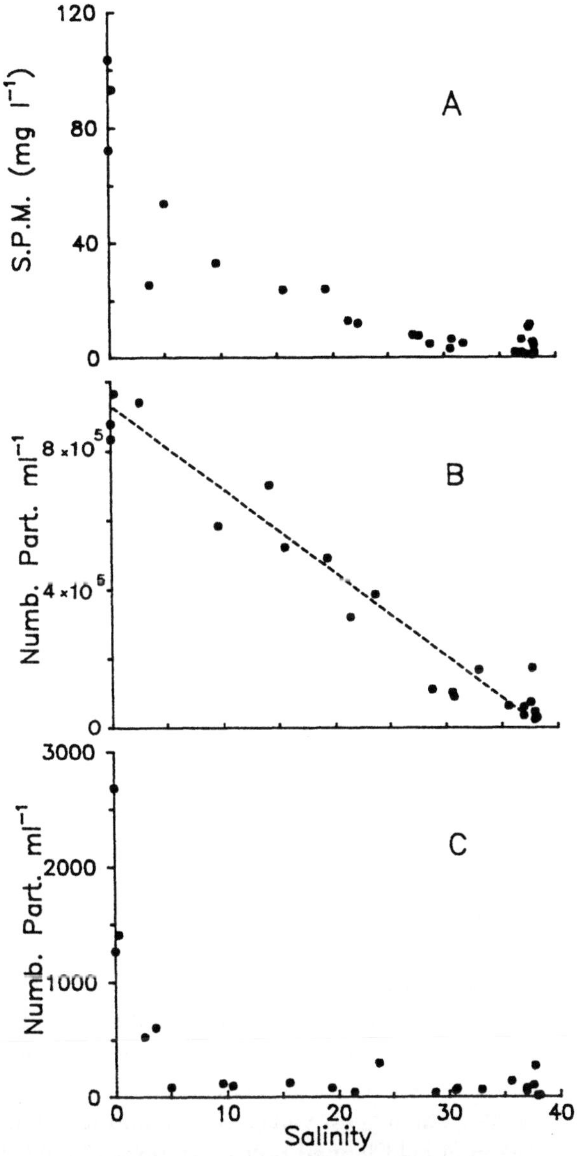

Fig. 5. SPM (A), 1.41 μm diameter particles (B), 28.5 μm diameter particles (C) *vs* Salinity diagrams for DYPOL III.

waters). Our measurements are consistent with those made by Arfi (1987) in placing NO_3-N values close to 120 $\mu mol\,l^{-1}$ in the river. This value exceeds the mean NO_3-N values for bottom waters in the Mediterranean by a factor of 15 ($\simeq 8$ $\mu mol\,l^{-1}$; Coste, 1987) and for unpolluted temperate rivers by a factor of 8.5 ($\simeq 14.3$ $\mu mol\,l^{-1}$; Meybeck, 1982).

The slight increase in NO_2-N noticeable in sea waters and at the freshwater/sea water interface

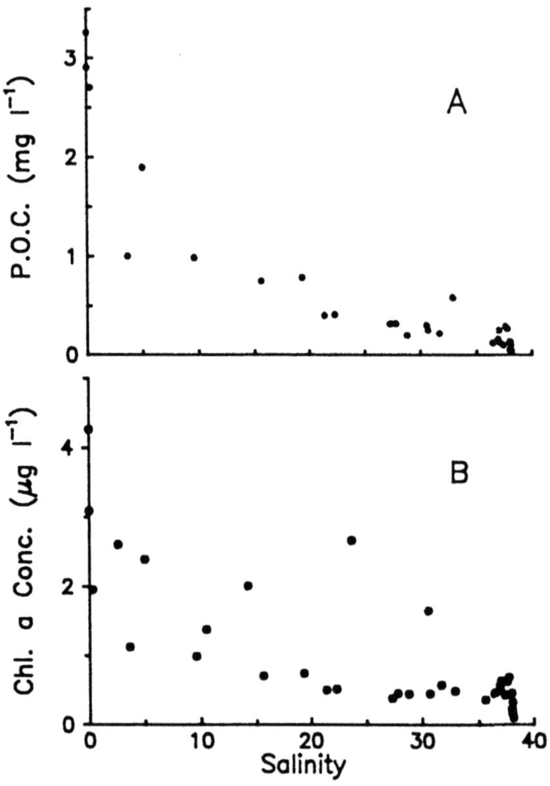

Fig. 6. POC (A), chlorophyll (B) *vs* Salinity diagrams for DYPOL III.

can be explained by the fact that nitrites can be produced by bacterial activity either through ammonium oxidation or nitrate reduction. Experimental evidence also suggests that phytoplankton taking up nitrate in a low light regime release nitrite (Vaccaro & Ryther, 1960; Fisher *et al.*, 1988). The highest values correspond to samples taken at light-limited bottom stations (05, 07, 13, 19) or at the interface, where the density of micro-heterotrophs increases.

Phosphorus concentrations in the Rhône, during winter, exceed the mean phosphate concentration for the world's rivers by a factor of 3 (Stallard & Edmond, 1983).

The presence of suspended particulate matter, the change in ionic balance and biological activity may affect phosphorus distribution. These phenomena can induce adsorption or coprecipitation with the colloidal fraction of the river waters (Morris *et al.*, 1981), or, conversely, desorption of phosphates present in the suspended particulate

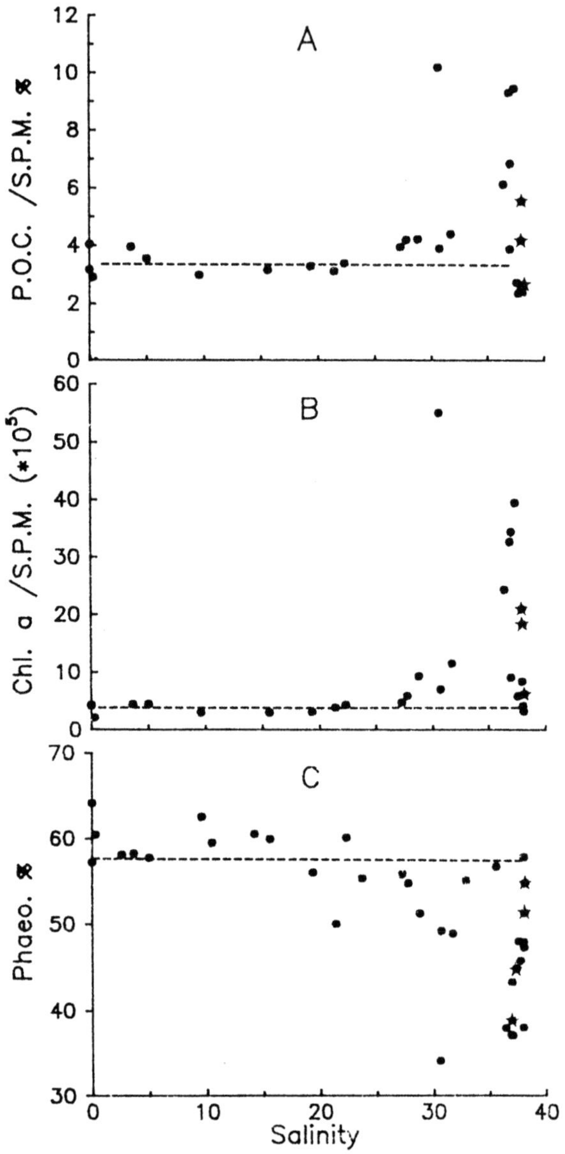

Fig. 7. Percentage of POC in the SPM (A), chlorophyll/SPM ratio ($\times 10^5$) (B), Phaeo/(Chl. *a* + Phaeo) $\times$ 100 (C) *vs* Salinity diagrams for DYPOL III (● surface and subsurface sampling, ★ deep sampling).

matter, as is the case for the Nile (Halim & Morcos, 1966) and the Amazon (Chase & Sayles, 1980). In the Rhône, we noted, in agreement with Coste (1974), that phosphates are relatively conservative from freshwater to sea water (Fig. 3A), as is the case for dissolved inorganic nitrogen. Deviations from conservative behaviour associated with the boundaries of the plume (salinity $\simeq$ 38) could reflect different origins of sea waters. As a matter of fact, the eastern boundary of the plume near the coast, may be under the influence of waters from the Gulf of Fos.

For silicon, it should be pointed out that a significant proportion of dissolved silicon in fresh and brackish waters may be precipitated upon freezing and our results may be underestimated. Concerning this nutrient, opinions differ, probably because of the possible influence of physical, chemical and biological factors. Some authors describe simple dilution of this nutrient according to salinity (Belzile & Lebel, 1983; Van Bennekom *et al.*, 1978), while others (Bien *et al.*, 1958; Liss & Pointon, 1973) indicate some removal of silicon from the solution in the estuarine area, which is seen as the result of adsorption and precipitation reactions between the silicon and the particulate matter, due to changes in the ionic balance. These latter authors suggest that the high concentrations of suspended matter and high silicon concentrations increase the rate of silicon removal from the dissolved fraction during freshwater – marine water mixing. In the Rhône, this phenomenon was not observed in keeping with Wollast's hypothesis (in Liss, 1976: 125), which maintains that the removal of soluble silica would take place only in the case of concentrations of more than 230 μmol l^{-1}, which is considerably higher than the values we found for the Rhône.

Results of DIN/PO_4 ratio variations are consistent with studies on nutrient limitation (e.g. Ryther & Dunstan, 1971; Bishop *et al.*, 1984), which have demonstrated that, for sea waters, nitrogen is usually the most limiting nutrient for potential phytoplankton production. Conversely, in limnology, the freshwater ecosystems are mostly phosphorus-limited (*e.g.* Schindler, 1981), and across salinity gradients there is often a change from phosphorus limitation in less saline waters to nitrogen limitation in more saline waters (Caraco *et al.*, 1987). The variance in the DIN/PO_4 ratio noted at the seawater end member of the graph (Fig. 4A) could be related to biological processes or could simply result from the influence of coastal waters rich in nitrates on the western side of the plume.

The non-linear relation between SPM and salinity indicates (Fig. 5) that differential settling is heavy in the estuary. In the river waters and at low salinity (S < 10), the SPM values show that as soon as the river enters the sea, heavy sedimentation takes place, especially for the coarser particles, resulting in a gradual loss of the coarser grains observed in particle size distribution further out to sea. Indeed, the freshwater masses spread out when they reach the sea, and the consequent reduction in speed results in varying particle sedimentation rates. As a result, the suspension become progressively finer offshore.

Based on the chlorophyll and SPM distributions it could be that some of the larger particles disappearing in the estuary were freshwater phytoplankton. This has lead us to conclude that the high phaeopigment concentrations in the plume reflected intensive zooplankton grazing on the large phytoplankton.

The value of POC expressed as a percentage of SPM (3.5%) in the river is consistent with the 3% mean value for the world's rivers for SPM in the 50 to 150 mg l^{-1} range (Meybeck, 1982) and with the figure of 4.7% found in the Zaire (Cadée, 1984). The seaward increase of organic content seems to be at least partly due to the settling of heavier and larger mineral particles in the plume and also to marine phytoplankton growth. The variations of POC/SPM ratio probably reflect organic matter along the perimeter of the plume.

There are therefore two distinct zones:

– The river plume, characterized by high concentrations of nutrients and particulate matter. In this zone, the conservative distribution of nutrients indicates that their use for primary production is slight (no uptake by phytoplankton). In addition, the high percentage of phaeopigments and the low POC/SPM ratio indicate that towards the end of the winter, the Rhône discharges considerable amounts of detritic plant matter, either resulting from runoff from the land during the spate, or consisting of freshwater phytoplankton in poor physiological condition or subject to intensive predation.

– The sea waters adjacent to the plume show lower chlorophyll *a* values than those found in the river, but with a lower percentage of phaeopigments in relation to total pigments. This indicates that phytoplankton are more photosynthetically active in these waters.

In relative terms, then, in the marine water column autochthonous living plant biomass is the predominant source of organic matter, while in the river plume terrestrial plant detritus and its associated microbiota is the main allochthonous source of organic matter.

Conclusion

Our investigations carried out in the river plume of the Rhône show that the majority of the dissolved compounds studied are conservative. Physicochemical (adsorption, desorption, precipitation, dissolution) and biological (photosynthesis) phenomena do not seem to have a great influence within the river plume. This might be due to the fact that the water masses are dispersed too rapidly at this period of spate. This implies that the nutrients discharged by the Rhone in winter can be directly used by marine phytoplankton and lead to the renewal of production observed in February (Jacques, 1970). This production can only occur in sea waters where salinity and light conditions are appropriate.

Acknowledgements

Thanks are due to C. Meynard for valuable assistance. The authors are also grateful to Dr G. Cauwet from the 'Laboratoire de Sédimentologie et Géochimie Marines de Perpignan' for providing POC and SPM data. We would also like to thank anonymous reviewers for helpful criticisms of the manuscript.

References

Arfi, R., 1987. Hydrologie et charge nutritive à l'embouchure du Rhône (saison froide). Bull. Ecol. 18: 123–130.

Bapst, A. & C. Beck, 1987. Structures néphéploïdales et thermiques sur le delta du Rhône. In: Coll. int. Océanol. Ecosystèmes de marges continentales. Comm. int. Mer Médit. (CIESM), Monaco (ed.), 26.

Belzile, N. & J. Lebel, 1983. Contribution à l'étude chimique du matériel dissous et particulaire de l'estuaire de la rivière Rimouski. Mar. Chem. 13: 15–34.

Bien, G. S., D. E. Contois & W. H. Thomas, 1958. Removal of soluble silica from fresh-water entering the sea. Geochim. Cosmochim. Acta 14: 35–54.

Bishop, S. S., K. A. Emmanuele & J. A. Yoder, 1984. Nutrient limitation of phytoplankton growth in Georgia nearshore waters. Estuaries 7: 506–512.

Blanc, F., M. Leveau & P. Kerambrun, 1975. Eutrophie et pollution: structure et fonctionnement du sous-écosystème planctonique. 10 th European Symposium on Marine Biology, Ostend, Belgium 2: 61–83.

Blanc, F., M. Leveau & K. H. Szekielda, 1969. Effets eutrophiques au débouché d'un grand fleuve (Grand Rhône). Mar. Biol. 3: 233–242.

Cadée, G. C., 1984. Particulate and dissolved organic carbon and chlorophyll *a* in the Zaïre river, estuary and plume. Neth. J. Sea Res. 17: 426–440.

Caraco, N., A. Tamse, O. Boutros & I. Valiela, 1987. Nutrient limitation of phytoplankton growth in brackish coastal ponds. Can. J. Fish. aquat. Sci. 44: 473–476.

Cauwet, G., 1984. Automatic determination of dissolved organic carbon in the sub ppm range. Mar. Chem. 14: 297–306.

Chase, E. M. & F. L. Sayles, 1980. Phosphorus in suspended sediments of the Amazon River. Estuar. coast. mar. Sci. 11: 383–391.

Correll, D. L. & D. Ford, 1982. Comparison of precipitation and land runoff as sources of estuarine nitrogen. Estuar. coast. Shelf Sci. 15: 45–56.

Coste, B., 1971. Circulation et évolution des composés de l'azote et du phosphore (en particulier des sels nutritifs) dans le bassin occidental de la Méditerranée. Influence sur la production organique. Thèse Doct. ès-Sci. Univ. d'Aix-Marseille II, 256 pp.

Coste, B., 1974. Rôle des apports nutritifs minéraux rhodaniens sur la production organique des eaux du Golfe du Lion. Téthys 6: 727–740.

Coste, B., 1987. Les sels nutritifs dans le bassin occidental de la Méditerranée. Rapp. Comm. int. Mer Médit. 30(3): 399–410.

Coste, B. & H. J. Minas, 1967. Premières observations sur la distribution des taux de productivité et des concentrations en sels nutritifs des eaux de surface du Golfe du Lion. Cah. océanogr. 19: 417–429.

Fisher, T. R., L. W. Jr Harding, D. W. Stanley & L. G. Ward, 1988. Phytoplankton, Nutrients, and Turbidity in the Chesapeake, Delaware, and Hudson Estuaries. Estuar. coast. Shelf Sci. 27: 61–93.

Halim, Y. & S. A. Morcos, 1966. Le rôle des particules en suspension dans l'eau du Nil en crue dans la répartition des

sels nutritifs au large de ses embouchures. Rapp. Comm. int. Mer Médit. 18(3): 733–736.
Holm-Hansen, O., C. J. Lorenzen, R. W. Holmes & J. D. H. Strickland, 1965. Fluorometric determination of chlorophyll. J. Cons. perm. int. Explor. Mer 30: 3–15.
Jacques, G., 1970. Aspects quantitatifs du phytoplancton de Banyuls sur mer (Golfe du Lion). IV. Biomasse et Production, 1965–1969. Vie Milieu. 21(1-B): 37–102.
Koroleff, F., 1969. Direct determination of ammonia in natural waters as indophenol blue. Int. Counc. Explor. Sea, C.M. 1969/C 9: 19–22.
Krey, J., 1964. Die mittler Tiefenverteibung von Seston, Mikrobiomasse und Detritus im nördlichen Nordatlantik. Kieler Meeresforsch. 20: 18–29.
Laws, E. A. & J. W. Archie, 1981. Appropriate use of regression analysis in marine biology. Mar. Biol. 65: 13–16.
Liss, P. S., 1976. Conservative and non-conservative behaviour of dissolved constituents during estuarine mixing. In J. D. Burton & P. S. Liss (eds), Estuarine chemistry, Academic Press, Lond.: 93–131.
Liss, P. S. & M. J. Pointon, 1973. Removal of dissolved boron and silicon during estuarine mixing of sea and river waters. Geochim. Cosmochim. Acta 34: 1073–1088.
Lochet, F., 1987. Etude de la structure frontale du Rhône grâce à la mesure en continu de descripteurs hydrobiologiques. In: Coll. int. Océanol. Ecosystèmes de marges continentales. Comm. int. Mer Médit. (CIESM), Monaco (ed.), 27.
Loeillet, C., 1984. Etude statistique des distributions de taille des particules en suspension dans deux zones soumises à l'influence Rhodanienne (Darse 1 de Fos, Golfe du Lion). Thèse Doct. 3eme Cycle, Univ. Aix-Marseille II, 103 pp.
Meybeck, M., 1982. Carbon, nitrogen, and phosphorus transport by world rivers. Am. J. Sci. 282: 401–450.
Morris, A. W., A. J. Bale & R. J. M. Howland, 1981. Nutrient distributions in an estuary: evidence of chemical precipitation of dissolved silicate and phosphate. Estuar. coast. Shelf Sci. 12: 205–216.
Ryther, J. H. & W. M. Dunstan, 1971. Nitrogen, Phosphorus, and Eutrophication in the coastal marine environment. Science 171: 1008–1009.
Schindler, D. W., 1981. Studies in lakes and their relevance to the estuarine environment. In: B. J. Neilson & L. E. Cronin (eds), Estuaries and nutrients. Humana Press, Clifton, 71–82.
Stallard, R. F. & J. M. Edmond, 1983. Geochemistry of the Amazon. 2. The influence of geology and weathering environment on dissolved load. J. Geophys. Res. 88: 9671–9688.
Tréguer, P. & P. Le Corre, 1975. Manuel d'analyse des sels nutritifs de l'eau de mer (Utilisation de l'Auto-Analyzer II Technicon R), 2nd édition, Laboratoire d'Océanologie chimique, Université de Bretagne Occidentale, Brest, 110 pp.
Vacarro, R. F. & J. J. Ryther, 1960. Marine phytoplankton and the distribution of nitrite in the sea. J. Cons. perm. int. Explor. Mer 25: 260–271.
Van Bennekom, A. J., G. W. Berger, W. Helder & T. P. De Vries, 1978. Nutrient distribution in the Zaire estuary and river plume. Neth. J. Sea Res. 12: 296–323.

Received 6 August 1989; in revised form 25 February 1990; accepted 15 June 1990

Hydrobiologia **207**: 105–114, 1990.
D. J. Bonin & H. L. Golterman (eds), Fluxes Between Trophic Levels and Through the Water–Sediment Interface.

Interpretation of the seasonal variations of nutrients in a mediterranean lagoon: étang de Thau

Bernadette Picot [1], Geneviève Péna [1], Claude Casellas [1], Daniel Bondon [2] & Jean Bontoux [1]
[1] *Département Sciences de l'Environnement et Santé Publique URA CNRS 1355, Faculté de Pharmacie, Av. Charles Flahault, 34060 Montpellier, France,* [2] *Centre de Recherches pluridisciplinaires Mèze Hérault Ceremher, 34140 Mèze, France*

Key words: nitrogen, phosphate, phytoplankton, Mediterranean lagoon

Abstract

A network of 63 stations was used on four occasions (June 1986, October 1986, February 1987, and May 1987) to study the spatio-temporal distribution of inorganic nutrients in Thau Lagoon ('l'étang de Thau'), which covers 7500 hectares on the French Mediterranean coast. Three environmental factors, revealed by multiple regression models, govern the distributions observed. Allochthonous inputs from the watershed enrich the environment with nitrogen and phosphorus compounds in the winter and autumn. Internal sources are essentially localized in the shellfish breeding zone of the lagoon. ln the summer, shellfish excretions and the rapid remineralization of organic deposits produce ammonium ions. Uptake by phytoplankton has a much larger impact on the seasonal variation of inorganic nitrogen than on that of phosphorus; the latter is present in excess in the lagoon waters. Thus, nitrogen appears to be the primary limiting nutrient for the development of the chorophyllous biomass.

Introduction

Coastal lagoons cover 13% of the world coastal zone. Their ecosystems are among the most productive in the biosphere. Inorganic nutrients, the first link in the trophic chain, contribute to this fertility. In a bibliographic review of the lagoons of the world, Nixon (1982) notes wide diversity in the geomorphology, hydrology, and biology determining the development of lagoon ecosystems. Nevertheless, he emphasized common points, specifically the similarities in the annual phosphate cycle, which shows fairly low variability. On the other hand, nitrogen levels tend to exhibit much more variability among lagoons. For instance, in certain lagoons (Gilmartin & Revelante 1978; Sfriso *et al.*, 1988), nitrogen tends to control the phytoplankton biomass, as is usually the case in the oceans (Hecky & Kilham, 1988), whereas in other lagoons (Okuda, 1981; Nowicki & Nixon, 1985) phosphorus may be the limiting factor, as is usually the case in bodies of fresh water (Hecky & Kilham, 1988). Dufour *et al.* (1981) reported wide spatio-temporal variability in Ebrié Lagoon (Ivory Coast), and showed that nitrogen is the primary limiting nutrient in the estuarine region, whereas phosphorus was more limiting in the continental region.

Thau Lagoon is one of the chain of lagoons extending along the Mediterranean coast. It is a site of intensive shellfish breeding, which produces 20000 tons of shellfish each year and has a breeding stock of 35 000 tons of oysters and 3500 tons of mussels. The development of shellfish breeding and its control are the basis for our knowledge of the nutrients and phytoplankton in this lagoon (Hénard, 1978; Tournier *et al.* 1982). These studies showed an inverse relationship between the high productivity of the lagoon and the low nitrogen concentrations observed, particularly in the summer. However, only the oxidized forms (nitrites, nitrates) were taken into account.

The present study concerns the spatio-temporal variability of nutrients throughout Thau Lagoon, and examines the different environmental and biological factors apparently responsible for this heterogeneity. Simultaneous analysis of the variations of nitrogen and phosphorus versus chlorophyll *a* allowed us to determine which nutrient is apt to control the phytoplankton biomass in the lagoon.

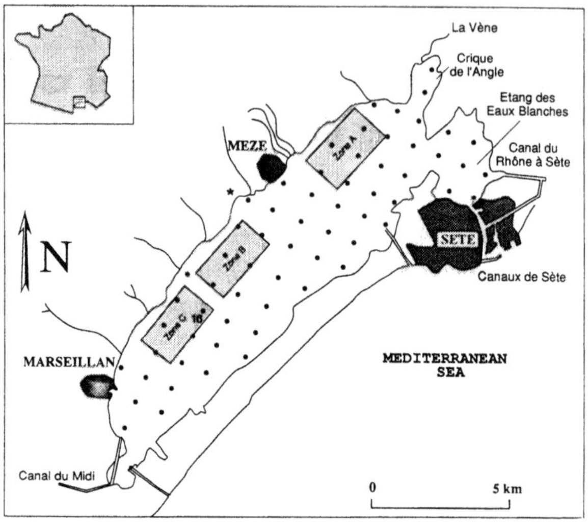

Fig. 1. Map of the sixty-three sampling stations (black dots) in the Thau Lagoon. Localisation of the three shellfish breeding zones (A,B,C) and the waste water treatment plant identified by an asterisk. Station 16 (temporal study) is situated in zone C.

Description of the site studied

The Thau Lagoon (Fig. 1) is located on the French Mediterranean coast (43° 24′ N, 3° 36′ E) with an area of 7500 hectares and an average depth of 3.5 m. The near-shore region has a substantial human population (Sète, Balaruc, Bouzigues, Mèze, Marseillan) of about 130 000 in summer, and the sewage from about one half of this population flows directly into the lagoon.

Freshwater enters from the Canal du Midi on the south-west, from the Rhône-Sète canal in the north-east and from the catchment area (30 000 hectares) north of the lagoon which is drained by a number of small streams and the Vène river. Sea water enters from the canals at Sète in the east and at Pisses-Saumes on the south-west. For a detailed description of the Thau Lagoon see Amanieu *et al.* (1989).

Materials and methods

Sixty three sampling sites one kilometre apart were established according to a systematic sampling plan (Fig. 1). Samples were taken in June and October 1986, February and May 1987 (Amanieu *et al.*, 1989). Sampling was completed in 4 to 6 hours. The samples were analysed within four hours of collection. Two liters of water were sampled 50 cm below the surface using a Niskin bottle. After the immediate fixing of a 100 ml aliquot for the determination of ammonium, the rest of each sample was kept cold (4 °C) and taken to the laboratory where determination of nutrients was carried out immediately, using the following techniques: ammonium (Solorzano, 1969), nitrates (Wood *et al.*, 1967) and phosphates (Murphy & Riley, 1962). After comparison of the reproducibility of the manual methods described by Aminot & Chaussepied (1983) and the automatic procedures described by Tréguer & Le Corre (1975), we chose (Canovas, 1987) a manual determination of ammonium and automatic determination of nitrates and phosphates using a Technicon autoanalyzer. The mapping of spatial distribution of the three nutrients was performed using a univariate kriging method. Using fragmentary sampling (here systematic sampling comprising 63 stations), this method of spatial analysis represents the distribution by estimation and interpolation of the variable at all points in the space. It is based on an experimental variogram (relation representing variance/distance between observations). A stepwise regression procedure was then used to examine the relations between the variable to be explained and the different explanatory variables of spatial distribution. The multiple regression technique requires that the data be distributed in a multi-normal manner. When the data deviated from normality, normalizing transformations were used that were simple functions such as $\sqrt{x}$ and log(x + 1). The final regression model is expressed as follows:

$$f(y) = \beta_1 f(x_1) + \beta_2 f(x_2) + \beta_3 f(x_3) + \ldots + \beta_0$$

where $f(y)$ and $f(x)$ are the normalized transformations of y and x.

Results

Simple statistical parameters such as means, standard deviations, and minimal and maximal values of distribution are used to express data on each ion, collected in the four campaigns (Table 1). This table shows the different seasonal patterns of nitrogen compounds and phosphates. The time course of ammonium ions and nitrates is characterized by minima in the spring and summer and maxima in the autumn and winter, whereas phosphates show minima in the winter and spring and maxima in the summer and autumn. A more detailed analysis of the spatial distributions obtained in each of the campaigns and for each nutrient revealed that: in June 1986, there were patchy areas characterized as low-nitrate zones (<1.5 μmol l^{-1}) throughout the shellfish breeding section (Fig. 3a), areas rich in ammonium ions (1.3 to 2.5 μmol l^{-1}) in zone A of the shellfish breeding section (Fig. 1 & Fig. 2a), and a zone with a high level of phosphates (3.2 to 10.4 μmol l^{-1}) downstream from the discharge of the ponding station at Mèze (Fig. 4a); in October 1986, a southwest-to-northeast gradient of increasing nitrate concentrations appeared; a large part of the lagoon showed high levels of ammonium ions and phosphates (10.3 to 14.4 μmol l^{-1} for NH_4^+ and 2.3 to 3.7 μmol l^{-1} for PO_4^{3-}); in February 1987, all the nutrients showed southwest-to-northeast gradients of increasing concentrations (Fig. 2b, 3b, 4b); in May 1987, no structure could be identified, and the nutrient levels (particularly nitrogen compounds) were low or very low.

The above distributions reflect the spatial heterogeneity of the lagoon, induced by biotic and

Table 1. Statistical parameters of the four periods sampling for the three nutrients (NH_4^+, NO_3^-, PO_4^{3-}) expressed in μmol l^{-1}, in the sixty-three sampling stations.

		June 86	October 86	February 87	May 87
NH_4^+	*Mean*	0.91	10.30	3.22	1.02
	Standard error	0.04	0.24	0.47	0.05
	Minimum	0.43	0.35	0.98	0.43
	Maximum	2.17	14.43	30.44	2.01
NO_3^-	*Mean*	1.97	10.45	15.22	0.16
	Standard error	0.15	0.65	1.47	0.02
	Minimum	0.22	2.69	2.32	0.00
	Maximum	7.00	21.85	69.67	1.24
PO_3^{3-}	*Mean*	1.82	2.09	0.86	0.42
	Standard error	0.22	0.06	0.04	0.02
	Minimum	0.52	1.17	0.36	0.11
	Maximum	10.39	3.69	1.76	0.90

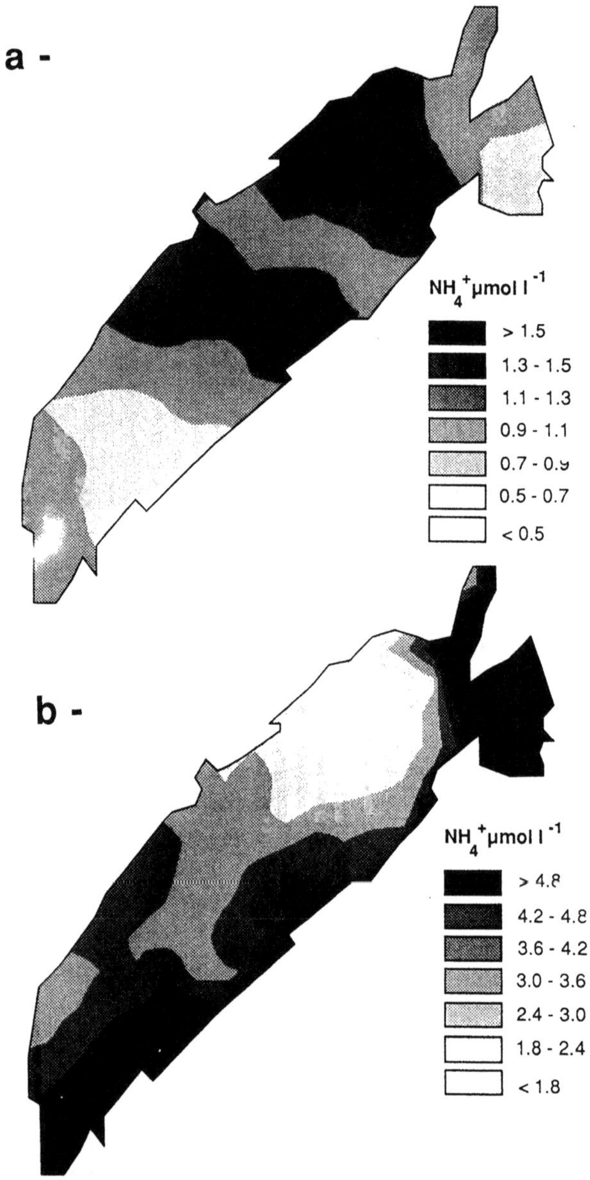

Fig. 2. Spatial distribution of NH_4^+. Interpolated map obtained by kriging. a – June 1986, b – February 1987.

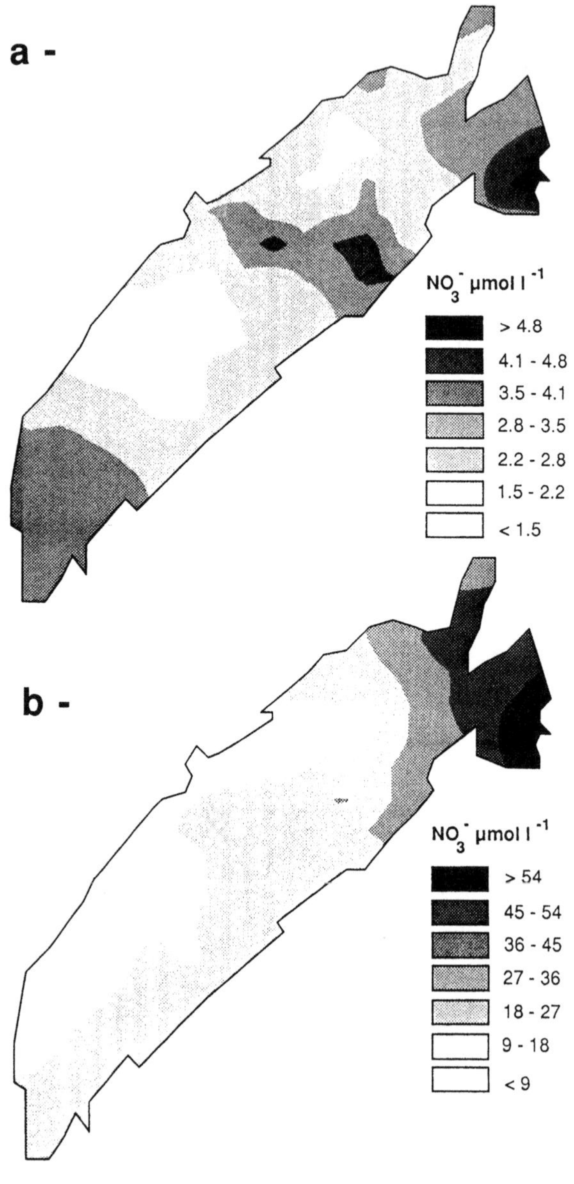

Fig. 3. Spatial distribution of NO_3^-. Interpolated map obtained by kriging. a – June 1986, b – February 1987.

abiotic factors. Multiple regression models were used to determine the factors or combinations of environmental and biological factors correlated with the large environmental disturbances and the spatial variability of nutrients. Models were constructed for each nutrient and each campaign, using the following variables: geographical distance from the shellfish breeding zone: synthetic variable representative of the shellfish biomass action (SZD: shellfish zone distance), salinity (S), temperature (TC), and phytoplankton biomass expressed indirectly by the chlorophyll *a* concentration in mg m^{-3} (Neveux & Panouse, 1987), as well as the level of ammonium ions, introduced as a variable in the model of nitrates, and the level of nitrates used in the model of ammonium ions. These models are summarized in Table 2. The regression coefficients are significant ($p < 0.01$).

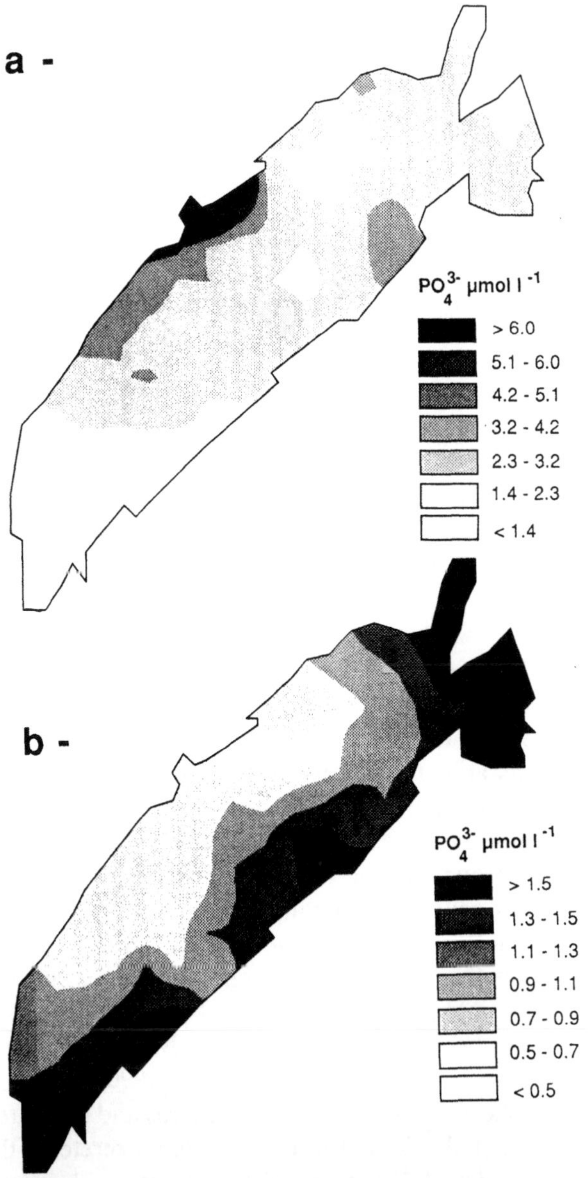

Fig. 4. Spatial distribution of PO_4^{3-}. Interpolated map obtained by kriging. a – June 1986, b-February 1987.

The results show the effect of the watershed (salinity), shellfish breeding zones (distance) and phytoplankton biomass (chlorophyll) on the spatial variability of nutrients in the lagoon.

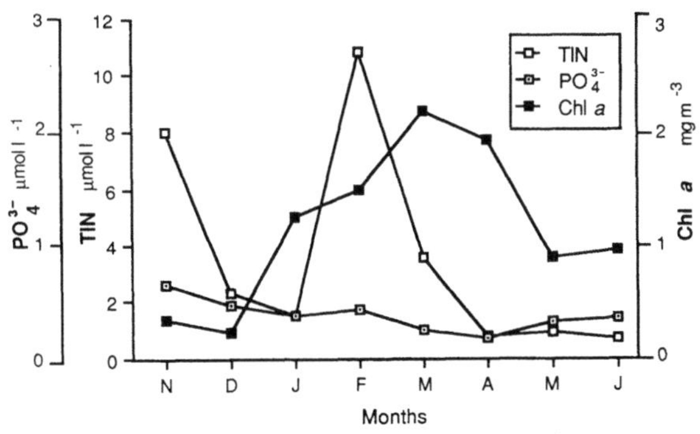

Fig. 5. Monthly variation of orthophosphate, total inorganic nitrogen and chlorophyll a at station n° 16 in a shellfish breeding zone.

Discussion

Effect of the watershed

In the autumn and winter, the Mediterranean climate is characterized by sudden showers. The salinity of the lagoon closely follows changes in the input of freshwater from land and atmosphere. In winter 1987, the zones of low salinity were located downstream from the outlet of the Vène, the main inflowing river. Under the influence of the movement of water masses, these inputs remained in the northwest quarter of the lagoon before diffusing along the western shore. The regression models show the strong influence of these inputs on the spatial distribution of each of the three nutrients. The zones of low salinity correspond to areas with high concentrations of nitrogen compounds and phosphates (Fig. 2b, 3b, 4b). External nutrient contributions in winter also determined high nutrient concentration variability in the Venice Lagoon waters (Facco *et al.*, 1986).

During the summer waste water input, rather than the drainage base, determined nutrient loading (Vaulot & Frisoni, 1986). These concentrated inputs are not large enough to decrease the salinity of the lagoon. In June 1986, the impact of discharges from the sewage ponds at Mèze can be seen on the map of the spatial distribution of phosphates (Fig. 4a). Driven by wind from the southeast, this input, which amounted to 500

Table 2. Stepwise regression model of the nutrients at differents seasons and order of the first three variables entered in the model. NH_4^+: ammonium ion, NO_3^-: Nitrates, PO_4^{3-}; Phosphate, TIN: Total Inorganic Nitrogen, SZD: Shellfish Zone Distance, S: Salinity, Chl *a*: Chlorophyll *a*, r^2: coefficient determination, p: level of significance, f(x): normalizing function.

Nutrients	Seasons	First variable	Second variable	Third variable	Regression model
NH_4^+	Summer	SZD	Chl *a*	NO_3^-	$f(NH_4^+) = -0.062f(SZD) + 0.15f(Chl\ a) - 0.016f(NO_3^-) + 0.396$ $r^2 = 0.36, p < 0.001$
	Winter	S	Chl *a*	–	$f(NH_4^+) = -0.036f(S) - 0.831\,f(Chl\ a) + 2.195$ $r^2 = 0.42, p < 0.001$
NO_3^-	Winter	S	Chl *a*	–	$f(NO_3^-) = -0.068f(S) + 0.74f(Chl\ a) + 2.743$ $r^2 = 0.76, p < 0.001$
PO_4^{3-}	Winter	S	–	–	$f(PO_4^{3-}) = -0.23f(S) + 0.966$ $r^2 = 0.74, p < 0.001$
TIN/PO_4^{3-}	Winter	Chl *a*	S	–	$f(TIN/PO_4^{3-}) = 0.38f(Chl\ a) - 0.01f(S) + 1.601$ $r^2 = 0.31, p < 0.001$

moles PO_4^{3-} per day, extended two to three kilometers along the western shore of the lagoon. In addition to this input of terrestrial origin, phosphate concentrations were elevated throughout the lagoon (mean concentration 1.8 μmol l^{-1}). Using path analysis, Ge (1988) explained 85% of the temporal variations of phosphate in the lagoon, measured from 1975 to 1982, and demonstrated the dominant role of temperature in the summer. Temperature increases favour anaerobic conditions in the sediments, which accelerate recycling (Golterman 1975, Seiki *et al.* 1989).

Effect of shellfish breeding zones

Shellfish breeding zones are special zones, with a dense biomass, the movement of water masses being slower here, and the levels of organic deposits in the sediments being high. In the summer, the shellfish area has a strong effect on the spatial distribution of the two nitrogen compounds. With regard to ammonium ions, the regression model indicated the importance of the factor of geographical distance between the sampling sites and the shellfish breeding zone (Table 2). High concentrations of ammonium ions were measured in zone A of the shellfish breeding area (Fig. 2a). This zone represents 50% of the total shellfish production of the lagoon (Hamon & Tournier 1986). The high concentrations of ammonium ions arise from different sources:

- direct regeneration by shellfish excretions; the nitrogen excretion is relatively constant throughout the year and consists essentially of dissolved organic nitrogen (urea, amino acids, etc.). Excretion by mollusks is extremely active in the summer, and ammonium ions are the dominant form excreted (Vincendeau & Robert, 1987; Boucher & Boucher-Rodoni, 1988). In Thau Lagoon, Outin (1990) measured the excretion of *Crassostrea gigas* and estimated that the 32000 tons of oysters excrete in the order of 10^7 moles of ammonium each year.
- indirect regeneration by degradation of faeces and pseudofaeces; while they remain in the water column, their composition can be considerably modified by solubilization of digestion products or unabsorbed products (Deslous-Paoli & Jousset, 1987). Once they have accumulated in the sediments, their organic nitrogen is mineralised by bacteria, to ammonium that is released into the water column. During this period, Grentz (1989) using benthic chambers in the shellfish breeding zone of Thau Lagoon showed that the flux of ammonium

ions is three to four times greater in this zone than outside it.

The low values of nitrates made it impossible to demonstrate significant spatial nitrate variations that could be used to construct a model. The study of spatial distribution in June 1986 (Fig. 3a) showed a nitrate deficiency throughout the shellfish breeding zone. In this period of the year, the depletion of nitrates in the water column does not appear to be compensated by the high concentrations in interstitial water. The diffusion of sediment to adjacent water may be masked by the immediate uptake of nitrates by benthic microflora, microphytes, and macrophytes (Feuillet-Girard *et al.*, 1988). The production of ammonium ions could also be a source of nitrates via nitrification. However, high temperature and the presence of high concentrations of organic matter produce anoxia in the sediments of the shellfish breeding zone, inhibiting chemical conversion of NH_4^+ to NO_3^-.

In the summer, the nitrate reserves are extremely depleted. Ammonium ions can represent as much as 40% of the total nitrogen, and at this time they constitute practically the only form of mineral nitrogen. These results are in agreement with those of Venice Lagoon (Sfriso *et al.* 1988). Nixon and Lee (1981) indicated that the sediments were a major source of ammonia in summer at Charlestown Green Hill Pond Lagoon complex on the south shore of Rhode Island. In the Thau Lagoon, this source is enhanced by the high production of faeces and pseudofaeces in shellfish breeding zones.

Effect of the phytoplankton biomass

Phytoplankton assimilation played a predominant role in the spatial distribution of nitrogen compounds. In February 1987, the chlorophyll *a* variable was in second position in the regression models for ammonium ions and nitrates. During this month, a monospecific phytoplankton bloom occurred. As the concentration of nitrates increased in the waters of the lagoon, the level of chlorophyll *a* increased: the regression coefficient is positive. In the same period, the association of a decrease in ammonium ion levels with an increase in chlorophyll *a* concentrations (negative regression coefficient) showed that this nitrogen compound was either preferentially taken up by phytoplankton or nitrified. In June 1986, the chlorophyll *a* variable was in second position only in the model of the ammonium ion. Since the nitrate reserves were practically inexistent, the only available nitrogen source was ammonium ions. The ratio of total inorganic nitrogen (ammonium + nitrate) to phosphate was very low (mean ratio 1.6), and the development of phytoplankton was related to the quantity of ammonium present in the environment.

The phytoplankton biomass did not appear to affect the distribution of phosphate. The phosphate levels in the lagoon did not fall below 0.1 μmol l^{-1}. The maxima generally measured in Mediterranean surface waters are in the order of 0.05 μmol l^{-1} (Hénard, 1978). Thus, Thau Lagoon contains reserves of mineral phosphorus.

A study of the ratio of total inorganic nitrogen to phosphate yielded a regression line that was significant for the campaign of February 1987, the principal factor corresponding to chlorophyll *a*. Due to the large allochthonous inputs of nutrients, the ratio showed wide variation, ranging from 7.9 to 33.6, which was similar to the distribution of chlorophyll *a* levels (1.3 to 7.3 mg m^{-3}). High ratios corresponded to high chlorophyll *a* levels. Since phosphate did not appear to be correlated with the phytoplankton biomass, total nitrogen appeared to be responsible for the distributions of phytoplankton biomass observed. This effect of nitrogen compounds on primary production appeared during February 1987, as well as in monthly variations at the station 16 (Fig. 1) where all the above-mentioned variables were measured.

The monthly variations of phosphates and chlorophyll *a* (Fig. 4) show a certain stability of phosphate which varied from 0.2 to 0.8 μmol l^{-1} and did not appear to be affected by the variations of chlorophyll *a*. On the other hand, the monthly variation of total nitrogen and chlorophyll *a* (Fig. 4) showed complete interdependence. The high levels of total nitrogen in November 1986 and February 1987 were followed by a large increase in phytoplankton biomass. This development of primary production eventually resulted in a sudden depletion of nitrogen concentrations.

Spatio-temporal study showed the predominant effect of nitrogen compounds on the chlorophyll *a* concentration. The high phosphate levels throughout the lagoon and during all months preclude determination of the effect of phytoplankton on this element. In contrast, when present in insufficient quantities, total nitrogen showed the effects of large uptake by primary production. Consequently, in Thau Lagoon, nitrogen is more apt to control the chlorophyll *a* concentration than phosphorus. Nitrogen appeared to be the limiting factor for algal growth during spring-summer in the Lagoon of Venice (Sfriso *et al.*, 1988) and in the Barrier Island lagoons of the Golf of California (Gilmartin & Revelante, 1978). Phosphate was often the primary limiting nutrient in other lagoons: north-eastern Golf of Mexico coastal estuaries (Myers & Iverson, 1981), lagoons of the Ebro Delta (Comin, 1982), coastal lagoons in Venezuela (Okuda, 1981) and in south-western Australia (Hodgkin & Birch, 1982); in these lagoons phosphate and nitrate decreased in summer when algal growth increased. Because of the great diversity of lagoon characteristics determining the development of lagoon ecosystems (i.e. shape and size, temperature and salinity, fresh water input, exchange with the sea, land use and human activities), account must be taken of all these data before the study of a specific ecosystem can be properly conducted.

Conclusion

In the winter, distributions of nutrients are governed by large inputs of freshwater from precipitation and output from the watershed. These inputs allow the formation of nitrogen reserves in the water column and in the sediments by absorption. In spring, phytoplankton blooms result in a sudden drop in nitrogen levels, including both ammonium, which is preferentially assimilated, and nitrates with concentrations of the former importantly affected by the large populations of molluscs in the shellfish breeding zone.

The variation of phosphate is governed by external inputs from precipitation in winter, as well as from domestic effluents in the summer, and by internal sources (essentially recycling from sediments) in the warm period, i.e. summer and autumn. Phosphate accumulates in the sediments of the shellfish breeding areas in particular (Sornin *et al.*, 1986). Small seasonal changes in total phosphate and large changes in total inorganic nitrogen, correlated with algae biomass, point to nitrogen as the element controlling primary production.

Despite seasonal fluctuations, the phosphate reserves remain relatively stable, making nitrogen the limiting factor in the development of the phytoplankton biomass. Oysters are bred throughout the year in the same area of Thau Lagoon, producing a certain accumulation of nitrogen in the sediments. The productivity of the lagoon is thus related to this accumulated sedimentation, which provides an ammonium source for a limiting period.

Acknowledgements

The authors thank Vincent Jarry for his contribution to the use of the Unimap program.

This study is the contribution n° 14 of the ECOTHAU program. This program is financed by PIREN, IFREMER and the Ministry of the Environment.

References

Amanieu, M., P. Legendre, M. Trousselier & G. F. Frisoni, 1989. Le programme Ecothau: Théorie écologique et base de la modélisation. Oceanol. Acta 12: 189–199.

Aminot, A. & M. Chaussepied, 1983. Manuel des analyses chimiques en milieu marin. Centre National pour l'Exploitation des Océans, Brest, 395.

Boucher, G. & R. Boucher-Rodoni, 1988. *In situ* measurement of respiratory metabolism and nitrogen fluxes at the interface of oyster beds. Mar. Ecol. Prog. Ser. 44: 229–238.

Canovas, S., 1987. Impact des effluents de la station de lagunage de Mèze sur l'étang de Thau. DEA National d'Hydrologie. Fac. de Pharmacie, Univ. Montpellier I, 144 p.

Comin, F. A., 1982. Seasonal changes of phytoplankton in three coastal lagoons of the Ebro delta in relation to environmental factors. Proc. int. Symp. on coastal lagoons SCOR/IABO/UNESCO. Bordeaux, France. Oceanol. Acta, Sp: 259–267.

Deslous-Paoli, J.M. & J. Jousset, 1987. Transformation du matériel fécal de *Crassostrea gigas*: évolution biochimique à court terme. Océanis 13: 581–584.

Dufour, P., L. Lemasson & J.L. Cremoux, 1981. Contrôle nutritif de la biomasse du seston dans une lagune tropicale de Côte d'Ivoire. Il Variations géographiques et saisonnières. J. exp. mar. Biol. Ecol. 51: 269–284.

Facco, S., D. Degobbis, A. Sfriso & A. A. Orio, 1986. Space and time variability of nutrients in the Venice Lagoon. In D. A. Wolfe (ed), Estuarine Variability. Academic Press: 307–317.

Feuillet-Girard, M., M. Héral, J. M. Sornin, J. M. Deslous-Paoli, J. M. Robert, F. Mornet & D. Razet, 1988. Eléments azotés de la colonne d'eau et de l'interface eau-sédiment du bassin de Marennes-Oléron: influence des cultures d'huitres. Aquat. Living Resour. 1: 251–265.

Ge, C., 1988. Contribution à la modélisation physico-chimique de l'étang de Thau. Thèse. Univ. Montpellier II, 163 p.

Gilmartin, M. & N. Revelante, 1978. The phytoplankton characteristics of the Barrier Island lagoons of the Gulf of California. Estuar. coast. mar. Sci. 7: 29–47.

Golterman, H. L., 1975. Physiological Limnology. An approach to the physiology of lake ecosystems. Developments in Water Science 2, Elsevier Scientific Publishing Company. 488 p.

Grentz C., 1989. Quantification et destinée de la biodéposition en zones de production conchylicole intensive en Méditerranée. Thèse. Univ. d'Aix-Marseille II, 145 p.

Hamon, P. Y. & H. Tournier, 1986. Evolution de la biomasse de mollusques en élevage dans l'étang de Thau de 1980 à 1984. Rev. Trav. Inst. Pêches Marit. 48: 33–44.

Hecky, R. E. & P. Kilham, 1988. Nutrient limitation of phytoplankton in freshwater and marine environments: A review of recent evidence on the effects of enrichment. Limnol. Oceanogr. 33: 796–822.

Hénard, D., 1978. Production primaire d'une lagune Méditerranéenne: l'étang de Thau (Hérault). Année 1976. Thèse Doct. Ing., Univ. Montpellier II, 85 p.

Hodghin, E. P. & P. B. Birch, 1982. Eutrophisation of a western Australian estuary. Proc. int. Symp. on coastal lagoons. SCOR/IABO/UNESCO. Bordeaux, France. Oceanol. Acta SP: 313–318.

Murphy, J. & J. P. Riley, 1962. A modified single solution method for the determination of phosphate in natural water. Anal. Chim. Acta 27: 31–36.

Myers, V. B. & R. I. Iverson, 1981. Phosphorus and nitrogen limited phytoplankton productivity in north-eastern Gulf of Mexico coastal estuaries. In B. J. Neilson & L. E. Cronin (ed), Estuaries and Nutrients. Humana Press, Clifton, New Jersey: 569–582.

Neveux, J. & M. Panouse, 1987. Spectrofluometric determination of chorophylls and pheophytins. Arch. Hydrobiol. 109: 567–581.

Nixon, S. W.,1982. Nutrients dynamics, primary production and fisheries yields of lagoons. Proc. int. Symp. on coastal lagoons. SCOR/IABO/UNESCO. Bordeaux. France. Oceanol. Acta 357–371.

Nixon S. W. & V. Lee, 1981. The flux of carbon, nitrogen and phosphorus between coastal lagoons and offshore waters. UNESCO. Tech. Pap. mar. Sci. 33: 325–348.

Nowicki, B. L. & S. W. Nixon, 1985. Benthic nutrient remineralization in a coastal lagoon ecosystem. Estuaries 8: 182–190.

Okuda, T., 1981. Inorganic and organic nitrogen in some coastal lagoons in Venezuela. UNESCO. Tech. Pap. mar. Sci. 33: 348–358.

Outin, V.,1990. Ecophysiologie de l'huitre *Crassostrea Gigas* Thunberg en milieu naturel. Rôle des populations d'une lagune méditerranéenne dans les transferts de matières particulaire et dissoute. Thèse. Univ. Paris VI, 152 p.

Seiki, T., H. Izawa & E. Date, 1989. Benthic nutrient remineralization and oxygen consumption in the coastal area of Hiroshima Bay. Wat. Res. 23: 219–228.

Sfriso, A., B. Pavoni, A. Marcomini & A. A. Orio, 1988. Annual variations of nutrients in the lagoon of Venice. Mar. Pollut. Bull. 19: 54–60.

Solorzano, L., 1969. Determination of ammonia in natural waters by the phenolhypochlorite method. Limnol. Oceanogr. 14: 799–801.

Sornin, J. M., M. Feuillet, M. Héral & J. C. Fardeau, 1986. Influence des cultures d'huitres Crassostrea Gigas sur le cycle du phosphore en zone intertidale: rôle de la biodéposition. Oceanol. Acta 9: 313–322.

Tréguer, P. & P. Le Corre, 1975. Manuel d'analyse des sels nutritifs dans l'eau de mer (utilisation de l'autoanalyseur II technicon®). 2nd édit. Lab. Océanol. Chim., Univ. Brest, France, 110 p.

Tournier, H., P. Y. Hamon & S. Landrein, 1982. Synthèse des observations réalisées par l'ISTPM sur les eaux et le plancton de l'étang de Thau de 1974 à 1980. Rev. Trav. Inst. Pêches Marit. 45: 283–318.

Vaulot, D. & G. F. Frisoni, 1986. Phytoplanktonic productivity and nutrients in five Mediterranean lagoons. Oceanol. Acta 9: 57–63.

Vincendeau M. L. & J. M. Robert, 1987. Variations qualitatives saisonnières des produits excrétés par deux mollusques bivalves *Crassostrea gigas* Thunberg et *Ruditapes philippinarum* (Adams et Reeve). Oceanis 13: 543–556.

Wood, E. D., F. A. Armstrong & F. A. Richards, 1967. Determination of nitrate in sea water by cadmium-copper reduction to nitrite. J. mar. Biol. Ass. U.K. 47: 23–31.

Received 7 August 1989; in revised form 1 March 1990; accepted 25 June 1990

Hydrobiologia **207**: 115–122, 1990.
D. J. Bonin & H. L. Golterman (eds), Fluxes Between Trophic Levels and Through the Water–Sediment Interface.

Effects of oxygen on denitrification in marine sediments

Patricia Bonin & Nathalie Raymond
Centre d'Océanologie de Marseille, Faculté des Sciences de Luminy case 901, 13288 Marseille, Cedex 9, France

Key words: oxygen, denitrification, nitrate reduction, marine sediment

Abstract

Experiments were carried out to determine the effect of oxygen on denitrification, simultaneously in sediment and in a bacterial strain isolated from this sediment. Descriptions of the apparatus used are given. In sediment, denitrification was found to occur when the oxygen concentration increased to 5 mg l^{-1}, but the activity was only 10% of that obtained under anaerobic conditions. The two first steps of denitrification measured in the sediment were less sensitive to oxygen. When the sediment was incubated in presence of 1,1 mg l^{-1} O_2, the denitrification rate was never fully restored by returning it to anaerobic conditions.

In cell suspensions of the denitrifying bacteria: *Ps. nautica* isolated from the sediment, denitrification was equally strongly influenced by oxygen. The different steps of denitrification, nitrate, nitrite and nitrous oxide reduction, were differently affected with respect to oxygen concentration. The nitrate reduction was less sensitive than nitrite and nitrous oxide reduction.

Introduction

Denitrification in marine and fresh water environment and sediments has been extensively studied (Payne, 1976; Knowles, 1982). Generally, denitrification takes place in aerobic-anaerobic interfaces, and some workers consider this process as a strictly anaerobic metabolism pathway (Payne, 1976). However, there has been a long controversy in literature concerning the lowest oxygen concentration that permits denitrification (See references herein and Robertson & Kuenen, 1984a). Recent reports have shown the occurrence of aerobic denitrification (Meiberg *et al.*, 1980; Hochstein *et al.*, 1984; Robertson & Kuenen, 1984b; Lloyd *et al.*, 1987; Christensen & Tiedje, 1988; Bonin *et al.*, 1989) and others have described how denitrification is blocked in the presence of low oxygen concentration, which inhibits and/or prevents the synthesis of the enzymes associated with denitrification (Payne, 1976; Kapralck *et al.*, 1982), or represses the active transport of nitrate into the cells (Alefounder *et al.*, 1981).

There has been frequent speculation that anoxic microsites may possibly support denitrification in oxic environment (Jannasch, 1960).

In this study, we have attempted to determine the effect of dissolved oxygen concentration on the two first steps of denitrification, nitrate and nitrite reduction and total denitrification (activity of the whole pathway), simultaneously in an experimental water-sediment open flow system and in a bacterial strain isolated from this sediment.

Description of site and bacterial strain studied

Sediment and sea water used in these experiments were sampled at Lavera, Gulf of Fos, French

Mediterranean coast. Undisturbed sediment cores were taken by hand in plexiglass tubes (inner diameter, 4 cm) at 5 m water depth. The sediment is sandy and highly porous.

The bacterium used was isolated from marine coastal sediment (Lavera) in our laboratory and identified as a denitrifier, *Pseudomonas nautica* (Bonin *et al.*, 1987a).

Materials and methods

Oxygen sensitivity of denitrifying activity in marine sediment

Experimental system with sediment:
The experimental system used to determine the effect of oxygen concentration on nitrate, nitrite, oxygen consumption rate and total denitrifying activities in sediment is illustrated in Fig. 1.

Sea-water supplemented with nitrate (about 200 μM) was placed in the reservoir (2). The reactor (4) containing 1 cm of sediment taken in the top two centimeters of the sediment core was supplied with sea-water by peristaltic pump (3) (Technicon), which was also used to control the flow (40 ml h^{-1}) and to maintain the dilution rate at 2.75 l. Dissolved oxygen concentrations in the influent were maintained constant by sparging the reservoir (2) with a gas mixture obtained with the aid of a proportional gas mixer (1) connected to cylinders of oxygen and nitrogen. The dissolved oxygen concentration was measured with a polarographic-type electrode (Orbisphere) continuously in the effluent (5) and in the reser-

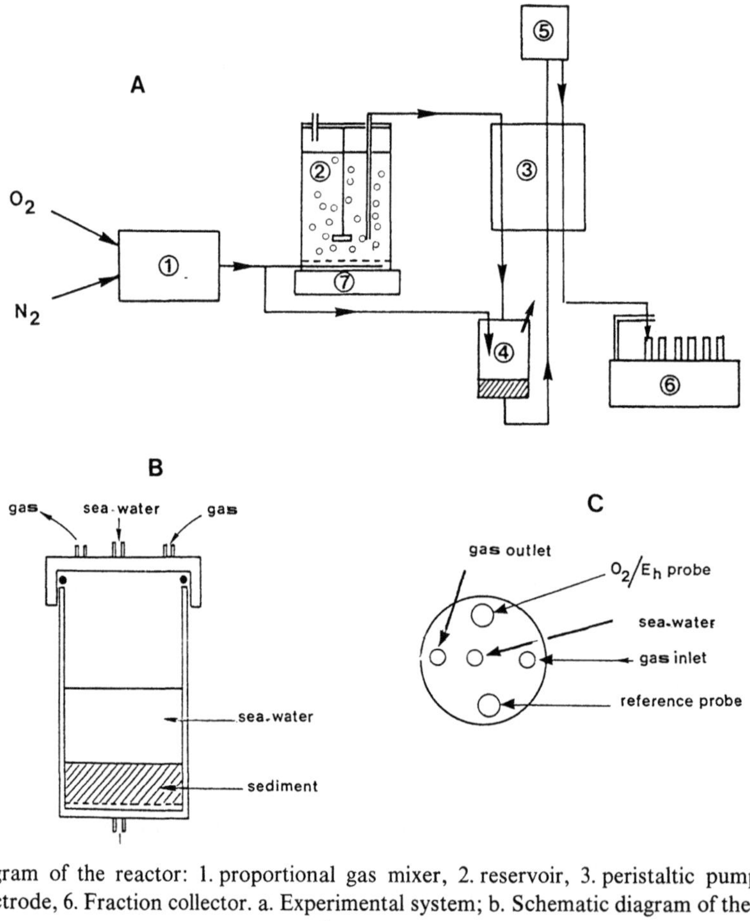

Fig. 1. Schematic diagram of the reactor: 1. proportional gas mixer, 2. reservoir, 3. peristaltic pump, 4. reactor, 5. oxygen polarographic type electrode, 6. Fraction collector. a. Experimental system; b. Schematic diagram of the reactor; c. Top view of the reactor.

voir (2). The outflow of the reactor was recovered in the test tubes of a fraction collector (6). The headspace of the reactor was purged continuously with nitrogen or gas mixture.

Measurement of microbial activity in sediment:

The rates of the two first steps of denitrification, nitrate and nitrite reductions, were estimated at steady state by measuring the nitrate and nitrite concentration in the outflow.

The measurement of denitrification was based on the acetylene inhibition technique (Sorensen, 1978).

4 cm^3 of sediment taken from the reactor at steady state were transferred to a rubber sealed 13 ml tube containing 4 ml of sea water supplemented with KNO_3 (200 μM). The headspace was gassed with N_2 or the gas mixture (O_2/N_2) used during the incubation period in the reactor and supplemented with 2 ml acetylene to inhibit the last step of denitrification. Vials were incubated for 30 min at 20 °C. After incubation, the vials were strongly agitated and nitrous oxide concentration was measured in the headspace and in the sea-water. The rate of nitrous oxide accumulation is used as a measure of total denitrification.

Oxygen sensitivity of denitrifying activity on bacterial strain

Cells of *Ps. nautica* were grown anaerobically in a serum flask containing Y.L.N. medium (Yeast extract 0.1% – Lactate 8.9 mM – Nitrate 10 mM) as described previously (Bonin *et al.*, 1987a; 1987b). At the beginning of the stationary phase of growth, the cells were harvested by centrifugation at 5000 × ***g*** for 15 min, washed twice with artificial sea-water (ASW) (Baumann & Baumann, 1981) without NH_4^+. Pellets were resuspended (3 mg of cells wet weight ml^{-1}) in ASW supplemented with chloramphenicol (100 $\mu g\ ml^{-1}$) (to prevent protein synthesis), electron donor (Na-lactate 133.5 mM) and electron acceptor (K NO_3 10 mM, or K NO_2 2 mM, or N_2O) in a 2 L reactor containing 600 ml of suspension.

The reactor was operated at 27 °C ± 1 °C and the medium was stirred at 700 rpm. The dissolved oxygen concentration was maintained constant during the experiment. A proportional gas mixer (Air Liquide CRDR) allowed adjustment of oxygen concentration in the inlet gas flow. The gas flow was 0.4 l min^{-1}. An oxygen polarographic electrode was used to monitor dissolved oxygen concentration. The rates of consumption of nitrate, nitrite and nitrous oxide in resting cell suspension were established at a particular dissolved oxygen concentration steady state (Bonin *et al.*, 1989).

Analytical assays

Nitrite and ammonium concentrations were determined colorimetrically by the method of Bendschneider & Robinson (1952) and Slawyk & Mac Issac (1972) respectively. Assays of nitrate were performed after reduction by Cu-Cd column according to Tréguer and Le Corre (1975).

Nitrous oxide was determined by a gas chromatograph Girdel series 30 equipped with an electron capture detector as previously described (Bonin *et al.*, 1987a).

Protein concentration was determined by the procedure of Lowry *et al.* (1951) using bovine serum albumin as standard, after heating (10 min at 90 °C) the cells with NaOH (5%).

Results

Oxygen sensitivity of denitrifying activity in marine sediment

The effect of dissolved oxygen concentration on denitrifying activity was studied using the open flow system previously described (Fig. 1).

The variations of nitrate, nitrite and ammonium concentrations were determined during experiments realized at different dissolved oxygen concentrations in the inflow sea-water: 0%, 3%, 5%, 7%, 10% and 21% (V/V, O_2/N_2 in 100% gas mixture). Figure 2 shows an example of results obtained under anaerobic conditions (0% O_2)

(Fig. 2a), or in the presence of oxygen (10% O_2) in the inflow sea water (Fig. 2b).

The experiment was continued for 24 hours. After 400 ml sea water percolation (for 10 hours) steady levels of dissolved oxygen, nitrate, ammonium in the effluent were achieved, and these levels remained constant throughout the experiment.

In steady state, under anaerobic conditions (Fig. 2a), the NO_3^- concentration in the effluent was lower than in the presence of oxygen (10%). But in this latter condition the nitrate concentration in the outflow was lower than in the inflow. Thus, the rate of nitrate reduction declined but was not completely inhibited in presence of 10% of oxygen in the gas mixture (Fig. 2b).

On the other hand, it is worth noting that the ammonium concentration in the effluent increased in presence of oxygen (10%).

At the end of each experiment we established the redox profiles in the sediment column, and the oxygen concentration in the influent and effluent of the reactor (Table 1). The redox values at the top and at the bottom of sediment were not strik-

Table 1. Oxygen concentration and redox potential values measured at the end of each experiment versus O_2 percentage in flow gas.

% O_2 v/v	O_2 in (a) $mg \cdot l^{-1}$	O_2 out (b) $mg \cdot l^{-1}$	Eh (c) mv	Eh (d) mv
0	0	0	86	26
3	0.73	$0.7\,10^{-3}$	90	32
5	1.1	0.6	131	89
7	1.4	1	135	106
10	2.2	0.54	180	135
21	5	0.3	180	180

a: O_2 concentration in the inflow.
b: O_2 concentration in the outflow.
c: Eh measured at the top of the sediment column in the reactor.
d: Eh measured at the bottom of the sediment column in the reactor.

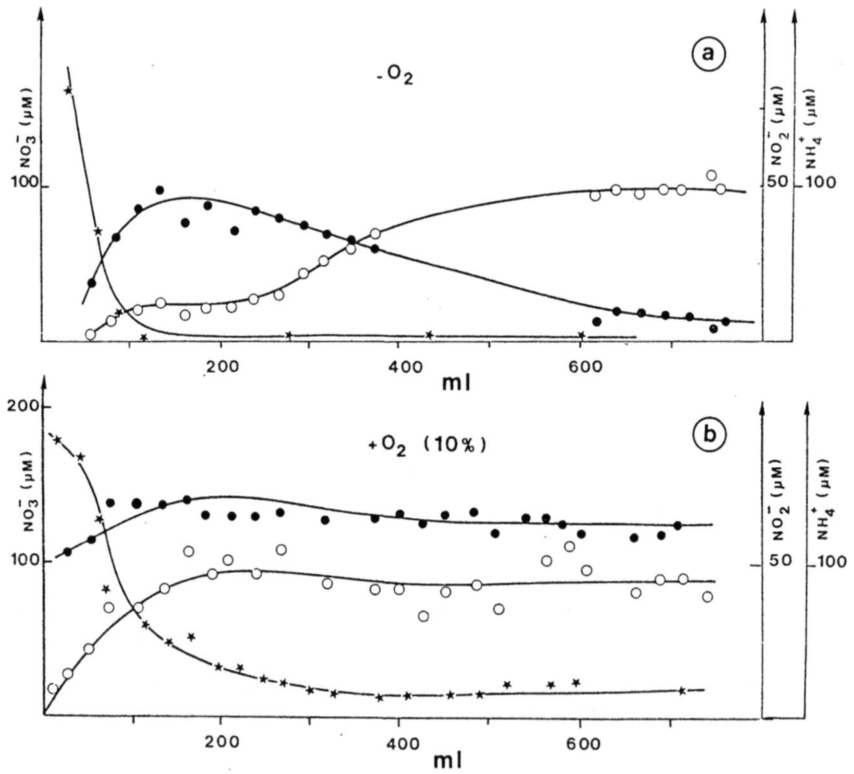

Fig. 2. Nitrate (●), nitrite (○), and ammonium (★) concentration measured under anaerobic conditions (a) or in presence of oxygen (10%) (b). The x axis is the volume (ml) of liquid percolated through the reactor.

ingly different. Furthermore, when the percentage of oxygen in the gas mixture was higher than 5% (O_2 in = 1.1 mg l^{-1}) the oxygen concentration was not negligible in the outflow. It seems that the bacteria in the sediment did not consume the whole quantity of oxygen supplied in the inflow.

The rate of the first step of denitrification, nitrate reduction, can be calculated at steady state by the following equation (Esteves *et al.*, 1986):

$$R\,NO_3^- = \frac{Q}{V}(C_i - C_e)$$

where Q is the flow rate (ml h^{-1}), V is the volume of sediment (ml) and C_i and C_e the nitrate concentration in the influent and effluent, respectively.

The nitrite concentration measured in the outflow is the result of the rates of production and consumption of nitrite in the sediment. The rate of nitrite reduction (RNO_2), the second step of denitrification, can be determined from the difference between the rate of nitrate consumption and the net rate of nitrite production (PNO_2):

$$RNO_2 = R\,NO_3 - P\,NO_2 \text{ (Esteves, 1984)}$$

The results concerning the influence of different oxygen concentrations in the effluent (0 to 5 mg l^{-1}) on the rates of nitrate and nitrite reduction are given in Fig. 3.

Nitrate and nitrite reduction rates were seriously affected when the dissolved oxygen concentration increased in the effluent. When the dissolved oxygen concentration was increased from 0 to 5 mg l^{-1} in the influent, the rates of nitrate and nitrite reduction decreased from 0.33 to 0.072 μmol NO_3^- reduced $cm^3\,h^{-1}$ and from 0.084 to 0.0041 μmol NO_2^- reduced $cm^3\,h^{-1}$ respectively. Such were our results for first two steps of denitrification.

At the end of each experiment, for different oxygen concentrations in the influent, the sediment was collected and the total denitrifying activity was determined by measuring the amount of nitrous oxide accumulated in sealed serum tubes in the presence of acetylene. Results related to the effect of oxygen concentration on denitrification are shown in Fig. 4. Activity was measured in anaerobic conditions (dashed bars) or in the presence of oxygen at the same concentration as that used during the incubation period in the reactor (blacked bars). Denitrifying activity is expressed in comparison to that estimated under completely anaerobic conditions. In the presence of 0.73 mg l^{-1} O_2, denitrifying activity was 63% of the reference level. When the oxygen concentration was further increased, up to 1 mg l^{-1} nitrous oxide production fell below 15%.

The rate of denitrification (activity of the whole

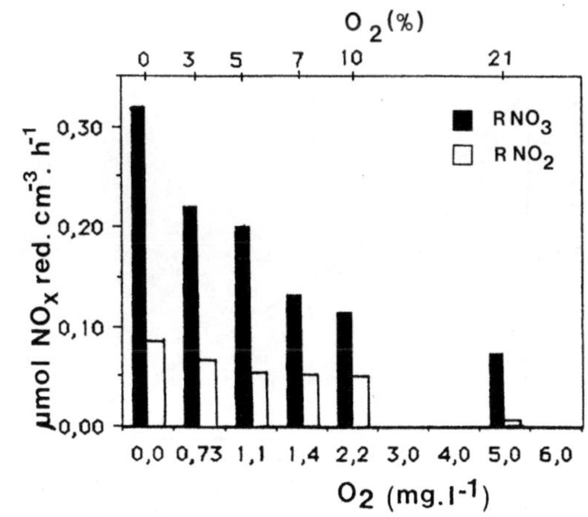

Fig. 3. The effect of oxygen concentration on the rates of nitrate and nitrite reduction in marine sediment. Activity is expressed in μmol NO_3^- or NO_2^- reduced, $cm^3\,h^{-1}$.

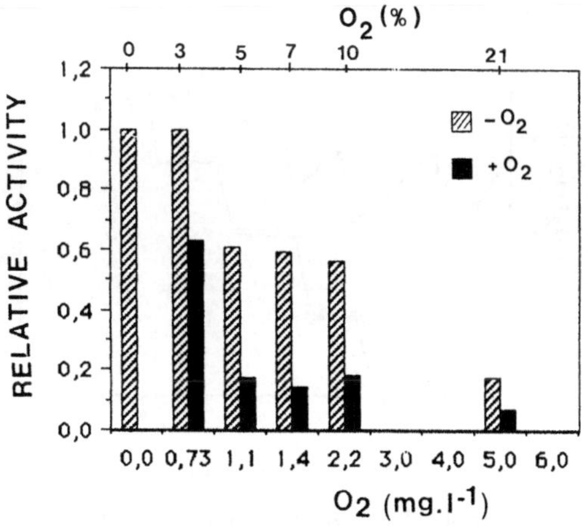

Fig. 4. The effect of dissolved oxygen concentration on the rate of denitrification in marine sediment. Activity was measured in presence of oxygen (■) or after return to anaerobic conditions (▨).

pathway) was more sensitive to oxygen than nitrate and nitrite reduction.

If the sediment was percolated in the reactor with sea-water containing 0.73 mg l^{-1} O_2, denitrifying activity was fully restored by returning to anaerobic conditions. Denitrifying activity was not fully recovered after the sediment was initially exposed to oxygen concentrations above 1 mg l^{-1} (Fig. 4).

Effect of oxygen on denitrification activity in resting cell suspension of Ps. nautica

Series of different levels of oxygen concentration were established in a 2 L reactor containing resting cell suspension of *Ps. nautica*. The kinetics of nitrate, nitrite or nitrous oxide reduction were determined for each dissolved oxygen concentration.

Figure 5 shows the results obtained using NO_3^- as electron acceptor. The nitrate was added before the establishment of each new oxygen concentration level to maintain an initial nitrate concentration of 10–11 mM. Increasing the levels of dissolved oxygen concentration progressively inhibited the rate of nitrate reduction.

In order to determine the effect of oxygen on nitrite and nitrous oxide reduction activities, similar experiments were carried out replacing NO_3^- by NO_2^- (2 mM) or N_2O (in the inflow stream).

In Fig. 6, the influence of oxygen on the relative rates of nitrate, nitrite and nitrous oxide reduction is shown. Activity is also given in comparison to that measured under anaerobic conditions (0.137 μmol mg^{-1} min^{-1}). The following results were obvious. First, denitrification takes place in the presence of oxygen. Second, enzymes associated with denitrification are differentially affected with respect to oxygen concentration. Nitrate reductase is the least sensitive and nitrous oxide reductase the most sensitive activity. Indeed, the nitrate reductase enzyme was completely inhibited at an oxygen concentration greater than 4.05 mg l^{-1}, compared with 2.15 and 0.25 for nitrite and nitrous oxide reductases respectively.

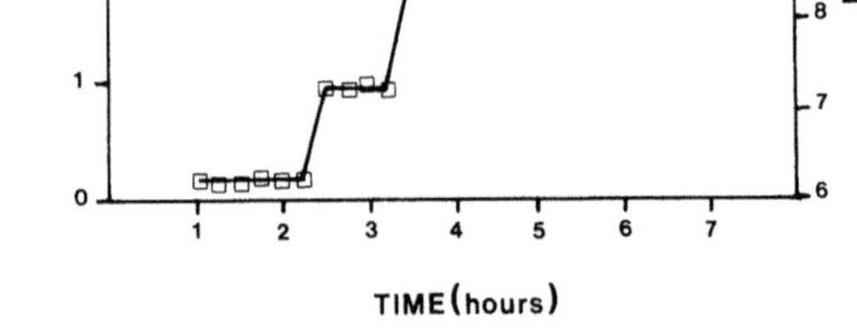

Fig. 5. The effect of dissolved oxygen concentration on the kinetics of nitrate consumption in resting cell suspension. Arrows indicate the shift of oxygen percentage and the increase of nitrate. (□) dissolved oxygen concentration (mg l^{-1}); (●) NO_3^- (mM).

Discussion – Conclusion

The partial pressure of oxygen is a fundamental parameter for determining whether conditions are favourable for denitrification.

The innovative feature of our experimental approach is the use of continuous open flow systems. This technique makes it possible to maintain aerobiosis conditions in the reactor containing sediment as well as well suspension of *Ps. nautica*.

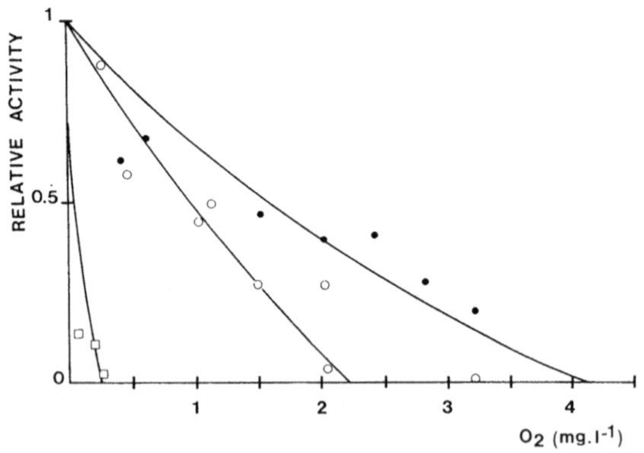

Fig. 6. The relative rate of nitrate (●), nitrite (○), and nitrous oxide (□) reduction versus dissolved oxygen concentration in resting cell suspension.

The results obtained in sediment, using the experimental apparatus previously described show that denitrification can take place in the presence of oxygen in the influent. The total denitrification rate in the sediment, assessed by N_2O production, decreased steeply when the oxygen concentration in the pore water increased from 0 to 1.1 mg l^{-1}. In the presence of 1.1 mg l^{-1} O_2, the denitrifying activity, was 15% of that measured under anaerobic conditions. When the oxygen concentration was further increased, up to 5 mg l^{-1}, denitrification declined slowly but was not completely blocked.

While denitrification is generally viewed as complete reduction of nitrate to molecular nitrogen:

$$NO_3^- \rightarrow NO_2^- \rightarrow NO \rightarrow N_2O \rightarrow N_2$$

part of the sequence can be abandoned by bacteria in response to oxygen.

The kinetics of the consumption of nitrate and nitrite, corresponding to the rates of the two first steps of denitrification, were measured at different oxygen concentrations. The results show that these steps were less sensitive to oxygen than nitrous oxide reduction.

The quantity of NH_4^+ in the effluent increased when the oxygen concentration increased. The appearence of ammonium in the effluent might be due to dissimilatory nitrate reduction pathway to ammonium in the presence of oxygen, rather than an increase in assimilatory nitrate reduction, since this route is generally blocked in the presence of ammonium. The dissimilatory nitrate reduction, called nitrate ammonification by Sorensen (1987), has been described as very important in marine sediments, accounting for as much as 52% of the total nitrate dissimilation. Furthermore, the appearence of ammonium can be related to an increase of mineralisation for higher oxygen level.

From results shown in Table 1, we may conclude that under these experimental conditions, the bottom of the sediment column in the reactor was never completely anoxic. It is difficult to explain the increasing of oxygen concentration in the outflow at higher oxygen concentration in the influent. Yet, this might be due to an increasing of the oxygen consumption by aerobic bacteria at higher initial oxygen concentration.

The data obtained on sediment, could be distorted by inadequate control of phenomena such as oxygen and nitrate diffusion through sediment, and obviously by formation of anoxic microsites with denitrification taking place only in these areas.

Experiments carried out on resting cell suspension of *Ps. nautica* isolated from this sediment, permit us to discount these ambiguities. Results show that the oxygen inhibition of denitrification was different at each step. The steps of nitrate and nitrite reduction were completely inhibited at oxygen concentrations greater than 4.05 and 2.15 mg l^{-1}, respectively, whereas the last step is more sensitive and totally blocked at 0.25 mg l^{-1} O_2. These data confirm that denitrification is not a strictly anaerobic process.

These results must not be generalized to all denitrifying strains, since the denitrification systems of some bacteria are inhibited by oxygen, while other species are able to denitrify at oxygen concentration approaching air saturation.

In conclusion, our results provide evidence that in the marine sediment studied, denitrification can occur in presence of oxygen. These findings are in agreement with those reported for cell suspension of *Ps. nautica*, isolated from this sediment.

Acknowledgements

This work was carried out in the framework of the programme 'Cycles biogéochimiques: évolution de la matière organique en milieu littoral' and supported by the Société Nationale Elf Aquitaine, the Centre National de la Recherche Scientifique and the Programme Interdisciplinaire de Recherche sur l'Environnement (PIREN).

References

Alefounder, P. R., J. E. G. Mc Carthy & S. J. Fergusson, 1981. The basis of the control of nitrate reduction by oxygen in *Paracoccus denitrificans*. FEMS Microb. Lett. 12: 321–326.

Baumann, P. & L. Baumann, 1981. The gram negative eubacteria: genus *Protobacterium*, *Beneckea*, *Alteromonas*, *Pseudomonas* and *Alcaligenes*. In: P. S. Mortimer, S. Heinz, H. G. Trüper, A. Ballows, H. G. Schlegel (eds.), The prokariotes. A handbook on habitats, york: pages 1302–1330.

Benschneider, K. & R. J. Robinson, 1952. A new spectrometric method for determination of nitrite in sea water. J. mar. Res. 11: 87–96.

Bonin, P., M. Gilewicz & J. C. Bertrand, 1987a. Denitrification by a marine bacterium *Pseudomonas nautica* strain 617. Ann. Inst. Past/Microbiol. 138: 371–383.

Bonin, P., J. C. Bertrand, G. Giordanno & M. Gilewicz, 1987b. Specific sodium dependence of a nitrate reductase in a marine bacterium. FEMS Microb. Lett. 48: 5–9.

Bonin, P., M. Gilewicz, J. C. Bertrand, 1989. Effect of oxygen concentration of each step of denitrification. Can. J. Microbiol. 35: 1061–1064.

Christensen, S. & J. M. Tiedje, 1988. Oxygen control prevents denitrifiers and barley plant roots from directly competing for nitrate. FEMS Microb. Ecol. 53: 217–221.

Esteves, J. L., 1984. Reduction du nitrate dans les sédiments marins. Thèse 3° cycle. Université d'Aix-Marseille II, 107 pp.

Esteves, J. L., G. Mille, F. Blanc & J. C. Bertrand, 1986. Nitrate reduction activity in a continuous flow through system in marine sediments. Microb. Ecol. 1: 283–290.

Hochstein, L. I., M. Betlach & G. Kritikos, 1984. The effect of oxygen on denitrification during steady state growth of *Paracoccus halodenitrificans*. Arch. Microbiol. 137: 74–78.

Jannasch, H. W., 1960. Denitrification as influenced by photosynthetic oxygen production. J. gen. Microbiol. 23: 55–63.

Kapralek, F., E. Jechova & M. Otavova, 1982. Two sites of oxygen control in induced synthesis of respiratory nitrate reductase in *Escherichia coli*. J. Bact. 149: 1142–1145.

Knowles, R., 1982. Denitrification. Microbiol. Rev. 46: 43–70.

Lloyd, D., L. Boddy & K. J. P. Davies, 1987. Persistence of bacterial denitrification under aerobic conditions: the rule rather than the exception. FEMS Microb. Ecol. 45: 185–190.

Lowry, O. H., N. J. Rosebough, A. L. Farr & A. J. Randall, 1951. Protein measurement with the folin phenol reagent. J. biol. Chem. 193: 265–275.

Meiberg, J. B. M., P. M. Bruinenberg & W. Harder, 1980. Effect of dissolved oxygen tension on the metabolism of methylated amines in *Hyphomicrobium X* in absence and presence of nitrate: evidence for aerobic denitrification. J. gen. Microbiol. 120: 453–463.

Payne, W. J., 1976. Reduction of nitrogenous oxides by microorganisms. Bact. Rev. 37: 409–452.

Robertson, L. A. & J. G. Kuenen, 1984a. Aerobic denitrification, old wine in new bottles. Antonie van Leeuwenhoek 50: 525–544.

Robertson, L. A. & J. G. Kuenen, 1984b. Aerobic denitrification, a controversy revived. Arch. Microbiol. 139: 351–354.

Slawyk, G. & J. J. Mac Issaac, 1972. Comparison of two automatic ammonium methods in a region of coastal upwelling. Deep Sea Res. 19: 521–524.

Sorensen, J., 1978. Denitrification rates in a marine sediment as measured by acetylene inhibition technique Appl. envir. Microbiol. 36: 139–143.

Sorensen, J., 1987. Nitrate reduction in marine sediment: Pathways and interactions with iron and sulfur cycling. Geomicrobiol. J. 5: 401–431.

Tréguer, P. & P. Le Corre, 1975. Manuel d'analyse des sels nutritifs dans l'eau de mer (Utilisation de l'autoanalyseur II – Technicon). Laboratoire d'Océanographie Chimique, Université de Bretagne Occidentale, Brest, France, 110 pp.

Received 25 July 1989; in revised form 10 January 1990; accepted 14 May 1990

Hydrobiologia **207**: 123–129, 1990.
D. J. Bonin & H. L. Golterman (eds), Fluxes Between Trophic Levels and Through the Water–Sediment Interface.

Oxygen profiles in intertidal sediments of Ria Formosa (S. Portugal)

Vanda Brotas,[1], Ana Amorim-Ferreira,[1] Carlos Vale,[2] Fernando Catarino[1]
[1] *Departamento de Biologia Vegetal, Faculdade de Ciências de Lisboa, Campo Grande, Bloco C2, 1700 Lisboa, Portugal;* [2] *Instituto Nacional de Investigação des Pescas, Av. Brasília, 1400 Lisboa, Portugal*

Key words: oxygen profiles, sediment-water interface, intertidal zones

Abstract

Microelectrode oxygen profiles were measured in intertidal sediments from Ria Formosa (S. Portugal), a very productive shallow coastal lagoon. Four intertidal sampling sites were selected according to different sediment characteristics. Individual profiles revealed a high degree of lateral variability on a centimeter spatial scale. Nevertheless, consistent differences were observed between oxygen profiles measured in atmosphere-exposed and inundated intertidal sediments: in organically poor sand oxygen-penetration depth varied from 3 mm in inundated cores to more than 7 mm in exposed ones, while in organically rich muddy sand and mud it remained between 0.5–2.0 mm. The oxygen input from inundated to exposed conditions was estimated for each sampling site. Semi-diurnal tidal fluctuation, leading to periodical atmospheric exposure of sediments plays a major role in the oxygenation process of intertidal zones of Ria Formosa.

Introduction

A thin oxygenated layer is commonly observed on the sediment surface of aquatic ecosystems (Revsbech *et al.*, 1980a; Lindeboom *et al.*, 1984; Baillie, 1986; Andersen & Helder, 1987; Silverberg *et al.*, 1987). At the sediment surface, oxygen is rapidly consumed, either by aerobic degradation of organic matter and respiration of benthic communities, or by reduced compounds produced in anaerobic layers and transported into the oxygenated zone. The oxygen required by all these reactions can be supplied by the water column, by the atmosphere and may also be produced by photosynthesis within the uppermost sediment layer.

In recent years, membrane-covered polarographic oxygen sensors have been developed to assay dissolved oxygen gradients near the sediment surface with a sub-millimeter spatial resolution (Revsbech *et al.*, 1980b; Helder & Bakker, 1985). Measurements performed in deep coastal sediments generally show an oxygen penetration of a few millimeters (Revsbech *et al.*, 1980a; Silverberg *et al.*, 1987). Mechanisms other than diffusion seem to dominate the oxygen supply across the sediment-water interface (Silverberg *et al.*, 1987). In intertidal sediments, however, several sources of oxygen production, oxygen uptake and oxygen transport are present, resulting in large variations in the oxygen profiles. Several studies have demonstrated that micro-algae photosynthesis increases oxygen concentration in the first sediment millimeter (Revsbech *et al.*, 1981, Revsbech & Jorgensen, 1983). More recently Baillie (1986), has emphasized the important effect of infusion associated with the tidal fluctuation on the oxygenation of intertidal sediments.

The present study reports oxygen profiles measured in intertidal sediments with different characteristics in air-exposed and inundated conditions. The results obtained point out the effect of tidal height fluctuations associated with periodical atmospheric exposure on the oxygenation processes of intertidal areas.

Methods

The study area was located in Ria Formosa, a shallow mesotidal coastal lagoon (mean depth 3.5 m), on the south coast of Portugal, permanently connected to the sea through several narrow channels and with no relevant freshwater input (Fig. 1). Salinity values average 36% all year round. The exposed intertidal area may reach 50 km^2 (*ca.* 1/3 of the total lagoon area) at low water during spring tides and almost 20% of this area is occupied by growth-banks of the clam *Ruditapes decussatus* L.. Large areas of the lagoon are covered by macrophytes (*Zostera* spp and macroalgae). To evaluate the spatial variability of sediment oxygenation in the intertidal zone, four sites were selected (Fig. 1): station 1 a sandy clam growth bank, station 2 a muddy sand clam growth bank, stations 3 and 4 both located on mudflats not used for aquaculture, but station 4 had a higher carbon content. These sites emerge for about two hours during spring tides. At sampling time (always between 10 and 12 a.m.) sediment temperature ranged from 17 °C in winter to 28 °C in summer.

From January 1988 to February 1989, sediment samples were collected three times during the winter period and once every month in the summer period. All samples were drawn while the sediment was atmosphere-exposed at low tide. In January 1990 both atmosphere-exposed and inundated sediments were collected at the same sites, at low and high tide. Samples were taken by hand with plexiglass tubes (15–20 cm length and 8 cm i.d.) which were immediately stoppered at the bottom with an hermetic lid and brought to the

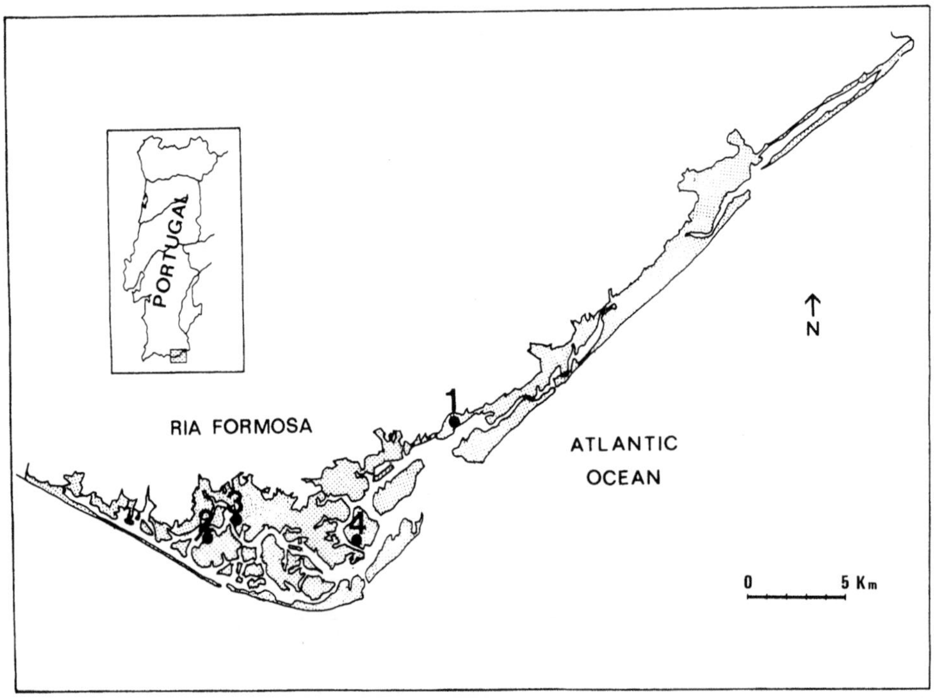

Fig. 1. Map of Ria Formosa with sampling stations.

laboratory. Inundated cores were transported in a basin containing water from the sampling locality several centimeters in depth. In January 1988, sediments from stations 1 and 2 were collected undisturbed and put inside transparent plexiglass box cores (40 × 40 × 30 cm). These were subsequently covered with water from the sampling locality (*ca.* 20 cm depth) and transported to the field laboratory where they were kept outdoors for two weeks under shade, with continuous stirring and aeration to maintain oxygen in the overlying water at saturation levels.

For each sampling station sediment was analyzed for grain size, organic carbon, chlorophyll *a*, phaeopigments and water content. The sand fraction was determined by sieving and the other fractions by means of a sedimentometer. Organic carbon were estimated by subtracting the inorganic carbon from total carbon content measured separately in a CHN-analyzer. Concentrations of chlorophyll *a* and phaeopigments in the upper centimeter were determined spectrophotometrically according to Lorenzen (1967). Water content was calculated by weighing the sediments before and after drying to constant weight at 105 °C. Porosity was expressed as the ratio of water volume to water volume plus volume of sediment sample.

The vertical distribution of oxygen was studied in the laboratory at room temperature (20–24 °C) and room light intensity (10 $\mu E\ m^{-2}\ s^{-1}$), with a polarographic needle sensor (2.5 cm length, 0.7 mm o.d.). Oxygen needle electrodes are indicated for coarse-grained sediments providing a good spatial resolution (0.25 mm), (Helder & Bakker, 1985). *In situ* measurements were attempted without success. The pore water movements created by the operator standing on the mud- and sand-flats made it impossible to make valuable measurements. The needle electrode (produt n° 760), the chemical microsensor (n° 1201) and Ag/AgCl reference electrode (n° 334) used in this study, were purchased from Diamond Electro-Tech Inc., Ann Arbour, Michigan, USA.

Before each experiment the electrode was allowed to stabilize at − 0.7 V at the measurement temperature in filtered water from the lagoon for at least 2 hours. The electrode was calibrated in oxygen saturated water from the lagoon (21% O_2) and in deep anoxic sediment (0% O_2). The reference electrode was always introduced in the calibration media as well as the measurement media (sediment or water). Stabilization was always confirmed with a strip chart recorder. When measuring the oxygen profiles the needle electrode was lowered into the sediment at 0.25 mm intervals using a manual micromanipulator. In inundated cores the overlying water above the sediment was 1–2 cm deep.

The oxygen profiles were measured on field samples in two situations: (i) undisturbed sediment cores one hour after collection (both atmosphere-exposed and inundated); (ii) atmosphere exposed sediment cores, which were subsequently immersed with water (oxygen saturated) from the sampling locality and allowed to stand with no stirring for 10 hours. Measurements were also performed in sediments collected from the box cores previously described, 14 days after field sampling.

Results

Oxygen concentration was measured in coarse- and fine-grained sediments containing low- and high-organic matter. Sediment characteristics within the upper 5 cm of the cores collected at the four sampling sites are presented in Table 1.

In order to evaluate the effect of time lag between collection and measurement of oxygen profiles in air-exposed cores, sequential profiles were measured in the laboratory at 10, 30, 60 and 120 minutes after collection (Fig. 2). During the first 60 minutes slight differences were recorded in the oxygen penetration depth which may be attributed to small-scale spatial heterogeneity. However, after this period, penetration depth decreased and surface value increased, defining a different oxygen profile. Therefore, we assume that only measurements within one hour were reliable.

Profiles of dissolved oxygen measured in air-

Table 1. Sediment characteristics of the sampled cores within the upper 5 cm; chlorophyll *a* and phaeopigment content represent annual average values of the upper sediment centimeter; water content was determined for inundated and air exposed sediments, there were no significant differences between these two conditions.

	St 1	St 2	St 3	St 4
Sand $\Theta > 60\ \mu m$ (%)	97.3	81.2	23.2	14.0
Silt $2\ \mu m < \Theta < 60\ \mu m$ (%)	2.4	16.0	67.3	76.6
Clay $\Theta < 2\ \mu m$ (%)	0.2	2.8	9.5	9.4
Water content	0.18	0.36	0.49	0.61
Porosity	0.34	0.56	0.72	0.80
Organic carbon (%)	0.20	1.46	1.68	2.80
Chlorophyll *a* $\mu g g^{-1}$ dw	4.39	7.42	6.19	2.96
Phaeopigments $\mu g g^{-1}$ dw	4.42	13.77	18.86	12.69

exposed sediments from stations 1 to 4, at two different sampling periods, are presented in Fig. 3. In spite of variations in the same core and between sampling periods, there were consistent differences between the stations in both the oxygen penetration depth and in the shape of the profiles. Oxygen penetrated down to 0.5–1.0 mm in mud (stations 3 and 4), to 1.0–1.8 mm in muddy sand (station 2) and to 6 mm or more in sand (station 1). Oxygen variation with depth was quite irregular at station 1, while exponential to almost linear distributions were found for the other stations. The oxygen gradients at stations 2, 3 and 4 varied from 105 to 395 $\mu M\ mm^{-1}$ while at station 1 they were always lower than 90 $\mu M\ mm^{-1}$.

Oxygen profiles were also measured in cores collected from the atmosphere-exposed and inundated intertidal areas of the same stations (Fig. 4). At stations 2, 3 and 4, oxygen penetration depths in exposed and inundated sediments were quite similar. A different picture was found at station 1: the erratic variation with depth observed in most exposed cores was not found in

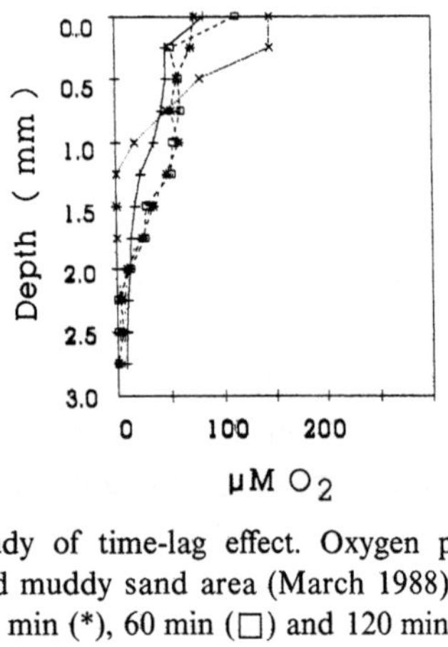

Fig. 2. Study of time-lag effect. Oxygen profiles in an exposed and muddy sand area (March 1988): 10 min (+), 30 min (*), 60 min (□) and 120 min (x).

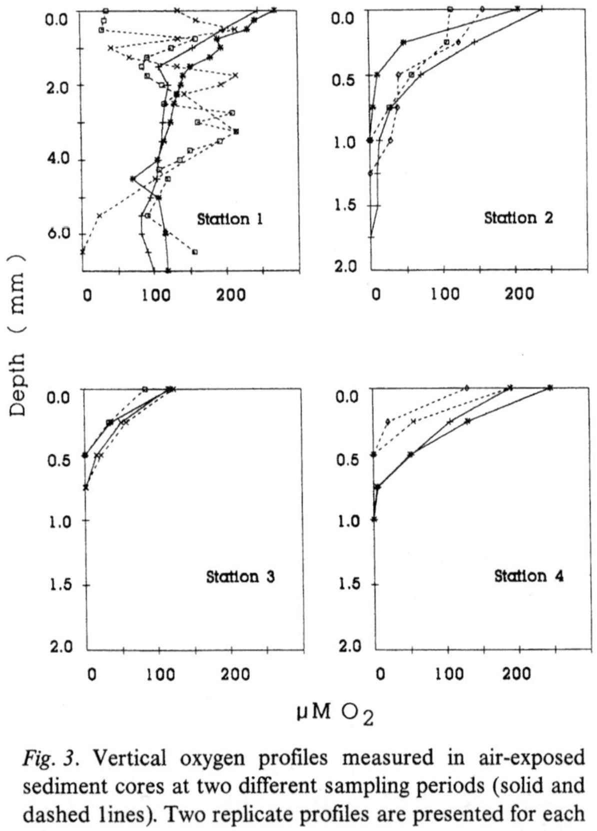

Fig. 3. Vertical oxygen profiles measured in air-exposed sediment cores at two different sampling periods (solid and dashed lines). Two replicate profiles are presented for each period. In spite of seasonal and small-scale spatial variations, consistent differences between cores in both depth penetration and shape of the profiles are evident.

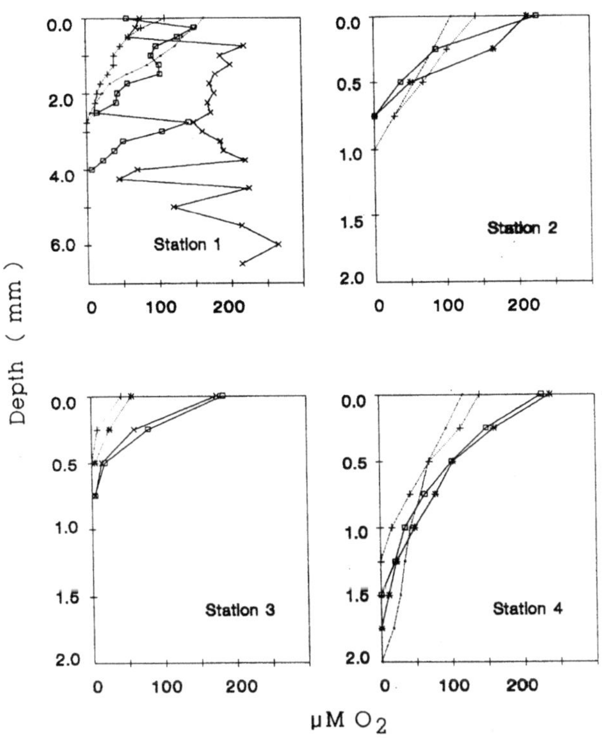

Fig. 4. Oxygen profiles in air-exposed (solid line) and inundated (dotted line) sediments collected in January 1990 at low and high tide, respectively. In the inundated cores oxygen concentration in the bulk of the overlying water was 180 μM for stations 1 and 2, 138 μM for station 3 and 170 μM for station 4. Oxygen saturation value at measuring temperature was 234 μM.

inundated conditions and penetration depth decreased substantially. At stations 2, 3 and 4 surface levels of oxygen in air-exposed sediments were higher than in inundated ones. For all sampling stations, oxygen gradients were 2–3 fold steeper in exposed cores than in inundated ones: 13 versus 55 $\mu M\ mm^{-1}$ at station 1, 125 versus 220 $\mu M\ mm^{-1}$ at station 2, 50 versus 180 $\mu M\ mm^{-1}$ at station 3, 130 versus 230 $\mu M\ mm^{-1}$ at station 4.

According to the Fick's law of diffusion, $J = \Theta\ Ds\ \Delta C/\Delta x$, the oxygen flux (J) at each core (for constant values of porosity, Θ, and sediment diffusion coefficient, Ds) increases as the gradient, $\Delta C/\Delta x$ is higher. This means that oxygen fluxes across the sediment-water interface are higher in the exposed cores than in the inundated ones.

The amount of oxygen (Q) present in the sediment, from the sediment-water interface (p0) to the maximum depth of oxygen penetration (pmax) may be calculated by unit area as follows:

$$Q = w \int C(p)\,dp \quad [p0, pmax]$$

where w is the water content and C the oxygen concentration at the oxygen penetration depth, p.

Based on oxygen profiles (Fig. 4) and water content measurements (Table 1), one may calculate the amount of oxygen at exposed (Qe) and inundated cores (Qi) of each site.

Assuming that oxygen profiles obtained are representative of the conditions encountered in these locations, the difference Qe – Qi, at each station, represents the oxygen input from inundated to exposed periods. The result of Qe – Qi was positive for all the stations: from 16 to 2 $nmol\ cm^{-2}$ at station 1, 0.8 $nmol\ cm^{-2}$ at station 2, 2.7 $nmol\ cm^{-2}$ at station 3, and 1.5 $nmol\ cm^{-2}$ at station 4. Except for station 1, where the irregularity of replicate profiles accounted for a large range of Qe – Qi, the other sites presented a comparable oxygen input.

Oxygen profiles measured in sediment cores from stations 1 and 2 artificially immersed in the laboratory for 10 hours are presented in Fig. 5. A diffusive boundary layer above the sediment surface was formed. Jorgensen & Revsbech (1985) relate the thickness of this layer to the rate of

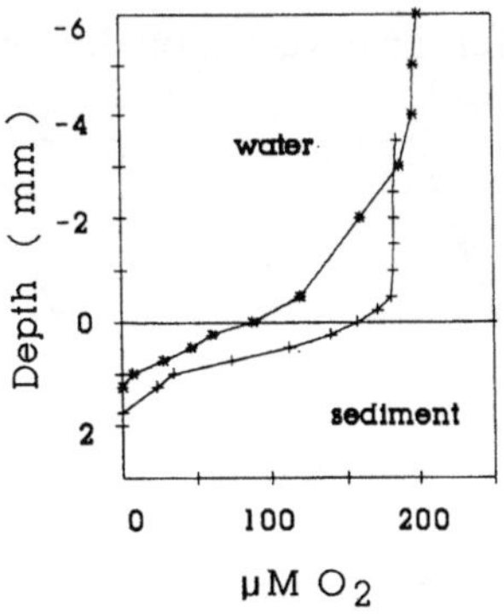

Fig. 5. Oxygen profiles measured in sediment cores from station 1 (+) and station 2 (x) artificially immersed in the laboratory for 10 hours. Oxygen saturation value at measuring temperature was 206 μM.

oxygen uptake of sediment surface: when steep gradients are established, the boundary layer may represent a major resistance to the transport of oxygen; whereas in sediments with a low respiration rate, diffusion through this layer does not play a significant role for oxygen uptake. Figure 5 indicates a higher rate of oxygen uptake for sediment of station 2, as the thickness of the diffusive boundary layer was 3 mm compared to 0.5 mm in the sediment from station 1. Furthermore, the sediment oxygenated layer recorded at station 2 was thinner (1.3 mm) than at station 1 (1.8 mm). In comparison with inundated cores (Fig. 4), the penetration depth was particularly reduced at station 1. This represents a reduction of oxygen flux from 20 nmol $cm^{-2} h^{-1}$ in exposed to 13 nmol cm − 2 h^{-1} in inundated sediments (Ds = 2×10^{-5} $cm^2 s^{-1}$, for 20 °C, Andersen & Helder, 1987).

Oxygen measurements in 14-day incubated sediments from stations 1 and 2, revealed almost linear profiles, oxygen dropping to 2.0–2.8 mm (Fig. 6). With stirring and continuous air-supply, oxygen penetration in sediment from station 1 is comparable to that recorded in inundated conditions, but in sediments from station 2, oxygen went 1.5 mm deeper (compare Figs. 4 and 6). Similarly, the comparison between 14-day and 10-hour incubations shows that at station 2 oxygen went 1 mm deeper, as oxygen is supplied to the overlying water, whereas at station 1 penetration depth remained roughly the same. The higher carbon content in sediments from station 2 accounts for the oxygenation differences between

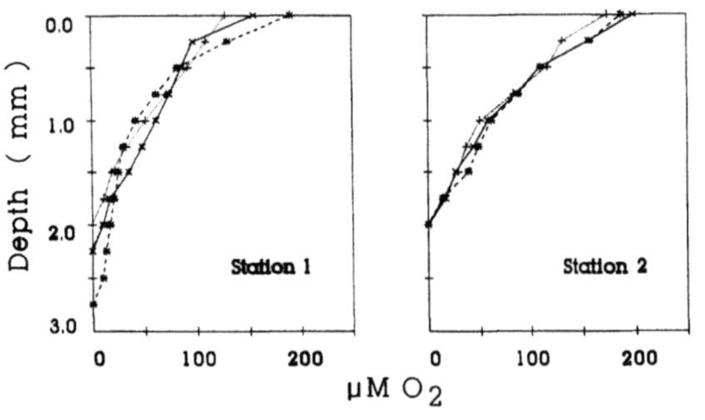

Fig. 6. Oxygen profiles measured in cores collected from sediment incubated with continuous stirring and aeration for 14 days. Three replicates are indicated for each core.

the experiments; Lindeboom *et al.* (1984) drew the same conclusion, from sediments with low and high POC content, in a shallow brackish lake.

Oxygen fluxes into the sediment were 35 and 59 nmol $cm^{-2} s^{-1}$ for stations 1 and 2, respectively, in the 14-day experiment; whereas in the 10-hour experiment they were 22 nmol $cm^{-2} s^{-1}$ for station 1 and 34 nmol O_2 $cm^{-2} s^{-1}$ for station 2 (see Figs. 5 and 6). Oxygen fluxes calculated in this study are comparable to those reported for shallow nearshore marine sediments (Silverberg *et al.*, 1987; Andersen & Helder, 1987).

Discussion

The results obtained in this study point out the contrasting oxygenation of sand/mud-exposed/inundated intertidal surface sediments of Ria Formosa. While in mud and muddy sand the oxygenated layer was thin (0.5–2.0 mm) under all the experimental conditions, a deeper and variable layer was found in sandflats (from 3 to more than 7 mm).

Exposure of sand to the atmosphere seems to be the major cause for the high oxygen penetration. The irregularities found in most oxygen profiles in sand, may be due to the presence of air bubbles which favor the diffusional exchange of oxygen with the deeper interstitial water retained around particle grains (Revsbech & Jorgensen, 1983). Revsbech *et al.*, (1981) demonstrated that oxygen levels in surface sediment may increase due to microalgal photosynthesis. In our measurements, autochthonous production of oxygen in exposed sand cannot alone justify mud-sand oxygen penetration differences. In fact, values of chlorophyll *a* were comparable in all the sampling locations and experiments were performed under similar light conditions.

Reduction in the thickness of the oxygenated layer in inundated sand probably results from the interruption of atmospheric oxygen supply to sand interstices with the rise of water level. Riedl & Machan (1972) studied the dynamics of interstitial water in intertidal sand and demonstrated the existence of a filling-draining process driven by the tidal height fluctuation. This pulsing mechanism may lead to the transport of oxygen from the atmosphere as well as the transport of reduced substances from deeper anaerobic layers (Fenchel, 1978). In mud, this pulsing mechanism tends to be less effective. Penetration depth was comparable in inundated and exposed conditions, but oxygen increased considerably in the surface of exposed mud.

The discrepancy of oxygen profiles at stations 1 and 2, both with a high percentage of sand (97% and 81%, respectively), may be related to the different organic matter content. Anderson & Helder (1987) showed that large differences in oxygen depth penetration may be due to both organic matter content and wave exposure. Infiltration of water between sand grains results in a faster turnover rate of easily degradable organic carbon at the exposed sites. In our study, station 1 is situated near the inlet channel of the lagoon, whereas station 2 is located in an inner area (see Fig. 1). Consequently sediments of station 2 have a minor wave exposure, which may result in a poorer renewal of interstitial water. This leads to a progressive accumulation of organic matter in the sediments.

The differences in oxygen penetration in sediments from stations 1 and 2 were drastically reduced in the 14-day incubation experiment. When tide-driven renewal of interstitial water is absent, the oxygenation of the sediment surface due to diffusional processes and net production of oxygen, is similar in both sediments.

Baillie (1986) pointed out the role of the tide and microalgae photosynthesis in the oxygenation process of sediment surface of intertidal areas. The experiments conducted in this study have simply emphasized the question of how the oxygenated layer varies from exposed to inundated conditions in sediments of Ria Formosa. Studies from intertidal (Riedl & Machan, 1972; Baillie, 1986) to coastal sediments (Silverberg *et al.*, 1987) concluded that mechanisms other than molecular diffusion must dominate the transport of oxygen across the sediment-water interface. In our study, the largest differences were observed in organically poor sands between atmosphere-exposed and inundated periods. These are mainly attributed to transport processes through pore water movement, which are driven by the large fluctuation of tidal height. Evapotransport, coupled with respiration, deplecting the oxygen in organically rich sediments, accounts for the thin oxygenated layer. These processes may be particularly intense in summer due to the high temperatures of the water lagoon. However, seasonal changes were not observed, probably because of the small-scale temporal variations of the oxygenation process. Further studies should therefore be conducted to understand the role of bottom primary producers coupled with the semi-diurnal effect of tide on the formation and maintenance of the oxygenated surface layer on the sediments of intertidal zones.

Acknowledgements

We thank M. Falcão for all help and facilities provided during this study, and Prof. S. Prates for the granulometry analysis. This work was supported by the Portuguese Junta Nacional de Investigação Científica e Tecnológica (JNICT) by contract n° 869.86.141.

References

Andersen F. O. & W. Helder, 1987. Comparison of oxygen microgradients, oxygen flux rates and electron transport system activity in coastal marine sediments. Mar. Ecol. Prog. Ser. 37: 259–264.

Baillie, P. W., 1986. Oxygenation of intertidal estuarine sediments by benthic microalgal photosynthesis. Estuar. coast. Shelf Sci. 22: 143–149.

Fenchel, T. M., 1978. The ecology of micro and meiobenthos. Ann. Rev. Ecol. Syst. 9: 99–121.

Helder, W. & J. F. Bakker, 1985. Shipboard comparison of micro-and minielectrodes for measuring oxygen distribution in marine sediments. Limnol. Oceanogr. 30: 1106–1109.

Jorgensen, B. B. & N. P. Revsbech, 1985. Diffusive boundary layers and the oxygen uptake of sediments and detritus. Limnol. Oceanogr. 30: 111–122.

Lindeboom, H. J., H. A. J. De Klerk & A. J. J. Sandee, 1984. Mineralization of organic carbon on and in the sediment of Lake Grevelingen. Neth. J. Sea Res. 18: 492–510.

Revsbech, N. P. & B. B. Jorgensen, 1983. Photosynthesis of benthic microflora measured with high spatial resolution by the oxygen microprofile method: capabilities and limitations of the method. Limnol. Oceanogr. 28: 749–756.

Revsbech, N. P., B. B. Jorgensen & T. H. Blackburn, 1980a. Oxygen in the sea bottom measured with a microelectrode. Science 207: 1355–1356.

Revsbech, N. P., B. B. Jorgensen & O. Brix, 1981. Primary production of microalgae in sediments measured by oxygen microprofile, H14CO-3 fixation and oxygen methods. Limnol. Oceanogr. 26: 717–730.

Revsbech, N. P., J. Sorensen, T. H. Blackburn & J. P. Lomholt, 1980b. Distribution of oxygen in marine sediments measured with microelectrodes. Limnol. Oceanogr. 25: 403–411.

Riedl, R. J. & R. Machan, 1972. Hydrodynamic patterns in lotic intertidal sands and their bioclimatological implications. Mar. Biol. 13: 179–209.

Silverberg, N., J. Bakker, K. M. Edenborn & B. Sundby, 1987. Oxygen profiles and organic carbon fluxes in Laurentian Trough sediments. Neth. J. Sea Res. 21: 95–105.

Received 8 August 1989; in revised form 14 March 1990; accepted 5 June 1990

Hydrobiologia **207**: 131–136, 1990.
D. J. Bonin & H. L. Golterman (eds), Fluxes Between Trophic Levels and Through the Water–Sediment Interface.

Daily cycles of dissolved oxygen and nutrient content in a shallow fishpond: The impact of water renewal

Vincent Escaravage
Institut de Biologie Marine, Université de Bordeaux, 33120 Arcachon, France

Key words: dissolved oxygen, nutrients, daily cycles, water renewal, fishponds

Abstract

This study was conducted in a shallow fishpond located on the eastern side of Arcachon Bay, France.

Dissolved oxygen, pH and nutrient content (NH_4, NO_2, NO_3 and PO_4) showed pronounced diel variations.

These phenomena seemed to be directly controlled by biological processes. Microbially mediated degradation of organic matter constitutes the main source of nutrients in the water column and primary producers consume the major part of these products. These processes are closely coupled to the light-dark cycle.

Man-induced water renewal caused low modifications in nutrient concentration and did not alter its temporal dynamics.

Introduction

Eutrophic lagoons and more specifically lagoonal ponds act as traps for organic materials of both autochthonous and allochtonous origin. In many fishponds, this phenomenon is enhanced by the hydrographic conditions resulting from stagnation periods interrupted by man-induced water renewals. During stagnation periods, the degradation processes acting in these organic rich sediments constitute the main source of vital nutrient for both benthic and pelagic primary productions (Martens, 1982). Man-induced water renewals, occurring regularly, aim to stabilize the trophic status of the ponds, in order to avoid the development of dystrophic crises. However, there is little information on the exact impact of this empirical practice on the ecosystem functioning.

In this paper, I focus on interpreting the short term variations of nutrients in relation to the community O_2 metabolism (a composite of gross productivity and respiration). Furthermore, the modifications induced by water renewal and their durability are studied.

Description of the site studied

The lagoonal fishponds of Arcachon Bay ('réservoirs à poissons') are old salt pans transformed into basins for extensive aquaculture (Amanieu, 1967; Lasserre & Gallis, 1975). These shallow reservoirs (< 0.3 m depth) are flooded monthly at high tide by seawater from the Bay through sluice gates established around the embankments. Between two water renewals, the lagoons are totally isolated. In summer, dystrophic crises occur in these anoxic stagnant waters. These events cause catastrophic diseases in the whole ecosystem (Labourg, 1976). The area studied is located near a sluice gate. Depth is 1.5 m, which is a relatively high value in comparison to the rest of the reservoirs. Moreover, this pond is subjected to a freshwater input from groundwater.

In order to reduce the macrophytic populations, the water level was maintained very low for three months before the start of the study. As a result, the water renewal caused the flooding of the desiccated and cracked banks.

Materials and methods

Hydrobiological and chemical data were collected during a 2-week survey from August 4 to August 16, 1986. Water renewal occurred from August 7 to August 16. The study was divided in two parts: before and after the water renewal.

- From August 4 to 6, preliminary measures were performed in order to characterize the initial state of the lagoon before water renewal.
- From August 11 to 16, the same parameters were measured in order to give information about the impact of the water renewal of the lagoonal characteristics.

Sampling and *in situ* measurements were performed between 8 a.m. and 8 p.m. Temperature and salinity (thermosalinometer), dissolved oxygen (polarographic oxymeter) and pH (pH meter) were recorded every 90 min. below the surface (– 0.10 m) and close to the bottom (– 1.25 m). The water samples were analysed by methods summarized by Aminot & Chaussepied (1983).

Water samples for nutrient analysis (NO_3 ions, NO_2 ions, NH_4 ions and PO_4 ions) and chlorophyll *a* were collected at – 0.75 m (mid depth) three times a day (8 a.m., 2 p.m. and 8 p.m.) before the water renewal, and every 120 min. during the 42 hours following the last closing of the sluice gate (August 11 at 9 a.m.).

Results

Salinity remained stable (from 28.5 to 29.0) throughout the study. This can be explained by the two following reasons: the relatively large depth (– 1.5 m) of the area studied and the fact that it receives freshwater from groundwater. These characteristics certainly reduced the effects of evaporation on salinity. As a result, the salinity was at the same level in the fishpond and outside.

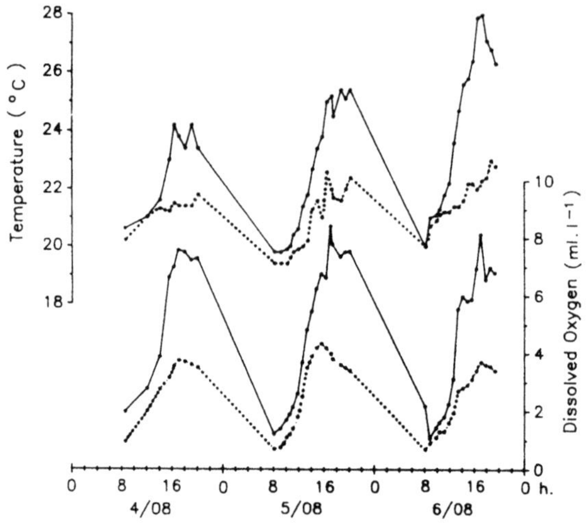

Fig. 1. Temporal variations of temperature and dissolved oxygen at surface (solid line) and bottom (dotted line) levels (from August 4 to 6).

Initial period (August 4 to August 8, 1986)

Temperature showed three peaks resulting from the daily exposure to the sun (Fig. 1). Changes of dissolved oxygen were synchronized with temperature. However, a close negative relation is known to link water temperature and oxygen solubility. We may therefore assume that the daily increase of dissolved O_2 concentrations is controlled by the photosynthetic activity itself, positively correlated with the light intensity.

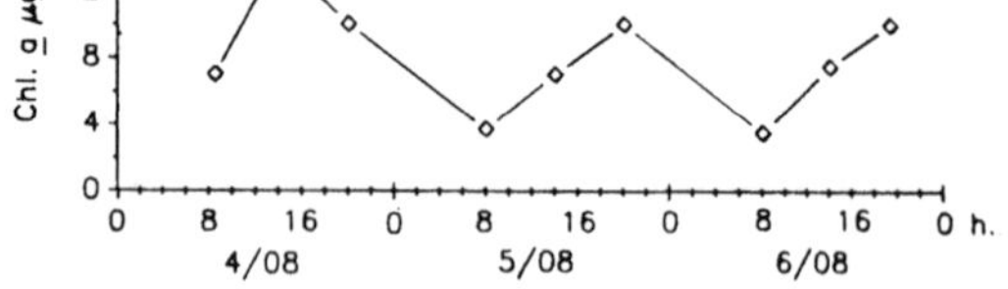

Fig. 2. Temporal variations of chlorophyll *a* concentrations at mid depth (from August 4 to 6).

The chlorophyll *a* concentrations are apparently linked to the photosynthetic activity and increased during the light period (Fig. 2).

The pH showed the same pattern as the previous variables (daily increase and nightly decrease) (Fig. 3) as pH is closely coupled with the assimilation of CO_2 by the photosynthetic organisms.

The nutrients showed equally a pronounced diel periodicity, which can be related to the whole ecosystem metabolism. During the night, the NH_4 and PO_4 concentrations increased, owing to degradation processes of benthic organic matter. These nutrients, not consumed, accumulated in the water column throughout the night. During the light period, the PO_4 and NH_4 concentrations strongly decreased, unlike the NO_3 concentrations which increased and showed peaks in evenings (Fig. 4).

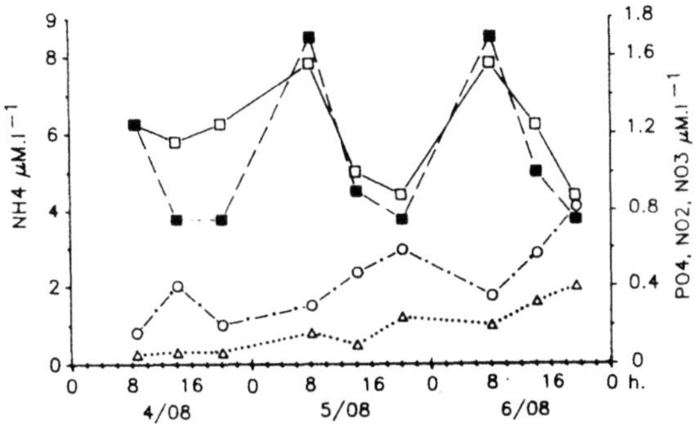

Fig. 4. Temporal variations of nutrient content at mid depth (from August 4 to 6): PO_4 = ■, NO_3 = ○, NO_2 = △, NH_4 = □.

Period following the water renewal (August 1 to August 16)

From August 7 to 11, the sluice gate was daily open at low and high tides for a renewal of the lagoonal water. The last opening of the sluice gate occurred on August 11, at high tide between 9 and 10 a.m. The seawater flooding the lagoon was characterized by 1) lower concentrations of phosphate and ammonium and 2) higher concentration of nitrate than the lagoonal water analysed just before the flooding (Table 1). Moreover, the seawater was well oxygenated (87% of saturation), while the concentration of dissolved oxygen did not exceed 40% of saturation at this time of day in the pond. The pH was higher in seawater (8.5) than in the pond (7.3).

The last seawater flooding was followed by an increase of NO_3 concentrations which reached 1.7 $\mu mol\, l^{-1}$ in mid afternoon on August 11 (Fig. 5). This value is higher than the maximum concentration obtained during the initial control period (0.8 $\mu mol\, l^{-1}$ on August 6). Moreover, this first post-renewal day was characterized by an intense increase in the concentration in ammonium. This increase went on until the next morning, when concentration reached 19 $\mu mol\, l^{-1}$.

From August 11 to 16, the mean concentrations of NH_4 and NO_3 progressively decreased to the levels observed during the control period. The PO_4 and NO_2 concentrations were not otherwise affected by the water renewal and remained simi-

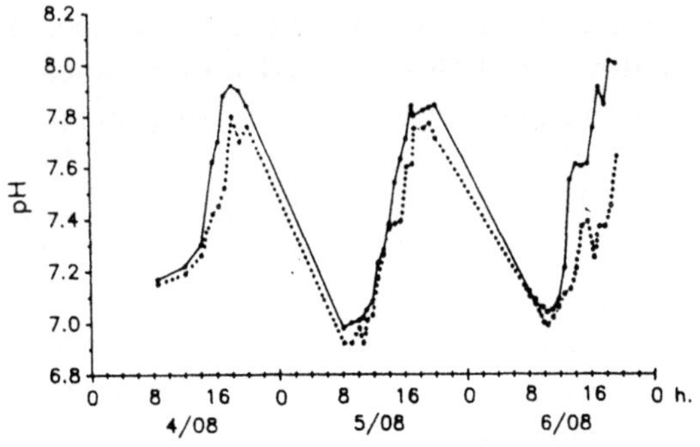

Fig. 3. Temporal variations of pH at surface (solid line) and bottom (dotted line) levels (from August 4 to 6).

Table 1. Characteristics of seawater and fishpond on August 11 at 8 a.m. (nutrient concentrations are expressed in $\mu mol\, l^{-1}$).

	Fish pond	Sea water
PO_4	1	0.25
NH_4	5.62	0.78
NO_3	0.30	1.14
NO_2	0.15	0.26
pH	7.3	8.5
O_2 (% of saturation)	35	87

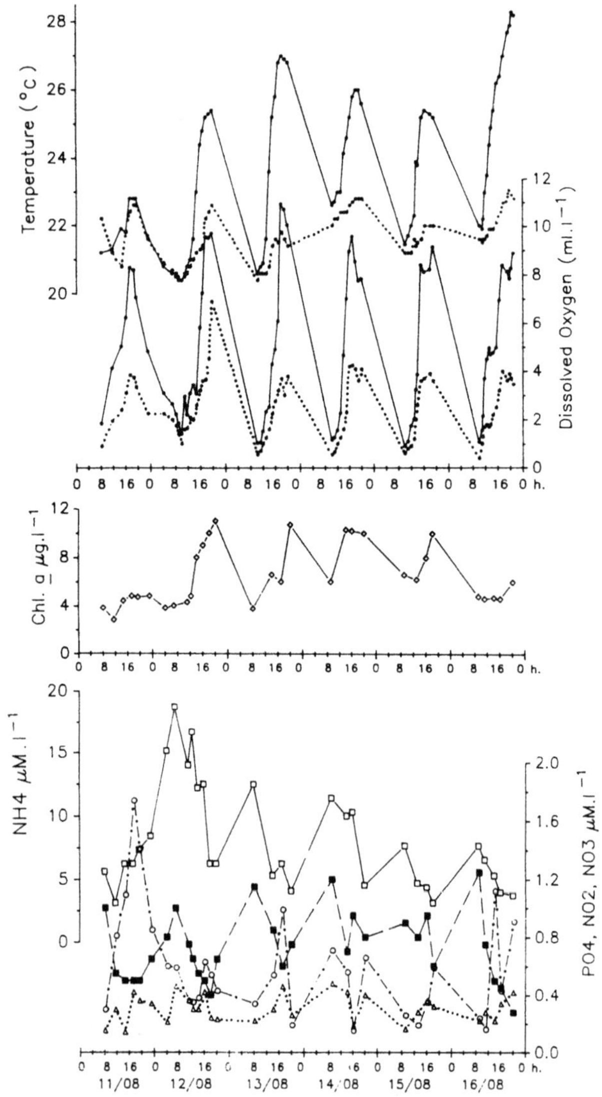

Fig. 5. Temporal variations of all the parameters measured the week following the water renewal (from August 11 to 16): PO_4 = ■, NO_3 = ○, NO_2 = △, NH_4 = □.

lar to those of the initial period. Moreover, the diel periodicity, observed during the initial period for all the variables measured, continued after the water renewal (for clarity, in the figures, pH is not represented).

Discussion

Before analysing the results themselves, let us first consider their representativity.

The first question to be addressed is the spatial representativity of the results. The station studied is located at the interface between the sea and the lagoon. Therefore, the characteristics of the station are presumably determined more by its juxtaposition to the sea than by its association with one or other fishpond systems. These characteristics are certainly common to all the ponds located at the sea : lagoon interface. Moreover, Gouleau (1987), who has studied a fishpond belonging to another hydraulic system (but similarly situated in the 'réservoirs à poissons' of Arcachon Bay), has obtained results similar to ours concerning phenomena occurring during the pond flooding.

A second question concerns the temporal representativity of the initial state, defined by a sampling series of three days during the first period of this experiment. This *a priori* short period seems, *a posteriori*, of minimal but sufficient duration to appraise the characteristics of the pond. Since atmospheric conditions remained stable throughout the study, the main source of disturbance for the ecosystem was the flooding of the pond (the phenomenon tested). The diel variations were similar over the three days of the first period, and the same pattern reappeared spontaneously during the second period, when the effects of the flooding has disappeared. Therefore, this pattern can reasonably be considered as typical of the steady state of the ecosystem at this period of the year (conditions of illumination, temperature, precipitation ...).

Photosynthetic activity seems to be the major parameter determining the daily pattern of O_2 concentration in the water column. Indeed, Kemp & Boynton (1980) observed, for Chesapeake Bay, that in shallow waters (< 3 m) diel changes of O_2 concentrations appeared to be generally dominated by biological metabolism. In the same way, Kenney *et al.* (1988) suggested that, in sufficiently shallow or fertile environments (both in our case), metabolic activity should be readily detectable as changes in the mass balance of the water column O_2 concentrations.

Chlorophyll *a* concentrations showed the same diel pattern as dissolved oxygen. Several hypotheses can be given to explain this phenomenon. Firstly, the pattern may be due to the ability of mobile phytoplankton to migrate vertically; Eppley *et al.* (1968) have described a daily climb and nightly drop of dinoflagellates within the water column. Secondly, concerning the occurrence of diel rhythms for cell division, Chisholm *et al.* (1978) indicated that phased cell division coupled with the light : dark cycle occurs in diatoms. A third possibility is the existence of diel cycles in the algal metabolism; Owens *et al.* (1980) have observed in cultures of *Skeletonema costatum*, periodicities of cellular chlorophyll *a* and *c* content which closely follow the light : dark cycle. They suggested that this periodicity results from subtle changes in rates of chlorophyll synthesis and degradation, which are coupled to the light : dark cycle and are unrelated to cell division. Our results to date do not allow us to choose between these possible explanations.

Photosynthetic organisms supposedly consume daily an important part of the night regenerated stock of nutrients (NH_4 and PO_4). According to Bulleid (1984) and Nowicki (1985), nutrient concentrations are closely controlled by photosynthetic organisms, which satisfy most of their needs in nitrogen with ammonium. In the same way, PO_4 constitutes the major source of phosphorus for autotrophic organisms. McCarthy *et al.* (1977) suggested a universally high phytoplankton preference for NH_4 and urea-N over NO_3 and NO_2. These authors indicate that, in Chesapeake Bay, NH_4 concentrations in excess of 0.5–1.0 $\mu mol\, l^{-1}$ almost totally suppress NO_3 utilization. In our case, NH_4 concentrations were always over 3 $\mu mol\, l^{-1}$. This can explain the apparent non-utilization of NO_3 by primary producers during the light period. One part of the NO_3 was probably released by biogeochemical processes in the oxic conditions of the light period, the other part proceeding from the oxydation of NH_4 (Henriksen *et al.*, 1984).

Water renewal had little effect on the characteristics of the fishpond. The nutrient release, recorded just after the water flooding, resulted probably from the moistening of the desiccated and cracked banks. These phenomena can be explained by a stimulation of mineralisation of organic matter. According to Arenas & De La Lanza (1981), dry and cracked sediments release high quantities of nutrients. Indeed, a sample taken in a submerged crack gave an exceptional NH_4 concentration of 30 $\mu mol\, l^{-1}$. Modifications induced by the water renewal disappeared very quickly and, a week later, the lagoonal system recovered similar hydrological characteristics to those recorded before the renewal.

Analogous results were obtained by Gouleau (1987), who studied nutrient fluxes at the water-sediment interface in another fishpond of the 'réservoirs à poissons' that dried before refilling. The use of dark : light bottles and bell jars allowed this author to evaluate the fluxes of NH_4 across the water-sediment interface. Hourly sampling was conducted during the four days following the flooding of the dry pond. During the first 24 hours, the flooded water underwent an intensive enrichment in NH_4 (up to 200 $\mu mol\, l^{-1}$), caused by its release from the desiccated sediment. During the three following days, the evolution of the NH_4 concentrations was quite similar to that described in the present paper: a diel cycle (day consumption / night regeneration) superimposed upon a day-by-day decrease. Four days after the flooding, the concentrations of NH_4 had fallen by more than 60%. Gouleau (1987) has suggested two processes to explain this decrease: on the one hand, a de-ammonification occurring both in the water column and at the water-sediment interface. The contribution of the interface to this phenomenon has been evaluated at between 30 and 60%. On the other hand, a kind of 'essoufflement' (*sic*) of the benthic ammonification processes could have induced a fall in the sediment-to-water flux and consequently caused a progressive decrease in NH_4 concentrations of the water column.

All these results illustrate the following thesis: in shallow and fertile environments, the greater part of the regulation of nutrient concentrations is controlled by biogeochemical phenomena (Martens, 1982). These mechanisms are hardly or not at all altered by the renewal of the water above.

In the fishponds, the principal result of water

renewal is to maintain salinity and temperature at levels compatible with biological activities. This is vital in most of the fishponds, which are generally shallower than the pond studied here. However, water renewal cannot radically transform other hydrological parameters directly implicated in biological processes.

References

Amanieu, M., 1967. Introduction à l'étude écologique des réservoirs à poissons de la région d'Arcachon. Vie Milieu 18: 381–446.

Aminot, A. & M. Chaussepied, 1983. Manuel des analyses en milieu marin. C.N.E.X.O. (ed.), Brest, 395 pp.

Arenas, V. & G. De La Lanza, 1981. The effect of dried and cracked sediments on the availability of phosphorus in a coastal lagoon. Estuaries 3: 206–212.

Bulleid, N. C., 1984. Deoxygenation and remineralization above the sediment-water interface; an *in situ* experimental study. Estuar. coast. Shelf Sci. 19: 15–25.

Chisholm, S. W., F. Azam & R. W. Eppley, 1978. Silicic acid incorporation in marine diatoms on light : dark cycles: Use as an assay for phased cell division. Limnol. Oceanogr. 23: 518–529.

Eppley, R. W., O. Holm-Hansen & J. D. H. Strickland, 1968. Some observations on the vertical migration of dinoflagellates. J. Phycol. 4: 333–340.

Gouleau, D., 1987. Etude de la mise à sec des marais endigués aquacoles. Rapport CNRS/CNEXO, G.I.S. 'Resources vivantes Sud Vilaine' N. 85/5 08 0581, IFREMER: 45 pp.

Henriksen, K., A. Jensen & M. B. Rasmussen, 1984. Aspects of nitrogen and phosphorus mineralization and recycling in the northern part of the Danish Wadden Sea. In R. W. P. M. Laane & W. J. Wolff (eds), Proceedings of the fourth international Wadden Sea symposium 'The role of organic matter in the Wadden Sea', 1–3 November 1983, Texel, The Netherlands, Netherland Institute for Sea Research, Publication Series n° 10: 51–69.

Kemp, W. M. & W. R. Boynton, 1980. Influence of biological and physical processes on dissolved oxygen dynamics in an estuarine system: Implications for measurement of community metabolism. Estuar. coast. mar. Sci. 2: 407–431.

Kenney, B. E., W. Litaker, C. S. Duke & J. Rasmus, 1988. Community oxygen metabolism in a shallow tidal estuary. estuar. coast. Shelf Sci. 27: 33–43.

Labourg, P. J., 1976. Les réservoirs à poissons du Basin d'Arcachon et l'élevage extensif des poissons euryhalins (muges, anguilles, bars, daurades). La pisciculture française 45: 35–42.

Lasserre, P. & J. L. Gallis, 1975. Osmoregulation and differential penetration of two grey mullets, *Chelon labrosus* (Risso) and *Liza ramada* (Risso) in estuarine fish ponds. Aquaculture 5: 323–344.

Martens, C. S., 1982. Biogeochemistry of organic rich coastal lagoon sediments. Proceedings International Symposium on coastal lagoons, SCOR/IABO/UNESCO, Bordeaux, France, 8-14 September 1981. Oceanol. Acta n° Sp.: 161–167.

McCarthy, J. J., W. R. Taylor & J. L. Taft, 1977. Nitrogenous nutrition of the plankton in the Chesapeake Bay. 1. Nutrient availability and phytoplankton preferences. Limnol. Oceanogr. 22 : 996–1011.

Nowicki, B. L., 1985. Benthic nutrient remineralization in coastal lagoon ecosystem. Estuaries 8: 182–190.

Owens, T. G., P. G. Falkowski & T. E. Whitledge, 1980. Diel periodicity in cellular chlorophyll content in marine diatoms. Mar. Biol. 59: 71–77.

Received 26 July 1989; in revised form 23 February 1990; accepted 19 March 1990

Hydrobiologia **207**: 137–146, 1990.
D. J. Bonin & H. L. Golterman (eds), Fluxes Between Trophic Levels and Through the Water–Sediment Interface.

Study of the Ria Formosa ecosystem: benthic nutrient remineralization and tidal variability of nutrients in the water

Manuela Falcao & Carlos Vale
Instituto Nacional de Investigação das Pescas, Centro Regional de Investigação Pesqueira, Av. 5 de Outubro, 8700 Olhão, Portugal

Key words: Nutrients, coastal lagoon, sediment-water exchanges, tide

Abstract

Exchange of nutrients across the sediment-water interfaces of Ria Formosa was studied in the laboratory. In the field, water samples were collected *(i)* fortnightly, at low and high tide over one year, and *(ii)* semi-diurnally, over three neap-spring tidal cycles in winter. Results from both laboratory experiments and field collection were quite variable. Higher liberation rates of silicates, phosphates and ammonium occurred in chambers whose bottom was formed by mixtures of mud and sand covered by vegetation with clams. The bottom took up nitrates from the overlying water. Nutrient variations in the lagoon water were controlled both seasonally and tidally. Higher concentrations of silicates were found at low tide independently of the season; phosphates in spring/summer at low tide; nitrates in winter/spring at high tide. Tidal flushing appears to be an important removal-mechanism in this lagoon.

Introduction

Coastal lagoons characterize more than 10% of the world coast line. Two important exchange processes have been recognized in these systems: sediment-water exchanges (Martens, 1982) and interactions between lagoon and coastal ocean (Nixon, 1980). Benthic-pelagic coupling is of particular interest in these systems, because the water volume is relatively small in comparison with the bottom area, and the influence of benthic processes on water chemistry may be correspondingly large (Nowicki & Nixon, 1985; Murphy & Kremer, 1985; Yamada & Kayama, 1987). In coastal lagoons where the tidal effect is very pronounced, the removal of dissolved constituents in the water is strongly influenced by tidal exchange with the sea (Millan-Nunez *et al.*, 1982; Ohtake *et al.*, 1984). In coastal systems with permanent connection to the sea, the so-called 'outwelling' hypothesis (Dame *et al.*, 1986) has give rise to controversy (Nixon, 1980) over whether it is a potential support to coastal ocean productivity.

The importance of Ria Formosa for the tertiary productivity of the adjacent coastal area has been evidenced by Monteiro (1989). Only recently have studies contributing to the understanding of nutrient cycling in this ecosystem been developed (Falcao, 1988). In the present work, results from microcosm experiments and statistical interpretation of nutrient fluctuations with tide and season are presented.

Description of the study area

Ria Formosa is a very productive shallow coastal lagoon permanently connected to the sea through several narrow channels (Fig. 1). Tidal amplitude varies from a maximum of 3.5 m at spring tide to 0.5 m at neap tide, which causes an important semi-diurnal and fortnightly tidal fluctuation of the water volume inside the lagoon. Its mean depth is approximately 3.5 m and the area reaches 160 Km^2, one third being intertidal. Growth-banks of clams *Ruditapes decussatus* occupy 10 Km^2 of the intertidal flats. At spring tide the banks are uncovered twice a day, around low tide, but at neap tide remain submerged. Large areas of the bottom are covered by vegetation (*Zostera spp*, *Enteromorpha spp* and *Spartina spp*). Grain size of bottom sediments vary from mud at inner areas to coarse sand in the inlet channels. Sand is transported by the aquaculturers from the beaches to the clam growth-banks located on the intertidal flats.

No relevant freshwater input to the lagoon takes place and salinity remained around 36‰ throughout the year, except during sporadic short run-off periods. Water temperature ranges annually from 12 to 26 °C, but dissolved oxygen in the water column remains close to or above saturation levels and no anoxic water layer developed. Untreated domestic sewage from about 10^5 inhabitants is discharged into the lagoon.

Sampling – Methods

In order to study the variability of nutrient exchanges across different sediment-water interfaces in Ria Formosa, several box-cores were collected from five representative substrates (Fig. 1): beach sand (S1); sand with clams, *Ruditapes decussatus*, (S2); sandy-mud with clams and vegetation, *Zostera spp*, (S3); muddy-sand with clams and algae, *Enteromorpha spp* (S4), and mud (S5). The density of clams was 9 individuals per 0.1 m^2, approximately the same as that in the growth-banks. Sediment samples were obtained during the autumn/winter: S2, S3 and S4 in November 1987; S1 and S5 in March 1988. Samples from S3 and S4 were collected again in December 1987, immediately after a run-off period. Water temperature in the experiments varied from 17 to 20 °C. *In loco*, sediments were placed inside

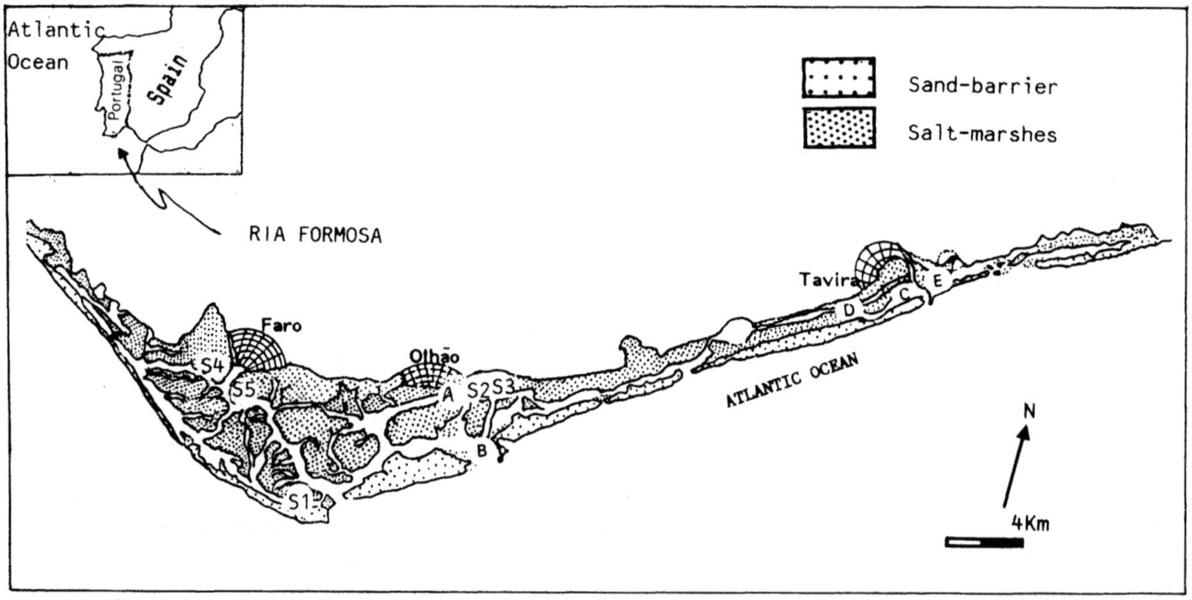

Fig. 1. Map of the lagoon Ria Formosa. Location of five stations where water collection was made: A, B, C, D, E. Location of sampling sites where benthic nutrients fluxes were measured: S1, S2, S3, S4, S5.

transparent plexiglass chambers (30 × 30 × 30 cm^3) and covered by lagoon water. The chambers were carefully transported to a field laboratory and during two weeks constant aeration was applied to maintain saturation levels of oxygen in the water. Overlying water was sampled daily and NO_3, NH_4, PO_4, and SiO_2 concentrations were determined.

In a shallow flat area, characterized by bottom vegetation cover (station E), water samples were collected daily over three neap-spring cycles at low and high tide. This sample schedule was followed in January, February and March 1988. Nutrients, temperature, pH, alkalinity and dissolved oxygen were determined.

In order to study the nutrient variations over longer time-scales, water was sampled fortnightly at low and high tide, including neap and spring tides, from September 1985 to September 1986, at four stations: A, B, C and D (Fig. 1). Two stations (A and B) were located near an inlet channel, and the other (C and D) at inner flat areas. Nutrients were determined with a Chemlab autoanalyser, alkalinity by potentiometric titration and pH, temperature and dissolved oxygen using a pH meter (Hanna) and a portable dissolved oxygen monitor (Nester instruments), respectively.

Results

Concentrations of silicate, phosphate, nitrate and ammonium during 15 days in the experimental chambers are presented in Fig. 2. Similar patterns of silicate and phosphate were found in several experiments: in chambers S1 (sand), S2 (sand + clams) and S5 (mud) concentrations in the water did not vary a great deal ($<20\ \mu M\ SiO_2$ and $<2\ \mu M\ PO_4$); in chambers with clams and vegetation (S3, S4) concentrations increased dramatically after the first three or four days, remaining at constant values after days 8 or 10. Concentrations reached higher values in cham-

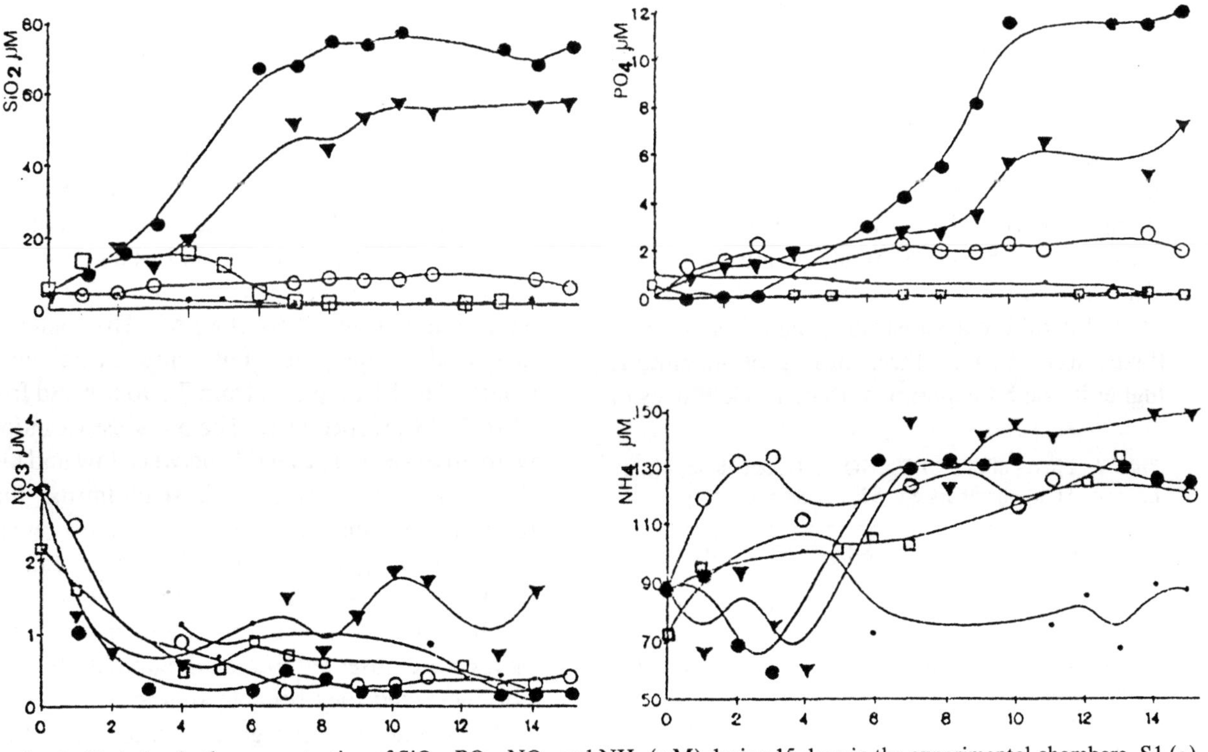

Fig. 2. Variation in the concentration of SiO_2, PO_4, NO_3 and NH_4 (μM) during 15 days in the experimental chambers, S1 (•), S2 (○), S3 (▼), S4 (●) and S5 (□).

ber S4 (70 μM SiO_2 and 11 μM PO_4) than in S3 (60 μM SiO_2, 6 μM PO_4). The major difference between these two chambers was the grain-size of the sediment: S4 sediment contained 52% of mud and S3 35% of mud. Ammonium concentrations increased with time to 130 μM in all the experiments, except for S1, where levels were kept lower (80 μM). Concentrations did not increase similarly in the several experiments. While in S5 levels increased progressively, sharp increases were recorded in S2 (in the first two days), in S3 (between days 4 and 7) and in S4 (between days 3 and 6). After eight days concentrations remained close to 130 μM in most experiments. For nitrates a quite opposite variation was recorded: in all the experiments values decreased in four days down to 1 μM.

Over a fortnightly tidal cycle, growth-clam banks located in the intertidal zone remain continuously submerged at neap tides, corresponding to approximately 8–10 days. Based on concentration variations over the first eight days, fluxes of nutrients (J) were determined by multiplying the slope of their concentrations versus day (m) by the water-volume/sediment-area ratio (v/s) of the chamber. Fluxes of silicates, phosphates, nitrates and ammonium (μmol m^{-2} d^{-1}) were calculated for each experiment (Table 1). These calculations emphasize the different nutrient exchanges across the sediment-water interfaces considered in this study. While nitrate fluxes were always negative and showed only slight differences for different substrates, the other nutrient fluxes were two to three orders of magnitude higher in the S4 experiment than in S1. Fluxes of silicates and ammonium were higher in S5 (mud) than in S1 (sand). In sand experiments (S1, S2) fluxes were higher in the chamber containing clams (S2). The sharpest nutrient increments were obtained in the S3 and S4 experiments which contained vegetative detritus and clams.

Table 1. Fluxes of SiO_2, PO_4, NO_3, NH_4 over 8 days (μmol $m^{-2} \cdot d^{-1}$), at chambers S1, S2, S3, S4 and S5.

Sediments	SiO_2	PO_4	NO_3	NH_4
Site S1 (sand)	-58	-5	-25	0
Site S2 (sand; clams)	29	25	-137	566
Site S3 (sandy mud; clams; vegetation)	522	47	-44	906
Site S4 (muddy sand; clams; vegetation)	761	96	-85	911
Site S5 (mud)	121	-6	-175	433

Freshwater discharge into the lagoon occurs mainly during run-off, and it may increase considerably the levels of silicates and nitrates (Falcao *et al.*, 1985). The comparison of nutrient concentrations in experiments S3 and S4 carried out at 'normal' hydrological conditions and immediately after run-off is presented in Fig. 3. Under these sporadic conditions, sharp decreases of silicates and nitrates in the water were recorded (from 180 to 40 μM SiO_2 in S4 and from 20 to 5 μM NO_3). Unlike 'normal' conditions, phosphates did not increase in the water, remaining at 1 μM. The irregular increase of ammonium continued to be observed. These results indicated that the additional amount of nutrients brought in by the freshwater run-off, tends to be taken up by the sediments.

In order to confirm the importance of bottom release on the nutrient concentration in the lagoon, nutrients were measured daily over three neap-spring tidal cycles at station E, located in a shallow muddy-sand area of vegetation cover and with benthic organisms, mainly clams. Nutrient levels varied within a broad interval. Nitrates ranged from 0.1 to 4.6 μM, phosphates from 0.1 to 3.5 μM, silicates from 1 to 165 μM and ammonium from 29 to 108 μM. The values of dissolved oxygen, pH and temperature varied from 5.8 to 18.1 mg l^{-1}, from 7.5 to 8.9 and from 11 to 24 °C, respectively. The broadest variations were observed at spring tide between low and high tide. Figure 4 illustrates the semi-diurnal fluctuations of silicates and phosphates which occurred along a neap-spring cycle in January 1988. At neap tide, water volume variation is very small and only slight low-tide/high-tide differences were observed, but as the amplitude of the tide increased they became sharper.

The occurrence of high concentrations of silicate and phosphate (contrary to other parameters) at periods of low-spring tide, over the three

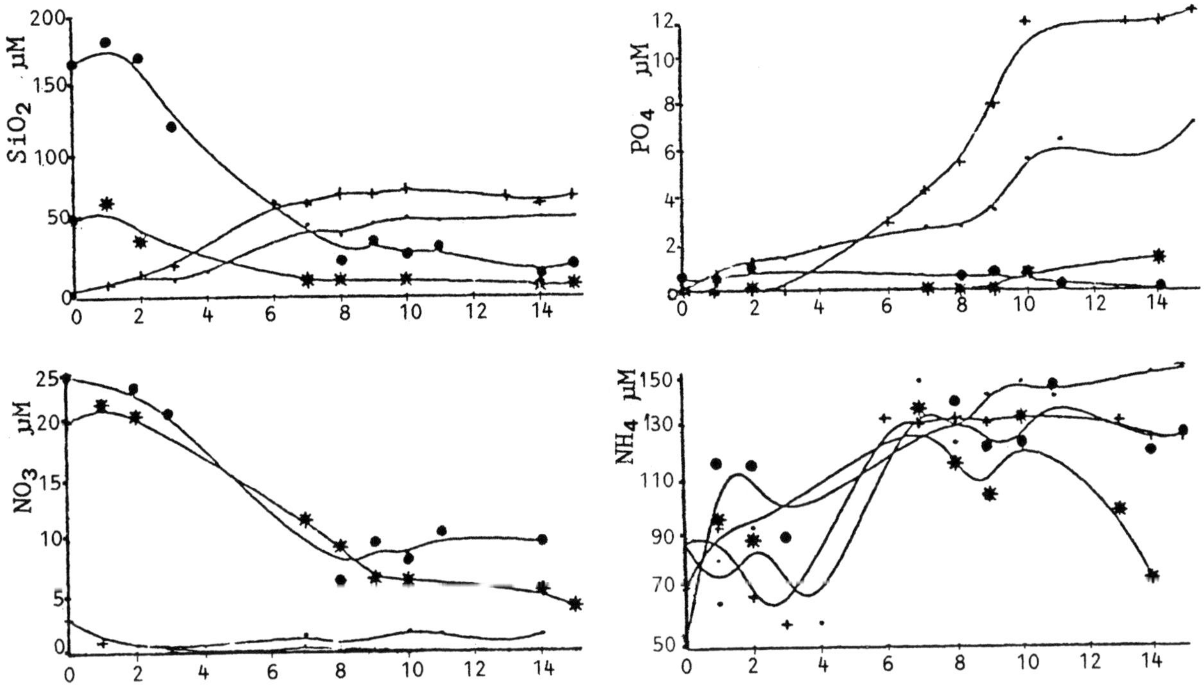

Fig. 3. Comparison of nutrient variation in experiments S3 and S4 carried out before (•, +) and immediately after run-off (*, ●).

fortnightly tidal cycles, may be shown using multivariate statistical methods (Legendre & Legendre, 1984). Two statistical methods were tested. In principal component analysis, inertia explained by the two principal axes was 59%, whereas in a correspondence factorial analysis it reached 95%. This means that nearly all the information is represented in the plane defined by axis A1, A2. So the latter representation was selected (Fig. 5). This plot clearly evidences the association of silicates and phosphates with low-spring tides in the three tidal cycles (1, 2, 3). Higher concentrations occurred in periods of higher physical perturbance of the bottom and lower water volume (minimum dilution with the incoming seawater). An opposite pattern was found for nitrates and ammonium as well as pH, alkalinity and dissolved oxygen. None of these parameters was preferentially related to those tidal conditions. Because results were obtained over short periods of time, no seasonal effect can be taken into account. Furthermore, extrapolation for other lagoon areas may be questionable, due to the variability of nutrient exchanges across the different sediment-water interfaces.

A tidal schedule sampling (water collection at neap and spring tide, both at low and high tide, every month over an annual period) was carried out at four stations, located near a sandy inlet channel (A, B) and at inner muddy-sand zones (C, D). Nutrient concentrations varied within a broad range of values: nitrates from 0.2 to 8.8 μM; phosphates from 0.05 to 5.6 μM and silicates from 9.2 to 14 μM. In order to condense the numerous data and to interpret the high variability of the results, principal component analysis was applied. The plane (A1, A2) for each parameter reached inertia higher than 80% (Fig. 6). The ordination of dissolved oxygen indicated its close association with the winter/spring period, when water temperature was lower (Fig. 6-a). However, while at station A and B (inlet channel), the higher values were recorded at high-spring tide, at stations C and D, located at inner areas, the highest levels were measured at neap tide. Values up to 180% saturation of dissolved oxygen were

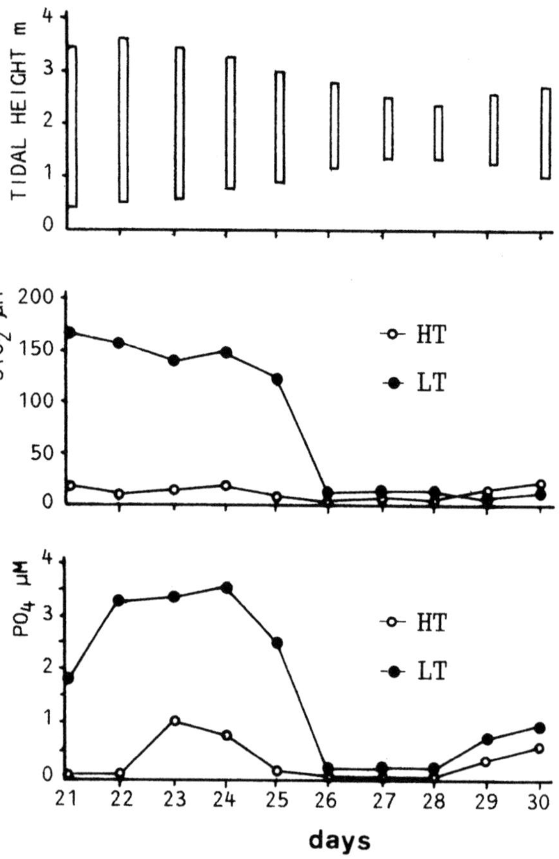

Fig. 4. High tide, low tide concentrations of SiO_4 and PO_4 (μM) over a neap-spring tidal cycle (21–30 January 1989).

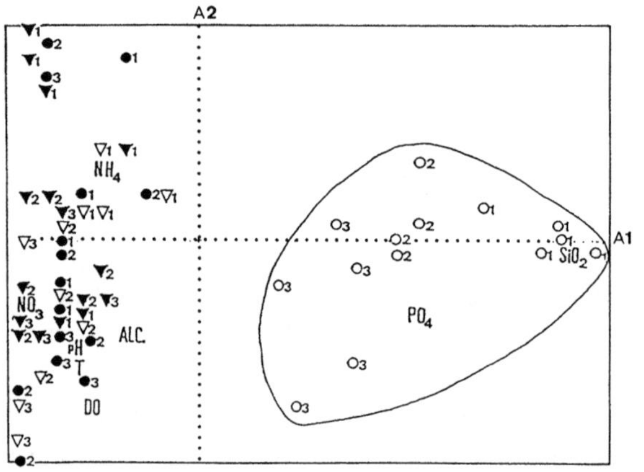

Fig. 5. Scattergram of parameters (T, pH, DO, NO_3, PO_4, SiO_2, Alc and NH_4) and tidal conditions (low spring tide o, low neap tide •, high spring tide ∇, and high neap tide ▼) at station E, in January (1), February (2) and March (3), in the plane defined by the two first principal axes of the correspondence factorial analysis, A1 and A2 (inertia explained: 95%).

measured in these stations, indicating the strong effect of photosynthesis. In summer/autumn the nitrates were very low (Fig. 6-b). In all stations nitrates were closely associated with winter/spring samples, although low nitrate concentrations were also measured during this period. For stations A and B, located in the inlet channel, nitrates appeared to be associated with high tides, both at neap and spring tides. Such tidal effect was obscured for the inner stations (C, D). Less clear seasonal patterns were found for phosphates and silicates. Although the highest phosphate concentrations occurred in spring/summer (Fig. 6-c), closer association was obtained for samples collected at spring tide. For silicates, the fluctuation with tide is superimposed on seasonal variations. In fact, a rod including only samples collected at low tide of both neap and spring tide may be delimited (Fig. 6-d).

Discussion

Results from microcosm experiments suggest that nutrient fluxes across the sediment-water interfaces of the Ria Formosa are quite variable. Fluxes were not calculated from measurements of *in situ* chambers, but such experiments with realistic flux values are difficult to perform in such a tidally dynamic lagoon. Chambers were incubated in a field laboratory, with sediments continuously covered with water. This means that results may be comparable only to periods of small variations in the lagoon water volume (neap tides, when water covers most of the intertidal areas). The maintenance of biological variables such as dissolved oxygen and stirring (Sundby *et al.*, 1986), lead us to assume that the flux variability obtained may illustrate the different exchanges between sediments and the overlying water column. Because the bottom had different characteristics, fluxes may vary greatly in magnitude and may even change direction. Probably sand takes up nutrients from the water, whereas most nutrients are released from mud sediments. Such a sand-mud take up-release contrast has also been reported by Nowicki & Nixon (1985) in a study of

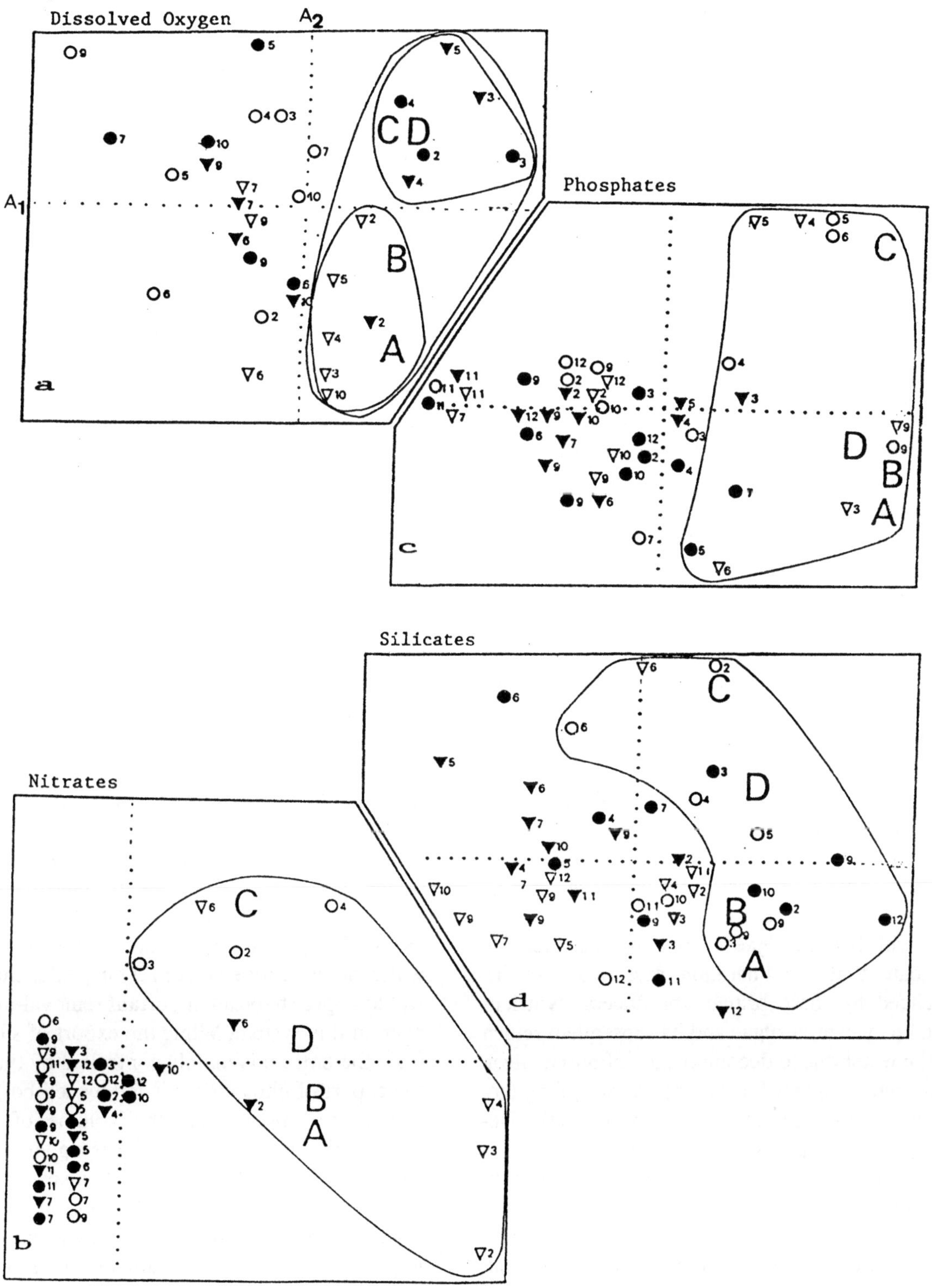

Fig. 6. Scattergram of stations A, B, C and D according to tidal conditions (low spring tide ○, low neap tide ●), high spring tide ▽ and high neap tide ▼) for various parameters (a) dissolved oxygen; (b) nitrates, (c) phosphates and d) silicates, over the year (months are indicated by numbers) in planes defined by the two principal components.

benthic nutrient remineralization in a coastal lagoon.

The presence of clams in sand (experiment S2) caused modifications in the exchanges with the overlying water: release of silicates and phosphates increased slightly and the liberation rate of ammonium was particularly sharp (Table 1). Lomstein *et al.* (1989) found that high sediment-water exchange of ammonium was correlated with a high macrofauna biomass. The increased amount of ammonium may depend on the biological effect of excretion and on liberation accelerated by bioturbation (Yamada & Kayama, 1987). These authors found that excretion by *Theora lata* was larger than that of bioturbation. The same dominant effect may also occur for *R. decussatus* in Ria Formosa, as ammonium fluxes in other growth-banks (S3 and S4) were about the same order of magnitude. Release of silicate and phosphate to the overlying water increased dramatically if, in addition to clams, the bottom was covered by vegetation and contained vegetative detritus. Silicate fluxes were increased to 7–20 times the level of that calculated in sand or mud alone. Increases of magnitude are in agreement with the literature. For example, Rutgers van der Loeff *et al.* (1984) reported that silicate liberation in the presence of organisms was much higher than those calculated from interstitial water profiles assuming one-dimensional molecular diffusion. The highest fluxes were calculated for sediments bioturbated and with benthic algae. Plants and their detritus indicate that the liberation rate was also influenced by these primary producers. Nutrient uptake by benthic algae and its subsequent return to the water due to decomposition of buried algae has been recognized as an important cycling mechanism in coastal marshes (Nixon, 1980). Decomposition appears also to be an important process influencing benthic nutrient regeneration in Ria Formosa. To what extent clams, and hence the presence of growth-banks in the intertidal area, stimulate the transport of nutrients from bottom sediments covered by vegetation (which utilize the nutrients) is not known. Murphy & Kremer (1985) have shown that nutrient cycling and its availability to phytoplankton depends on the fate of the benthic algae.

Field measurements corraborated the importance of benthic nutrient regeneration for nutrient concentration in the water column. However, nutrient concentrations in Ria Formosa are quite time-dependent, as they are greatly influenced by the tide. Indeed, we found that concentrations varied in different time-scales: seasonally with the rise of temperature and all its induced processes; and with the tide, due to the large water volume exchanged with the sea. Higher concentrations of phosphates in spring/summer particularly at low-spring tide, suggest that exportation of phosphorus from sediments is faster in periods of higher temperature and maximum physical perturbation of the bottom. These observations agree with the analysis of phosphorus dynamics in marsh sediments made by Nixon (1980), which attributes the phosphorus release to the decomposition of Spartina materials. For silicates, the sediment-water exchange appears to be the major factor regulating its concentration in the water. In fact, silicates were always higher at low tide over the annual survey, that is, at situations of lower water volume in the lagoon, independently of the season. In the microcosms particularly high liberation of silicates was found when organisms living in sediments modified the physical structure of the bottom. The occurrence of high silicate levels at distinct areas of the lagoon (intertidal mud- and sand-flats, sandy channels near the entrance) may be interpreted as benthic silicate regeneration influencing the water of the entire lagoon. Thus, tidal flushing would appear to be an important removal-mechanism in this system. While the export of silicates from the bottom is mainly controlled by the tide, the export of phosphorus is influenced both seasonally and tidally. As the volume of water entering the lagoon at spring tide is higher than the subtidal volume (Monteiro, 1989), an exchange of the nutrients regenerated in the bottom with the adjacent coastal water should be expected. The importance of this mechanism on export of nutrients to the sea has been recognized in several coastal environments (Nixon, 1980; Ohtake *et al.* 1984; Dame *et al.*, 1986).

A different behavior should be expected for nitrates, as minimum concentrations in the inlet channel occurred in periods of minimum water volume. The exchange of nitrates with the coastal water apparently followed a pattern opposite to that observed for silicates and phosphates. The import of nitrates is observed in most marsh systems, and its disappearance is attributed to high rates of denitrification in the anaerobic sediments (Nixon, 1980; Yamada & Kayama, 1987). In the microcosm experiments, nitrate fluxes were always negative, which also indicates rapid denitrification in the sediment. On the other hand, higher nitrate concentrations in winter and spring may be due to higher rates of nitrification. Increase of nitrates with upland run-off, which is rather sporadic, probably does not account for the enhancement observed in particular at high tide. Unlike phosphates and silicates, production and uptake of nitrogen species are affected by a larger number of processes, such as nitrification, nitrate reduction and denitrification in sediments, all of which may occur simultaneously at different rates and depths in the sediment (Smethie, 1987; Lomstein *et al.*, 1989).

In spite of this complexity the levels of ammonium were around 50 times higher than the nitrate concentrations, except for the run-off periods. Metabolic activities of animals and decomposition processes in the system may transform large quantities of particulate nitrogen into dissolved reduced forms (Lomstein *et al.*, 1989). The enrichment of ammonium is also reflected in the ratio of inorganic nitrogen to orthophosphates, which may reach 800. However, in periods of minimum water volume (low-spring tide), it decreased to 20, which is about the ratio obtained in the microcosm by dividing the respective fluxes. Intermediate values were obtained at intermediate tidal conditions, which means that most of the time the N/P ratio is much higher than the ratio of average plankton materials proposed by Redfield (16 : 1). If those inorganic ratios are realistic for the entire year, Ria Formosa tends to show a phosphorus limitation in the water rather than nitrogen limitation commonly associated with marine systems (Dufour *et al.*, 1981). Nowicki & Nixon (1985) observed a lack of available phosphate in the water of a coastal lagoon, and attributed the lack of remineralized phosphate to a take up by benthic microflora at the sediment surface.

Ria Formosa seems to be similar to other marsh systems, with nutrients being transformed within the lagoon and eventually escaping to the adjacent coastal waters (Nixon, 1980; Dame *et al.*, 1986). Nutrients are internally recycled by the benthic primary producers, and those released to the water are utilized by phytoplankton which presumably enters the lagoon in larger amounts during the warmer periods, as one may infer from higher clorophyll *a* in this period (Falcao, 1988). Tidal flushing leads to exchanges of internally recycled nutrients with the adjacent coastal water. Therefore a feed-back loop between Ria Formosa and coastal ocean should take place.

Acknowledgements

We thank M.L. Inácio, A.J. Figueiredo and A. Figueiredo for help with field and laboratory work. Support for this work was provided by the PIDR-Ria Formosa on the behalf of CCRA.

References

Dame, R., T. Chrzanowski, K. Bildstein, B. Kjerfve, H. McKellar, D. Nelson, J. Spurrier, S. Stancyk, H. Stevenson, J. Vernberg & R. Zingmark, 1986. The outwelling hypothesis and North inlet, South Carolina. Mar. Ecol. Prog. Ser. 33: 217–229.

Dufour, P., L. Lemasson & J. L. Cremoux, 1981. Contrôle nutritif de la biomasse du seston dans une lagune tropicale de côte d'Ivoire. II. Variations géographiques et saisonnières. J. Exp. Mar. Biol. Ecol. 51: 269–286.

Falcao, M. M., 1988. Alguns aspectos químicos, físicos e biológicos da zona lagunar Ria Formosa. Proc. V Congresso do Algarve, Albufeira. 2: 699–705.

Falcao, M., J. Pissarra & M. M. Cavaco, 1985. Características físicas e químicas da água na zona de Tavira: (1984/1985). Relat. Inst. Nac. Invest. Pescas, Lisboa, 60: 19 pp.

Legendre, L. & P. Legendre, 1984. Écologie numérique. 2. Le traitement multiple des données écologiques. Masson, Paris, 335 pp.

Lomstein, B. A., T. H. Blackburn & K. Henriksen, 1989. Aspects of nitrogen and carbon cycling in the northern Bering Shelf sediment. I. The significance of urea turnover in the mineralization of NH_4^+. Mar. Ecol. Prog. Ser. 57: 237–247.
Martens, C. S., 1982. Biogeochemistry of organic-rich coastal lagoon sediments. Proc. inter. Symp. on coastal lagoons, SCOR/IABO/UNESCO, Bordeaux, France, 8–14 September, 1981. Oceanol. Acta. N° SP: 161–167.
Millan-Nunez, R., S. Alvarez-Borrego & D. M. Nelson, 1982. Effects of physical phenomena on the distribution of nutrients and phytoplankton productivity in a coastal lagoon. Estuar. coast. Shelf. Sci. 15: 317–335.
Monteiro, C. C., 1989. La faune ichthyologique de la lagune Ria Formosa (sud Portugal). Thèse doct. Univ. Montpellier: 219 pp.
Murphy, R. C. & J. N. Kremer, 1985. Bivalve contribution to benthic metabolism in a California lagoon. Estuaries. 8: 330–341.
Nixon, S. W., 1980. Between coastal marshes and coastal waters – a review of twenty years of speculation and research on the role of salt marshes in estuarine productivity and water chemistry. in P. Hamilton, K. B. MacDonald eds. Estuarine and wetland process. Plenum Publ. Corp., New York: 437–525.
Nowicki, B. & S. Nixon, 1985. Benthic nutrient remineralization in a coastal lagoon ecosystem. Estuaries. 8: 182–190.
Ohtake, H., S. Kato & Y. Date, 1984. The role of tides in removing phosphorus from lake Nakanoumi, a coastal lagoon in Japan. Arch. Hydrobiol. 99: 443–462.
Rutgers van der Loeff, M. M., L. G. Anderson, P. O. J. Hall, A. Iverfeldt, A. B. Josefson, B. Sundby & S. F. G. Westerlund, 1984. The asphyxiation technique. Limnol. Oceanogr. 29: 675–686.
Smethie, W. M., 1987. Nutrient regeneration and denitrification in low oxygen fjords. Deep-Sea Res. 34: 983–1006.
Sundby, B., L. G. Anderson, P. O. J. Hall, A. Iverfeldt, M. M. Rutgers van der Loeff & S. F. G. Westerlund, 1986. The effect of oxygen on release and uptake of cobalt, manganese, iron and phosphate at the sediment-water interface. Geochim. Cosmochim. Acta 50: 1281–1288.
Yamada, H. & M. Kayama, 1987. Liberation of nitrogenous compounds from bottom sediments and effect of bioturbation by small bivalve *Theora lata* (Hinds). Estuar. coast. Shelf. Sci. 24: 539–555.

Received 11 July 1989; in revised form 29 January 1990; accepted 4 May 1990

Hydrobiologia **207**: 147–152, 1990.
D. J. Bonin & H. L. Golterman (eds), Fluxes Between Trophic Levels and Through the Water–Sediment Interface.

Distribution of microaerophilic bacteria through the oxic-anoxic transition zone of lagoon sediments

Maria J. Ferrara-Guerrero & Armand Bianchi
Microbiologie Marine CNRS UPR 223 – Faculté des Sciences de Luminy – 70, route Léon Lachamp – 13288 Marseille cedex 09, France

Key words: seawater, marine sediments, bacteria, anaerobes, microaerophiles, oxygen concentration

Abstract

In marine sediments, where soluble gases diffuse only very slightly, many organisms struggle for molecular oxygen. Microaerophilic bacteria, able to grow at reduced pO_2 between 0.2 and 12%, have an advantage. Distribution of aerobes, microaerophiles and anaerobes was compared with the oxygen gradient in seawater and sediment samples collected in a northern Mediterranean lagoon. In the near bottom seawater and in the 0–10 mm upperlayer of sediment, the microaerophilic counts were less than 1% of aerobe densities. In the 10–15 mm zone, these two groups were equivalent in density (1×10^5 cells ml^{-1}). As expected, the microaerophiles took advantage of their low oxygen tension requirements in the subsurface sediments, between the well aerated zone (0–5 mm depth) and the low redox potential zone. Then, beyond a depth of 20 mm, the anaerobes prevailed in this sandy clay.

Introduction

Most of the microbial ecosystems are organized into heterogeneous microniches, in which microbial proliferation is controlled by the interactive relationships of many physical and chemical parameters. In the water-sediment boundary layer and upper sediment layers (0–5 cm depth), in which biomasses of macroorganisms and microorganisms are very extensive, oxygen concentration and redox potential can vary drastically from oxygen oversaturation to anoxic conditions.

Electron acceptor availability is an important parameter for bacterial community distribution within an ecosystem (Marty *et al.*, 1988a). In spite of this, most of the data concerning bacterial densities in the marine environment refer to total count, without any reference to the physiological status of the microscopically counted bacterial cells and to counts of viable obligate or facultative aerobes. Few data on the distribution of anaerobes are available (Marty, 1981; Winfrey *et al.*, 1981; Buck, 1982; Marty & Garcin, 1987; Marty *et al.*, 1988b), and counts concerning microaerophiles in the marine or even the terrestrial environments are very scarce (Okon *et al.*, 1977; MacFarlane *et al.*, 1984; Barrequio *et al.*, 1988). Microaerophilic bacteria, which use molecular oxygen as terminal electron acceptor for their respiratory metabolism, are not able to grow under atmospheric oxygen concentrations. Because such bacteria need an oxygen-depleted environment (0.2–12.4%), the upper layers of marine sediments could be an advantageous habitat for them.

The aim of the present study was to compare the distribution of microaerophilic bacteria with the distribution of aerobes and anaerobes, at the water-sediment boundary layer in a Mediterranean lagoon, and to relate the oxygen distribution and Eh profiles.

Study area

Samples were collected in Le Brusc lagoon, a 320 ha lagoon formation, with a dominant seagrass flora of *Posidonia, Cymodocea* and *Zostera* (Aillaud & Pellegrini, 1970), on the northern coast of the western Mediterranean Sea. At the sampling station, the shallow waters (< 50 cm depth) had an average temperature and salinity of 17 °C and 31‰, respectively. Sediments were sandy clays. *Cymodocea* and *Zostera* were the dominant seagrasses, and the macrofauna was characterized by various boring polychaetes and a large colony of *Upogebia pusilla*, a very active boring crustacean (Vicente, 1973).

Materials and methods

Sediment sampling and characteristics

Undisturbed sediment and its overlying seawater were sampled using plexiglass tubes (3 cm inner diameter, 13 cm long). Along the tube we bored a series of 0.5 cm i.d. holes at 1.0 cm intervals. Before coring, the holes were sealed with a thin layer of silicone sealant. Cores were brought to the laboratory in an ice bow and processed within 1 hour.

Eh and pH profiles were measured with combined minielectrodes (Ingold, respective diameters 5 mm and 4 mm). Oxygen profiles were measured with a needle electrode (0.07 o.d., 2.5 cm l., Diamond Electro-Tech. Inc. type 760–45 °), modified according to Helder and Bakker (1985) and covered with a DepeX membrane (Revsbech, 1983). The reference electrode was an Ag/AgCl minielectrode (Diamond Electro-Tech. Inc., type 334). These electrodes were connected to a Chemical Microsensor Transidyne 1201.

For oxygen concentration measurements the core was kept in the dark and in a cooling water bath using local seawater at *in situ* temperatures. The electrodes were introduced into the vertically handled core by steps of 5 mm to 25 mm sediment depth. Then the core was kept in a horizontal position to permit electrode penetration into the preexisting holes at 10 mm intervals to a 50 mm sediment depth.

Two cores were fractionated into 10 mm segments for organic carbon content and porosity measurements. Two replicates were made on each segment. Organic carbon content was calculated from loss of dry weight at 550 °C for 24 h (Gallardo *et al.*, 1987).

Bacterial counts

Most Probable Number (MPN) was determined for microaerophilic, aerobic and anaerobic bacteria in the overlaying water and at 10 mm intervals into the sediment to 50 mm depth. Viable heterotrophs were determined by using 2216E liquid medium (Oppenheimer & ZoBell, 1952) in screw-capped vials for aerobes, and in rubber-stoppered vials with a controlled atmosphere (80/20: CO_2/N_2) for anaerobes (Hungate, 1969).

For microaerophilic viable counts, the SA500 semi-solid agar (Ferrara-Guerrero & Bianchi, 1989) was used in screw-capped tubes (15 × 140 mm) incubated at 20 °C for 15 days. The tubes in which growing rings occurred at 5 to 20 mm depth (the microaerophilic zone), were counted as positive for MPN determination.

For total direct counts, 1 ml samples were suspended in 9.0 ml of 0.2 μm freshly filtered water preserved with 2% formalin (final concentration) and stored at 4 °C until processing. To facilitate separation of bacterial aggregates, 1 ml of this suspension was diluted into 9.0 ml of 2 mM $Na_4P_2O_7$ (Albright *et al.*, 1986) and 20 min later sonicated on ice. Sonication was carried out three times for 1 min (in an MSE Ultrasonic probe at low power), with intervals of 30 s. Aliquots of this suspension (or aliquots of an appropriate dilution) were stained with acridine orange, filtered through a 0.2 μm porosity black Nuclepore membrane. Bacterial cells were counted using epifluorescence microscopy (Hobbie *et al.*, 1977).

Results

For each parameter presented, the following results describe the mean values obtained for three different sampling operations processed in different hydrodynamic conditions.

Environmental parameters

Oxygen concentration and Eh values decreased regularly from the near sediment/water interface to 50 mm depth into the sediment core, describing a progressive change from an aerobic to an anaerobic ecosystem (Fig. 1). The porosity decreased also with increasing depth in the sandy clay, while the pH values increased very slightly from 5 to 40 mm depth in the core (Table 1). The organic carbon content showed erratic fluctuations (Fig. 2).

Bacterial densities

Direct counts of bacterial cells exhibited small variations along the whole sediment depth. Viable counts of the three respiratory groups of bacteria showed larger fluctuations (Fig. 3).

The density of aerobes was maximal at the water-sediment boundary layer (1.4 × 10^6 cell ml^{-1}), and then decreased with depth. Microaerophilic bacteria were usually detected in the water as well as in the sediment samples. The

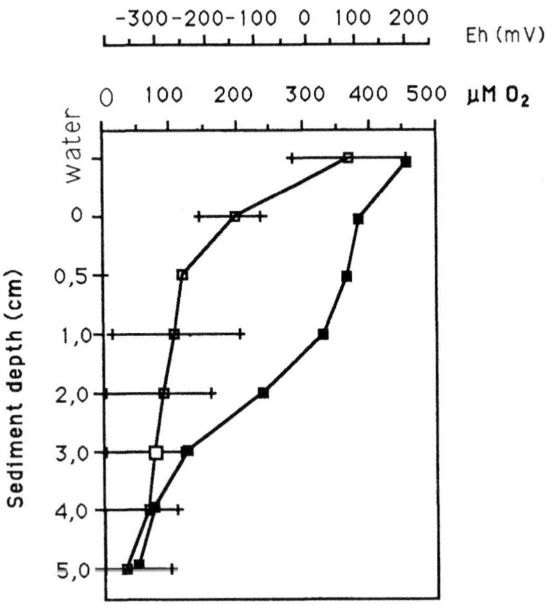

Fig. 1. Redox potential in mV (■) and oxygen concentration in μM (□) vertical profiles in the near bottom water and sediment samples. Mean values of three sampling operations in diverse hydrological conditions. Horizontal bars indicate oxygen ranges. 0 cm depth = sediment surface.

Table 1. Vertical distribution of physical, chemical and bacteriological parameters in the near bottom seawater and the sandy clay sediment in Le Brusc lagoon.

Depth (cm)	pH	Eh (mV)	Porosity % (v/v)	Org. carbon % dry wt.	μM O_2	Viable bacteria: Microaer × 10^3 ml^{-1}	Viable bacteria: Aer × 10^5 ml^{-1}	Viable bacteria: Anaer × 10^6 ml^{-1}	Direct count 10^8 bact. ml^{-1}
Water	7.9	+205			367.9	0.22	2.60	0.08	00.18
Surface	7.9	+108	26.7	5.5	197.7	0.73	14.00	0.04	10.30
0.5–1.0	7.5	+083	22.5	4.0	116.4	0.25	2.00	0.02	15.00
1.0–1.5	7.6	+036	24.7	5.3	104.4	91.80	1.13	3.80	10.20
2.0–2.5	7.6	−085	21.3	4.2	90.3	0.46	0.20	1.50	7.03
3.0–3.5	7.7	−237	19.4	7.3	76.9	6.82	0.43	55.00	4.03
4.0–4.5	7.7	−303	21.7	4.7	67.3	3.28	4.82	10.00	6.00
5.0–5.5	7.6	−337	14.2	8.3	33.0	0.68	0.20	1.50	11.50

Mean values of three sampling operations in diverse ecological conditions.

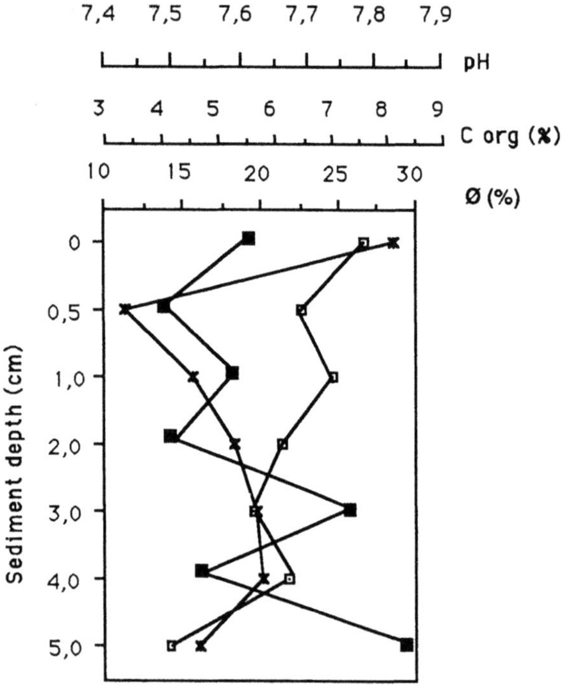

Fig. 2. Profiles of vertical abundance of organic carbon (■) referring to vertical evolution of porosity (□) and pH (×).

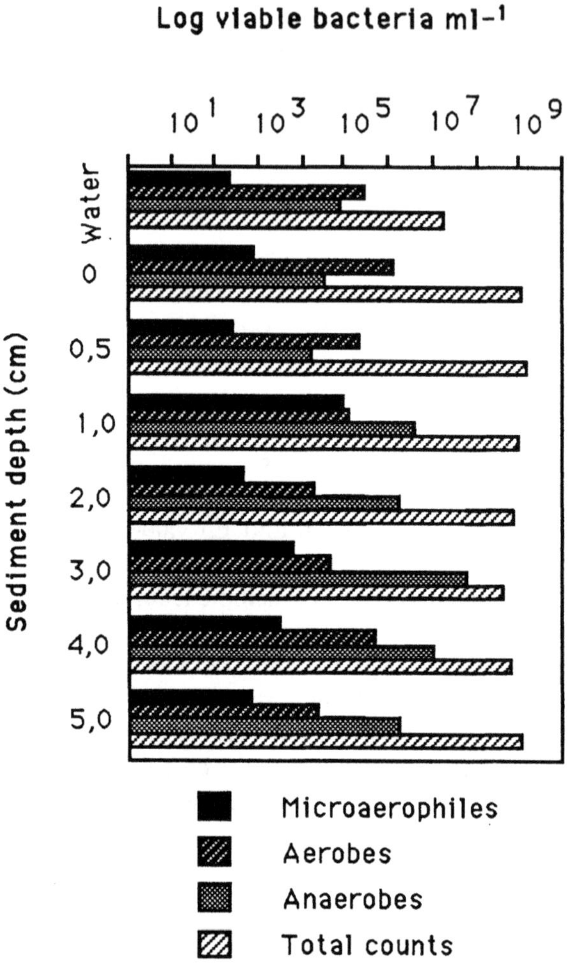

Fig. 3. Viable and total bacterial counts. Comparison of microaerophile, aerobe and anaerobe viable counts with the total counts.

highest density of these bacteria (9.1×10^5 cell ml^{-1}) was measured at 10 mm depth (Table 1), while they appeared less concentrated in the well-oxygenated upper sediment layers (few hundred cells per ml). The anaerobes were also present in both seawater and sediments. Their greatest concentration (5.5×10^7 cell ml^{-1}) was noted at 30 mm depth in sandy clay, and their density decreased lower in the profile.

Discussion and conclusion

Water mass exchanges between the lagoon and offshore seawater are probably responsible for the variations in oxygen concentration (283–455 μM O_2) observed in the near bottom seawater samples. In this lagoon, seagrasses provide a large quantity of oxygen and the resulting highly oxygenated seawater does not constitute a favourable environment for microaerophilic bacteria, not able to take advantage of their reduced oxygen tension requirements. Thus they were lacking in some seawater samples, or were present only at low concentrations (maximum value 2.5×10^3 bacteria ml^{-1}, mean density 2.2×10^2 bacteria ml^{-1}).

In the seawater samples, anaerobic counts appeared higher than the microaerophilic bacterial counts (mean density 8.0×10^4 cells ml^{-1}), but this difference could be due to the development of aerobic-anaerobic consortial pairs in the culture medium used to count anaerobes (Sieburth, 1988). Moreover, microaerophiles, better known as oligocarbophilic and oligonitrophilic bacteria (Stanier *et al.*, 1976; Okon *et al.*, 1977;

Krieg, 1981), were counted using the oligotrophic culture medium, which appeared in a previous study as the most efficient for marine microaerophile counts (Ferrara-Guerrero & Bianchi, 1989). On the contrary, anaerobes were counted in a nutrient rich broth. Thus, the difference in densities could also be due to differences in the nutritional requirements of these two groups.

In the upper layer of sediment, oxygen concentration was high (198 μM). In this layer, aerobes were the major group, but anaerobes were able to proliferate in anoxic microniches (Jørgensen, 1977; Bianchi, 1979). Oxygen gradients between these microniches could be favourable for microaerophilic bacteria. Their counts, however, appeared very low (few hundred cells per ml), probably for the same nutritional reasons suggested above. The oxygen concentration dropped to 116 μM at 5–10 mm depth in the sediment. This decrease in oxygen content resulted in a decrease of viable aerobe counts (Table 1). Deeper in the sediment the oxygen concentration decreased slightly to 33 μMO_2 at 50 mm depth, where the redox potential was as low as − 350 mV. This distribution of molecular oxygen in sediment rich in organic matter (5–7% organic C, w/w), could have several causes:

(i) this organic material resulting mostly from seagrass decomposition, is refractory in nature, due to lignocellulosic fibres which are not a good substrate for bacterial growth;
(ii) the high porosity (14% at 50 mm depth) of these sandy clays facilitates both water and oxygen circulation;
(iii) bioturbation processes due to diverse excavating polychaetes and crustaceans may have induced a flow of oxygenated supernatant seawater down into the sediment.

Thus, in spite of a high oxygen uptake both by macrofauna and microbial communities in the upper layer of sediments, we can see that in the 50 mm layer, where the redox potential was as low as − 250 or − 350 mV, the oxygen concentration (100–30 μMO_2) was sufficient to support an aerobic microflora, probably in oxic microchannels maintained in these conditions by sediment porosity or by bioturbation processes.

As expected, aerobes in the surface sediment layer and in the 5–10 mm upper layer showed the highest counts and the densities of microaerophiles were less than 1% of the aerobes. A notable presence of microaerophilic bacteria, with a density equivalent to that of aerobes, was observed in the 10–15 mm layer, concomitantly with the beginning of increased anaerobic counts and decreased aerobic counts. In this layer, the combination of low oxygen concentration unfavourable for aerobic growth, and positive values of Eh not advantageous for anaerobic growth, created environmental conditions of which the microaerophiles were able to take advantage. Nevertheless, the anaerobic microflora was here the major group. Increases of anaerobes and microaerophiles in this oxygen depleted zone could be favoured by metabolic products resulting from organic material degradation by aerobic microbial communities in the uppermost layer (Marty *et al.*, 1988a).

Just below this zone, microaerophilic counts decreased drastically to a few hundred cells per ml. At 30–35 mm depth, where the carbon content of sediment was high (7%) and the Eh low (− 237 mV), anaerobes exhibited the highest densities of 5.5×10^7 bacteria ml^{-1}, higher than the counts of aerobes and microaerophiles by three orders of magnitude. Within this level, the viable anaerobic density/total count ratio was almost 14%, whereas it remained less than 1% in all other levels.

In conclusion, the aerobes in this lagoon appeared to be the major group of bacteria only in the seawater and in the upper layer of the sandy clay sediment. Microaerophiles were present in all the seawater and sediment samples, but were high in number (almost 1×10^5 cells ml^{-1}) only in a very thin layer in the near surface sediment. This area corresponded to the boundary layer between the well aerated superficial layer (0–5 mm depth) and the low redox potential area, appearing at 20 mm depth in this sandy clay, where anaerobes were more numerous than other respiratory groups. This localization corresponded to the pattern expected, given the metabolic requirements of microaerophilic organisms.

Acknowledgements

We thank P. Rundel for his valuable comments on the manuscript; Danièle Mazaudier for assistance in field and laboratory work, Fernando Garcia (I.M.P., Mexico) and Michel Denis (Centre d'Océanologie de Marseille) for their assistance with the oxygen minielectrode utilization. M. J. Ferrara-Guerrero thanks the French Government and CONACyT (Mexico) for fellowships.

References

Aillaud, G. & M. Pellegrini, 1970. Etude écologique et biocénotiques dans la baie du Brusc (Var). Contribution à l'étude de l'ambiance climatique de la lagune de Brusc. Bull. Inst. océanogr. Monaco 69: 1–24.

Albright, L. J., S. K. McCrae & B. E. May, 1986. Attached and free-floating bacterioplankton in Howe Sound British Columbia, a coastal marine Fjord-Embayment. Appl. envir. Microbiol. 51: 614–621.

Barrequio, W. L., A. Dumont & R. Knowles, 1988. Enumeration of a Free-Living Aerobic N_2-Fixing H_2-oxidizing bacteria by using a heterotrophic semisolid medium and Most-Probable-Number Technique. Appl. envir. Microbiol. 54: 1313–1317.

Bianchi, A., 1979. Distribution qualitative et qualitative des populations bactériennes à l'interface eau-sédiment. In R. Daumas (ed.), Biogéochimie de la matière organique à l'interface eau-sédiment marin. Editions du C.N.R.S., Paris: 269–273.

Ferrara-Guerrero, M. J. & A. Bianchi, 1989. Comparison of culture methods for enumeration of microaerophilic bacteria in marine sediments. Res. Microbiol. (Inst. Pasteur/Elsevier) Paris 140: 255–261.

Gallardo, J. F., J. Saavedra & T. Martin-Palino, 1987. Solid organic matter determination. Commun. in Soil. Sci. Plant. Anal. 18: 699–707.

Helder, W. & J. F. Bakker, 1985. Shipboard comparison of micro-and minielectrodes for measuring oxygen distribution in marine sediments. Limnol. Oceanogr. 30: 1106–1109.

Hines, M. E. & J. D. Buck, 1982. Distribution of methanogenic and sulfate-reducing bacteria in near shore marine sediments. Appl. envir. Microbiol. 43: 447–453.

Hobbie, J. E., R. J. Daley & S. Jasper, 1977. Use of Nuclepore filters for counting bacteria by fluorescence microscopy. Appl. envir. Microbiol. 33: 1225–1228.

Hungate, R. E., 1969. A roll tube methods for cultivation of strict anaerobes. In J. R. Norris & D. W. Ribbon (eds), Methods in microbiology. Academic Press, London, 3B: 117–132.

Jørgensen, B. B., 1977. The sulfur cycle of a coastal marine sediment (Limfjorden, Denmark). Limnol. Oceanogr. 22: 814–832.

Krieg, N. R., 1981. The Genera *Spirillum*, *Aquaspirillum* and *Oceanospirillum*. In M. P. Starr, H. Stolp, H. G. Trüper, A. Balows & H. G. Schlegl (eds), The Prokaryotes a handbook on habitats, isolation and identification of bacteria. Springer-Verlag, Berlin. 1: 596–608.

Marty, D. G., 1981. Distribution of different anaerobic bacteria in Arabian Sea sediments. Mar. Biol. 63: 277–281.

Marty, D. G. & J. E. Garcin, 1987. Présence de bactéries méthanogènes méthylotrophes dans les sédiments profonds du détroit de Makassar (Indonésie). Oceanol. Acta 10: 249–251.

Marty, D. G., J. C. Bertrand & P. Caumette, 1988a. Les métabolismes bactériens dans les systèmes sédimentaires marins. In M. Bianchi, D. G. Marty, J. C. Bertrand, P. Caumette & M. Gauthier (eds), Micro-organismes dans les écosystèmes océaniques. Masson, Paris: 101–151.

Marty, D. G., J. C. Bertrand, P. Caumette & A. Bianchi, 1988b. Mineralization of organic material in coastal eutrophic ecosystems according to available terminal electron acceptors. Prog. Oceanogr. 21: 167–176.

MacFarlane, G. T., M. A. Russ, S. M. Keith & R. A. Herbert, 1984. Simulation of microbial processes in estuarine sediments using gel-stabilized systems. J. gen. Microbiol. 130: 2927–2933.

Okon, Y., S. L. Albrecht & R. H. Burris, 1977. Methods for growing *Spirillum lipoferum* and for counting it in pure culture and in association with plants. Appl. envir. Microbiol. 33: 85–89.

Oppenheimer, C. H. & C. E. ZoBell, 1952. The growth and viability of sixty-tree species of marine bacteria as influence by hydrostatic pressure. J. mar. Res. 11: 10–18.

Revsbech, N. R., 1983. In situ measurement of oxygen profiles of sediment by use of oxygen microelectrodes. In E. Gnaiger & H. Forstner (eds), Polarographic oxygen sensors. Aquatic and physiological applications. Springer-Verlag, Berlin, 265–273.

Sieburth, J. McN., 1988. The trophic roles of bacteria in marine ecosystems are complicated by synergistic-consortia and mixotrophic-cometabolism. Prog. Oceanogr. 21: 117–128.

Stanier, R. Y., E. A. Adelberg & J. L. Ingraham, 1976. Microbial World. Prentice-Hall, Inc.; Englewood Cliffs, New Jersey: 30–58.

Vicente, N., 1973. Intérêt écologique de la Lagune du Brusc. Fondation Scientifique Ricard. Bulletin de l'Observatoire de la mer. Le Brusc (Var, France). 1–12.

Winfrey, M. R., D. G. Marty, A. Bianchi & D. M. Ward, 1981. Vertical distribution of sulfate reduction, methane production, and bacteria in marine sediments. Geomicrobiol. J. 2: 341–362.

Received 31 July 1989; in revised form 8 January 1990; accepted 20 April 1990

Hydrobiologia **207**: 153–160, 1990.
D. J. Bonin & H. L. Golterman (eds), *Fluxes Between Trophic Levels and Through the Water–Sediment Interface.*

In situ biochemical and bacterial variation of sediments enriched with mussel biodeposits

Christian Grenz[1], Marie-Noelle Hermin[2], Dominique Baudinet[1] & Raoul Daumas[2]
[1] *Centre Océanologie de Marseille, Station Marine d'Endoume, 13007 Marseille, France;* [2] *Laboratoire de Microbiologie Marine (CNRS), Faculté des Sciences de Luminy, Case 907, 13288 Marseille Cedex 9, France*

Key words: mussel deposits, bacterial activity, exoenzymes, degradation rate

Abstract

In order to estimate the qualitative variation of sedimented mussel deposits, biochemical and microbial measurements were undertaken after a sediment enrichment with fresh faeces and pseudo-faeces collected in a mussel farming area (Carteau, Gulf of Fos).

During two months, cores sampled were investigated at three stations: a first station enriched with mussel deposits, a second considered as a reference station without mussel deposits and occasionally a third corresponding to a continuously enriched sediment under a rope culture.

Carbohydrate, organic carbon and nitrogen contents in the sediments gave evidence of a short term variation after sedimentation.

Bacterial production increased rapidly at the enriched station and returned to its initial level 8 days later. At this station, exoglucosidasic activity of bacteria, low at the beginning, presented a maximum two weeks after enrichment whereas exoproteolytic activity, which was high in the biodeposits, decreased in the course of the first week. These exoenzymatic activities were significantly higher at the enriched station than at the reference stations. Carbohydrate measurements were in agreement with these results.

Degradation rate of biodeposits is discussed on the basis of exoenzymatic activity, organic carbon, nitrogen and free amino acid content at the three stations.

Introduction

A number of works dealing with coastal ecosystem functioning have demonstrated the dominant role played by filter feeding animals in sedimentation of organic matter (Kuentzler, 1961; Dame *et al.*, 1984; Kautsky & Evans, 1987). Biodeposits have then an influence on the recycling process of nutrients after mineralization. These processes are mainly due to physical and chemical dissolution, bacterial activity as well as invertebrate trophic activity leading to sediment particle reworking (*e.g.* bioturbation).

Average mussel biodeposit production in suspended culture areas reaches enormous quantities, up to 345 kg m^{-2} y^{-1} (ash free dry weight) (Grenz, 1989). Since huge quantities of organic matter are contained in faeces and pseudo-faeces, an estimation of biodegradation times is essential for the understanding of the role played by biodeposits in matter and energy transfers in these specific areas.

Stuart *et al.* (1982) compared degradation of seaweed fragments and deposits from a mollusc (*Aulocomya ater*). Results showed that for a temperature of 10 °C, biodeposits were mineralized in 33 days with a maximal efficiency in the first 3 days. Bacterial biomass increased within the first 3 days of the experiment and slowly returned to the initial level. These experimental studies performed under controlled conditions suggested that organic matter degradation processes occurred over a relative small time scale. According to Jacobsen & Azam (1984) and Kirchman (1983), complete decomposition of copepod faecal pellets occurred in 50 days at 18 °C.

The purpose of this study was to follow, *in situ*, degradation states of organic matter in a sediment enriched with mussel biodeposits, compared to a standard sediment, during a 60 day period. Observations deal mainly with biochemical and microbiological changes in sediments. Results presented here are concerned with:

- Carbon and nitrogen content, as an estimation of organic matter,
- Carbohydrates, which are the main component of mussel biodeposits,
- Dissolved free amino acids in pore water,
- Bacterial exoenzymatic activity and production.

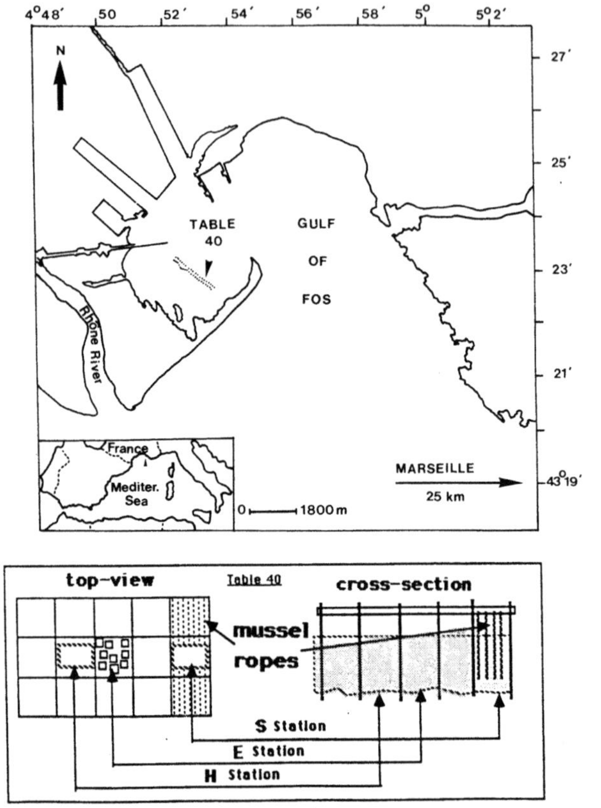

Fig. 1. Geographical localisation of the experimental stations: E (artificially enriched with mussel biodeposits), H (reference site) and S (sediment under mussel ropes) inside the rope hanging structure called 'Table 40' in the Gulf of Fos.

Materials and methods

A sediment free from mussel biodeposits was contaminated with freshly collected faeces and pseudo-faeces. Qualitative variations were measured during a 60 day period at the end of spring 1988 in Carteau Cove (Gulf of Fos). The study site was located under a partially colonized mussel rope hanging structure *e.g.* 'Table 40' (Fig. 1) where no mussel biodeposition occurred. The enrichment experiment was performed in eight 30 cm^2-quadrates. In each quadrate, about 400 ml freshly collected biodeposits were placed at the interface to a height of 1 to 2 cm on day 4. Day 0 corresponds to point zero. Cores were sampled by scuba divers following the sampling schedule (Table 1), at three stations: station E (enriched with biodeposits), station H (reference site without biodeposits) and station S (continually enriched sediment under a rope culture).

The cores were either pre-treated (exoenzymatic activity and bacteria production) or frozen. At the lab they were cut into one cm sections, freeze-dried, ground and sieved on a 63 μm mesh for subsequent biochemical analysis. Temperature was measured with the Horiba sensor.

- Chlorophyll *a* in seawater was estimated as follows: samples (250 ml) were filtered out on GFC glass fiber filters, extracted with acetone 90% (6 ml) and absorbance was tested at 665 nm before and after acidification. Calculations were made using the Lorenzen formula (1967).

Table 1. Survey calendar during 1988 experiment and corresponding parameter analyses (1: carbohydrate analyses; 2: the other analyses) at station E (enriched), station H (ref. without mussel biodeposits) and station S (under a mussel 'table').

Date	04/21	04/25	04/27	05/03	05/09	05/24	05/30	06/13	06/18
Days	0*	4**	6	12	18	33	38	53	58
Analysis 1	H-S	E-H	E-H	E-H	E-H	E-H	E-H	E-H	E-H
Analysis 2	H-S	E	E	E	E	E	H-S	E	E

* Point zero.
** Biodeposit enrichment.

- Organic carbon (OC) and nitrogen (N) contents were measured using the Leco 800 CHN analyser. OC was estimated according to Kristensen & Andersen (1987). Results were expressed as percentage of total dry matter. The coefficient of variation on triplicate analyses of sediment samples was 1.5% (OC) and 40.0% (N).
- Soluble and insoluble carbohydrates were extracted according to Khripounoff's techniques (1979). Analyses were carried out using the Dubois *et al.* (1956) method. Results were expressed as μg glucose equivalent per mg dry matter.

Acidic hydrolysis kinetics: we adapted methods described by Jocteur-Monrozier & Jeanson (1979) to carbohydrates in order to give an estimation of sugar degradability. T_{75} was the time required for 75% of carbohydrates to be released by hydrolysis with H_2SO_4 (1 N). T_{75} increased with hydrolytic resistant material. Thus velocity of hydrolysis processes appeared to correspond to the degree of resistance of organic matter.

- Dissolved free amino acids. Identification of amino acids was done with HPLC using pre-column derivatization with OPA (Lee & Drescher, 1978; Lindroth & Mopper, 1979; Mopper & Lindroth, 1982; Hermin *et al.*, 1986). HPLC was a LCS 620 Kontron Compact system with a Beckman column (ODS Ultrasphere 5 μm 4.6 × 250 mm).
- Exoenzymatic activity. Using hydrolysis of chromogenic substrata, it was possible to measure exoenzymatic activity of bacteria living in the environment. Substrata were as follows:
  - L-Leucine-4-methyl-coumarinyl-7-amide (MCA-Leu) specific to amidopeptidases
  - 4-methylumbelliferyl-β-D-glucoside (MUF-Glu) specific to β-glucosidase.

Substrata were injected (MCA-Leu and MUF-Glu 5nM dissolved in sterile sea water) into the sediment. Incubation duration in darkness is 1 hour at sampling temperature. Reactions were stopped by freezing the samples. Measurements were done using a spectrofluorimeter (SFM 23/B) as described by Hermin (1989). In order to assess exoenzymatic activities in the sediments, the estimated hydrolysis rates were recalculated on the basis of real substratum content (Leu and Glu) measured *in situ*. Results were expressed as nmol of hydrolysed substratum per g dry sediment and per hour.

- Bacterial production was estimated using ^{3}H-methyl-Thymidine (concentration 4 μCi/10 μl) (Moriarty & Pollard, 1981; 1982; Hermin, 1989). The conversion factor used for the calculation of increase in bacteria cells was 1.30 10^{18} cell mol^{-1} incorporated ^{3}H-Tdr. Production was expressed as the number of cells developed in 1 hour and in 1 cm^3. Taking the carbon content in the cell as 2.37 10^{-11} mg C Cell^{-1} (Moriarty, 1982; Fallon *et al.*, 1983; Stanley *et al.*, 1987), bacterial production was calculated and expressed in mg C m^{-2} d^{-1} at each depth investigated.

Results

Chlorophyll *a* (Chl*a*) variation in water column during the experimental period is shown in Fig. 2.

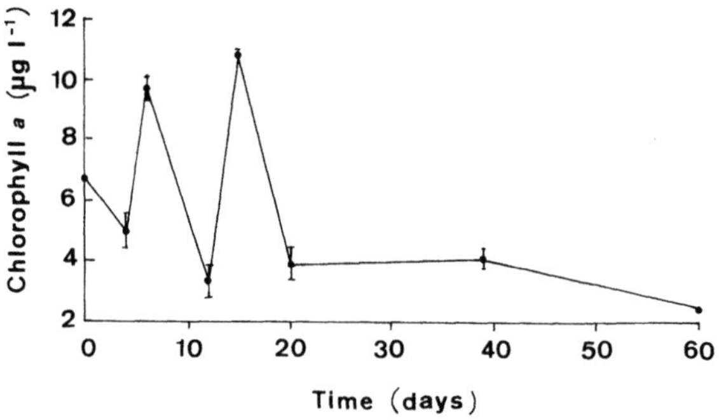

Fig. 2. Variation with time in chlorophyll a content in the water column during the experiment. Error bars are $\pm t_{95\%}$ standard error.

The classic features of spring bloom may be noted. Chl*a* reached values of 9.8 $\mu g\ l^{-1}$ and then dropped in a few days to 3.3 $\mu g\ l^{-1}$. Temperature was stable between day 0 and day 12 (14.8–15.0 °C) and then increased to 18 °C at day 53.

The time variation of organic carbon (OC) and nitrogen (N) content versus sediment depth is shown in Figs. 3a and b. Day 0 (point zero) corresponds to first sampling date. At that time, OC and N contents were respectively 0.54% and 0.08% in the first cm of sediment at station E. After enrichment (day 4), OC content reached 1.36 to 1.56% and N content 0.13 to 0.19% within a few days in the upper levels. A significant decrease can be seen at day 12, 1 week after enrichment. Beyond day 12, a second increase may be observed in the upper levels. On days 38 and 53, OC and N content at stations E and H were quite similar.

Figures 4a, b and c show respectively the variation of total carbohydrate content, insoluble/

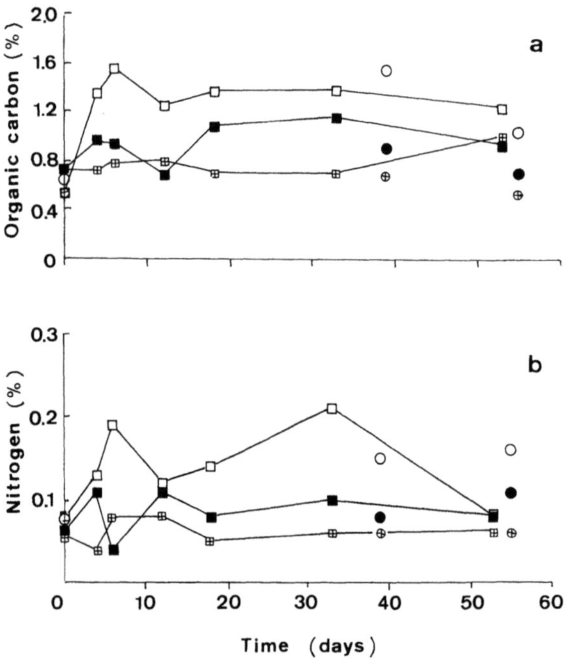

Fig. 3. Variation with time of organic carbon (a) and nitrogen (b) content in the 3 sediment depths (□: 0–1 cm, ■: 1–2 cm, ⊞: 4–5 cm) at station E (continuous line) and station H (○: 0–1 cm, ●: 1–2 cm, ⊕: 4–5 cm; isolated points).

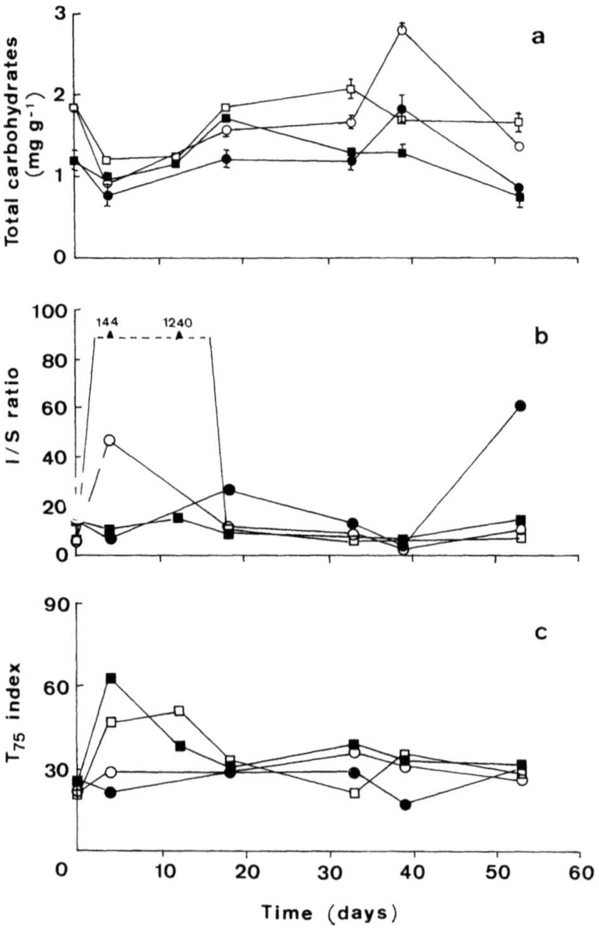

Fig. 4. Variation with time of a: Total carbohydrate content, b: Insoluble/Soluble ratio (I/S), c: Time of 75% hydrolysis index (T_{75}) in 2 sediment depths at station E (□: 0–1 cm, ■: 2–3 cm) and station H (○: 0–1 cm, ●: 2–3 cm). Error bars are $\pm t_{95\%}$ standard error.

soluble sugar ratio (I/S) and T_{75} index in the first cm (0–1 and 2–3 cm) of sediments sampled in both stations. Regarding total carbohydrates, no significant difference between stations, related to the biodeposit enrichment, could be determined. Highest content was measured on day 33 in the upper level. It is noteworthy that the I/S ratio increased enormously in the two first cm of the sediment in station E. It returned to its original value within 12–18 days. T_{75} index increased on day 4 and returned to its original value on day 18. At station S on day 0, a T_{75} value of 84 was recorded.

The sum of 23 DFAA in pore water was higher at station S than at the other stations (> 150 nmol g^{-1}). In the upper layer of sediment at station E, an increase of 108 nmol g^{-1} occurred on day 4 followed by a decrease within the next two days (Fig. 5). On day 18, DFAA content returned to its initial value. A slight increase was noticed on day 33 only at station E. DFAA contents at station H were lower than at the other stations during the whole period, except on the last sampling day.

In superficial sediment at station E, enrichment with faeces and pseudo-faces has a clear effect on exoproteolytic activity (EPA) (Fig. 6a). EPA increased from 87 to 318 nmol hydrolysed MCA-Leu g^{-1} h^{-1} on day 4, then dropped drastically to 100 nmol g^{-1} h^{-1} in the next sample. Induction of β-D-Glucosidase (EGA) differed from

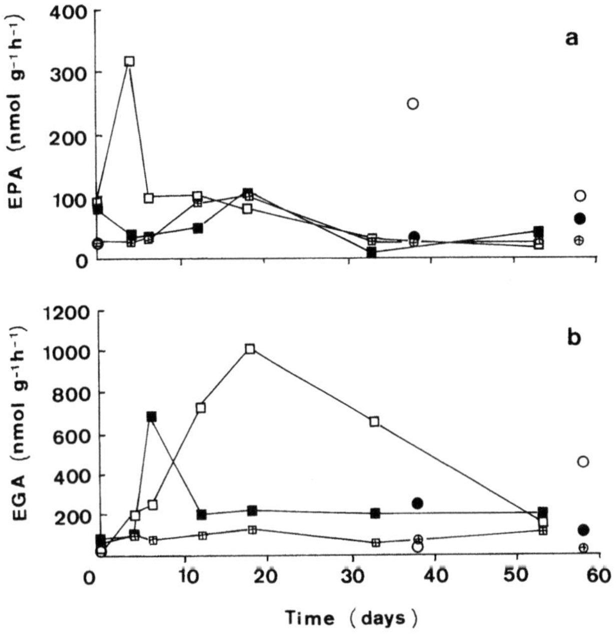

Fig. 6. Variation with time of a: Exoproteolytic activity (EPA), b: Exoglucosidasic activities (EGA) in sediments at station E (continuous line; □: 0–1 cm, ■: 1–2 cm, ⊞: 2–3 cm) and station H (○: 0–1 cm, ●: 1–2 cm, ⊕: 2–3 cm; isolated points).

Protease induction (Fig. 6b). In fact, at station E, EGA, induced by biodeposit enrichment, was weak at the beginning, increased from 200 to 1000 nmol g^{-1} d^{-1} until day 18, and then decreased progressively. EPA and EGA were generally higher at station S than at station H (> 200 nmol g^{-1} d^{-1}).

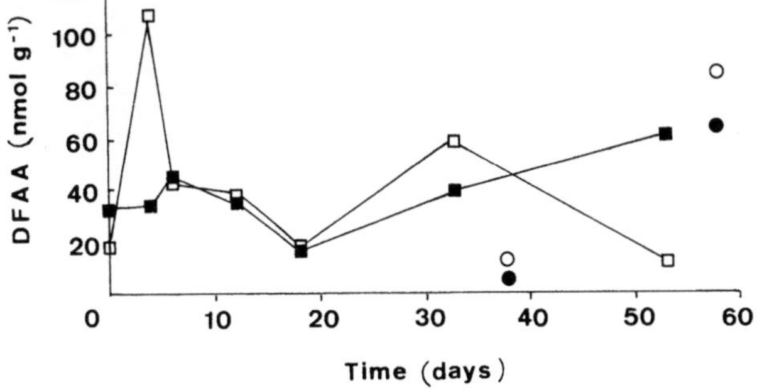

Fig. 5. Variation with time of the sum of 23 DFAA in interstitial water from sediments at station E (□: 0–1 cm, ■: 2–3 cm) and station H (isolated points; ○: 0–1 cm, ●: 2–3 cm).

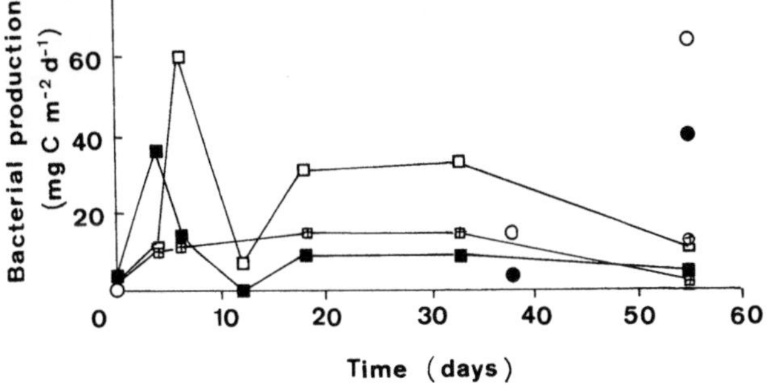

Fig. 7. Variation with time of bacterial production measured at station E (continuous line; □: 0–1 cm, ■: 1–2 cm, ⊞: 2–3 cm) and station H (○: 0–1 cm, ●: 1–2 cm, ⊕: 2–3 cm; isolated points).

Moriarty, D. J. W., 1982. Bacterial biomass and productivity in sediments, stromatolites and water of Hamelin Pool, Shark Bay, Western Australia. Geomicrobiol. J. 3: 121–133.

Moriarty, D. J. W. & P. C. Pollard, 1981. DNA synthesis as a measure of bacterial productivity in seagrass sediments. Mar. Ecol. Prog. Ser. 5: 151–156.

Moriarty, D. J. W. & P. C. Pollard, 1982. Diel variation of bacterial productivity in Seagrass (*Zostera capricorni*) beds measured by rate of thymidine incorporation into DNA. Mar. Biol. 72: 165–173.

Stanley, S. O., K. G. Boto, D. M. Alongi & F. T. Gillan, 1987. Composition and bacterial utilization of free amino acids in tropical mangrove sediments. Mar. Chem. 22: 13–30.

Stuart, V., R. C. Newell & M. I. Lucas, 1982. Conversion of kelp debris and faecal material from mussel *Aulocomya ater* by marine microorganisms. Mar. Ecol. Prog. Ser. 7: 47–57.

Received 2 September 1989; in revised form 20 March 1990; accepted 27 March 1990 after 20 days at the earliest and probably depending on temperature conditions and other environmental parameters. In our experiment, post-bloom sedimentation occurring at mid-period, interfered with biodeposit qualitative variation, so that clear interpretation of such long term degradation phenomena was not possible.

Nevertheless it is clear that under rope hanging structures where mussel biodeposition occurs continuously, first degradation steps are of particular importance when considering the fast recycling of the labile fraction of organic matter in such an ecosystem.

Acknowledgements

We would like to thank anonymous reviewers for critical comments on an earlier version of this manuscript. This project was supported in part by funds provided by the Groupement Scientifique (G.S.): 'Cycles biogéochimiques – Devenir de la matière organique en milieu littoral' (GS: PIREN – CNRS – Société Nationale Elf-Aquitaine – Université d'Aix-Marseille II).

References

Alberic, P. & A. Khripounoff, 1984. Relations entre les compositions en acides aminés des particules en voie de sédimentation, du contenu intestinal des holothuries abyssales et du sédiment environnant. Mar. Chem. 14: 379–394.

Alberic, P., J. P. Feral & M. Sibuet, 1987. Les acides aminés libres, reflet de l'activité bactérienne dans les contenus digestifs des holothuries: différence entre zones abysalle et littorale. C.r. Acad. Sci., Paris, 305, Série III: 203–206.

Daatselaar, M. C. C. & W. Harder, 1974. Some aspects of the regulation of the production of extracellular proteolytic enzymes by a marine bacteria. Arch. Mikrobiol. 101: 21–34.

Dame, R., R. Zingmark & E. Haskin, 1984. Oyster reefs as processors of estuarine materials. J. exp. mar. Biol. Ecol. 83: 239–247.

Dubois, M., K. A. Gilles, J. K. Hamilton, P. A. Rebers & F. Smith, 1956. Colorimetric method for determination of sugars and related substances. Analyt. Chem. 28: 350–356.

Fallon, R. D., S. Y. Newell & C. S. Hopkinson, 1983. Bacterial production in marine sediments: will cell-specific measures agree with whole-system metabolism? Mar. Ecol. Prog. Ser. 11: 119–127.

Fontigny, A., G. Billen & J. Vives-Rego, 1987. Some kinetic characteristics of exoproteolytic activity in coastal seawater. Estuar. coast. and Shelf Sci. 25: 127–133.

Grenz, C., 1989. Quantification et destinée de la biodéposition en zones de production conchylicole intensive en Méditerranée. Thèse Doct. Univ. Aix-Marseille II, 144 pp.

Halemejko, G. Z. & R. J. Chróst, 1986. Enzymatic hydrolysis of proteinaceous particulate and dissolved material in an eutrophic lake. Arch. Hydrobiol. 107: 1–21.

Hermin, M. N., 1989. Dégradation microbienne de la matière organique à l'interface eau-sédiment en milieu marin. Thèse Doct. Univ. Aix-Marseille II, 202 pp.

Hermin, M. N., R. A. Daumas & N. Drogue, 1986. Transferts d'acides aminés dans un système eau-sédiment: étude préliminaire en milieu expérimental. In GERBAM. Deuxième Colloque internationnal de Bactériologie Marine-CNRS, Brest, 1–5 octobre 1984, IFREMER. Actes de colloques 3: 61–72.

Jacobsen, T. R. & F. Azam, 1984. Role of bacteria in copepod fecal pellet decomposition: colonization, growth rates and mineralization. Bull. mar. Sci. 35: 495–502.

Jocteur-Monrozier, L. & P. Jeanson, 1979. L'azote organique et ammoniacal. Combinaison et stabilité chimique. In M. Arnould et R. Pelet (eds), Géochimie organique des sédiments marins profonds ORGON III, Mauritanie, Sénégal, Iles du Cap Vert. Editions du CNRS, Paris: 441 pp.

Kautsky, N. & S. Evans, 1987. Role of biodeposition by *Mytilus edulis* in the circulation of matter and nutrients in a baltic coastal ecosystem. Mar. Ecol. Prog. Ser. 38: 201–212.

Khripounoff, A., 1979. Relations trophiques dans l'écosystème abyssal atlantique: descriptions et bilan énergétique. Thèse 3ème cycle. Univ. Paris VI: 132 pp.

Kirchman, D., 1983. The production of bacteria attached to particles suspended in a freshwater pond. Limnol. Oceanogr. 28: 858–872.

Kristensen, E. & F. Andersen, 1987. Determination of organic carbon in marine sediments: a comparison of two CHN-analyser methods. J. exp. mar. Biol. Ecol. 109: 15–23.

Kuentzler, E. J., 1961. Phosphorus budget of a mussel population. Limnol. Oceanogr. 6: 400–415.

Lee, K. S. & D. G. Drescher, 1978. Fluorimetric amino acids analysis with orthophtaldialdehyde (OPA) Int. J. Biochem. 9: 457–468.

Lindroth, P. & K. Mopper, 1979. High performance liquid chromatographic determination of subpicomole amounts of amino acids by precolumn fluorescence derivatization with O-phtaldialdehyde. Analyt. Chem. 51: 1667–1674.

Lorenzen, C. J., 1967. Determination of chlorophyll and phaeopigments. Spectrophotometric equations. Limnol. Oceanogr. 12: 343–346.

Mopper, K. & P. Lindroth, 1982. Diel and depth variations in dissolved free amino acids and ammonium in the Baltic Sea determined by shipboard HPLC analysis. Limnol. Oceanogr. 27: 336–347.

Bacterial production (Fig. 7) increased at station E from 2 to 10 and even 60 mg C $m^{-2} d^{-1}$, following enrichment. Then a rapid decrease in a few days was observed in the first 2 cm of sediment. On day 18 and day 34, samples exhibited values of 30 mg C $m^{-2} d^{-1}$, which suggests new bacterial production. All values at station S were higher than the other stations.

Discussion

Two enrichments with organic matter were observed in the course of our experiment. The first, artificial enrichment, due to mussel faeces and pseudo-faeces deposited on day 4 at station E, was only recorded at this station. A further enrichment due to bloom deposition was recorded at all the three stations (E, S and H). The main impact of post-bloom sedimentation was on bacterial activity, resulting in some difficulty in interpreting bacterial phenomena after day 18. However, with regard to biochemical variation, it seems to be possible to estimate degradation velocity of mussel biodeposits. Variation of carbon and nitrogen content occurred quite rapidly, within a period of up to ten days. These results are in good agreement with the data of Kirchman (1983) and Jacobsen & Azam (1984). These authors observed full degradation of copepod faecal pellets within 50 days at 18 °C. Total carbohydrates did not exhibit wide variations, when measurement of variability was taken into account. I/S ratios and T_{75} index gave evidence of catabolic activity at the water-sediment interface even if the strong decrease of I/S value between day 12 and day 18, was only insufficiently reflected by total carbohydrate contents and EGA values. Nevertheless, potential exoglucosidasic activity increased after enrichment and decreased after day 18, which suggested that substratum became limiting. As no drastic temperature variation occurred during this period, increase in bacterial activity could be due to organic matter supply. For exoproteasic activity, results were quite different: activity was very strong when enrichment took place, and then faded out in a few days. This might suggest the hypothesis that bacteria from the digestive tract of mussels were probably deposited with biodeposits during sedimentation. Subsequent increase in proteasic activity and bacterial production could explain the first step of organic matter degradation. This appeared to be confirmed by observations made at station S where biodeposition was continuous. DFAA pool in particular was very high at this station. Alberic & Khripounoff (1984) and Alberic *et al.* (1987) found similar values for DFAA in gut content of Holothuria. Significant decrease of bacterial production within a few days might be explained by the disappearance of intestinal bacterial populations when in contact with sediment. However, another reason might be the lack of adequate substratum to sustain bacterial production, because of the low protein content versus carbohydrate content of mussel biodeposits (Grenz, 1989). The existence of enzymatic inhibitors in the environment is worth mentioning (Daatselaar & Harder, 1974; Halemejko & Chróst, 1986; Fontigny *et al.*, 1987). A new bacterial induction in sediment around day 18 may be deduced from the results concerning exoglucosidasic activities. Later on, it is difficult to distinguish between the effects of artificial enrichment and post-bloom sedimentation. Furthermore temperature increases were recorded during this second step.

Conclusion

Sediment enrichment with mussel biodeposits induced a strong reaction of bacterial (autochthonous or allochthonous) compartment at the sediment-water interface. This was consistent with an enzymatic hydrolysis of sedimented material. Duration of these effects was about 10 days. We might hypothesize that degradation of faeces and pseudo-faeces occurs in two steps. First, degradation of the less hydrolytic resistant fraction during the first week was due to bacteria issued from mussel intestinal tracts. The second step was slower and involved degradation of the most refractory fraction of organic matter, occurring

Hydrobiologia **207**: 161–166, 1990.
D. J. Bonin & H. L. Golterman (eds), Fluxes Between Trophic Levels and Through the Water–Sediment Interface.

Variation of nutrient stocks in the superficial sediments of Lake Geneva from 1978 to 1988

Daniel Span, Didier Arbouille, Hélène Howa & Jean-Pierre Vernet
Institute F.-A. Forel, 10 route de Suisse, CH-1290 Versoix, Switzerland

Key words: nutrients, speciation of phosphorus, superficial sediments, Lake Geneva

Abstract

Two sediment sampling campaigns were conducted in 1978 and 1988 in Lake Geneva (Switzerland). Organic carbon, total nitrogen, total phosphorus and its various forms were analyzed. Results indicate a stability of organic carbon and nitrogen mass, and a significant increase of phosphorus. The variation of phosphorus mass is related to the increase of nonapatite inorganic phosphorus. This study attempts to quantify the phosphorus exchanges at the water sediment interface. The dissolved oxygen level in the bottom water determines the exchange direction. In aerobic conditions, sediments accumulate the excess of phosphorus, while in anaerobic conditions, they constitute an internal source.

Introduction

The eutrophication of lakes is caused largely by excessive loading of phosphorus into the lakes (Vollenweider *et al.*, 1979).

At the beginning of this century, Lake Geneva showed the typical characteristics of an oligotrophic lake. By 1976, Lake Geneva was identified as phosphorus-limited and classified as mesotrophic. The total phosphorus concentration in the lake water was then about 90 $\mu g\,l^{-1}$ but, in recent years (1986-1987), that concentration has decreased to approximately 60 $\mu g\,l^{-1}$ (Blanc *et al.*, 1986).

A large fraction of this nutrient load is retained by the lake sediments. Since 1950, the vertical distribution of phosphorus and nitrogen concentrations has increased exponentially towards the sediment surface (Ravera & Pelletier, 1980). Under appropriate circumstances, the sedimented phosphorus can be released as biologically available phosphorus.

Since 1957, CIPEL (Commission Internationale pour la Protection des Eaux du Léman) has conducted systematic studies on the lake and its watershed. The aim of this contribution is to analyse the recent trends of nutrient loading in superficial sediments of Lake Geneva (Arbouille *et al.*, 1989).

Description of the site studied

Lake Geneva is the largest of the alpine lakes with a surface area of about 582 km^2, a volume of 89 km^3 and a maximum depth of 310 m (Fig. 1).

By differences in bathymetry, morphology and geochemical facies of superficial sediments, Lake Geneva can be divided into four distinct areas (Jaquet *et al.*, 1982):

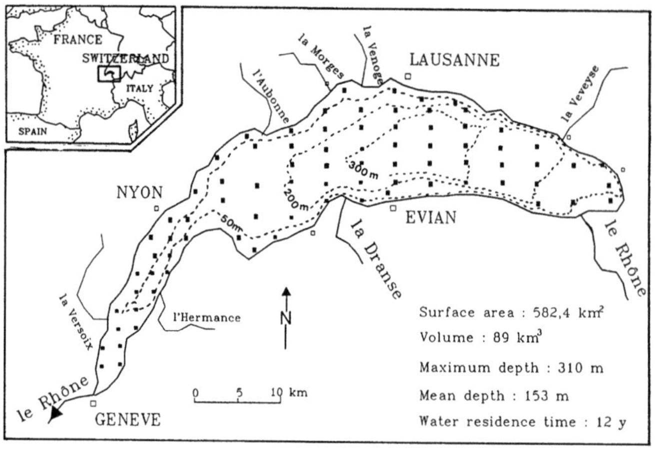

Fig. 1. Location of sampling stations and principal tributaries of Lake Geneva.

- the Rhone and Dranse fan areas where sediments are siliciceous (Rhone) or dolomitic (Dranse) allochtonous materials.
- the central plain: the deepest and largest area where clastic sediments are mixed with autochtonous carbonates.
- the 'Petit Lac' from Nyon to Geneva: this basin, the shallowest part of the lake is separated from the central plain by a morphological ridge off Nyon. Most of the sediments are autochtonous carbonates.
- the bay of Lausanne where the sediments are particularly contaminated by heavy metals.

The Rhone, the main river, supplies about 75% of the water and about 80% of the solids in the lake. The river Dranse's contribution to sediment input is about 10%. Remaining inputs result from numerous Swiss tributaries.

The major phosphorus input to the lake comes from the Rhone but only 15% is potentially bioavailable (Burrus *et al.*, 1989). Total phosphorus loading from the Rhone is therefore not the essential factor controlling the trophic state of Lake Geneva.

The nutrient input of anthropogenic origin is discharged primavility from 136 sewage treatment plants (STEP) located on the lake shore (near Lausanne) or on the lake tributaries.

Materials and methods

This work relies on the study of 170 samples collected with a Shipek grabber during two spring campaigns in 1978 and 1988 (85 samples for each campaign). All the different sedimentary and geochemical environments were sampled at depths varying from 5 to 310 meters (Fig. 1). The top few millimeters of each sample were wet sieved and dried. Nutrient concentration was determined only in the fraction less than 63 μm.

Chemical analyses were performed using the routine methods of the F.-A. Forel Institute (University of Geneva).

A Redox titration method was used to analyse organic carbon (%): sediment samples with excess $K_2Cr_2O_7$ were heated at 130 °C for one and half hours, then titrated with $FeSO_4$.

The forms of phosphorus in samples were considered as three fractions: apatite phosphorus (AP), nonapatite inorganic phosphorus (NAIP) and organic phosphorus (OP). These three forms were obtained using a procedure derived from Williams *et al.* (1976, 1980). Total phosphorus (TP) was estimated by perchloric acid digestion. Concentrations in digested solutions were measured using a colorimetric procedure (ascorbic acid method). The results are reported relative to dry sediment weight (mg kg^{-1}).

Total nitrogen in sediments was measured using the Kjedahl method.

Table 1. Evolution of nutrient contents. Means are calculated with a water content of 78.5%.

	O.C. in %	T.N. in %	T.P. analysed in mg kg^{-1}
1978	1,07	0,09	334
1988	1,11	0,11	344

Results and discussion

Nutrient variations

The average content of the sediments in nutrient elements over the last 10 years (1978 and 1988) is given in Table 1.

The quantities of carbon, nitrogen and phosphorus stored in the sediments vary with the oxydative capacity of the bottom water. The organic carbon (OC) and total nitrogen (TN) contents are low and indicate that mineralization of organic matter is efficient in the lake. Organic carbon and total nitrogen concentrations show little variation over the ten years of the survey. For these two nutrients, the lowest concentrations are observed near the Rhone and the Dranse fans (Fig. 2). In those areas, the dilution effect by sediment supplies is the most important. This phenomenon seems to extend along the active channel of the Rhone fan.

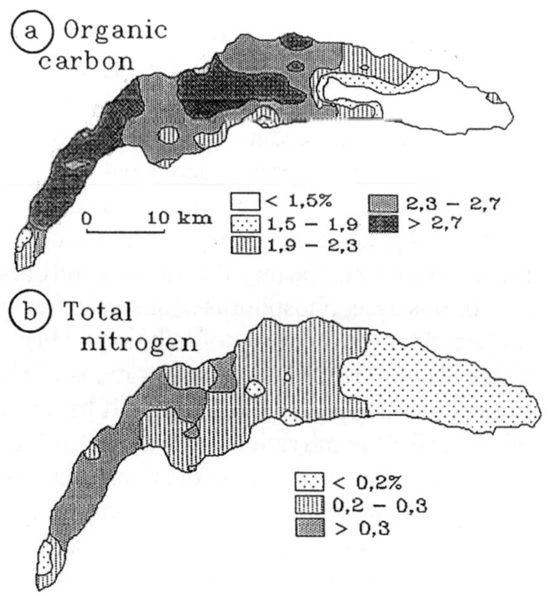

Fig. 2. Distribution of nutrients in superficial sediments (dry sediment and size less than 63 μm) of Lake Geneva (results of campaign 1988).
a) Organic carbon in %
b) Total nitrogen in %

Speciation of phosphorus

The analysis of different forms of phosphorus is as important as the quantity of total sediment phosphorus. From the results of previous analyses, it appears that apatite phosphorus has a detrital origin, whereas nonapatite inorganic phosphorus is anthropogenic (detergents...). Organic phosphorus is due to biological activity.

Apatite phosphorus (AP): This from consists of orthophosphates which are present in the crystal lattices of apatite. Apatite is natural and detrital. It represents more than 50% of total phosphorus in lake sediments. Burrus *et al.* (1989) found that

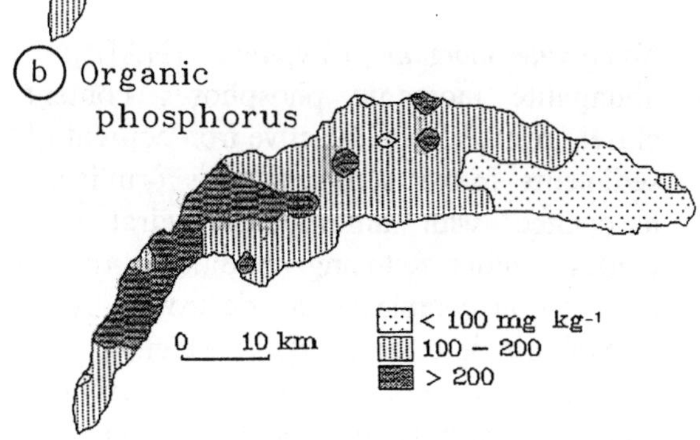

Fig. 3. Distribution of the forms of phosphorus in superficial sediments (dry sediment and size less than 63 μm) of Lake Geneva (results of campaign 1988).
a) Apatite phosphorus (A.P) in mg kg^{-1}
b) Organic phosphorus (O.P) in mg kg^{-1}

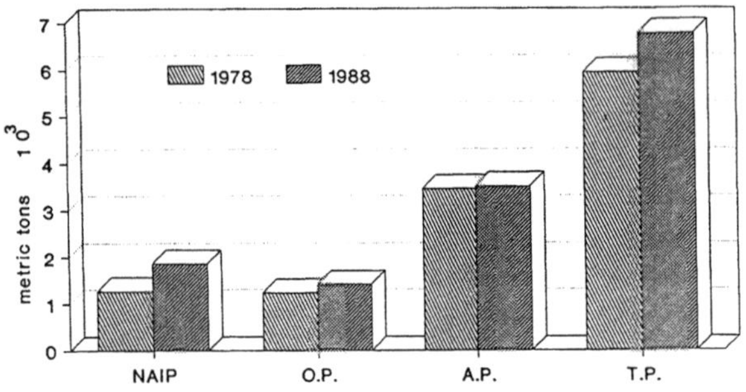

Fig. 4. Evolution of stocks of forms of phosphorus in superficial sediments of Lake Geneva. Stocks are calculated for sediments (2 cm of thickness) with a water content of 78.5%.

more than 80% of the total phosphorus of the Rhone river was associated with sediments. The apatite phosphorus content of the superficial sediments decreases gradually from the mouth of the Rhone towards the central plain (Fig. 3a). Between 1978 and 1988, the concentrations and the stock of this form have not changed (Fig. 4). However, its extension can vary with the seasonal trends of the Rhone sediment fluxes.

Organic phosphorus (OP): Organic Phosphorus (OP) has its origin in autochtonous organic matter and human activity in the watershed. A part of the organic phosphorus can be remobilized during organic matter decomposition. The spatial distribution (Fig. 3b) conforms to that of the organic carbon. A small increase of the stock occurred between the two surveys (Fig. 4).

Nonapatite inorganic phosphorus (NAIP): The nonapatite inorganic phosphorus content is closely related to the reactive iron content of the sediments. In oxic sediments, this form is mainly associated with amorphous hydrated ferric oxides. Under reducing conditions, the phosphorus-iron complexes are destroyed, releasing phosphorus into the overlying water.

In contrast to other species of phosphorus, a sharp increase in nonapatite inorganic phosphorus content was found over the last 10 years (Fig. 4). The spatial distribution of nonapatite inorganic phosphorus in 1978 was very different from that in 1988 (Fig. 5a and b). In 1988, the highest concentration of nonapatite phosphorus is observed in the central plain and near the sewage input from Lausanne and Evian. The evolution of dissolved oxygen in the bottom water of the central plain can explain this phenomenon (Fig. 5c). During the period 1970 to 1978, seasonal mixing of the bottom water did not occur due to relatively mild winters. The dissolved oxygen contents in the bottom water of the lake (Fig. 5c) decreased to a minimum of 0.3 mg l^{-1} (Blanc *et al.*, 1986). Under these near-anaerobic conditions, PO_4^{3-} diffuses into the water column. The orthophosphate concentration in the bottom water reached 230 $\mu g\, l^{-1}$. A major part of the nonapatite inorganic phosphorus was remobilized. Thus, the concentration of these species of phosphorus in the sediment was less important. This phenomenon is most noticeable in the central plain where the conditions were more reducing. In this area (depth greater than 300 m with a surface of 79 km^2), the mean phosphorus flux from sediments varied from 3 to 6 mg $m^{-2}\, d^{-1}$ under anoxic conditions (Rapin *et al.*, 1984). Thus, in 1978 about 90 to 180 metric tons of phosphorus were released from superficial sediments.

In the southern part of the central plain, another phenomenon can explain the small concentration of phosphorus observed in 1978. During the period preceding the first campaign, a 10 cm thick sand layer was deposited by a turbid current derived from the Dranse River (Vernet *et al.*, 1984). In 1978, the layer deposited before the turbidite could be mixed with this sandy material. In this case, phosphorus concentrations were underestimated, because of dilution. This turbidite also influenced the concentrations in phosphorus in the superficial sediments. It behaved as a barrier for the migration towards the surface of the phosphorus dissolved in the interstitial water.

Since 1978, winter mixing has improved the oxygenation of the bottom water with concentrations of dissolved oxygen greater than 4 mg l^{-1} (Fig. 5c). The orthophosphate contents in the bottom water are now lower than 90 $\mu g\, l^{-1}$ (Blanc *et al.*, 1986). Under these improved conditions, nonapatite inorganic phosphorus is retained in

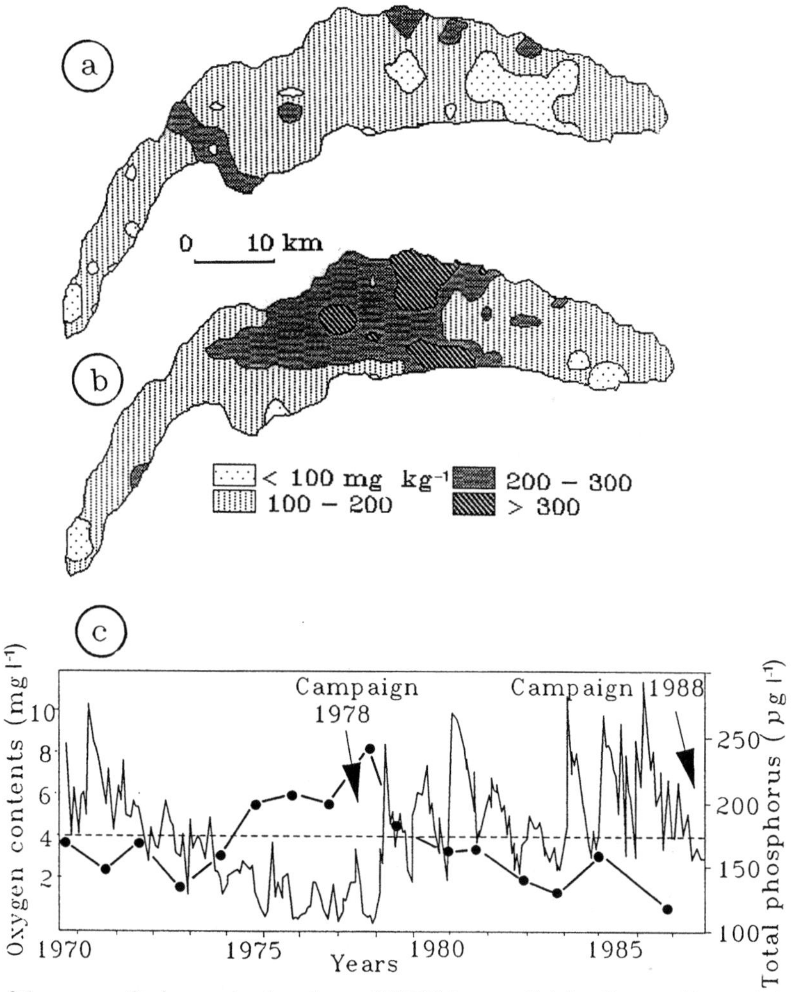

Fig. 5. Comparison of the nonapatite inorganic phosphorus (NAIP) in superficial sediments (dry and size less than 63 μm) of Lake Geneva, related to the evolution of dissolved oxygen contents in the bottom water of central plain, from 1970 to 1986.
a) NAIP in 1978 in mg kg^{-1}.
b) NAIP in 1988 in mg kg^{-1}.
c) Dissolved oxygen concentration and maximum total phosphorus (•) measured at 308 meters depth (data from Blanc *et al.*, 1986). The broken line indicates the limit of dissolved oxygen concentration for phosphorus diffusion from sediment.

superficial sediments. In the central plain, under oxic conditions the diffusive phosphorus flux can be estimated to be between 1.0 and 1.5 mg m^{-2} d^{-1} (Rapin *et al.*, 1984). Thus, only 29 to 43 metric tons of phosphorus were released from sediment to bottom water in 1988.

Conclusion

Superficial sediments provide time-integrated and informative data on the most dramatic changes within a lake. The comparison of the results from two survey campaigns on Lake Geneva does not indicate significant quantitative variations of organic carbon and total nitrogen, but shows an increase of phosphorus. This increase of total phosphorus in superficial sediments (5900 metric tons in 1978 to 6700 metric tons in 1988) is concomitant with the decrease of total dissolved phosphorus in lake water (8000 tons in 1978 to 6000 tons in 1986).

The speciation of phosphorus allows a qualitative interpretation of the content variation in the sediment.

Only the nonapatite inorganic phosphorus has increased distinctly over the last ten years. This phenomenon is directly related to the intensity of the mixing of the bottom water. The nonapatite inorganic phosphorus can be recycled at the sediment-water interface when the dissolved oxygen level is below 4 mg l^{-1} and lead to increase in dissolved phosphorus concentrations in the bottom water.

Thus, despite improvements in waste water treatment and the phosphorus ban in detergents since 1986 in Switzerland, eutrophication remains a serious problem because of the remobilisation potential of a part of the sedimentary phosphorus resulting from previous pollution.

Acknowledgements

This study was supported by the 'Commission Internationale pour la Protection des Eaux du Léman' (CIPEL) in Lausanne. We are most thankful for the technical and scientific help provided by B. Gallerini. The manuscript has benefited greatly from review and comments by Dr. J. Dominik.

References

Arbouille, D., H. Howa, D. Span & J.-P. Vernet, 1989. Etude générale de la pollution par les métaux et répartition des nutriments dans les sédiments du Léman. In: Rapport Commission internationale pour la protection des eaux du Léman contre la pollution, campagne 1988, Lausanne: 139–172.

Blanc, P., C. Corvi & F. Rapin, 1986. Evolution physico-chimique des eaux du Léman. In: Rapport Commission internationale pour la protection des eaux du Léman contre la pollution, campagne 1986, Lausanne: 21–42.

Burrus, D., R. L. Thomas, J. Dominik & J.-P. Vernet, 1989. Recovery and concentration of suspended solids in the upper Rhone river by continuous flow centrifugation. Hydrological Processes, Wiley & Sons Ltd., Sussex, England 3: 65–74.

Jaquet, J.-M., E. Davaud, F. Rapin & J.-P. Vernet, 1982. Basic concepts and associated statistical methodology in the geochemical study of lake sediments. Hydrobiologia 91: 139–146.

Rapin, F., J. Dominik, F. Giovanoli, H. El-Ghobary, P.-Y. Favarger & J.-P. Vernet, 1984. Evolution du phosphore dans l'eau interstitielle et dans les sédiments. In: Rapport Commission internationale pour la protection des eaux du Léman contre la pollution, campagne 1984, Lausanne: 81–96.

Ravera, O. & J.-P. Pelletier, 1980. Short description of the involved lakes. OECD Eutrophication Program, Regional Project, Alpine Lakes: Swiss Federal Board for Environmental Protection, Bern: 19–41.

Vernet, J. P., J. Dominik & P.-Y. Favarger, 1984. Texture and sedimentation rates in Lake Geneva. Envir. Geol. 5: 143–149.

Vollenweider, R. A., W. Rast & J. Kerekes, 1979. The phosphorus loading concept and Great Lakes eutrophication. Phosphorus management strategies for lakes, Proceedings of the 1979 Conference, Ann Arbor Sci Publ.: 207–234.

Williams, J. D. H., J. M. Jaquet & R. L. Thomas, 1976. Forms of phosphorus in the surficial sediments of Lake Erie. J. Fish. Res. Bd Can. 33: 413–429.

Williams, J. D. H., H. Shear & R. L. Thomas, 1980. Availability to *Scenedesmus quadricauda* of different forms of phosphorus in sedimentary materials from the Great Lakes. Limnol. Oceanogr. 25: 1–11.

Received 28 June 1989; in revised from 20 December 1989; accepted 23 February 1990

Hydrobiologia **207**: 167–177, 1990.
D. J. Bonin & H. L. Golterman (eds), Fluxes Between Trophic Levels and Through the Water–Sediment Interface.

Phosphorus mobility in interstitial waters of sediments in Lake Kinneret, Israel

Barbara Staudinger[1], Stefan Peiffer[1], Yoram Avnimelech[2] & Thomas Berman[3]
[1] *Limnologische Station der Universität Bayreuth, P.O. Box 101251, 8580 Bayreuth, F.R.G.;* [2] *Technion – Israel Institute of Technology, Technion City, Haifa 32000, Israel;* [3] *Kinneret Limnological Laboratory, P.O. Box 345, Tiberias 14102, Israel*

Key words: phosphorus-solubility, calcium-phosphate-complexes, hydroxyapatite, iron sulfides

Abstract

Monthly samples of sediment cores from maximum depth (~42 m) in Lake Kinneret were taken from May 1988 until January 1989. The chemical composition of the interstitial and overlying water was investigated with respect to phosphate, Fe^{2+}, Fe^{3+}, Ca^{2+}, alkalinity and electric conductivity. pH, pH_2S and pe (electron-activity) were measured by microelectrodes inserted directly into the sediment core immediately after sampling.

Ion activity products of vivianite, siderite, ironsulfides, Ca-P complexes and Ca-P solid phases were calculated; in addition, Ca/P ratios for the overlying and pore water were obtained by using the potential diagram technique. Despite the fact that anoxic conditions prevail for most of the year, no control of phosphate solubility by a Fe-P relationship could be found. Determination of IAPs, together with calculated molar Ca/P-ratios, suggests that hydroxyapatite as well as surface complexes like dicalcium-phosphate are the solubility-controlling species in pore water. For the overlying water a $Ca_3(HCO_3)_3PO_4$ surface complex is assumed to fix the phosphorus, accompanied by a subsequent transformation of the bound P into apatite.

Introduction

Much effort has gone into the investigation of the phosphate release from lake sediments, because of its strong influence as a limiting nutrient on trophic status. This is especially significant for Lake Kinneret, which is a major water supply source for Israel.

The sediment-water interface may react either as a sink or as a potential source for phosphate (Boström *et al.*, 1982; Enell & Löfgren, 1988). Classic works by Mortimer (1941, 1942), Ohle (1937, 1938), Einsele (1936, 1937, 1938), Einsele & Vetter (1938), confirmed by subsequent studies, reported that the exchange of phosphate from sediments is related to redox-dependent iron chemistry. Phosphate is adsorbed on FeOOH or hydrated Fe^{3+}-complexes which then form solid phases. Under reduced conditions, which develop during stagnant periods, phosphate as well as Fe^{2+} is released into solution.

Furthermore, with increasing pH-value, processes such as ligand exchange mechanisms also liberate phosphate adsorbed to hydrated Fe or Al complexes, due to competition of OH^- and PO_4^{3-} ions (Lijklema, 1977). For calcareous systems, Stumm & Leckie (1971) reported that with decreasing pH there was a release of P adsorbed on calcite, with a formation of surface complexes.

A useful way of evaluating the Ca-P solid phase, which controls the solubility of orthophosphate, o-PO_4 in interstitial waters, is the application of ion activity products and calculation of molar Ca/P ratios. These ratios are characteristic of certain Ca-P complexes, which may be determined by the potential diagram technique, PDT, (McGregor & Brown, 1965; Avnimelech, 1980; 1983).

The purpose of this work was to investigate the overlying- and sediment pore water of Lake Kinneret with respect to the processes that control phosphate solubility. In this lake the sediment pore-water has been investigated by Serruya *et al.* (1974) for the nutrient composition. These authors investigated two layers of sediment: 0–2 cm and >2 cm. Concerning o-PO_4 their results show that most of the sediment phosphorus was bound as calciumphosphate, but that the exchangeable fraction belonged to the iron-bound phosphorus. They noted further that this Fe/P relationship seemed to be responsible for the o-PO_4 release into the pore water in summer at low redox potentials, accompanied by the formation of iron sulfides. The observed phosphorus solubility in the pore water and hypolimnion seemed to be caused by the formation of calcium phosphates.

Potential-Diagram-Technique: PDT

As mentioned above, calculation of ion activity products and comparison with the corresponding solubility products allow the determination of the P-mineral or -complex responsible for the o-PO_4--solubility of the solution.

PDT can also be used to investigate these determinants. This procedure is used when the stoichiometry of forming Ca-P-complexes is not known. PDT has been applied by McGregor & Brown (1965), Cole & Olsen (1959) and Avnimelech (1980, 1983). McGregor & Brown (1965) assume that a compound with the generalized empirical formula $Ca_xH_y(PO_4)\ zH_2O$ with 2x + y = 3 and y having possibly negative values, forms according to

$$\times\ Ca(OH)_2 + H_3PO_4 + (z - 2x)\ H_2O \rightarrow Ca_xH_y(PO_4)\ zH_2O \qquad (1)$$

Providing that there is equilibrium, stable or metastable, between the solid phase (Ca_xH_y-$(PO_4)\ zH_2O$) and the saturated solution, the sum of the chemical potentials, μ, on the left hand side of this equation, equals the characteristic free energy of formation of the solid phase $F^0_{(s)}$.

$$\times\ \mu Ca(OH)_2 + \mu H_3PO_4 + (z - 2x)\ \mu H_2O = F^0_{(s)}\,. \qquad (2)$$

The chemical potential of H_2O in solute solutions may be considered constant and may be combined with $F^0_{(s)}$ to a new constant F^*. If we express the chemical potentials by their activities and by

$$K_1 = \frac{1}{R * T * 2.303} * (\mu^\circ Ca(OH)_2 + \mu^\circ H_3PO_4) + F^* \qquad (3)$$

it follows that:

$$-\log\{H^+\}^3 * \{PO_4{}^{3-}\} = -x * -\log\{Ca^{2+}\} * \{OH^-\}^2 + K_1\,. \qquad (4)$$

Regarding a carbonate-phosphate-surface complex, $Ca_x(HCO_3)_3PO_4(H)_{3-x}$, a similar equation was obtained by Avnimelech (1980).

$$-\log\{H^+\}^3 * \{PO_4{}^{3-}\} = -x * -\log\{Ca^{2+}\} * \{CO_3{}^{2-}\} + K_2 \qquad (5)$$

where K_1 and K_2 are constants specific for the calciumphosphate under consideration. { } denotes activities and x is the molar Ca/P ratio:

$$\Delta(-\log\{H^+\}^3 * \{PO_4{}^{3-}\})/$$
$$\Delta(-\log\{Ca^{2+}\} * \{OH^-\}^2$$

and

$$\Delta(-\log\{H^+\}^3 * \{PO_4^{3-}\})/$$
$$\Delta(-\log\{Ca^{2+}\} * \{CO_3^{2-}\})$$

which is given by the stoichiometry of the considered Ca-P-complex and carbonate-phosphate-surface complex respectively. The molar Ca/P ratio corresponds to the slope when $-\log\{H^+\}^3 * \{PO_4^{3-}\}$ is plotted against $-\log\{Ca^{2+}\} * \{OH^-\}^2$ or $-\log\{Ca^{2+}\} * \{CO_3^{2-}\}$, respectively. Theoretical molar Ca/P values for equation (4) are: $CaHPO_4$ (dicalciumphosphate: 1.0; $Ca_3(PO_4)_2$ (β-tricalciumphosphate): 1.5; $Ca_4(PO_4)_3$ (octacalciumphosphate): 1.3 and $Ca_5(PO_4)_3OH$ (hydroxyapatite): 1.67; for equation (5): $Ca_3(HCO_3)_3PO_4$ surface complex: 3.0.

Methods and sampling site

Lake Kinneret is a warm monomictic lake located in the northern part of the Jordan Rift Valley, Israel (for general description, see Serruya, 1978).

Sediment and overlying water samples were taken at maximum water depth (42 m). Perspex tubes with holes drilled laterally (diameter: 5 mm) at 1 cm intervals were used to obtain cores with a gravity corer described by Tessenow *et al.* (1977). Immediately after sampling, the cores were covered with aluminium foil, to avoid stimulation of the growth of photosynthetic H_2S-consuming bacteria, and transported to a 17 °C room in order to reduce gassing out of H_2S, CO_2 and possibly CH_4. On the same day, pH, pe and pH_2S measurements were performed with rH and pH_2S microelectrodes, inserted laterally into the sediment core. Since it was not possible to carry out all investigations immediately on the day of sampling, a second core, which was used for pore-water extraction, was stored at 4 °C in a dark room overnight. Although oxidation of Fe^{2+} to Fe^{3+} may occur at 4 °C, it seems quite impossible that it occurred in these samples, because of the strong smell of H_2S that escaped from the overlying water removed. A pressure filtration apparatus (Reeburgh, 1976) was used to separate pore water from sediment. The pore water was filtered twice, through a GF/C filter and then through a 0.45 μm membrane filter. Both filters were washed with 200 ml of distilled water before use. Analyses of Fe^{2+}, Fe^{3+} and phosphate were performed immediately after extracting the pore water. The remaining pore water was stored in airtight flasks with septa at 4 °C in the dark. The extraction, filtration and determination of Fe^{2+}, Fe^{3+} and phosphate were carried out under anoxic conditions in an anaerobic hood to avoid oxidation of Fe^{2+} to Fe^{3+} and a subsequent coprecipitation of phosphate. The cores were examined down to a depth of 14 cm, with sample slices taken every cm down to 4 cm, then every 2.5 cm. o PO_4 was analyzed by the molybdenum blue method. Fe^{2+}-, Fe^{3+}-concentrations were determined by the bathophenanthroline method. APHA-Standard-Methods (1981) were used for the analyses of SO_4^{2-} and Ca^{2+}, referring to the turbidimetric method with $BaCl_2$ as reagent and the EDTA titrimetric method with murexide as indicator, respectively. Alkalinity was obtained by titration with a 0.01 M HCl solution.

All methods described above were optimized for 1 ml sample and standard volume, except for sulfate and electrical conductivity (5 ml) and alkalinity (2 ml) determinations. Duplicates were analyzed, but due to the small total sample volume, only single measurements of SO_4^{2-}, Ca^{2+} and alkalinity were made.

Results and discussion

Three different kinds of phosphate concentration profiles in the sediment pore waters were observed, corresponding to: i) the 'mixed' period in May 88, and subsequently from November 88 to January 89; ii) the 'transition' period in June and July 88 and iii) the 'stagnant' period from August until October 88. These periods are roughly correlated with the mixed and stratified lake periods, indicated by homothermy and thermal stratification respectively. (Fig. 1a–c)

The stratification of the lake in summer causes

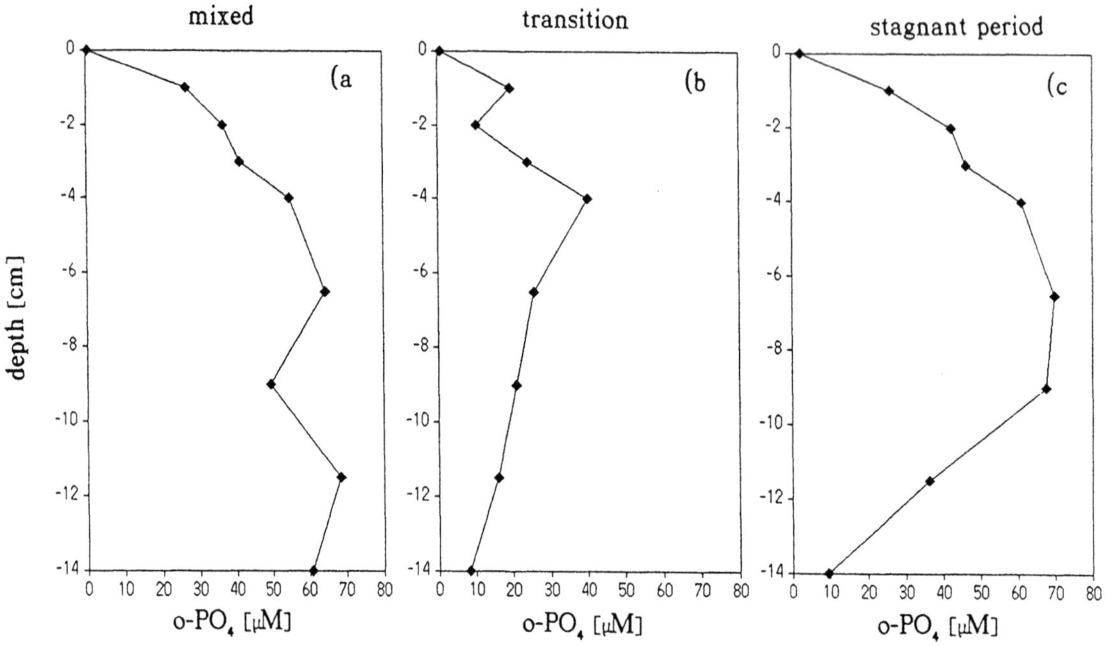

Fig. 1. Phosphate concentration patterns with sediment depth shown at three periods of different sediment conditions (see text), which could be roughly correlated to lake mixing and lake stratification: a) 'mixed'; b) 'transition'; c) 'stagnant'.

the onset of reducing conditions in the sediment. Fig. 1b shows a typical o-PO_4-concentration profile in the pore water for this period, where a zone of o-PO_4-production could be found at ~6.5 cm depth. From this zone o-PO_4 diffuses into the overlying water and deeper sediment horizons, following the concentration gradient.

Since o-PO_4-concentrations were always higher in the sediment pore waters than in the overlying water (42 m) throughout the whole sampling period (Fig. 1a–c), o-PO_4-fluxes between the overlying water and the first cm of the sediment pore waters were calculated with the assumption of molecular diffusion using Fick's first law. This assumption is valid for the 'stagnant' period. During the 'mixed' and 'transition' periods, turbulent diffusion has to be considered. Thus the calculated o-PO_4-fluxes represent the minimum values for the o-PO_4-fluxes from the pore water into the overlying water. They ranged from a minimum value of 49 $\mu M * m^{-2} * d^{-1}$ in June 88 to a maximum value of 70 $\mu M * m^{-2} * d^{-1}$ in January 89. The diffusion coefficient, $D_s = (3.6 \pm 1.1) * 10^{-6}\ cm^2 * s^{-1}$, was taken from Krom & Berner (1980a).

The turnover of the lake in winter is responsible for the profile of the mixed period (Fig. 1a). The lowering of the thermocline and the following strong convectional currents, which were confirmed by hydrodynamic measurements (Serruya *et al.*, 1974), destroyed the sediment-water interface built up in summer. These bottom currents are responsible for a certain mixing of the upper layers of the sediment pore-waters in winter. Serruya *et al.* (1974) estimated a mixing down to 15 cm sediment depth. Resuspension of a few mm of the sediment may also occur (Serruya, 1976). In winter the dissolved species maintain a more or less constant concentration between 4 and 14 cm depth; o-PO_4 concentration ranged between 60 and 70 μM o-PO_4 (Fig. 1a). The pattern of the upper 4 cm might be caused by a persistent concentration gradient.

The transition period is characterized by the development of stable conditions, which prevail throughout the summer (Fig. 1b).

To examine the variability of the sediment and

Table 1. Sediment pore-water concentrations of o-PO_4, Fe^{2+}, Ca^{2+}, SO_4^{2-}, carbonate-alkalinity, ionic strength as well as pH- and pH_2S-values at station A of Lake Kinneret, 5 Oct. 1988 ('stagnant' period).

Depth [cm]	o-PO_4 [μM]	Fe^{2+} [μM]	Ca^{2+} [mM]	Alk.* [meq l^{-1}]	SO_4^{2-} [μM]	I [M]	pH	pH_2S
0**	2.4	1	1.2	2.93	317	0.018	7.35	3.70
1	26	1	1.2	3.83	196	0.018	7.16	3.42
2	42	0	1.3	4.50	218	0.018	7.07	3.45
3	46	0	1.6	5.33	170	0.021	7.03	3.66
4	61	0	1.6	5.70	90	0.021	7.00	3.77
6.5	70	0	1.4	5.95	84	0.022	6.96	4.10
9	68	0	1.7	6.40	84	0.023	6.93	4.45
11.5	36	0	1.7	6.50	82	0.023	6.90	5.60
14	10	0	1.9	6.85	78	0.025	6.84	6.57

* Alk. corresponds to carbonate-alkalinity (HS^--concentration has been substracted).
** 0 cm depth corresponds to the overlying water.

pore water composition, three sediment cores were taken at the same time at station A. We observed differences in the pore water concentrations of several dissolved species between the three sediment cores, which however did not exceed 13%. We therefore assume that the patchiness of the sediment cannot account for the observed seasonal variations of the concentration profiles.

The measured o-PO_4, Fe^{2+}, Ca^{2+}, SO_4^{2-} and alkalinity concentrations, calculated ionic strength-values (Griffin & Jurinac, 1973), as well as pH- and pH_2S-values for two characteristic measurements representing the 'mixed' and 'stagnant' period (5.01.89 and 5.10.88, respectively), are given in Tables 1 and 2.

The redox potentials (pe-values) were always < 150 mV. Since the pe-value represents a mixed potential of several redoxactive substances present in the sample solution, which react with the platinum electrode cell, it is difficult to interpret these values qualitatively. The determination of pH_2S-values in H_2S-bearing waters is therefore much more informative with respect to the redox state of the sample solution (Peiffer, 1989).

The first 15 centimetres of the sediment are characterized by high biological activity. Sulphate-reducing bacteria dominate in the first 1–3

Table 2. Sediment pore-water concentrations of o-PO_4, Fe^{2+}, Ca^{2+}, SO_4^{2-}, carbonate-alkalinity, ionic strength as well as pH- and pH_2S-values at station A of Lake Kinneret, 5 Jan. 1989 ('mixed' period).

Depth [cm]	o-PO_4 [μM]	Fe^{2+} [μM]	Ca^{2+} [mM]	Alk.* [meq l^{-1}]	SO_4^{2-} [μM]	I [M]	pH	pH_2S
0**	0.5	0.7	1.00	2.1	500	0.016	7.80	22.7
1	27	2.8	1.60	4.9	194	0.019	7.09	4.8
2	37	0.8	1.60	4.8	135	0.018	7.11	5.7
3	41	0.8	1.87	6.3	58	0.022	6.93	7.4
4	55	0.9	1.70	7.3	0	0.022	6.88	7.7
6.5	64	3.6	1.96	8.2	0	0.022	6.86	11.7
9	49	1.6	2.14	8.4	0	0.025	6.83	11.9
11.5	68	2.8	2.30	9.3	0	0.026	6.80	14.0
14	61	1.3	2.20	9.3	0	0.026	6.78	15.6

* Alk. corresponds to carbonate-alkalinity (HS^--concentration has been substracted).
** 0 cm depth corresponds to overlying water.

centimetres, reducing SO_4^{2-} to H_2S as indicated by a decrease in SO_4^{2-}-concentration with depth and a H_2S-activity maximum at 1–3 cm depth. Similar to the pattern of the o-PO_4-concentration, H_2S diffuses from this maximum into the overlying water and into deeper sediment horizons. SO_4^{2-}-concentration decreases from 270–500 μM in the overlying water to 50–190 μM at 4 cm depth in the sediment and to values of 0–80 μM down to 14 cm depth, causing an increase in carbonate-alkalinity with depth. Fermenting and probably methane-producing bacteria dominate below 4 cm depth, causing a decrease in the pH-value. During the sampling period, the pH-value decreased from 7.2–7.8 in the overlying water to 6.2–6.85 at 15 cm sediment depth. Carbonate-alkalinity increased from 2.1–3.8 meq l^{-1} in the overlying water to 6.6–9.3 meq l^{-1} at 15 cm depth. As can be seen from Tables 1 and 2, Ca^{2+} and HCO_3^- (representing carbonate alkalinity) diffuse into the overlying water because of the existing concentration gradient. Only a slight increase with depth could be observed for Ca^{2+}-concentration.

In the sediment pore-waters Fe^{3+} could not be detected, in contrast to Fe^{2+} which was measurable. Furthermore, the presence of H_2S at high concentrations (Table 1a, b) and the low EMF values (< 150 mV) of the platinum electrode cell (except for the one overlying water sample at turnover in January 1989) exclude the existence of dissolved Fe^{3+} and Fe^{3+}-hydroxides in the overlying- and sediment pore-water of Lake Kinneret. Formation of iron sulfides in presence of H_2S was probably responsible for the observed Fe^{2+}-concentrations (0–8 μM, exception: June 88: 15–29 μM Fe^{2+} from 6.5 to 14 cm depth) in the overlying- and sediment pore-waters. These Fe^{2+}-concentrations are low in comparison to other freshwater systems (Emerson & Widmer, 1978). Soluble bisulfide complexes with ferrous iron: $Fe(HS)^+$, $Fe(HS)_2$, $Fe(S)(HS)^-$ and $Fe(HS)_3^-$ (Landing & Westerlund, 1988) have to be considered as causing the free Fe^{2+}-concentration to be lower than the measured Fe^{2+}-concentration. As no data were available for $Fe(HS)^+$, $Fe(HS)_3^-$ and $Fe(S)(HS)^-$, only $Fe(HS)_2^0$ has been considered for the calculation of the free Fe^{2+}-concentration and the iron activity product of the iron sulfides. The calculated free Fe^{2+}-concentration as well as the calculated ion activity products might therefore be overestimated.

$$Fe^{2+} + 2HS^- \rightarrow Fe(HS)_2 \quad K = 10^{8.9} \tag{6}$$

and

$$Fe_{tot} = Fe^{2+} + Fe(HS)_2 \tag{7}$$

with

$$Fe_{tot} = \text{measured } Fe^{2+}$$

Because of the difficulty in obtaining reliable values for the second dissociation constant of the $H_2S/HS^-/S^{2-}$ system, Davison (1980) proposes the use of the following equations for the calculation of the ion activity products (IAP) of iron sulfides.

$$H^+ + MS \rightleftarrows HS^- + M^{2+} \tag{8}$$

where M = Fe and

$$IAP = \frac{\{Fe^{2+}\} * \{HS^-\}}{\{H^+\}} \tag{9}$$

{ } denotes the activity of the metal-, hydrogen-, hydrogen sulfide-ions etc. HS^--activities can be obtained from H_2S-activities. Values for the ionic strength range between 0.016 and 0.025 M for all samples. Since the Davies-approximation is valid for I < 0.5 M, we used this approximation to calculate activity coefficients γ.

$$-\log \gamma_{m\pm} = A * Z_{m\pm}^2 \left\{ \frac{\sqrt{I}}{\sqrt{I}+1} - 0.2 * I \right\} \tag{10}$$

with I = ionic strength [M]
A = constant for 25 °C: 0.508
Z = charge of ion

From equation (6) and (7) it follows that:

$$[Fe^{2+}] = \frac{[Fe_{tot}]}{(1 + 10^{8.9}\{HS^-\})} \quad (11)$$

Transformation into activities,

$$\{Fe^{2+}\} = [Fe^{2+}]\,\gamma_{Fe^{2+}} \quad (12)$$

and combining equation (12) with (9) gives

$$IAP = \frac{[Fe_{tot}]}{(1 + 10^{8.9}\{HS^-\})}\gamma_{Fe^{2+}} * \frac{\{HS^-\}}{\{H^+\}} \quad (13)$$

The relation between IAP and solubility constant K_{so} for the following equation

$$FeS \rightarrow Fe^{2+} + S^{2-} \quad (14)$$

is given by

$$pIAP = pK_{so} - pK_2$$

where K_2 is the second dissociation constant. The values obtained for pIAP (denoting $-\log IAP$) in the sediment pore-waters range between 2,9 and 4.9 in the first 4 cm and between 4.0 and 6.1 down to 14 cm depth. Iron sulfide solubility products (K_{sp}) following equation (8) reported by Landing & Westerlund are: amorphous FeS pK_{sp}: 3.6, greigite Fe_3S_4 pK_{sp}: 4.2, mackinawite pK_{sp}: 4.9 and pyrrhotite pK_{sp}: 6.53. Comparison of the pIAP's with the solubility products of the several iron sulfides indicates the possible formation of amorphous FeS, greigite Fe_3S_4, mackinawite $Fe_{(1+x)}S$ and pyrrhotite (Emerson, 1976) with depth.

Analysis and comparison of the sediment Fe_{tot}--concentration (16–30 mg g^{-1} dry substance) and of the acid soluble sulfide representing the iron fraction, which is bound as iron sulfide but not as pyrite, FeS_2, (20–36 mg g^{-1} dry substance) showed that most of the iron in the first 3–4 cm of the sediment is bound as iron sulfide. Acid soluble sulfide decreases with depth to about 2–3 mg g^{-1} dry substance, whereas total iron stays constant. This difference between total and acid soluble sulfide below 4 cm is probably due to the formation of pyrite.

The results obtained by the calculation of ion activity products of vivianite, $Fe_3(PO_4)_2$, and siderite $FeCO_3$, respectively, indicate undersaturation of the pore-waters with respect to both vivianite and siderite. The metastable iron sulfides, amorphous iron sulfide, greigite, mackinawite and pyrrhotite, therefore control Fe^{2+}-solubility in the sediment pore-water.

Carbonate

The ion activity product of $CaCO_3$ has been calculated as

$$IAP = \gamma_{Ca^{2+}}[Ca^{2+}]\,\gamma_{CO_3^{2-}}[CO_3^{2-}]$$

It is necessary to consider a $CaHCO_3^+$-complex, since this can lower the free Ca^{2+}-concentration (Plummer & Busenberg, 1982). CO_3^{2-} has been calculated from pH-values and carbonate-alkalinity concentrations. The results indicate that the sediment pore-waters are undersaturated concerning $CaCO_3$. The pIAP lies between 8.46 and 8.85 from May 88 until January 89. Chemisorption of impurities, which are attached at sites of $CaCO_3$ dissolution can greatly retard this process (Berner & Morse, 1974). In the case of $CaCO_3$ such impurities are mainly o-PO_4 (Berner & Morse, 1974) and organic matter (Suess, 1970). This suggests that $CaCO_3$ dissolution in Lake Kinneret sediments may have been inhibited by adsorption of o-PO_4 and organic matter.

Phosphorus

As discussed above, we could exclude the presence of Fe^{3+} in the sediment pore-waters throughout the sampling period. Furthermore the sediment pore-waters were undersaturated with respect to ferrous phosphate complexes like $FeH_2PO_4^+$ and $FeHPO_4$. (formation constants: $pK = 2.7$ and $pK = 3.7$ respectively. Nriagu 1972). Also vivianite, a mineral which is often present in freshwater lakes (Emerson & Widmer,

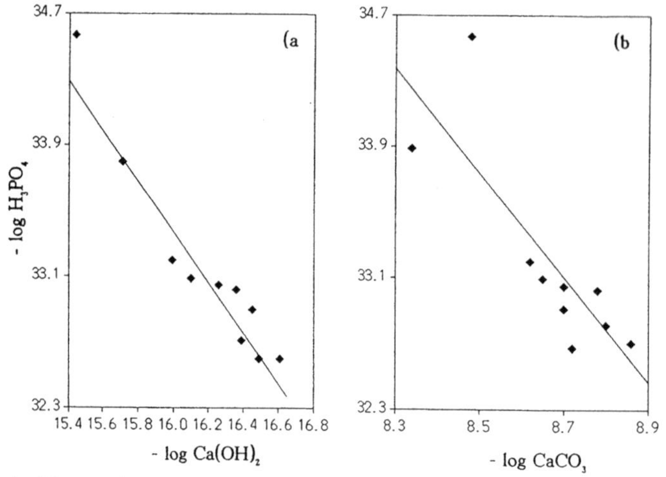

Fig. 2. Ca/P ratios calculated with equation (4) and (5) (see text) for the overlying water throughout the sampling period. The calculated equations are: a) $-\log H_3PO_4 = -1.59 * -\log Ca(OH)_2$, $r = -0.96$; b) $-\log H_3PO_4 = -3.18 * -\log CaCO_3$, $r = -0.83$; indicating the presence of apatite and a $Ca_3(HCO_3)_3PO_4$ surface complex, respectively.

1978), does not determine o-PO_4- and Fe^{2+}--solubility in the sediment pore-waters of Lake Kinneret. Thus, a direct Fe-P relationship, responsible for the o-PO_4-solubility in the sediment pore-waters, may not exist here (Serruya *et al.*, 1974) We therefore wished to demonstrate that a Ca-P relation within these periods could have determined the solubility of o-PO_4 in the sediment pore- and overlying-waters. The potential diagram technique, together with the calculation of ion activity products of several Ca-P species was used to evaluate such a possible relationship.

The calculated molar Ca/P ratios (Fig. 2a, b) indicate the presence of apatite (Ca/P = 1.59, equation (4)) and a $Ca_3(HCO_3)_3PO_4$ surface complex (Ca/P = 3.18, equation (5)) in the overlying water. Such a clear relationship cannot be found for the pore water, if we consider each discrete layer (depth) throughout the whole sampling period. (We excluded the anomalous Nov. 88 sample, since the first 3 cm of the sediment were apparently advected by currents). Fig. 3 shows a plot of these molar Ca/P pore-water ratios versus sediment depth. A linear relationship could be found for 1–9 cm depth (correlation coefficients $r = -0.62$ to -0.93) but not for 11.5 and 14 cm depth ($r = -0.26$ and -0.36). Theoretical values could be obtained for 3 cm and 6.5 cm depth, whereas the molar Ca/P ratios for 1,5 and 4 cm depth lie between mono-calcium-phosphate and di-calciumphosphate, and for

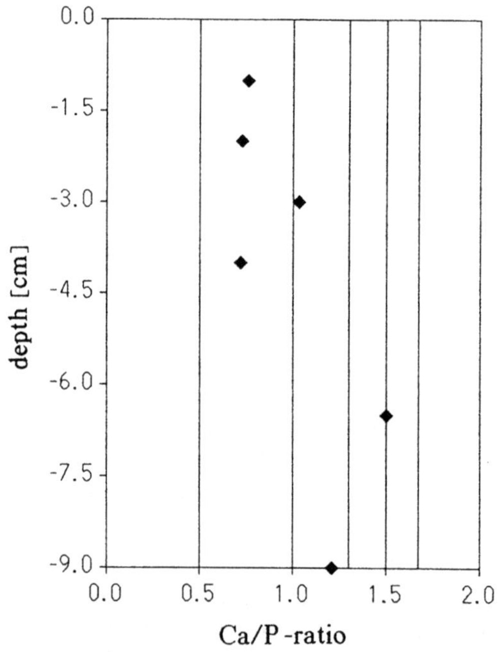

Fig. 3. Calculated Ca/P ratios with depth for single layers of the sediment pore water throughout the sampling period, using equation (4) (see text). The vertical lines represent the theoretical values of the calcium phosphates. $Ca(H_2PO_4)_2 = 0.5$; $CaHPO_4 = 1.0$; $Ca_3(PO_4)_2 = 1.5$; $Ca_4H(PO_4)_3 = 1.3$; $Ca_5(PO_4)_3OH = 1.67$.

9 cm depth between di-calciumphosphate and octa-calciumphosphate.

The experimental determination of solubility products for calcium phosphates is complicated because of their metastable character. It is therefore difficult to clearly define a solution as under-, over- or saturated with respect to a distinct calcium phosphate mineral.

Solubility products for di-calciumphosphate, $CaHPO_4$, are given as $2.77 * 10^{-7}$ mole $* l^{-1}$ (Moreno *et al.*, 1960) and $2.56 * 10^{-7}$ mole $* l^{-1}$ (Greogry *et al.*, 1970), for β-tri-calciumphosphate, $Ca_3(PO_4)_2$ as $3.16 * 10^{-29}$ mole $* l^{-1}$ (Duff 1971) and for octa-calciumphosphate, $Ca_4H(PO_4)_3$ as $1.45 * 10^{-47}$ mole $* l^{-1}$ (Moreno *et al.*, 1960). The solubility product for hydroxyapatite, $Ca_5(PO_4)_3OH$ (or $Ca_{10}(PO_4)_6(OH)_2$), ranges from 10^{-50}, calculated from data from the river Rhine and Rhone (Golterman & Meyer, 1985), to many experimentally determined solubility products, e.g. 10^{-57}, Stumm & Morgan (1981), $6.3 \pm 2.1 * 10^{-59}$ Avnimelech & Moreno (1973). In this article, we refer to the experimentally determined data.

Our calculated activity products indicated that the sediment pore-waters were undersaturated with respect to octa-calciumphosphate (pIAP = 48.4–55.7) at all depths, and saturated from 1–9 cm depth with respect to di-calcium phosphate (pIAP = 7.7–8.1) and β-tri-calcium-phosphate (pIAP = 28.3–30.0), taking into account the variability of the solubility products and the difficulties of their determination. Di-calciumphosphate and β-tricalciumphosphate were undersaturated in the overlying water and in the sediment pore-water at 11.5 and 14 cm depth (pIAP < 8.1 and < 30, respectively). The pIAP's of hydroxyapatite lie between 51.3 and 54.1 at 1–9 cm depth. pIAP values between 52.1–54.4 in the mixed period and 54.3–56.9 in the transition and stagnant periods have been found at 11.5 and 14 cm depth.

The sediment pore-waters can be interpreted as slightly oversaturated with respect to hydroxyapatite from 1–9 cm, whereas from 11.5 to 14 cm depth the sediment pore-waters were at saturation during the transition- and stagnant-, but oversaturated during the mixed-period concerning hydroxyapatite.

The overlying water was always saturated with respect to hydroxyapatite.

The results obtained for the sediment pore-waters by PDT and by calculation of ion activity products do not agree very well. Possibly, the adsorption of o-PO_4 on clay minerals (montmorillonite), uptake of o-PO_4 by bacteria (Gächter *et al.*, 1988), or formation of mixed phosphate phases (see below) could distort the results obtained by these two methods. Nevertheless, we suggest that di-calciumphosphate, β-tri-calciumphosphate and probably intermediates of both, as indicated by the molar Ca/P ratios, could be responsible for the o-PO_4-solubility in the sediment pore-waters.

Oversaturation of solutions with respect to hydroxyapatite has been widely reported (Murray *et al.*, 1978; Krom & Berner, 1980b; Matisoff *et al.*, 1980), and could be caused by the presence of surface complexes on hydroxyapatite.

Smith *et al.* (1974) reported that di-calciumphosphate exists as a chemical compound on the surface of dissolving apatite. Furthermore, Cole *et al.* (1953) reported that, at high o-PO_4-concentrations, di-calciumphosphate precipitates and that this compound is an intermediate of hydroxyapatite, whereas at low o-PO_4-concentrations o-PO_4 will adsorb on $CaCO_3$.

Since interstitial waters represent multicomponent systems, the existence of precipitates consisting of pure minerals or single solid phases is unlikely. Thus, the computation of activity products in such systems frequently gives unreliable results with respect to solubility. For these systems, mixed phases (e.g. Fe-Mn phosphates, Fe-Ca phosphates or Mg-Ca carbonates) have to be considered (Suess, 1979; Matisoff *et al.*, 1980). In our case no evidence can be given for the existence of mixed phases, but the prevailing geochemical conditions strongly imply the possibility of their presence.

These results, together with the difficulties in determining the solubility product of metastable calcium phosphates and stable hydroxyapatite, suggest that, besides the possible formation of mixed phases, surface complexes, such as di-calciumphosphate and β-tri-calciumphosphate existing on the hydroxyapatite surface, determine the o-PO_4-solubility in the sediment pore-waters. These metastable calcium phosphates will be transformed into the stable hydroxyapatite.

The results of the Ca/P ratios and the determination of the IAP's for the overlying water agree well, indicating the presence of hydroxyapatite and a carbonate-phosphate-surface complex, $Ca_3(HCO_3)_3PO_4$, as described by Avnimelech (1980).

The profiles of o-PO_4 (Fig. 1a–c) and the Ca^{2+}-concentrations (Table 1 and 2) in the upper sediment layers throughout the sampling period were mainly determined by diffusion of o-PO_4 and Ca^{2+} into the overlying water because of the concentration gradient. After being released from the sediment pore-water, o-PO_4 can be transported into the epilimnion and there, with increasing pH during stratification, coprecipitate or interact with calcite to form a surface complex, $Ca_3(HCO_3)_3PO_4$. According to Stumm & Leckie (1971), o-PO_4 can be adsorbed immediately on calcite and may subsequently be transformed into an amorphous carbonate-hydroxy-apatite growing epitaxially in a pH-range of 6.8–8.3. We assume that this process occurs in Lake Kinneret.

Finally, we emphasise that other mineralogical methods (e.g. SEM, TEM), which were beyond the scope of this study, would be desirable to verify our present conclusions.

Acknowledgement

This work was financially supported by a Minerva scholarship granted to B. Staudinger, who thanks Drs. W. Eckert, A. Nishri, O. Hadas, Y. Salingar and the staff of the Kinneret Limnological Laboratory for their advice and help.

References

A.P.H.A., 1981. Standard Methods for the Examination of Water and Wastewater. 15th Edition, Washington D.C., 1134 pp.

Avnimelech, Y., 1980. Calcium-carbonate-phosphate surface complex in calcareous systems. Nature 290: 255–257.

Avnimelech, Y., 1983. Phosphorus and calcium carbonate solubilities in lake Kinneret. Limnol. Oceanogr. 28: 640–645.

Avnimelech, Y. & E. C. Moreno, 1973. Solubility and surface properties of finely divided hydroxyapatite. Journal of Research of the National Bureau of Standards (U.S.) 77a Phys. and Chem. 1: 149–155.

Berner, R. A. & J. W. Morse, 1974. Dissolution of calcium carbonate in sea water: IV. Theory of calcite dissolution. Am. J. Sci. 27: 108–134.

Boström, B., M. Jansson & C. Forsberg, 1982. Phosphorus release from lake sediments. Arch Hydrobiol. Beih. Ergebn. Limnol. 18: 5–59.

Cole, C. V., S. R. Olsen & C. O. Scott, 1953. The nature of phosphate sorption by calcium carbonate. Soil Sci. Am. Proc. 17: 352–356.

Cole, C. V. & S. R. Olsen, 1959. Phosphorus solubility in calcareous soils: I. Dicalciumphosphate activities in equilibrium solutions. Soil Sci. Soc. Proc. 159: 116–118.

Davison, W., 1980. A critical comparison of the measured solubilities of ferrous sulphide in natural waters. Geochim. Cosmochim. Acta. 44: 803–808.

Duff, E. J., 1971. Orthophosphates Part IV. J. Chem. Soc. A. 1971: 921–926.

Einsele, W., 1936. Über die Beziehung des Eisenkreislaufs zum Phosphatkreislauf im eutrophen See. Arch. Hydrobiol. 29: 664–686.

Einsele, W., 1937. Physikalisch-chemische Betrachtung einiger Probleme des limnischen Mangan- und Eisenkreislaufs. Verh. int. Ver. Limnol. 5: 69–84.

Einsele, W., 1938. Über chemische und kolloidchemische Vorgänge in Eisen-Phosphat-Systemen unter limnochemischen und limnogeologischen Gesichtspunkten. Arch. Hydrobiol. 33: 361–387.

Einsele, W. & H. Vetter, 1938. Untersuchungen über die Entwicklung der physikalischen und chemischen Verhältnisse im Jahreszyklus in einem mäßig eutrophen See (Schleinsee bei Langenargen). Int. Rev. Hydrobiol. 36: 285–324.

Emerson, S., 1976. Early diagenesis in anaerobic lake sediments: chemical equilibria in interstitial waters. Geochim. Cosmochim. Acta 40: 925–924.

Emerson, S. & G. Widmer, 1978. Early diagenesis in anaerobic lake sediments – 2. Themodynamic and kinetic factors controlling the formation of iron phosphate. Geochim. Cosmochim. Acta 42: 1307–1316.

Enell, M. & S. Löfgren, 1988. Phosphorus in interstitial water: methods and dynamics. Hydrobiologia 170: 103–132.

Gächter, R., J. S. Meyer & A. Mares, 1988. Contribution of bacteria to release and fixation of phosphorus in lake sediments. Limnol. Oceanogr. 33: 1542–1558.

Golterman, H. L. & M. L. Meyer, 1985. The geochemistry of two hard water rivers, the Rhine and the Rhone, Part 4:

The determination of the solubility product of hydroxyapatite. Hydrobiologia 126: 25–29.
Gregory, T. M., E. C. Moreno & W. E. Brown, 1970. Solubility of $CaHPO_4 * 2H_2O$ in the system $Ca(OH)_2$-H_3PO_4-H_2O at 5, 15, 25, and 37.5 °C. Journal of Research of the National Bureau of Standards – A. Physics and Chemistry 74: 461–475.
Griffin, R. A. & J. J. Jurinak, 1973. Estimation of activity coefficients from electrical conductivity of natural aquatic systems and soil extracts. Soil Sci. 116: 26–30.
Krom, M. D. & R. A. Berner, 1980a. The diffusion coefficient of sulfate, ammonium and phosphate ions in anoxic marine sediments. Limnol. Oceanogr. 25: 327–337.
Krom, M. D. & R. A. Berner, 1980b. Adsorption of phosphate in anoxic marine sediments. Limnol. Oceanogr. 25: 797–806.
Landing, W. M. & S. Westerlund, 1988. The solution chemistry of iron (II) in Framvaren Fjord. Mar. Chem. 23: 329–343.
Lijklema, L., 1977. The role of iron in he exchange of phosphate between water and sediments. – In H. L. Golterman (ed.) Interactions between sediments and fresh water. Dr. W. Junk B.V., The Hague: 313–317.
MacGregor, J. & W. E. Brown, 1965. Blood: Bone equilibrium in calcium homoeostasis. Nature 4969: 359–361.
Matisoff, G., A. H. Lindsay, S. Matis & F. M. Soster, 1980. Trace metal mineral equilibria in Lake Erie sediments. J. Great Lakes Res. 6: 353–366.
Moreno, E. C., W. E. Brown & O. Osborn, 1960. Stability of dicalcium phosphate dihydrate in aqueous solutions and solubility of octacalciumphosphate. Soil Sci. Am. Proc. 24: 94–98.
Mortimer, C. H., 1941. The exchange of dissolved substances between mud and water in lakes. I. J. Ecol. 29: 280–329.
Mortimer, C. H., 1942. The exchange of dissolved substances between mud and water in lakes. II. J. Ecol. 30: 147–201.
Murray, J. W., V. Grundmanis & W. M. Smethie Jr., 1978. Interstitial water chemistry in the sediments of Saanich Inlet. Geochim. Cosmochim. Acta 42: 1011–1026.
Nriagu, J. O., 1972. Stability of vivianite and ion-pair formation in the system $Fe_3(PO_4)_2$–H_3PO_4–H_2O. Geochim. Cosmochim. Acta 36: 454–470.
Ohle, W., 1937. Kolloidgele als Nährstoffregulanten der Gewässer. Naturwissenschaften 25: 471–474.
Ohle, W., 1938. Die Bedeutung der Austauschvorgänge zwischen Schlamm und Wasser für den Stoffkreislauf der Gewässer. Vom Wasser XIII: 87–97.
Peiffer, S., 1989. Biogeochemische Regulation der Spurenmetallöslichkeit während der anaeroben Zersetzung fester kommunaler Abfälle. Ph.D. thesis, Univ. Bayreuth, Dept. Hydrology: 197 pp.
Plummer, L. N. & E. Busenberg, 1982. The solubilities of calcite, aragonite and vaterite in CO_2–H_2O solutions between 0 and 90 °C, and an evaluation of the aqueous model for the system $CaCO_3$–CO_2–H_2O. Geochim. Cosmochim. Acta 46: 1011–1040.
Reeburgh, W. S., 1976. An improved interstitial water sampler. Limnol. Oceanogr. 12: 163–165.
Serruya, C., 1976. Rates of sedimentation in Lake Kinneret. – In H. L. Golterman (ed.) Interactions between sediments and fresh water. Dr. W. Junk B.V., The Hague: 48–56.
Serruya, C., 1978. Lake Kinneret. Dr. W. Junk B.V., The Hague, 501 pp.
Serruya, C., M. Edelstein, U. Pollingher & S. Serruya, 1974. Lake Kinneret sediments: Nutrient composition of the pore water and mud water exchanges. Limnol. Oceanogr. 19: 489–507.
Smith, A. N., A. M. Posner & J. P. Quirk, 1974. Incongruent dissolution and surface complexes of hydroxyapatite. J. Colloid Interface Sci. 48: 442–449.
Stumm, W. & J. O. Leckie, 1971. Phosphate exchange with sediments; its role in the productivity of surface waters. Adv. Water Poll. Res. 5; 1970, III-26: 1–16.
Stumm, W. & J. J. Morgan, 1981. Aquatic chemistry. An introduction emphasizing chemical equilibria in natural waters. John Wiley & Sons, New York: 780 pp.
Suess, E., 1979. Mineral phases formed in anoxic sediments by microbial decomposition of organic matter. Geochim. Cosmochim. Acta 43: 339–352.
Tessenow, U., T. Frevert, W. Hofgärtner & A. Moser, 1977. Ein simultan schließender Serienwasserschöpfer für Sedimentkontaktwasser mit fotoelektrischer Selbstauslösung und fakultativem Sedimentstecher. Arch. Hydrobiol. Suppl. 48: 438–452.

Received 26 July 1989; in revised form 8 March 1990; accepted 4 May 1990

Hydrobiologia **207**: 179–186, 1990.
D. J. Bonin & H. L. Golterman (eds), Fluxes Between Trophic Levels and Through the Water–Sediment Interface.

Porewater chemistry of inorganic nitrogen compounds in the eastern Skagerrak (NE North Sea)

M. Wilken, K. K. Anton & G. Liebezeit[1,*]
Geologisch-Paläontologisches Institut, Bundesstraße 55, D-2000 Hamburg 13, Germany; [1]Present address: Nationalparkverwaltung, Virchowstraße 1, D-2940 Wilhelmshaven, Germany (author for correspondence)*

Key words: Skagerrak, postindustrial sediments, porewater chemistry

Abstract

Porosity, Eh and interstitital nitrogen compounds profiles are given for four locations in the eastern Skagerrak. Sediments down to 30 cm depth, i.e. postindustrial in age, were analysed. The results together with a detailed age determination (^{210}Pb) suggest a high degree of both lateral and vertical variability as a result of local bottom topography and currents as well as event-controlled sedimentation processes.

Introduction

With the exception of a few isolated areas in the German Bight and the central North Sea, the Skagerrak is the main area for deposition of particulate material in the North Sea. Based on ^{210}Pb analyses van Weering *et al.* (1987) estimated the annual deposition at 28.4 × 10^6 tons. From published age determinations (Erlenkeuser & Peterstad, 1984; Müller & Irion, 1984; Erlenkeuser, 1985; van Weering *et al.*, 1987) sedimentation rates appear to range from 0.8 to 11 mm yr^{-1}. Nevertheless, a general trend of high rates is evident for the southern slope with an increase from SW to NE.

Anton (1988) and Wirth & Wiesner (1988) have suggested that the majority of sedimenting inorganic material is derived from southern sources, although locally contributions from the Scandinavian mainland manifest themselves. Sediments on the southern slope grade from sands in shallow waters to silty clays in the deep basin (van Weering, 1981; Anton, 1988). Three facies can be distinguished: [1] a well sorted fine sand with median values <4 Ø in water depths up to 150 m, [2] a badly sorted silt with a fine sand content of up to 30% and median values between 4 and 6 Ø and [3] a badly sorted clay-rich (20–40%) silt with median values >6 Ø occurring in water depths >300 m. These differences exert a strong influence on the distribution of organic carbon in Skagerrak sediments. Generally, increasing C_{ORG} contents are found with increasing water depth (and fine fraction content). It can be expected that the amount and nature of organic matter control the concentrations of remineralisation products such as phosphate or inorganic nitrogen compounds (Lerman, 1975).

Despite its recognised importance as the major sink for particulate material produced in or introduced into the North Sea via e.g. riverine input or dumping, surprisingly few studies deal with postindustrial sediments in the Skagerrak. Shallow seismic profiling (van Weering, 1982; Salge & Wong, 1988a,b) generelly lacks the depth resolution required to elucidate sedimentation processes within the last 150 years. Other studies such as those by Jørgensen *et al.* (1981) among others use a rather coarse spatial resolution, which obscures short term variability. Here we present data on physical and chemical variables in cores taken along a downslope profile in the eastern Skagerrak with a spatial resolution of 2 cm roughly corresponding to a temporal resolution of < 5 years.

Material and methods

Cores were taken during cruise 57 of RV 'Valdivia' in May 1987. Sampling locations are given in Figure 1. A 100 × 8 cm gravity corer was used (Meischner & Rumohr, 1974), allowing us to recover undisturbed sediments, since it operates without a core catcher. Resuspension of superficial sediment is restricted to the 'fluff' layer *i.e.* the top millimetre(s). At each station up to four

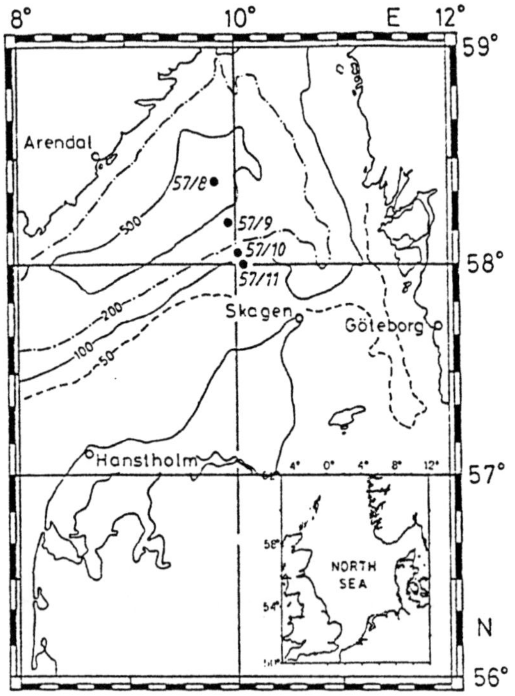

Fig. 1. Core locations in the Skagerrak.

parallel cores were taken. In the case of cores intended for porewater analysis, most of the resuspended sediment was allowed to settle. The supernatant was then siphoned off and stored frozen in PE bottles. Sediments were extruded and cut into 2 cm slices. After Eh and pH determinations (WTW pH-Digi 88 with combined glass electrode Ingold PT 405-85), these were squeezed under 10 bar nitrogen overpressure in a modified version of the apparatus described by Hartmann (1965). Cellulose acetate filters (0.45 μm) were used. Porewaters were collected and stored deep-frozen in PE bottles prior to analysis.

Inorganic nutrients (see below) were determined with the techniques described in Grasshoff *et al.* (1983) with the modifications suggested by Liebezeit & Velimirov (1984) on samples diluted 1 : 10 with bi-distilled water. Due to the small volumes obtained only single determinations could be carried out. Standard deviations calculated from replicate analysis of a combined sample were $\pm 1.3\%$ (ammonia), $\pm 3.3\%$ (nitrate), $\pm 34.4\%$ (nitrite).

Continuous porosity profiles were obtained from high resolution (2 mm) water content determinations using gamma-ray densitometry (Erlenkeuser & Metzner, unpubl. manuscript).

^{210}Pb analyses were carried in cooperation with G. W. Berger at the Netherlands Institute of Sea Research, Texel. The technique used has been described by Heijnis *et al.* (1987) and van Weering *et al.* (1987). Results used in the calculations are excess ^{210}Pb corrected for natural background resulting from the radioactive decay of ^{226}Ra.

Results and discussion

Physical environment

Porosity generally showed the expected exponential decrease reflecting sediment consolidation with depth (Fig. 2). Superimposed on this trend are deviations caused by bioturbation (non-parallel excursions), and changes in grain size composition (parallel excursions). Bioturbation

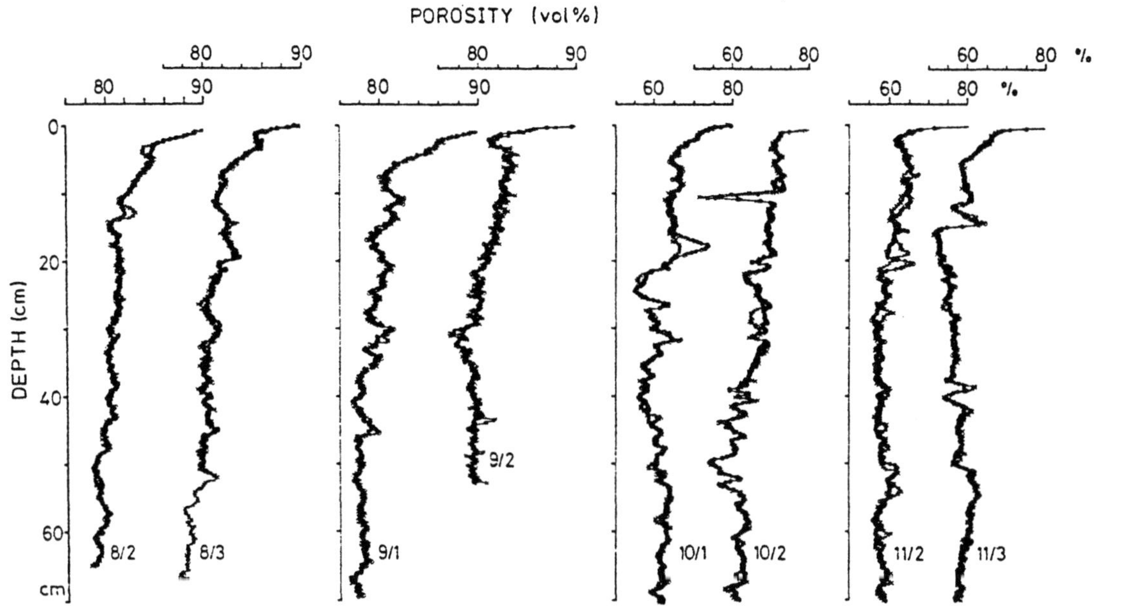

Fig. 2. Continuous porosity profiles of parallel cores calculated from gamma ray densitometric determination of water content at 2 mm depth resolution.

effects are particularly noticeable in the shallow cores 10 and 11, while they are either absent or only weakly pronounced in the deeper cores 8 and 9.

From the porosity data, cores 10 and 11 can be classified as facies 2 sediments, whereas 8 and 9 fall into facies 3 (see above). This is consistent with the distribution pattern given by van Weering (1981) for surface sediments.

In all cores large variability in grain size composition was found being more pronounced in the shallow cores. This is particularly evident in 10/1 and 10/2: in the latter a porosity minimum occurs at 10 cm depth (Fig. 2). Grain size analysis indicates for this layer an increase in the sand fraction from a background value of 14.0 ± 4.3% (n = 9) to 42.7% (Fig. 3). These phenomena presumably result from high variability in local micromorphology and bottom currents.

Organic matter in marine sediments is closely associated with the fine fraction (*e.g.* Jocteur-Monrozier *et al.*, 1983). Hence, variations in grain size composition should give rise to similar variations in the distribution of remineralisation products, at least on the time scale considered here.

Heterogeneity in sedimentation processes is further illustrated by a ^{210}Pb profile for core 10/2 (Fig. 3). Here, three layers of different activities were found: [1] a surface layer with a relatively rapid decrease terminating at the sand layer mentioned above, [2] a layer with a relatively uniform activity (11.5–18.5 cm) and [3] a layer with a slower decrease in ^{210}Pb activity. A benthic mixed layer (BML) commonly observed in shelf sediments (*e.g.* Nittrouer *et al.*, 1984), if present at all, would then be restricted to the upper 2 cm of the sediment column. This is supported by the ^{210}Pb profiles given by van Weering *et al.* (1987), which showed the BML to be either of variable thickness or completely absent.

Layer 2 can be described as highly bioturbated, while for layers 1 and 3 standardised sedimentation rates of 3.8 (<20 μm) and 3.9 (<2 μm) mm yr^{-1}, respectively, and 7.9 (<20 μm) and 8.6 (<2 μm) mm yr^{-1} can be calculated. A standardisation is necessary since lead, like other trace metals, is closely associated with the fine fraction (Förstner & Wittmann, 1979).

Using these sedimentation rates, the age of the sand layer at 10 cm depth is about 25 years. Since

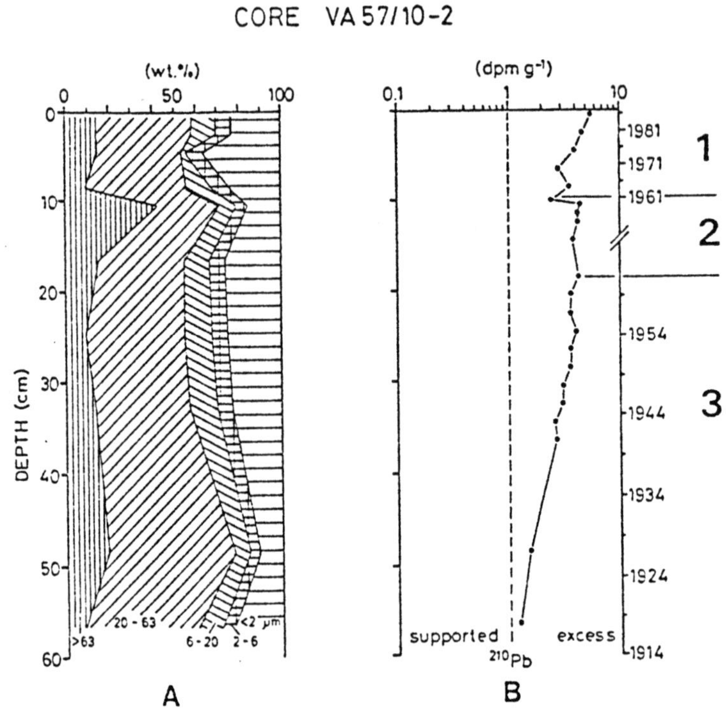

Fig. 3. Core VA 57/10-2. A: Cumulative grain size composition. B: ^{210}Pb profile. Supported ^{210}Pb activity represents approximately 1 dpm g^{-1}. 1 to 3 represent intervals with different sedimentation rates as discussed in the text.

one of the most disastrous storm floods in the North Sea occurred in 1962 this supports the hypothesis (Liebezeit, 1988) that sedimentation processes on the southern slope of the Skagerrak are to a large extent event controlled, in this case by storm floods. However, the sand layer in core 10/2 is of local origin and not the result of long range transport, as postulated by Liebezeit (1988) for the fine fraction.

Thus, local variability -both vertical and lateral- is a prominent feature of Skagerrak sediments. This can be found in variables as diverse as grain size composition/porosity (Fig. 2, 3A), ^{210}Pb activity (Fig. 3B, Erlenkeuser & Peterstad, 1984; Erlenkeuser, 1985; van Weering *et al.*, 1987) and carbohydrate contents (Liebezeit, 1988). Although corresponding high resolution measurements are lacking at present, variability in bottom currents and micromorphology can be assumed to be responsible for the observed irregular distributions. In addition, sedimentation events will leave an imprint upon the sedimentary record, as illustrated by the x-ray radiographs presented by Jørgensen *et al.* (1981).

Chemical variables

The redox potential in the cores used for pore-water sampling showed a rapid decrease from oxic to suboxic conditions in the deeper cores 8 and 9 and suboxic conditions throughout in 10 and 11 (Fig. 4). In one of the cores taken at station 11 anoxic conditions were found at 25 cm sediment depth again pointing to a high lateral variability over short distances.

Measurements using an oxygen microprobe (Helder & Bakker, 1985; Sørgensen & Jørgensen, 1987) showed oxygen penetration not much deeper than 10 mm at two stations in the Skagerrak at water depths of 387 and 520 m, confirming the results for cores 8 and 9. The data of the latter authors also suggest that at 520 m microbial reductions of oxygen, nitrate, Mn(IV),

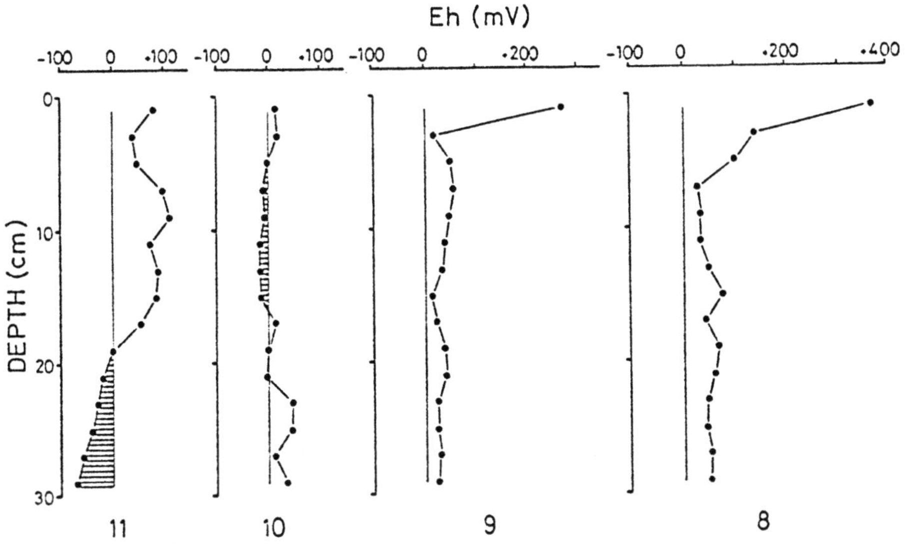

Fig. 4. Eh profiles of cores used for porewater sampling.

Fe(III) and sulphate are not spatially separated. Thus Eh values around 0 mV interpreted above as reflecting suboxic conditions, appear to be the result of a complex interplay of diverse microbial populations.

Nitrate/nitrite concentrations decrease with depth in all four cores (Fig. 5). Contents higher than in the overlying waters are restricted to the upper few centimetres and result from oxidation of ammonia during oxygen reduction. In deeper layers, concentrations of nitrate/nitrite remain essentially constant with the exception of cores 8 and 11 where a second maximum is to be noted at 15 and 13 cm depth, respectively. This might be related to bioturbation, which transports oxygen rich bottom water and material from the sediment surface into deeper layers. During coring operations using a 50 x 50 cm spade box corer, *Callianassa* burrows either still occupied or filled with finer sediment were noted to occur in deeper parts of the Skagerrak and the Norwegian Trough at sediment depths down to 30 cm (Liebezeit & Wirth, unpubl. observations).

The presence of nitrate/nitrite at depths of up to 30 cm is in stark contrast to the report of Sørensen & Jørgensen (1987) who did not detect nitrate at depths >2 cm at their deep Skagerrak station (see above). Both sediment and porewater sampling and analytical techniques are comparable, the only difference being storage of samples. Deep freezing, however, does not alter nitrate concentrations, at least for standards (A. Wenck, pers. comm., 1985). As nitrate depletion at shallow sediment depths was also observed in our extensive sampling programme of Skagerrak porewaters (Liebezeit & Wilken, unpubl. results), the explanation for the observed discrepancy compatible with the data at hand lies in local variability, as discussed above.

The fact that nitrate/nitrite are detectable throughout all cores might be related to the susceptibility of the organic material towards heterotrophic attack, since high contents of organic carbon are generally present in deeper sediment layers. In core 10/2, for example, C_{ORG} decreases from 1.34% at the surface to 0.72% at 10 cm depth (sand layer), followed by an increase to 1.06% at 32.5 cm (Anton, 1988). From mass balance calculations it has been suggested by van Weering *et al.* (1987) that a significant amount of refractory carbon is present in Skagerrak sediments. This has been recently confirmed by a number of independent techniques, such as maceral analysis (Wiesner *et al.*, in press), NMR spectroscopy of humic acid isolates (Fengler, 1989) and carbohydrate analysis (Liebezeit,

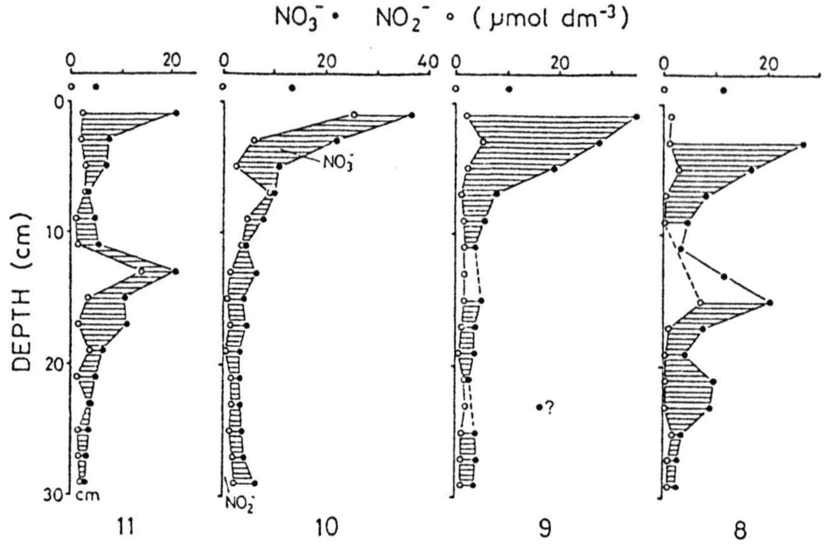

Fig. 5. Nitrate and nitrite profiles in Skagerrak cores. Data points on top refer to overlying water concentrations.

1988). A comparison between flux and burial data suggests that between 50 and 70% of the organic carbon pool do not stem from autochthonous primary production, although some import of 'fresh' organic matter has to be taken into account (Liebezeit, unpubl. results).

These data allow us to differentiate between two organic carbon pools in Skagerrak sediments: an easily degradable pool, which is remineralised within the upper (few) centimetre(s) of the sediment, and a refractory pool. The latter, which is obviously not attacked by heterotrophic organisms, survives for longer time spans as is evident from the analysis of a 3 m core from the eastern Skagerrak (Meyenburg, 1989), where a rapid decrease within the upper 10 cm is followed by virtually constant values throughout the rest of the core. These data are in line with the oxygen penetration data discussed above. The different methodological approaches used by Sørensen & Jørgensen (1987) and in the present communication will, however, inevitably result in a different picture of the processes operating within Skagerrak sediments. While our data reflect the bulk porewater chemistry and suggest a spatial separation of nitrate production (oxygen reduction) and reduction, the data of Sørensen & Jørgensen (1987) suggest parallel reduction of oxygen, nitrate, etc. as the result of microniche formation.

Ammonia in all cores increases with sediment depth, although an initial decrease is to be seen in core 11 (Fig. 6). This would suggest ongoing degradation of organic matter. However, neither iron monosulphides nor hydrogen sulphide were found to be present in all four cores, thus excluding a marked reduction of sulphate from being noticeable in the bulk samples analysed here. Neither iron nor manganese were determined for the present series of porewater samples. From the data of Sørensen & Jørgensen (1987) it has, however, to be assumed that these compounds play a significant role in the microbially mediated redox chemistry of Skagerrak sediments.

Using a molecular diffusion coefficient of 5.8×10^{-6} cm^2 sec^{-1} and a modified version of Fick's 1st law (Dicke, 1986), diffusive fluxes of 30 (core 8), 71 (core 9) and 81 (core 10) µmol m^{-2} d^{-1} can be calculated. These values fall in the range of data given by Dicke (1986) for Kiel Bight sediments (57–136 µmol m^{-2} d^{-1}) and by Berner (1974) for Santa Barbara Basin (137 µmol m^{-2} d^{-1}). Sediments with comparable organic carbon contents, however, gave considerably higher rates of diffusive transport, e.g. in Somes Sound, Maine, a value of 712 µmol m^{-2} d^{-1} was

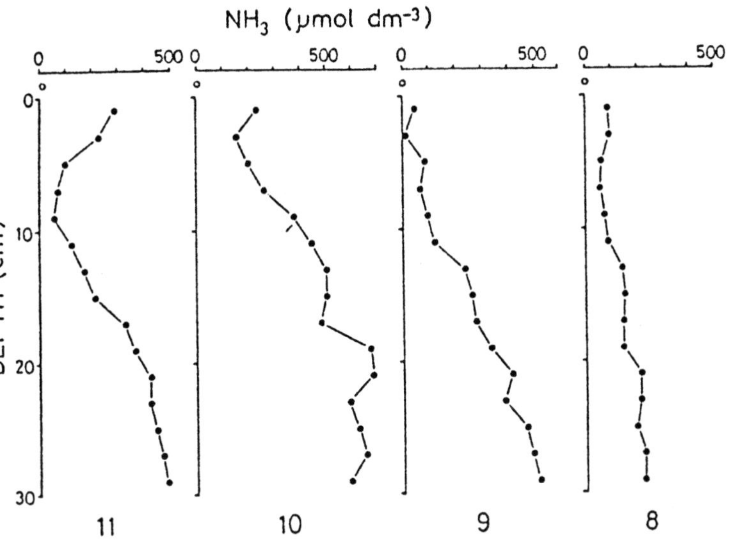

Fig. 6. Ammonia profiles for Skagerrak cores. Data points on top refer to overlying water concentrations.

found (Berner, 1974) whereas for Saanich Inlet, British Columbia, a diffusive flux of 49 x 103 μmol m^{-2} d^{-1} has been reported (Lerman, 1975). This again supports the notion that the degradability of organic matter largely controls porewater nutrient chemistry, as discussed above.

Acknowledgments

We are indebted to captain and crew of R/V 'Valdivia' for expert assistance in core recovery. G. W. Berger, Texel, kindly provided ^{210}Pb data. Porosities were determined in the laboratory of H. Erlenkeuser, Kiel. This work was supported financially by the Bundesministerium für Forschung und Technologie under grant MFU 05438. Responsibility for the contents of this paper lies with the authors.

References

Anton, K. K., 1988. Sedimentationsprozesse am SE-Hang des Skagerraks. Diploma Thesis, Universität Hamburg, 1–68.

Berner, R. A., 1974. Kinetic models for the early diagenesis of nitrogen, sulfur, phosphorous and silicon in anoxic marine sediments. In E. D. Goldberg (ed.), The Sea. New York, Wiley, 5: 427–450.

Dicke, M., 1986. Vertikale Austauschkoeffizienten und Porenwasserfluß an der Sediment/Wasser-Grenzfläche. Ber. Inst. Meeresk. Kiel 155: 1–164.

Erlenkeuser, H., 1985. Distribution of ^{210}Pb with depth in core GIK 15330-4 from the Skagerrak. Norsk Geol. Tidskr. 65: 27–34.

Erlenkeuser, H. & K. Peterstad, 1984. Recent sediment accumulation in Skagerrak as depicted by ^{210}Pb dating. Norsk. Geol. Tidskr. 64: 135–152.

Fengler, G., 1989. Huminstoffe in holozänen Sedimenten des Skagerraks (NE Nordsee) und der Elbe. Diploma Thesis, Universität Hamburg, 1–118.

Förstner, U. & G. Wittmann, 1979. Metal Pollution in the Aquatic Environment. Berlin, Springer Verlag, 1–486.

Grasshoff, K., M. Ehrhardt & K. Kremling, eds., 1983. Methods of Seawater Analysis. Weinheim, Verlag Chemie, 2nd edition, 1–419.

Hartmann, M., 1965. An apparatus for the recovery of interstitial water from recent sediments. Deep-Sea Res. 12: 225–226.

Heijnis, H., G. W. Berger & D. Eisma, 1987. Accumulation rates of estuarine sediment in the Dollard area: Comparison of ^{210}Pb and pollen influx methods. Neth. J. Sea Res. 21: 295–301.

Helder, W. & J. F. Bakker, 1985. Shipboard comparison of micro- and minielectrodes for measuring oxygen distribution in marine sediments. Limnol. Oceanogr. 30: 1106–1109.

Jocteur-Monrozier, L., M. Benijoly, P. Pillou, F. Andreux & B. Souchier, 1983. Distribution of organic matter in grain-size fractions of some recent sediments. In M. Bjoroy *et al.*

(eds.), Adv. Org. Geochem. 1981. New York, Wiley, 323–327.

Jørgensen, P., H. Erlenkeuser, H. Lange, J. Nagy, J. Rumohr & F. Werner, 1981. Sedimentological and stratigraphical studies of two cores from the Skagerrak. Spec. Publ. int. Ass. Sedim. 5: 397–414.

Lerman, A., 1975. Migrational processes and chemical reactions in interstitial waters. In E. D. Goldberg (ed.), The Sea. New York, Wiley, 6: 695–738.

Liebezeit, G., 1988. Early diagenesis of carbohydrates in the marine environment. II. Composition and origin of carbohydrates in Skagerrak sediments. In L. Mattavelli & L. Novelli (eds.) Adv. Org. Geochem. 1987, Org. Geochem. 13: 387–391.

Liebezeit, G. & B. Velimirov, 1984. Distribution of inorganic and organic nutrients in a sandy beach at Ischia, Bay of Naples. Oceanis 10: 437–447.

Meischner, D. & J. Rumohr, 1974. A light-weight, high-momentum gravity corer for subaqueous sediments. Senckenberg. marit. 6: 105–117.

Meyenburg, G., 1989. Sedimentologie und Geochemie eines Sedimentkernes aus dem östlichen Skagerrak. Diploma Thesis, Universität Hamburg, 1–83.

Müller, G. & G. Irion, 1984. Chronology of heavy metal contamination in sediments from the Skagerrak (North Sea). Mittlg. Geol.-Paläontol. Inst. Univ. Hamburg 56: 413–421.

Nittrouer, C. A., D. J. DeMaster, B. A. McKnee, N.H. Cutsall & I.L. Larsen, 1984. The effect of sediment mixing on ^{210}Pb accumulation rates for the Washington continental shelf. Mar. Geol. 54: 201–221.

Salge, U. & H. K. Wong, 1988a. Seismic stratigraphy and Quaternary sedimentation in the Skagerrak (NE North Sea). Mar. Geol. 81: 159–174.

Salge, U. & H. K. Wong, 1988b. The Skagerrak: A depo-environment for recent sediments in the North Sea. Mittlg. Geol.-Paläontol. Inst. Univ. Hamburg 65: 367–380.

Sørensen, J. & B. B. Jørgensen, 1987. Early diagenesis in sediments from Danish coastal water: Microbial activity and Mn-Fe-S geochemistry. Geochim. cosmochim. Acta 51: 158–1590.

van Weering, T. C. E., 1981. Recent sediments and sediment transport in the northern North Sea: surface sediments of the Skagerrak. Spec. Publ. int. Ass. Sedim. 5: 335–359.

van Weering, T. C. E., 1982. Shallow seismic and acoustic reflection profiles from the Skagerrak: implications for recent sedimentation. Proc. Konin. Acad. Wetensch. B85: 129–154.

van Weering, T. C. E., G. W. Berger & J. Kalf, 1987. Recent sediment accumulation in the Skagerrak, northeastern North Sea. Neth. J. Sea Res. 21: 177–189.

Wiesner, M. G., B. Haake & H. Wirth, in press. Organic facies in the North Sea. Org. Geochem.

Wirth, H. & M. G. Wiesner, 1988. Sedimentary facies in the North Sea. Mittlg. Geol.-Paläontol. Inst. Univ. Hamburg 65: 269–287.

Received 28 June 1989; in revised form 30 May 1990; accepted 28 June 1990

Hydrobiologia **207**: 187–196, 1990.
D. J. Bonin & H. L. Golterman (eds), Fluxes Between Trophic Levels and Through the Water–Sediment Interface.

Incidence of mussel culture on biogeochemical fluxes at the sediment-water interface

Dominique Baudinet, Elizabeth Alliot, Brigitte Berland, Christian Grenz, Marie-Reine Plante-Cuny, Raphaël Plante & Chantal Salen-Picard
Centre d'Océanologie de Marseille, Station Marine d'Endoume, 13007 Marseille, France

Key words: nutrient fluxes, sediment-water interface, mussel culture, biodeposit

Abstract

Upward nutrient fluxes at the sediment-water interface were studied in a mussel farming zone (Carteau, Gulf of Fos, France) in order to estimate the impact of organic matter input from biodeposition. Nitrate, nitrite, ammonia, silicate, phosphate and oxygen were measured. Fluxes were estimated by means of polyacrylate benthic chambers placed at sites located under (UM) and outside (OM) the rope hanging structures. Transformation of biodeposited organic matter increases phosphate, silicate and ammonia fluxes. No variation in nitrite fluxes could be detected and only minor differences were observed in nitrate and the oxygen production/consumption equilibrium at the two stations. Phosphate and silicate fluxes, which were always higher at the UM than at the OM site, decreased from spring to winter. Ammonia fluxes were very high under mussel cultures in May and September and lower in November. The fact that ammonia flux was always higher at the UM than at the OM sites might be explained by degradation of mussel biodeposit, as well as by benthic macrafauna excretion. Discrepancies between fluxes of the nutrients studied at the UM and OM sites increased as organic particulate matter in the water column decreased. Variations of oxygen flux followed a different pattern, since they were correlated with presence and abundance of photosynthetic microphytes on the bottom and in the water. Bottom respiration exceeded production of oxygen only in May 1988 at the UM station.

As it now stands, biodeposit input into the sediment under mussel ropes does not affect the ecosystem, although the flow of nutrients towards the water column is higher than in other areas.

Introduction

Because light conditions are optimal and supply of exogenous nutrients abundant, coastal estuarine and lagoon-like environments are characterized by high pelagic and benthic microphyte biomasses. These features make these areas ideal locations for shellfish farming. However, intense culture of filter-feeding molluscs (e.g. mussel ropes) may alter the equilibrium of the original ecosystem by removing particulate organic matter (POM) from the water column while increasing the sedimentation rate (Foster-Smith, 1975; Dahlbäck & Gunnarsson, 1981; Sornin *et al.*, 1983; Rodhouse & Roden, 1987; Kautsky & Evans, 1987; Kaspar *et al.*, 1985).

Degradation of organic matter at the water-sediment interface modifies the redox potential which influences direction and intensity of nutrient fluxes. Fisher *et al.* (1982) estimated that in the estuaries of North Carolina, between 28 and 35% of the nitrogen and phosphorus requirements of the water column are supplied by regeneration from sediments. Nixon (1981) claimed that 50% of the deposited organic carbon is oxidized and released back into the overlying water in mineral form. This remineralization process has been studied in French estuaries where tides play an important hydrodynamic role (Boucher & Boucher-Rodoni, 1988 and Feuillet-Girard *et al.*, 1988). To our knowledge, however, no data are available for the Mediterranean sea, where hydrodynamic conditions are different. Weak tidal variations but great wind variability result in more sporadic resuspension of biodeposited material. Since these factors affect both the metabolism of the molluscs and biodeposition, it is difficult to transpose data from other regions to our study site and to assess the fate of nutrients produced by remineralization of POM.

The purpose of this study was to evaluate the impact of POM on exchange processes at the water-sediment interface and the contribution of biodeposit remineralization to nutrient fluxes in the ecosystem. The enhancement of the nutrient content of the water column by sediment rich in organic matter can be detrimental for two reasons:

1. An overenrichment of the water column in nutrient salts by large biodeposit input could lead to extensive eutrophication, which could upset the equilibrium of the ecosystem, although high densities of molluscs may control eutrophication (Officer *et al.*, 1982) and mussel culture represents a more or less self-regulated extensive aquaculture system (Folke & Kautsky, 1989).

2. A disequilibrium in the kinetics of release of major components, i.e. nitrogen, phosphorus, and silicates, could change nutrient ratios and consequently the taxonomic composition of the phytoplankton community.

Study site

The Gulf of Fos (France) opens widely (3 km) onto the Mediterranean Sea (Fig. 1). Two rivers (the Rhone and the Durance) drain fresh water into

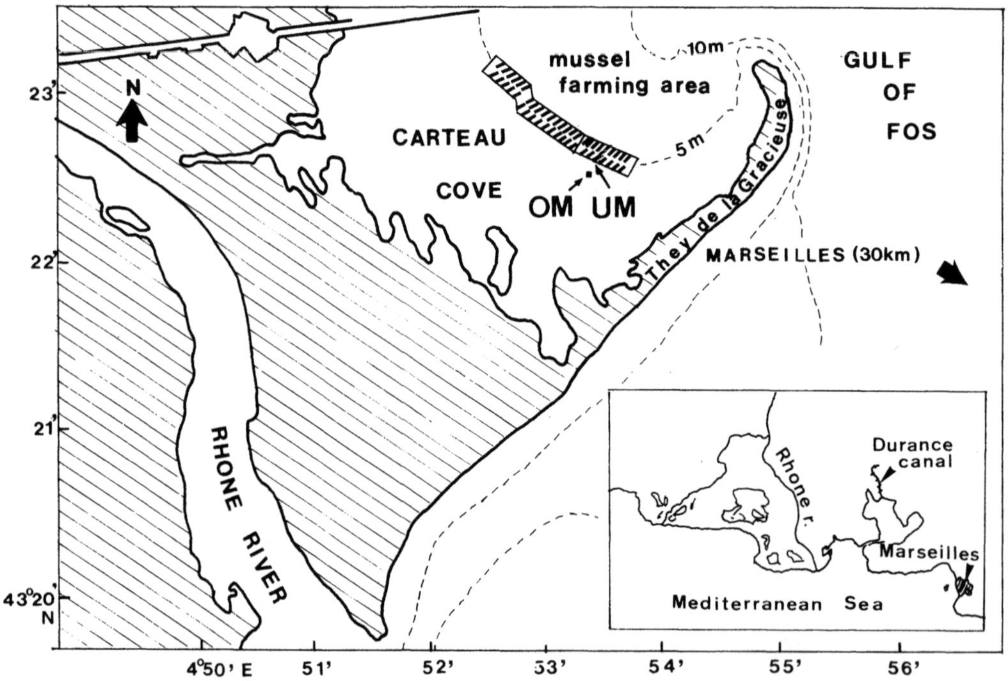

Fig. 1. Location of Carteau Cove with the mussel farming area. Outside mussel station is labelled OM and under mussel station UM.

the Gulf. The study site, Carteau Cove, has been the site of intense mussel farming for 5 years. In 1984, 70 mussel rope hanging structures were placed in zigzag fashion across the isobath situated at − 5 m. Each 15 × 50 m table can support 1000 to 1500 mussel ropes 3 to 4 m long. The cultivation area covers 0.0525 km^2 i.e. 0.4% of the whole Carteau Cove (13 km^2). The zone is sheltered to some degree by a natural sand bank (They de la Gracieuse).

Sediments outside the culture site (OM) are rich in organic matter, but mixing is intense, as shown by the depth of redox break line (more than 10 mm). It is classified as 'muddy sand' (25 to 50% of grains less than 63 μm). The macrofauna (adult invertebrates larger than 0.5 mm) was almost entirely comprised of detritus-feeding polychaetes (40% Cirratulidae, 30% Capitellidae mainly *Mediomastus sp.*) with densities of up to 50 000 per m^2. This is a poorly diversified fauna (Shannon index between 2 and 3).

Installation of mussel cultures (*Mytilus galloprovincialis*) and the resulting heavy production of faecal matter led to substantial physical and chemical changes. The under-mussel sediment (UM) must be classified as 'sandy mud' (75% of grains less then 63 μm). The original infauna has been overwhelmingly replaced by polychaetes characteristic of high-grade organic pollution. This new population is paucispecific (Shannon Index between 1.5 and 2), e.g. 10 000 *Capitella capitata* per m^2 and 60 000 *Ophryotrocha sp.* per m^2. These small species live in the top layer of the sediment, consuming and stirring the biodeposits. Under these conditions the redox beak line is observed within only 2 mm from the surface of the sediment.

Material and methods

Five surveys (as shown in Table 1) were carried out between 1987 and 1988 under various hydrological conditions.

Various physical, chemical, and biological indicators were monitored in the water immediately above the sediment in six transparent hemispheric 17-liter polyacrylate bell jars ($S = 0.13\ m^2$) placed under the mussel cultures (UM) on the first day of sampling period and at a reference site (OM) on the following day. On the day before each of the experiments, 6 PVC embasements per station were pushed randomly into the sediment; then sediment structure could be restored during the night. On the following day, the bell jars were clipped on the embasement. Water in the jars was continuously homogenized by a magnetic stirring bar. Water sampling and oxygen measurements were made in the chamber through a hole fitted with a soft rubber stopper. Samples were collected with a syringe by scuba divers every two hours over 6-hours (9 a.m. till 3 p.m.). Salinity was measured by means of a Beckman salinometer. Temperatures and dissolved oxygen were recorded *in situ* using an oximeter featuring automatic salinity and temperature correction (Orbisphere System 2609, 2112, 2120). Surface and underwater quantum flux were measured with a quantameter (Lambda instrument quantameter LICOR 550 equipped with a light sensor LI 192 SB). Nutrients, i.e. NO_2^-, NO_3^-, PO_4^{3-} and $Si(OH)_4^-$, were assayed in a Technicon autoanalyzer according to the protocol described by Tréguer & Le Corre (1975). NH_4^+ was determined by the Koroleff method (1969). Particulate suspended matter (PSM) and certain of its components were recorded monthly in the water column (− 2 m): particulate organic matter (POM) by the Krey method (1950), chlorophyll *a* by the Lorenzen method (1967), glucids by the Dubois *et al.* method (1956), proteins by the method of Lowry *et al.* (1951), and lipids by the Marsh & Weinstein method (1956). The available energy content of the POM ($j.l^{-1}$) was estimated by totalizing glucids, proteins and lipids, to which energy coefficients were assigned (Brody, 1945).

Nutrient fluxes at the water-sediment interface were estimated by the average slope of concentration plotted against time observed in the 6 chambers.

Macrofauna density expressed in terms of number per m^2 was calculated from 10 cm sediment cores after sieving through a 1 mm mesh screen and preservation in 5% formaldehyde.

Table 1. Sampling schedule and parameters measured in the water column. Minimal and maximal values are given when available.

Date	UM station				OM station				Nutrient concentrations in the water column (μmol.dm^{-3})				
	Temperature (°C)	Salinity	Time Total	Time Step	Temperature (°C)	Salinity	Time Total	Time Step	NH_4^+	NO_2^-	NO_3^-	PO_4^{3-}	$Si(OH)_4^-$
22/07/1987	–	–	–	–	19.7	37.5–38.0	24 h	var.	1.01–1.32	0.20–0.39	0.28–4.06	0.30–0.72	6.37– 8.33
22/09/1987	18.5–20.0	36.9	24 h	4 h	18.5–20.0	36.9	24 h	4 h	1.31–3.83	0.26–0.40	0.83–2.29	0.32–0.55	1.52– 3.19
25–26/05/1988	15.9–16.4	33.0–36.5	6 h	2 h	16.5	35.5–35.0	6 h	2 h	0.70–1.70	0.20–0.29	0.51–1.10	0.20–0.50	7.10–14.50
25–26/09/1988	18.2–18.6	37.9	6 h	2 h	18.5–19.0	37.8	6 h	2 h	0.99–2.00	0.20–0.30	0.65–1.77	0.24–0.41	7.18–12.20
24–25/11/1988	7.6– 8.4	33.2–36.4	6 h	2 h	8.1–10.4	36.6–36.9	6 h	2 h	2.00–2.90	0.35–0.48	3.02–4.93	0.43–0.76	11.72–15.63

Table 2. Ranges of dissolved nutrient fluxes (μmol m^{-2} h^{-1}) from nearshore sediments. Minimal and maximal values are given when available.

Authors		Location	Observed fluxes: μmol m^{-2} h^{-1}					
			NH_4^+		PO_4^{3-}		$Si(OH)_4^-$	
Aller & Benninger	1981	Long Island Sound (USA)	– 41.7;	+ 333.3			+ 833.0	
Aller *et al.*	1985	Yangtse River (China)	– 108.3;	+ 458.3			+ 1.2 ;	+ 550.0
Balzer *et al.*	1985	Maribago (Philippines)					+ 85 ;	+ 145.0
Boucher & Boucher-Rodoni	1988	Bay of Morlaix (France)	+ 51.0;	+ 369.0				
Callender & Hammond	1982	Potomac River (USA)	– 129.2;	+ 1083.3	– 25.0;	+ 166.7	+ 792.0	
Elderfield *et al.*	1981	Narragansett Bay (USA)	+ 37.5;	+ 750.0	+ 6.7;	+ 320.8	+ 6.6 ;	+ 320.8
Fisher *et al.*	1982	North Carolina (USA)	0.0;	+ 458.3	– 2.0;	+ 45.8		
Gouleau	1988	Salt marshes (French Atlantic coast)			+ 100.0;	+ 1270.0		
Helder & Andersen	1987	Bay of Fundy (Canada)			+ 92.0;	+ 288.0		
		Texel – Wadden Sea (Holland)			– 50.0;	+ 887.0		
Ullman & Sandstrom	1987	Bowling Green Bay (Australia)	– 6.6;	+ 73.1	– 1.0;	+ 1.2	– 41.25;	+ 73.1
Present work	1989	Gulf of Fos (France)	+ 3.5;	+ 366.0	– 3.2;	+ 116.2	+ 15.5 ;	+ 1206.0

Results

In the water column, seasonal fluctuations in PSM weight were strictly consistent with a coastal estuarine environment. In winter, there were periods of resuspension, and chlorophyll *a* pattern showed alternating periods of high and low phytoplanktonic biomass (Fig. 2). Figure 3 shows variations in energy content of POM, as ascertained by totalizing glucids, lipids and proteins in the water column, as well as the population density of *Capitella capitata* in sediment under mussel cultures. A lag was noted between the bloom and the development of polychaetes. Nutrient fluxes (as shown in Fig. 4) were generally positive. Ammonia, phosphates and silicates showed high and significant values. Other fluxes were low and not always significant. For example, nitrites varied from 0.7 to 2.8 μmol m^{-2} h^{-1}. Nitrate fluxes, which were almost always negative, displayed larger and more significant variations than nitrites (i.e. -59 to 5 μmol m^{-2} h^{-1}). The oxygen balance was generally in equilibrium with consumption. The concentration values ranged from 6 to 8.5 mg l^{-1}.

1. Fluxes were highest at both sites in May 1988 (PO_4^{3-} flux OM 35.7 and UM 116.2 μmol m^{-2} h^{-1} and especially $Si(OH)_4^-$ flux OM 376.2 and UM 1206.1 μmol m^{-2} h^{-1}, highest observed during the 2-year study). Although it was also high

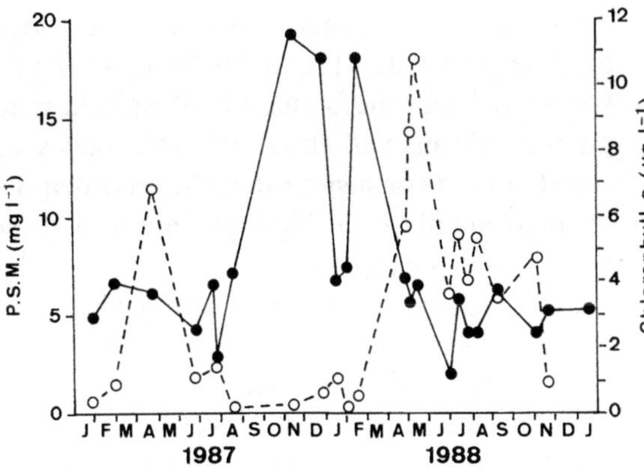

Fig. 2. Seasonal variations of Particulate Suspended Matter (PSM represented by a solid line) and chlorophyll *a* (dashed line) during 1987–1988.

in May 1988, NH_4^+ flux was about the same in September 1988 (UM site 339.4 vs. 366 μmol m^{-2} h^{-1}; OM site 73.4 vs. 64.4 μmol m^{-2} h^{-1} respectively). Spring was also characterized by a significant consumption of oxygen at the UM site (-63.8 mg O_2 m^{-2} h^{-1}).

2. In November 1988, water temperature was the lowest recorded in the cove (7.6 to 10.4 °C). The fluxes for all components except nitrates were also among the lowest observed. Only on this occasion were significant fluxes of nitrates detected either under culture or at the reference site (UM -18.5 and OM -59 μmol m^{-2} h^{-1}). Release of other nutrients by the sediment was clearly higher under mussel cultures than at the reference site. Slightly significant absorption of phosphates by the sediment was in fact noted at the reference site (-3.2 μmol m^{-2} h^{-1}). NH_4^+ and $Si(OH)_4^-$ fluxes at the reference site were particularly low (3.5 and 15.5 μmol m^{-2} h^{-1} respectively).

3. As mentioned above, NH_4^+ flux observed in September 1988 was the highest value recorded during the study. PO_4^{3-} and $Si(OH)_4^-$ were mid-range between those recorded in May and November 1988. The latter values were, however, higher under mussel cultures than at the reference site (UM 39.2 μmol m^{-2} h^{-1} for PO_4^{3-} and 163.5 μmol m^{-2} h^{-1} for $Si(OH)_4^-$). NO_3^- flux was low and negative at both sites, while nitrite release was significant under mussel cultures. To summarize, our data indicate that a decrease in PO_4^{3-} and $Si(OH)_4^-$ fluxes occurs at both the UM and OM stations from spring to winter, whereas absorption of nitrates by the sediment increases. It should be noted that in May 1988 phosphate and silicates fluxes recorded at the reference site were higher or equal to those recorded under the mussel cultures in September and November.

4. Although data from September 1987 and September 1988 were recorded under different conditions (24 h vs. 6 h period respectively; cf Table 1), comparison shows similar PO_4^{3-} and $Si(OH)_4^-$ fluxes at both sites, over the first 6 hours.

5. Data from July 1987 were recorded with

variable time intervals (Table 1) and only at the reference station. Fluxes were generally low.

Discussion – conclusion

1. On the whole the silicate fluxes reported here are similar to those reported in literature (Table 2) for coastal areas. Silicates generally diffuse towards the water column. The maximum flux observed under mussel cultures in May 1988 (1206 μmol $m^{-2} h^{-1}$) is among the highest reported to date. Silicate fluxes are 3 times higher under the mussels than at the reference site.

Dissolution of silica, like other components of organic matter, is highly dependent on temperature, salinity, intensity of bioturbation from macrofauna (Rutgers van der Loeff *et al.*, 1984; Helder & Andersen, 1987) and on oxidoreduction conditions (Yamada & D'Elia, 1984; Aller & Yingst, 1985). In Carteau Cove, the salinity of the water on the bottom varies from 33 to 38‰. This variability, which is low for an estuarine site cannot explain the great seasonal variations in $Si(OH)_4^-$ fluxes. In certain estuaries summertime regeneration may exceed allochtonous input from the water column (Wilke & Dayal, 1982; D'Elia *et al.*, 1983), with ratios between sediment and other sources ranging from 2 to 7.5 (Yamada & D'Elia, 1984). Even if the low temperature (7.6–10.4 °C) may explain the low silicate fluxes measured in November 1988 (lower dissolution and bioturbation), other hypotheses are needed to account for the low summertime fluxes (July 1987). They may be due to rapid silicate remineralization (<2 months) after the spring bloom at the end of April 1987 (6.9 $\mu g\, l^{-1}$ of chl *a*) because at this time chlorophyll *a* in the overlying water remains low ($1.1 \pm 0.5\ \mu g\, l^{-1}$). In May 1988 high silicate fluxes were recorded, following sedimentation of the bloom that took place at the beginning of the month (see Fig. 3). The diatoms *Chaetoceros curvisetus* and *C. debilis* were the main components of this bloom (Travers, Aix-Marseille II Univ., pers. comm.) which lasted for more than 15 days. Silica flux would thus appear to be highly dependent on the quantity and quality (silica-rich diatoms) of POM deposited on the bottom.

It is noteworthy that correlations between phosphate and silicate fluxes measured during the last three surveys were highly significant ($y = 9.603x + 0.886$; $r = 0.987$ and $p \leq 0.001$ for OM; $y = 10.8x - 0.979$; $r = 0.992$ and $0.001 < p \leq 0.025$ for UM, $n = 3$, where y: silicate fluxes and x: phosphate fluxes). Smith *et al.* (1987) reported a similar correlation in the Tomales Bay (California). This similarity is probably due to identical remineralization and release processes.

2. In this study phosphate exchanges were almost always positive, i.e. from the sediment to the water column. In literature, negative fluxes have been reported in some cases (Table 2). The only phosphate flux from the water column to the sediment was recorded in November 1988 at the OM station. One explanation for this negative flux might be absorption or a chemical reaction with iron hydroxides or calcite. An alternative explanation might be assimilation by microorganisms under winter conditions. The flux values reported here are of roughly the same magnitude as those mentioned in previous reports (as shown in Table 2 and Nixon *et al.*, 1980). There is, therefore, a strong increase in phosphate flux under mussel cultures. The ratios of fluxes between dissolved $Si(OH)_4^-$ and PO_4^{3-} were relatively constant throughout the year. During the last 3 surveys at both sites, these ratios were between 6 and 15. This is much lower than ratios calculated for the concentrations of these ions in the overlying waters (18 to 70). Smith *et al.* (1987) speculated that the optimal Si/P ratio in natural phytoplankton was about 10. The values measured at Carteau suggest that the recorded fluxes resulted mainly from biodegradation of recently sedimented phytoplankton.

3. Ammonia is the main nitrogen-containing mineral component released by sediment, since NO_3^- fluxes were two orders of magnitude lower. Our results concur with previous data from coastal environments (Table 2). However, unlike previous authors, we never observed negative NH_4^+ fluxes. As for PO_4^{3-} and $Si(OH)_4^-$, a high

NH_4^+ flux was observed in May 1988, a period in which the rate of mussel biodeposition was maximal. Bioturbation by a large population of Capitellidae in the sediment accelerated the release of components and their excretion could increase ammonia fluxes (see Fig. 3). Cerco (1989) indicates that ammonia release is enhanced up to 70% in anoxic conditions. The high NH_4^+ fluxes observed in September of the same year can be attributed either to these causes (polychaete population under the mussel ropes smaller than at reference site), nor to an enrichment of the water column in protein, nor to prevailing temperatures. A possible explanation could be deduced from the composition of macrobenthic populations. At this time the proportion of Nassidae, necrophagous gasteropods, was unusually high (up from 50 to 293 per m^2), while the density of polychaetes declined. These changes resulted from the lethal effect of high temperature combined with sedimentation of the summer bloom.

4. Nitrate flux variations followed a different pattern from other nutrients (Fig. 4). Under mussel ropes nitrates were generally absorbed, although the rate was sometimes low. At the reference site absorption tended to be higher. These observations indicate that the additional POM had no direct effect on nitrate fluxes. It should be

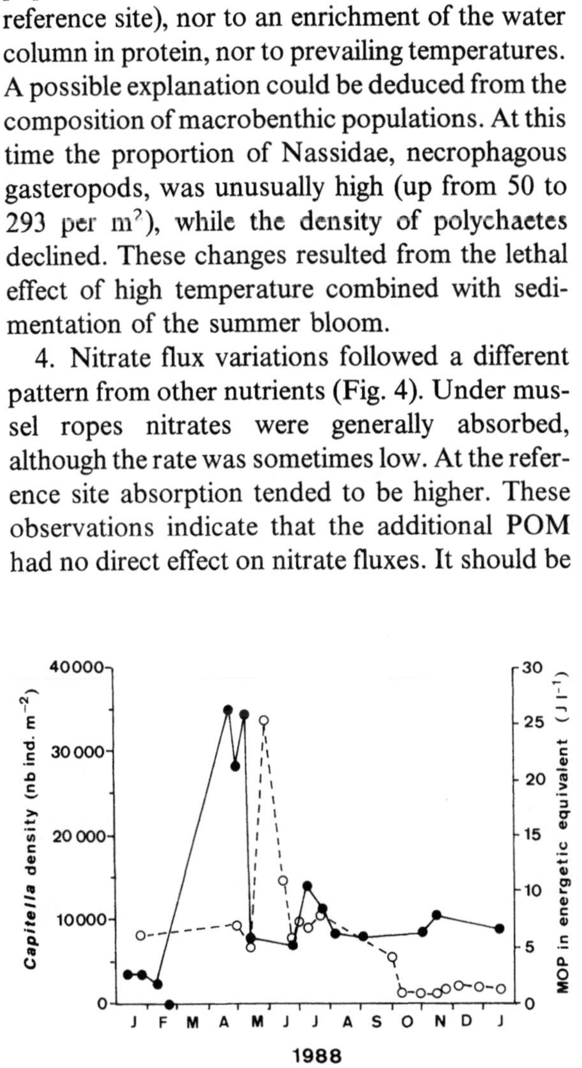

Fig. 3. Variations of *Capitella capitata* density under a mussel table (dashed line), and Particulate Organic Matter (sum of glucids, lipids and proteins expressed as energetic values, represented by solid line) in the water column, during 1988.

pointed out that absorption was highest when the seston load was lowest (November 1988). Seitzinger *et al.* (1980) showed that increasing nitrate concentrations in the water column enhanced denitrification and thus NO_3^- consumption. Similarly, Boynton *et al.* (1980) and Boynton & Kemp (1985) found an increase of sediment uptake when NO_3^- concentration increases in the water column. In Carteau Cove an accumulation of this ion was noted in the water column during winter, when the sediment uptake at the UM and OM sites are high and become significant. The N/P ratio varied from 2 to 9. According to Blackburn & Henriksen (1983) this low ratio is also a marker of denitrification.

5. Changes in oxygen concentrations on the bottom suggest that during daytime photosynthetic activity is counterbalanced by consumption. The depth in the study area being around − 5 m, light reaching the bottom accounts for 6 to 30% of the incidental energy, depending on the turbidity of the water. During our surveys, the lowest incidental light energy, recorded in July 1987 in overcast weather, was strong enough to sustain the growth of an autotrophic system. Photosynthesis by microphytobenthos played a direct role in oxygen production during the day and in absorption of nutrients. The only time that a negative O_2 balance was noted was in May when the sediment was particularly rich in organic matter under the mussel cultures. This tendency to anoxia may probably explain higher phosphate release (Sundbäck & Granéli, 1988; Cerco, 1989).

Table 3. Impact of mussel farming area (0.0525 km^2) fluxes on the whole Carteau Cove area (13 km^2). Average input values to the water column are estimated from fluxes of the last three surveys.

	NH_4^+	PO_4^{3-}	$Si(OH)_4^-$
A-Nutrient input by OM sediment (mol h^{-1})	612.3	169.0	2012
B-Nutrient input by UM sediment (mol h^{-1})	14.0	3.0	28.7
Influence of mussel cultivation zone (B/A %)	2.3	1.8	1.4

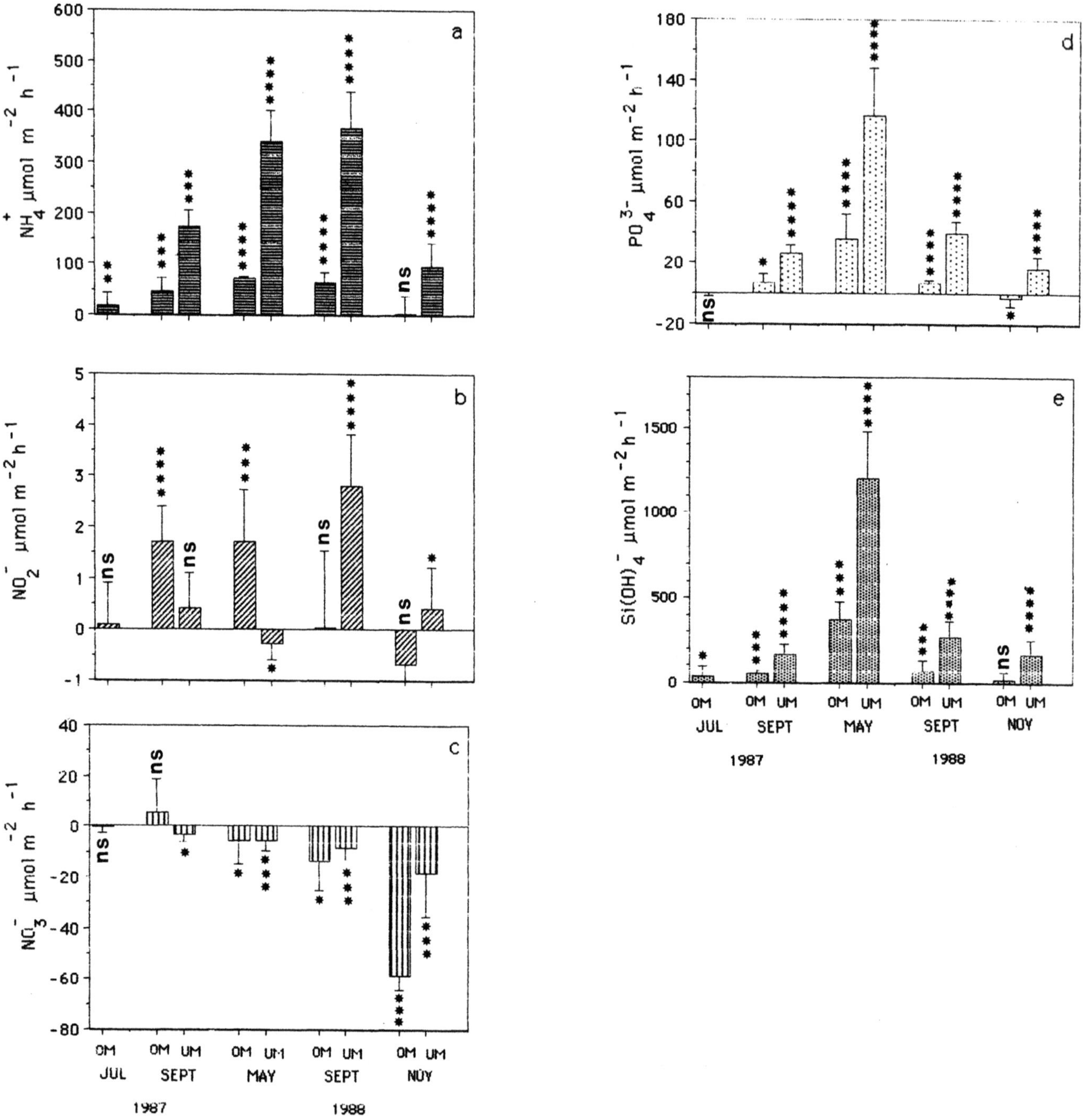

Fig. 4. Nutrient fluxes at both stations (UM and OM) measured during 5 surveys. Mean values for UM and OM stations are presented with 95% confidence interval. Rank statistical test of fluxes p is symbolized as: (****) $p \leq 0.001$; (***) $0.001 < p \leq 0.005$; (**) $0.005 < p \leq 0.1$; (*) $0.1 < p \leq 0.25$; (n.s.) $p > 0.25$.

6. In conclusion, the impact of mussel farming on nutrient enrichment of the area may be estimated by comparing the average total release of three nutrients (NH_4^+, PO_4^-, $Si(OH)_4^-$) with and without considering the mussel tables. The mussel zone (0.0525 km^2) covers no more than 0.4% of the total area (13 km^2) of the cove, whereas the presence of the mussel zone increases nutrient releases from 1.4 to 2.3% of the whole production of these elements (Table 3).

In spite of this phenomenon, the mussel cultivation area is not at present endangering the equilibrium of the whole ecosystem in Carteau Cove.

Acknowledgements

This work was supported by the 'Groupement scientifique CNRS–Université-Société nationale Elf-Aquitaine: Cycles biogéochimiques. Devenir de la Matière organique en milieu littoral.'

We thank anonymous reviewers for critical help.

References

Aller, R. C. & L. K. Benninger, 1981. Spatial and temporal patterns of dissolved ammonium, manganese and silicon fluxes from bottom sediments of Long Island Sound, U.S.A. J. mar. Res. 39: 295–314.

Aller, R. C. & Y. Yingst, 1985. Effects of the marine deposit-feeders *Heteromastus filiformis* (Polychaeta), *Macoma balthica* (Bivalvia) and *Tellina texana* (Bivalvia) on averaged sedimentary solute transport, reactive rates and microbial distributions. J. mar. Res. 43: 615–645.

Aller, R. C., J. E. Mackin, W. J. Ullman, C. H. Wang, S. M Tsai, J. C. Jisi, Y. N. Sui & J. Z. Hong, 1985. Early chemical diagenesis, sediment-water solute exchanges and storage of reactive organic matter near the mouth of the Changjiang, East China. Sea cont. Shelf Res. 4: 227–251.

Balzer, W., B. Von Bodungen & F. Pollehne, 1985. Benthic degradation of organic matter and regeneration of nutrients in shallow water sediments of Mactan, Philippines. The Philippine Scientist 22: 30–41.

Blackburn, T. H. & K. Henriksen, 1983. Nitrogen cycling in different types of sediments from Danish waters. Limnol. Oceanogr. 28: 477–493.

Boucher, G. & R. Boucher-Rodoni, 1988. *In situ* measurements of respiratory metabolism and nitrogen fluxes at the interface of oyster beds. Mar. Ecol. Prog. Ser. 44: 229–238.

Boynton, W. & W. Kemp, 1985. Nutrient regeneration and oxygen consumption by sediments along an estuarine salinity gradient. Mar. Ecol. Prog. Ser. 23: 45–55.

Boynton, W., W. Kemp & C. Osborne, 1980. Nutrient fluxes across the sediment water interface in the turbid zone of a coastal plain estuary. In V. Kennedy (ed.), Estuarine Perspectives. Academic Press, New-York, 93–109.

Brody, S., 1945. Bioenergetic and growth. Reinhold publishing Corp., New York, 1023 pp.

Callender, E. & D. E. Hammond, 1982. Nutrient exchange across the sediment-water interface in the Potomac River Estuary. Estuar. coast. Shelf Sci. 15: 395–413.

Cerco, C., 1989. Measured and modelled effects of temperature, dissolved oxygen and nutrient concentration on sediment water exchange. Hydrobiologia 174: 185–189.

Dahlbäck, B. & L. A. H. Gunnarsson, 1981. Sedimentation and sulfate reduction under a mussel culture. Mar. Biol. 63: 269–275.

D'Elia, C. F., D. M. Nelson & W. R. Boynton, 1983. Chesapeake Bay nutrient and plankton dynamics (III): the annual cycle of dissolved silicon. Geochim. Cosmochim. Acta 47: 1945–1955.

Dubois, M., K. A. Gilles, J. K. Hamilton, P. A. Rebers & F. Smith, 1956. Colorimetric method for determination of sugars and related substances. Analyt. Chem. 28: 350–356.

Elderfield, M., N. N. Luedtke, R. J. McCaffrey & M. Bender, 1981. Benthic flux studies in Narragansett Bay. Am. J. Sci. 288: 768–787.

Feuillet-Girard, M., M. Héral, J. M. Sornin, J. M. Deslous-Paoli, J. M. Robert, F. Mornet & D. Razet, 1988. Eléments azotés de la colonne d'eau et de l'interface eau-sédiment du bassin de Marennes-Oléron: influence des cultures d'huîtres. Aquat. living Resour. 1: 251–265.

Fisher, T. R., P. R. Carlson & R. T. Barber, 1982. Carbon and nitrogen productivity in three North Carolina estuaries. Estuar. coast. Shelf Sci. 15: 621–644.

Folke, C. & N. Kautsky, 1989. The role of ecosystems for a sustainable development aquaculture. Ambio 18: 234–243.

Foster-Smith, R. L., 1975. The effect of concentration of suspension on the filtration rates and pseudofaecal production for *Mytilus edulis* L., *Cerastoderma edule* L. and *Venerupis publastia* Montagu. J. exp. mar. Biol. Ecol. 17: 1–22.

Gouleau, D., 1988. Cycles journaliers de la silice dissoute dans les eaux libres d'un bassin aquacole. Rôle respectif des diatomées et du sédiment. J. Rech. océanogr. 13: 55–58.

Helder, W. & F. Ø. Andersen, 1987. An experimental approach to quantify biologically mediated dissolved silicate transport at the sediment-water interface. Mar. Ecol. Prog. Ser. 39: 305–311.

Kaspar, H. F., P. A. Gillepsie, I. C. Boyer & A. L. MacKensie, 1985. Effect of mussel aquaculture on the nitrogen cycle and benthic communities in Kenepuru

Sound, Marlborough sounds, New Zealand. Mar. Ecol. Prog. Ser. 38: 201–212.
Kautsky, N. & S. Evans, 1987. Role of biodeposition by *Mytilus edulis* in the circulation of matter and nutrients in Baltic coastal ecosystem. Mar. Ecol. Prog. Ser. 38: 201–212.
Koroleff, F., 1969. Direct determination of ammonia in natural as indophenol blue. Int. Coun. Explor. Sea C.M.C.: 9: 1–6.
Krey, I., 1950. Eine neue Methode zur quantitativen Bestimmung des Planktons. Kieler Meeresforsch. 7: 58–75.
Lorenzen, C. J., 1967. Determination of chlorophyll and phaeopigments. Spectrophotometric equations. Limnol. Oceanogr. 12: 343–346.
Lowry, O. H., N. I. Roseborough, A. L. Farrand & R. J. Randall, 1951. Protein measurement with the Folin phenol reagent. J. biol. Chem. 193: 263–275.
Marsh, J. B. & D. B. Weinstein, 1956. Simple charring method for determination of lipids. J. Lip. Res. 7: 574–576.
Nixon, S. W., 1981. Remineralization and nutrient cycling in coastal marine ecosystems. In B. J. Nelson & L. E. Cronin (eds), Estuaries and nutrients. The Humana Press, Clifton, New Jersey: 111–138.
Nixon, S. W., J. R. Kelley, B. N. Furnas, C. A. Oviatt & S. S. Hale, 1980. Phosphorus regeneration and the metabolism of coastal marine bottom communities. In K. K. Tenore & B. C. Coull, Marine benthic dynamics. University of South Carolina Press: 219–242.
Officer, C. B., T. J. Smayda & R. Mann, 1982. Benthic filter feeding: a natural eutrophication control. Mar. Ecol. Prog. Ser. 9: 203–210.
Rodhouse, P. G. & C. M. Roden, 1987. Carbon budget for a coastal inlet in relation to intensive cultivation of suspension feeding bivalve molluscs. Mar. Ecol. Prog. Ser. 36: 225–236.
Rutgers van der Loeff, M. M., L. G. Anderson, P. O. J. Hall, A. Iverfeldt, A. B. Josefson, B. Sundby & S. F. G. Westerland, 1984. The asphyxiation technique: an approach to distinguish between molecular diffusion and biologically mediated transport at the sediment-water interface. Limnol. Oceanogr. 29: 675–686.
Seitzinger, S., S. Nixon, M. E. Q. Pilson & S. Burke, 1980. Denitrification and N_2O production in nearshore marine sediments. Geochim. Cosmochim. Acta. 44: 1853–1860.
Smith, S. V., W. J. Wiebe, J. T. Hollibaugh, S. J. Dollar, S. W. Hager, B. E. Cole, G. W. Tribble & P. A. Wheeler, 1987. Stoichometry of C, N, P, and Si fluxes in a temperate climate enbayment. J. mar. Res. 45: 427–460.
Sornin, J. M., M. Feuillet, M. Héral & J. M. Deslous-Paoli, 1983. Effet des biodépôts de l'huître *Crassostrea gigas* (Thunberg) sur l'accumulation de matières organiques dans les parcs du bassin de Marennes-Oléron. J. mollusc. Stud. (Suppl. 12A): 185–197.
Sundbäck, K. & W. Granéli, 1988. Influence of microphytobenthos on the nutrient flux between sediment and water: a laboratory study. Mar. Ecol. Prog. Ser. 44: 229–235.
Tréguer, P. & P. Le Corre, 1975. Manuel d'analyse des sels nutritifs dans l'eau de mer (utilisation de l'autoanalyseur Technicon II R), 2ème éd., Laboratoire d'océanographie chimique, Brest, 110 pp.
Ullman, W. J. & M. W. Sandstrom, 1987. Dissolved nutrient flux from the nearshore sediments of Bowling Green Bay, Central Great Barrier Reef Lagon (Australia). Estuar. coast. Shelf Sci. 24: 285–303.
Wilke, R. J. & R. Dayal, 1982. The behavior of iron, manganese and silicon in the Peconic River estuary, New York. Estuar. coast. Shelf Sci. 15: 577–586.
Yamada, S. S. & C. F. D'Elia, 1984. Silicic acid regeneration from estuarine sediment cores. Mar. Ecol. Prog. Ser. 18: 113–118.

Received 31 July 1989; in revised form 10 January 1990; accepted 19 March 1990

Hydrobiologia **207**: 197–207, 1990.
D. J. Bonin & H. L. Golterman (eds), Fluxes Between Trophic Levels and Through the Water–Sediment Interface.

Modifications in habitat use patterns and trophic interrelationships in the fish fauna of an oligotrophic artificial lake: Sainte Croix (Provence, France)

Guy Brun, Rémi Chappaz, Georges Olivari
Laboratoire d'Hydrobiologie, Université de Provence, 1 Place Victor Hugo, 13331 Marseille Cedex 3, France

Key words: Artificial lake, Fish, Niche overlap

Abstract

Lake Sainte-Croix is a hydro-electric reservoir on the river Verdon that came into operation about 16 years ago. The area of the lake is about 22 km^2 and the volume about 767 Mm^3. The water level drops by as much as 16 m in winter. Sainte-Croix is a warm monomictic oligotrophic lake. The indigenous fish fauna of the Verdon originally included 8 species. Successive stocking has raised this number to 17. In catches made in 1976–1977 with 27 mm mesh nets, 80 to 90% of the fish with a body size of between 15 and 25 cm were of the four main species indigenous to the river before the dam was built: *Leuciscus cephalus*, *Barbus fluviatilis*, *Chondrostoma toxostoma*, *Salmo trutta*. Catches made at same period with 14 mm mesh nets gave only *Chondrostoma toxostoma* and *Leuciscus cephalus*.

Changes in the fish fauna were apparent from catches made in 1984 and 1987 with nets of the same mesh size. There was a marked demographic increase of roach, and more recently of bleak and gudgeon, all three species introduced after the construction of the dam. At the same time, there was a decrease in catch frequency of *Ch. toxostoma*, *L. cephalus* and *B. fluviatilis* in the downstream area of the lake. Stomach content analysis carried out in different seasons on the six most abundant species of the lake (roach, toxostome, chub, bleak, perch and gudgeon) provided evidence on the trophic organisation of the fish fauna. We have measured niche breadth for each species by the Simpson diversity index and niche overlap by the Pianka index. Bleak is the only species to feed on zooplankton in open water. Roach is a particularly opportunistic feeder, sometimes taking plankton, but at Sainte-Croix it is exclusively herbivorous and benthophagous. Niche overlap between these two species is slight or null, depending on the season. On the other hand, there is a high degree of overlap between roach and chub and between roach and toxostome. The decrease in toxostome and chub frequency in the downstream area of the lake, where roach is most abundant, would appear to be a competitive exclusion phenomenon. There is high niche overlap between bleak and young perch, which partly explains the low abundance of the latter species; other contributory factors are the scarcity of zooplankton inshore, and the scarcity of zoobenthos resulting from the drop in the water level in winter.

Introduction

the abrupt changes in the physical and chemical characteristics of the environment that result from the building of a dam on a river are inevitably accompanied by more or less immediate modifications in the distribution and composition of the fish fauna. This process can be controlled or accelerated by human intervention, notably by the introduction of new species. The lake of Sainte Croix on the river Verdon (Fig. 1) is an interesting example of this process. The reservoir was filled in 1974, and has been the subject of continuous scientific investigation since 1976. The lake and modifications in its physical and chemical parameters have been the subject of several studies: Champeau *et al.* (1979) and Champeau (1982). Bressac & Champeau (1985) and Pont *et al.* (1989) have described developments in the zooplankton. CEMAGREF (1977), Champeau *et al.* (*op. cit.*), Chappaz *et al.* (1987, 1988, 1989a, b) have compiled a list of fish species, and described the ecology of certain fish species in the lake. The running water ecosystem originally included 8 species of indigenous fish: *Salmo trutta* L., *Barbus barbus* (L.), *Chondrostoma toxostoma* (Vallot), *Leuciscus cephalus* (L.), *Telestes souffia* (Risso), *Cottus gobio* L., *Anguilla anguilla* (L.); 10 other species were introduced by stocking in the years following the filling of the reservoir: *Salmo gairdneri* Richardson, *Alburnus alburnus* (L.), *Gobio gobio* (L.), *Rutilus rutilus* (L.), *Scardinius erythrophtalmus* (L.), *Tinca tinca* (L.), *Cyprinus carpio* L., *Lepomis gibbosus* (L.), *Perca fluviatilis* L., *Esox lucius* L. the latest lists (Chappaz *et al.*, 1987, 1988, 1989a, b)) mention 17 species: *Cottus Gobio* has not been caught recently in the main part of the lake. The present article analyzes modifications in the trophic and spatial interrelationships of the fish fauna that have become apparent since the filling of the reservoir, the current spatial distribution of species and seasonal variations. We suggest a possible interpretation of these phenomena based on changes in physical and chemical parameters, and on new competitive relationships that have arisen with the development of new species introduced with varying degrees of success.

Materials and methods

Lake Sainte-Croix (Fig. 1) is an artificial reservoir on the river Verdon in the French Southern Alps

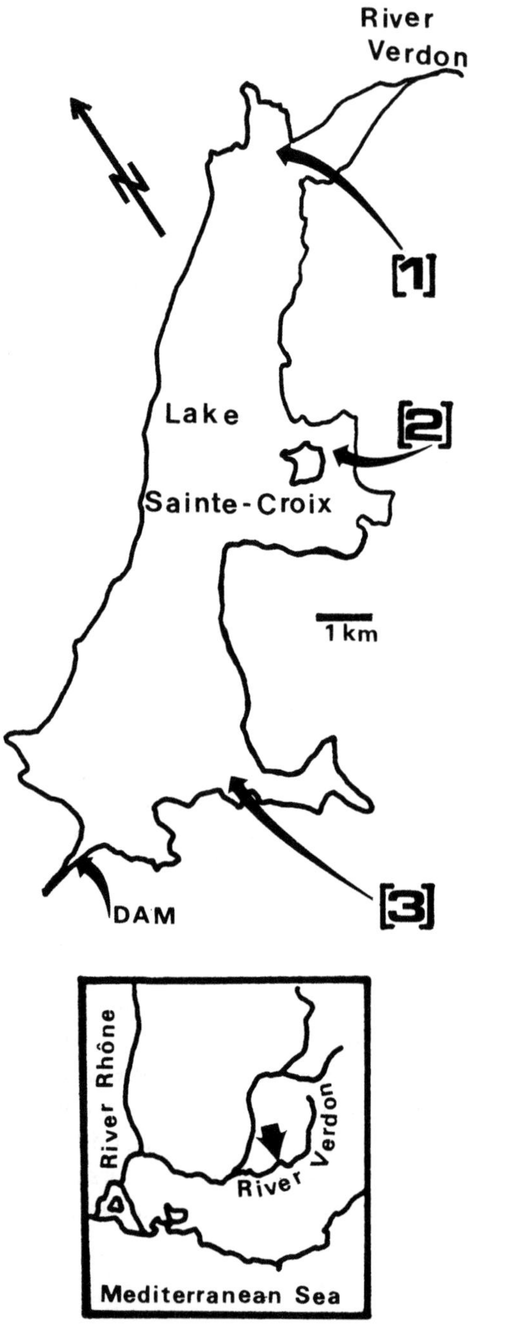

Fig. 1. Map of Lake Sainte-Croix showing the three areas studie: (1) upstream area, (2) central area, (3) downstream area.

(Alpes du Sud). The normal maximum water level of the lake is 477 m above sea level. When full, the lake covers an area of 2200 ha and has a volume of 767 Mm^3. Maximum depth is 83 m. The water level is subject to seasonal fluctuations, depending on the functioning of the hydroelectric power station, with a maximum range of 16 m. The lowest water level occurs in late winter, before the snow melts. Detailed temperature records are available thanks to measurements carried out by the electricity company (Electricité de France) at a permanent site. Temperatures are recorded every three hours at intervals of 3 m from 1 m to 80 m depth. This parameter determines to a large extent the growth rate of fish. In Lake Sainte-Croix, the surface water temperature varies between 6 °C and 7 °C in winter and 24 °C in summer, and remains above 20 °C in the first 10 to 15 m for more than 3 months. The thermocline is very well-defined in June. The period of mixing begins in November. Below 30 m, temperature remains fairly constant throughout the year at between 6 °C and 12 °C.

Fish were collected with gill nets and by electric fishing. Gill nets were placed at nightfall and taken up the following morning. To catch fish for census purpose, for the study of the growth rate of different species and for the study of diet, we used the following range of mesh sizes: 10, 14, 18, 22, 27, 32, 50 mm. The first three were in monofilament nylon, 25 m long and 1,5 m high; the other four were in cotton, 100 m long and 2 m high. There were two sets of gill nets: the first were placed at the surface, the second on the bottom. Small-bodied fish, or certain species whose behaviour makes them difficult to catch with a gill net, were caught by electric fishing from a boat. This method was used in the marginal areas, in *myriophyllum* or pondweed beds, in submerged terrestrial vegetation and among rocks. The method is identical to that used in running water, and involves an anode operator and one or two catch-net holders. It is much less efficient in very open or deep water where it is easier for the fish to escape.

The stomach, or the whole digestive tube, is removed *in situ* and fixed in a 10% formalin solution.

The field data are shown in the form of a general term table (species × time × food type): s_{ij}.

$$ps_{ij} = s_{ij} \Big/ \sum_{j=1}^{r} s_{ij} \qquad (1)$$

s_{ij} = number of individuals of species (S) utilizing the resource (j) at a given time (i). Food types = $(1, 2, \ldots, j, \ldots, r)$

ps_{ij} is the frequency of utilization of the resource (j) by the species (S) at a given date (i). The trophic niche breadth of the species (S) at a given date (i) is calculated by the standard Simpson Index:

$$Is = Bi - 1/N - 1 \qquad (2)$$

where

$$Bi = 1 \Big/ \sum_{j=1}^{r} s_{ij}$$

N = number of food types consumed by the fish.

Quantification of competition between species utilizing the same resources requires knowledge of the resource availability in the environment, and the frequency of use of different resources by each species. We have considered only the second factor by calculating the degree of niche overlap, using Pianka Index $O_{(A, B)}$:

$$O_{(A, B)} = \frac{\sum_{j=1}^{r} pa_{ij} \cdot pb_{ij}}{\left[\sum_{j=1}^{r} pa_{ij}^2 \cdot \sum_{j=1}^{r} pb_{ij}^2\right]^{1/2}} \qquad (3)$$

Where pa_{ij} and pb_{ij} are the frequency of use of food type (j) by species (A) and (B) at a give date (i).

The use of spatial or trophic resources varies in time according to the physiological condition of organisms and seasonal environmental changes. Seasonal migrations and changes of prey are among the factors likely to reduce niche overlap and, consequently, potential competition. These variations have been monitored by calculation of the diet overlap $O_{(A, B)}$ index for each sampling date. We have characterized overall time and diet overlap over a seasonal cycle by a generalisation of the Pianka Index. To this end we use the standard scalar product of the 2 matrices A and B defining the trophic spectra in species (A) and (B) over a seasonal cycle is. Following the same conventions, the degree of overlap is expressed as:

$$P_{(A,B)} = \frac{\sum_{j,i} pa_{ij} \cdot pb_{ij}}{\left[\sum_{j,i} pa_{ij}^2 \cdot \sum_{j,i} pb_{ij}^2\right]^{1/2}} \tag{4}$$

Results

Modifications in the habitat use patterns of the fish fauna: the samplings we carried out from 1983–1988 as well as those caried out of other workers in 1976 (CEMAGREF 1977) and in 1976–77 (Champeau *et al.*, 1979) show a decreasing gradient of fish population density from the edges towards the center of the lake. The axial zone of the lake is colonised only during the warm period, and then only by a single species, the bleak. Echo-sounding surveys carried out in 1987 and 1988 (Guillard *pers. com.*) gave a population density gradient decreasing from upstream to downstream.

From the comparison of the data obtained since 1976, it was possible to describe the distribution and the structure of the fish fauna over an interval of 8 years and to determine what changes had occurred. The comparative study held for areas of the lake surveyed at the same time of year and using the same equipment.

27 mm mesh gill nets: these nets catch fish of a body length of 15 to 25 cm. The comparison holds for three areas of the lake sampled in 1976–77 and again in 1984–85 (Fig. 1). The upstream sampling area is situated at the outlet of the gorges that bring the waters of the Verdon into the lake; the central and downstream areas are situated in bays on the eastern shore. The three sampling sites are 5 km apart. Fishing operations were carried out in September. In the three areas investigated, three years after the filling of the dam, 80 to 90% of the fish caught in a 27 mm gill net belong to the 4 major species indigenous to the Verdon (Fig. 2): trout, chub, barbel and toxostome. The highest frequency for the last species was recorded in the upstream area, whereas barbel was more frequent downstream. The newly introduced species represented only a low percentage of the fish caught (perch, roach and carp).

Table 1. Fry and small body-sized fish caught by electric fishing in the three areas of Lake Sainte-Croix in September and October 1984. N = number of fish, m = minimum length, M = maximum length in the sample.

Species	Upstream			Central			Downstream		
	N	Length		*N*	Length		*N*	Length	
		m	*M*		*m*	*M*		*m*	*M*
L. cephalus	48	4	14						
C. toxostoma	131	2	6	2	4	9	2	4	9
A. alburnus				100	5	10	316	5	14
G. gobio	17	5	11	36	3	9	5	3	11
S. erythrophtalmus				5	3	5			
R. rutilus				131	3	6	58	4	7
T. tinca	6	4	8	27	5	8	11	4	8
P. fluviatilis				2	8	9	13	3	8
L. gibbosus	15	3	6				3	8	10
R. lucius				2	10	15			

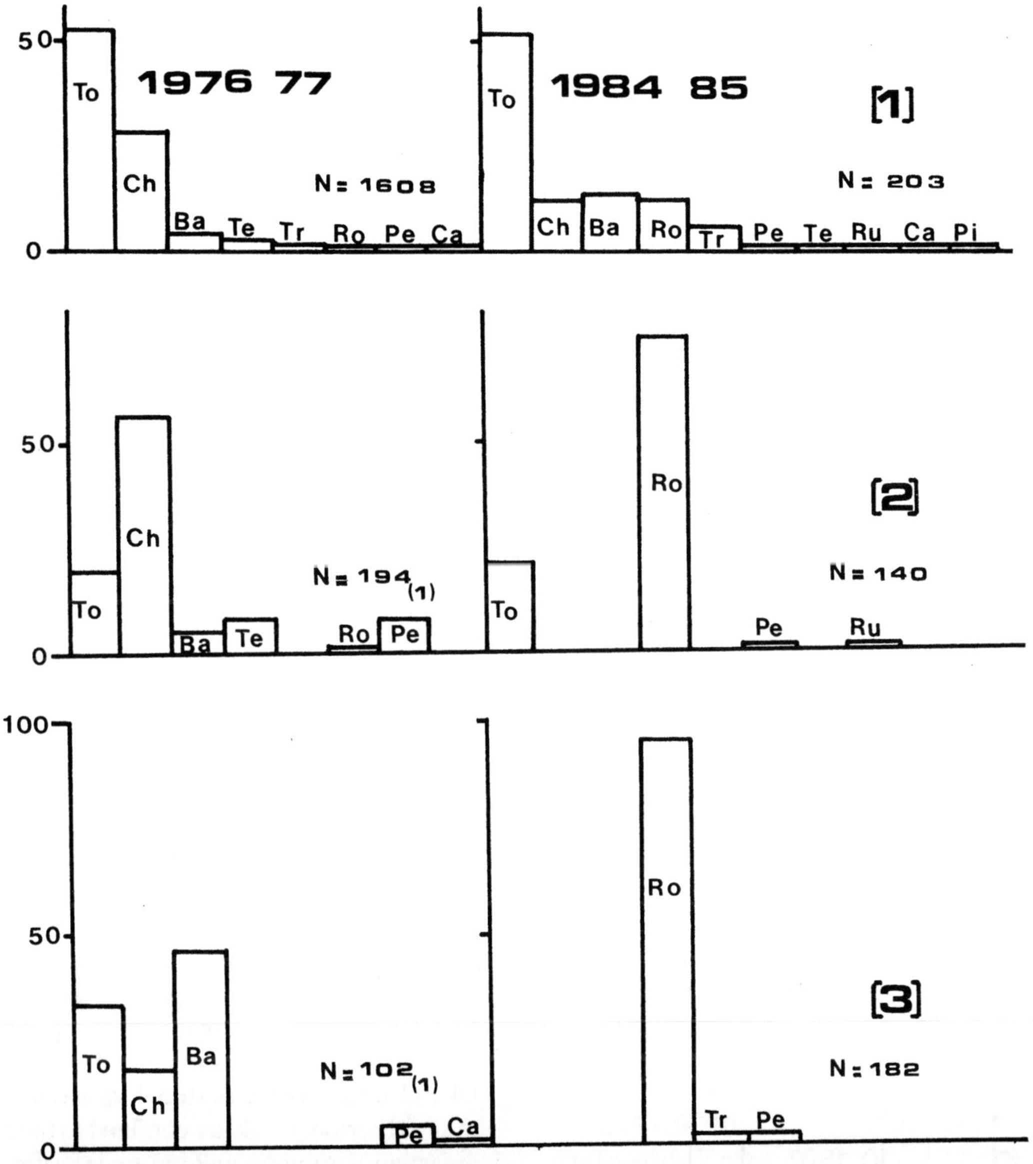

Fig. 2. Evolution in the habitat use patterns of the fish fauna in Lake Sainte-Croix. Frequency by number of each species caught with 27 mm size mesh net. Comparison between catches made in September 1976 & 1977 and catches made in September 1984 & 1985 in three areas of the lake (see the map). (1) From CEMAGREF 1977. To: toxostome, Ch: chub, Ba: barbel, Te: tench, Tr: trout, Ro: roach, Pe: perch, Ca: carp, Ru: rudd, Pi: pike.

The samplings of 1984 and 1985 showed considerable changes. In the upstream area, toxostome still represented 50% of the catch, but chub showed a decrease: 12% instead of 36%. On the other hand, the number of roach had increased, with 13% against 1%. In the central area of the lake, and in the downstream area, the changes were more marked. They concern firstly roach which represented 75% and 95% of the catch in 1984 and 1985 respectively, against 2% and 0% in 1976–1977. Toxostome had maintained the same frequency levels in the central area, but had disappeared downstream. Chub now represented only about 1% of the catch in the central area, and had also disappeared downstream, as had the barbel. The frequency of other species was too low to permit comparison between the two series of samplings. Mention should nonetheless be made of the appearence of pike.

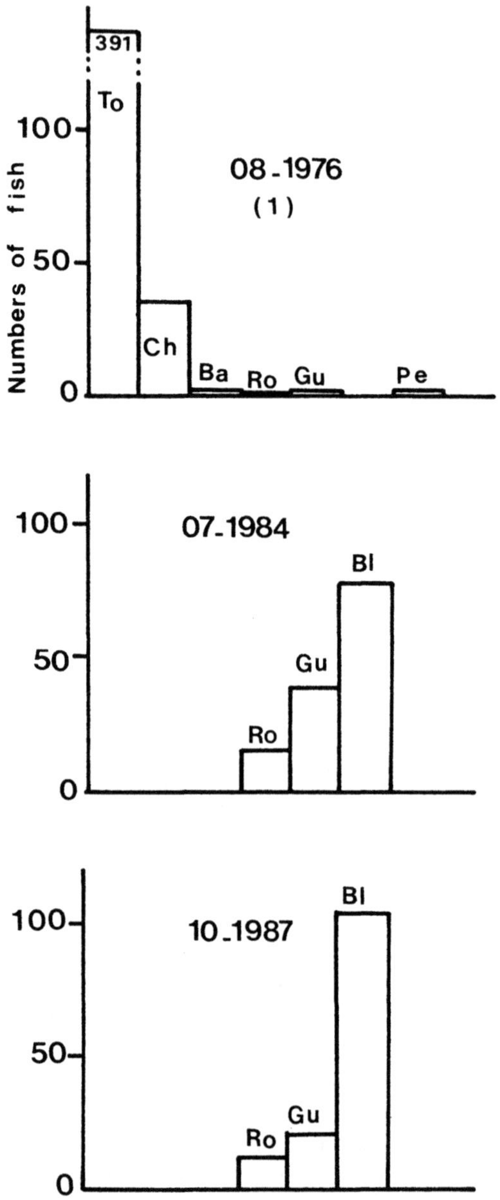

Fig. 3. Numbers of fish caught in the central area of Lake Sainte-Croix with 14 mm size mesh net placed during one night. (1) From CEMAGREF 1977. To: toxostome, Ch: chub, Ba: barbel, Ro: roach, Gu: gudgeon, Pe: perch, Bl: bleak.

Using 14 mm mesh gill nets, comparison with previous data is only possible for the central area. Such nets catch fish with a body length between 9 and 14 cm, and are thus suitable for catching small-bodied species such as gudgeon and bleak, as well as juveniles of large-bodied species (Fig. 3). In 1976, chub and toxostome constituted virtually the whole catch. Both these species had disappeared in 1984 and 1987, in favour of roach, gudgeon and bleak, which were introduced in 1975–76.

Using electric fishing, fry or small body size can be caught in places that would be largely inaccessible with nets. We have grouped the results obtained in September and October 1984 (Table 1) in the three lake areas described above. Fry of toxostome and chub were more abundant in the upstream area, whereas they were very rare elsewhere. Roach, bleak, gudgeon, tench and perch are the most conspicuous species in the central and downstream areas.

We analysed seasonal fish migrations by monitoring changes in the relative frequency of catches of roach and toxostome using a 27 mm mesh gill net. Other species were grouped together (Fig. 4). In the upstream area, there is little variation in the structural patterns of the fish fauna. Toxostome constitutes 50% of the catch throughout the year. Elsewhere, this species is absent at certain periods: May–June in the central area, and April to September downstream. It reappears during the cold period. The distribution of bleak in the lake also undergoes seasonal variations. During the cold period, bleak are confined to the marginal areas, and are not caught in open water. During the warm period from June to October, when the water is distinctly stratified, shoals of bleak occur several hundred metres offshore. This is the only species to occupy the open water.

Trophic niche breadth and overlap

Of the 17 fish species currently represented in lake Sainte-Croix, six account for more than 90% of the total biomass caught and are present in sufficiently large numbers in our samples for an

Table 2. Stomach content of fish of Lake Sainte-Croix S_{ij} = number of individuals of species (S) using the resource (j) at the date (i). Winter = W, Spring = Sg, Summer = Sr, (A) = adults, (J) = juveniles.

Species (S)	Roach			Toxostome			Chub			Gudgeon			Bleak			Perch (A)		Perch (J)	
Season (i)	W	Sg	Sr	W	Sg	Sr	W	Sg	Sr	W	Sg	Sr	W	Sg	Sr	W	Sr	W	Sr
Taxon (j)																			
Crustacea																			
Cladocera	10	2						3	5				22	14	22	4		38	2
Copepoda		14	1										1	1	2	4		30	8
Ostracoda										12	14	13				1		5	
Isopoda																1	1	4	1
Amphipoda																8	1	1	
Decapoda																3			
Insecta																			
Ephemeroptera											19	6		14	7	15	8	21	4
Tricoptera	1	6						2	6	2	25	3				5	3	5	
Odonata											2	2				3	3	5	1
Heteroptera	1	2					2									4	2	34	8
Coleoptera										1	11	4				3		4	1
Diptera																			
Chironomidae	2	7			1		1	7	6	15	52	6	18		20	18	18	41	8
Alter	1	3	1		1					3	22	12				8	2	5	4
Hymenoptera	3						5	3	7										
Mollusca																			
Gasteropoda	21	13	5	10	9	8				31	35	26							
Oligocheta		1	1					6		20	2					1		1	
Pisces		1	2				7	8	13							9	9	4	2
Algae																			
periphyton	19	12	11	25	17	26	2	8	4										
Vaucheria	25	15	36	4	2	20		15	16										
Spirogyra	1	13	1				6			4	8								
Chara	14	1	4																
Monocotyledons	2		24			6													
Dicotyledons	2		6			7													
$\sum_{j=1}^{r} S_{ij}$	106	93	92	39	30	67	23	52	57	88	190	70	48	37	51	87	57	199	39
Number of stomachs	66	76	67	53	30	56	37	22	21	37	88	41	26	26	27	45	22	75	17
Average size of fish in mm	188	194	212	231	220	228	230	315	245	131	126	130	124	130	129	185	180	< 100	

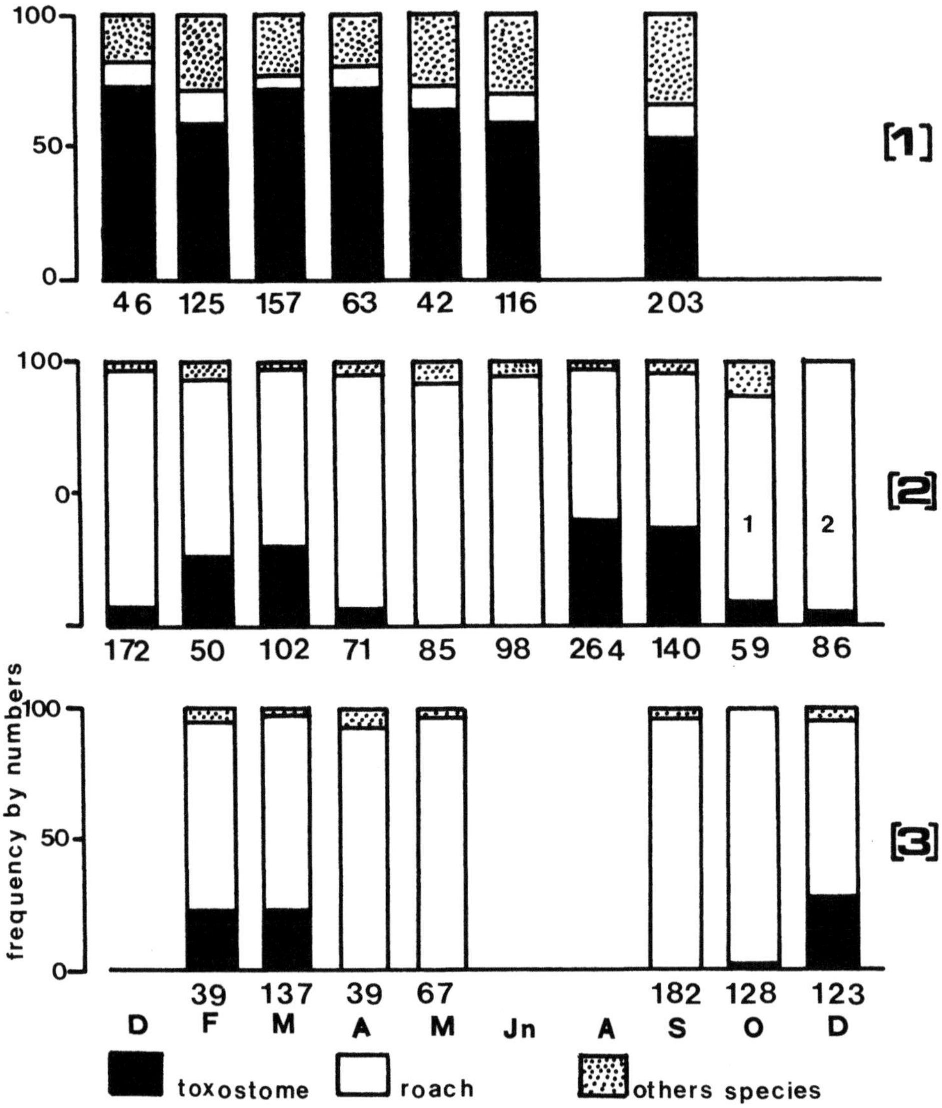

Fig. 4. Frequency by number of fish caught in lake Sainte-Croix throughout the years 1983–84, with 27 mm mesh net. Letters and numbers show respectively month and numbers of catches. (1) October 1985, (2) October 1987.

Table 3. Trophic niche breadth (eqn. 2) for the comon fish species of Lake Sainte-Croix. Toxos = toxostome, (A) = adult, (J) = juvenile.

Season	Species						
	Roach	Toxos	Chub	Bleak	Gudgeon	Perch (A)	Perch (J)
Winter	0.25	0.05	0.17	0.08	0.18	0.42	0.20
Spring	0.42	0.08	0.29	0.11	0.31		
Summer	0.16	0.14	0.24	0.08	0.19	0.22	0.23

Table 4. Trophic niche overlap for the common fish species of Lake Sainte-Croix: winter, spring and summer (eqn. 3); general (eqn. 4). (A) = adult, (J) = juvenile, Toxos = toxostome.

	Roach	Toxos	Chub	Bleak	Gudgeon	Perch (A)	Perch (J)
Roach	winter	0.66	0.67	0.30	0.37	0.10	0.20
	spring	0.57	0.71	0.19	0.47	–	–
	summer	0.77	0.63	0.01	0.09	0.03	0.02
	general	0.65	0.65	0.15	0.28	0.06	0.10
Toxos			0.41	0.08	0.27	0.00	0.00
			0.47	0.17	0.26	–	–
			0.56	0.00	0.18	0.00	0.00
			0.45	0.11	0.22	0.00	0.00
Chub				0.15	0.40	0.27	0.14
				0.15	0.39	–	–
				0.30	0.28	0.29	0.30
				0.17	0.35	0.27	0.20
Bleak					0.38	0.47	0.86
					0.33	–	–
					0.14	0.51	0.80
					0.27	0.48	0.82
Gudgeon						0.26	0.19
						–	–
						0.36	0.22
						0.30	0.20
Perch (A)							0.73
							–
							0.62
							0.65

analysis of their stomach content to be made. These are roach, toxostome, chub, perch, bleak, and gudgeon. We grouped the samplings into three periods: winter (December to February), spring (April to June) and summer (July to September), (Table 2). Given the diversity of prey types, and the fact that the cyprinids mash their food with their pharyngeal teeth, identification of prey was in the case of animal prey necessarily rudimentary, and often went no further than the order.

By means of the standard Simpson Index (eqn. 2) calculation, two patterns of feeding behaviour can be distinguished (Table 3): specialised feeding for toxostome and bleak, and generalised feeding for chub, roach, gudgeon and perch. Toxostome is a herbivorous substrate-scraper, which can also take in a wide range of plant debris. Bleak is essentially a plankton-feeder in the lake, but also takes chironomid pupae during the period of their emergence. The percentage of chironomid pupae among the preys in bleak stomach contents 8% in summer and 20% in winter. Among the generalist species, two are carnivorous and benthophagous, gudgeon and perch, and two are essentially herbivorous, chub and roach. The opportunistic feeding patterns of the last two species mean that they also take zooplankton, zoobenthos and even fish fry.

The dietary pattern of roach shows a high level of similarity to those of toxostome and chub (eqn. 3) (Table 4). There is also a high level of similarity of dietary patterns between bleak and young perch, and between young and adult perch which both prey on aquatic insect larvae. Seasonal variations in diet overlap are slight, whatever the pair of species considered. The general index values (eqn. 4) are slightly lower than the arithmetical mean of indices calculated for each season.

Discussion

Lake Sainte-Croix is conspicuously oligotrophic (El Hafa, 1987). The population density of zooplankton is low: 11 000 ind/m^3 on average (Pont *et al.*, 1987). The benthic fauna of the sediment is poor, consisting of Oligochaeta in the deep zone (average 6 900 ind/m^2) and chironomids in the littoral zone (average 2 800 ind/m^2) (Bressac & Champeau, 1985). The inputs of nutrients, plant debris and drifting invertebrates, washed into the lake by the river Verdon explain the increase in population density of fish fauna from downstream to upstream.

Modifications of habitat use patterns, which have resulted in the scarcity of the indigenous species throughout most of Lake Sainte-Croix, are not exclusively due to the abrupt change in the physical characteristics of the environment. Three years after the filling of the reservoir, chub, toxostome and barbel were represented all over the lake by adults, as well as juveniles in large numbers, which suggests that reproduction might take place *in situ*. The most profound upset to the ecological balance has resulted from the growth of populations of certain species introduced by man, in particular roach and bleak.

The present situation result therefore from competition for space and food between species characteristic of a running water ecosystem and those characteristic of a stagnant water ecosystem. Initially, in the absence of stagnant water species, the indigenous Verdon species colonised the whole lake, which was then unoccupied. The roach population, which was till insignificant three years after the original stocking of the lake in 1974, had reached sufficient size for competitive relationships to exist by 1982. This notoriously opportunistic species takes planktonic crustaceans when they are abundant (Lesmark, 1983; Vollestad, 1985; Ponton & Gerdeaux, 1987) although it is generally herbivorous and benthophagous in Lake Sainte-Croix. In this lake, its trophic niche overlaps considerably with toxostome and chub. Chappaz *et al.* (1989b) have shown that, in Lake Sainte-Croix spawning of toxostome occurs at the same time as roach (end of May or beginning of June). Spawning took place in the central area of the lake during first years of the existence of the lake. Since 1984 no adult toxostome has been caught downstream or in the middle of the lake during the reproduction period of this species. Electric fishing operation carried out in the marginal areas have resulted in the capture of only toxostome and chub fry in the upstream area of the lake which is directly influenced by the river and where the roach is poorly represented. Spatial competition has resulted in the toxostome spawning sites being moved upstream towards the river water outlet.

Trophic niche overlap between two species results in a competitive relationship only if the amount of resources available is a limiting factor for the growth of populations. This is indeed the case for zooplankton that are preyed on by both bleak and perch, essentially at the fry stage. Niche overlap between these two species is high. Pont *et al.* (1989) have demonstrated the impact of fish predation on zooplankton in lake Sainte-Croix. Over the past few years, this predation has resulted in the scarcity of large-bodied Cladocera, especially during the warm period. It is at the expense of large-bodied Cladocera that bleak and perch fry are fed. In Lake Sainte-Croix, perch have a low growth rate and poor fecundity (Chappaz *et al.*, 1989a). Perch population, which was considerable and well balanced (CEMAGREF, 1977), appears to have declined. A similar phenomenon has been described by Holcik (1970). The growth of the bleak population is also undoubtedly one of the factors that determine the herbivorous diet of the roach.

Two factors suggest that the fish fauna of Lake Sainte-Croix is not yet stable: 1) The zooplankton-fish system is still in the process of development (Pont *et al.*, *op. cit.*); 2) The increasing number of catches of young pike by electric fishing in the weed beds shows that this species in on the increase. The possibility cannot be excluded that predation on small fish prey might eventually limit the growth of small-bodied fish populations.

References

Bressac, Y. & A. Champeau, 1985. Contribution à l'étude du réseau trophique des lacs de barrage: le zooplancton de Serre Poncon et Sainte-Croix. In D. Gerdeaux & R. Billard, (Eds). Gestion piscicole des lacs et retenues artificielles, INRA Paris. 67–77.

C.E.M.A.G.R.E.F., 1977. Inventaire piscicole de la retenue de Sainte-Croix-Fontaine l'Evéque. Ministère de l'Agriculture. Groupement d'Aix-Marseille: 1–52.

Champeau A., 1982. Les retenues hydroélectriques du Verdon. Impact sur la rivière. Conséquences du marnage. Bull. Ecol. 13: 203–239.

Champeau, A., A. Grégoire & G. Brun, 1979. Le peuplement piscicole des lacs artificiels du Verdon. Ann. Limnol. 14: 245–271.

Chappaz, R., G. Brun & G. Olivari, 1987. Mise en évidence de différences de régime alimentaire dans une population d'ablettes *Alburnus alburnus* (L) dans le Lac de Sainte-Croix. Conséquences sur la croissance et la fécondité. Ann. Limnol. 23: 245–252.

Chappaz, R., G. Brun & G. Olivari, 1988. Stratégie de développement du goujon *Gobio gobio* (L) dans un lac oligotrophe: la retenue de Sainte-Croix (Provence, France). C.R. Acad. Sci. Paris, 307, série 3: 493–497.

Chappaz, R., G. Brun & G. Olivari, 1989a. Contribution à l'étude du régime alimentaire, de la croissance et de la fécondité de la perche (*Perca fluviatilis* L.) dans un lac oligotrophe de région tempérée. Rev. Sci. Eau 2: 95–107.

Chappaz, R., G. Brun & G. Olivari, 1989b. New data about biology and ecology of a little-studied fish: *Chondrostoma toxostoma* (Vallot, 1836). Comparison with Nase, *Chondrostoma nasus* (L., 1766). C.R. Acad. Sci. Paris, 309, série 3: 181–186.

El Hafa, M., 1987. Bilan des apports de sels nutritifs au lac de Sainte-Croix (Provence) et leur utilisation par le phytoplancton de la retenue. Thèse 3ème cycle, Univ. Provence, Aix-Marseille I, 130 pp.

Holcik, J., 1970. Klicava reservoir: an ecological study. Biologicke Prace Bratislava, 15: 5–94.

Lessmark, O., 1983. Competition between perch *Perca fluviatilis* and roach *Rutilus rutilus* in south swedish lakes. Doct. dissert. Univ. Lund, 172 pp.

Pont, D., R. Chappaz, G. Brun & A. Champeau, 1989. Interactions zooplancton-poissons dans une retenue oligotrophe de mise en eau récente (Sainte Croix, Provence, France.) Rev. Sci. Eau 2: 777–792.

Ponton, D. & D. Gerdeaux, 1987. La population de gardon *Rutilus rutilus* (L.) du lac Léman. Structure en âge, déterminisme du recrutement, analyse de la croissance. Bull. fr. Piscic. 305: 43–53.

Vollestad, L. A., 1985. Resource partitioning of roach *Rutilus rutilus* and bleak *Alburnus alburnus* in two eutrophic lake in SE Norway. Holarct. Ecol. 8: 88–92.

Received 31 August 1989; in revised form 10 January 1990; accepted 23 March 1990

Hydrobiologia **207**: 209–219, 1990.
D. J. Bonin & H. L. Golterman (eds), Fluxes Between Trophic Levels and Through the Water–Sediment Interface.

Metal exchanges in the fauna-sediment system. The case of *Nereis diversicolor* and *Scrobicularia plana* in the Bou Regreg estuary (Morocco)

Mohamed Cheggour[1], Hervé Texier[2*], Gérard Moguedet[3] & Bernard Elkaïm[4]
[1] *Département de Biologie, E.N.S. Marrakech, B.P. 541, Marrakech, Morocco;* [2] *Laboratoire de Géologie marine (chimie marine), Université de Caen, 14032 Caen Cedex, France; (* author for correspondence);* [3] *Laboratoire de Géologie, Université d'Angers, 49045 Angers Cedex, France;* [4] *Laboratoire d'Hydrobiologie, Université P. et M. Curie, 12, rue Cuvier 75005 Paris, France*

Key words: Biogeochemistry, Metals, Estuary, Morocco, Polychaete, Mollusc

Abstract

This study identifies the types and quantities of metals retained by *Nereis diversicolor* and *Scrobicularia plana* in a predominantly arid estuary environment.

Certain abiotic parameters and a number of physiological processes linked to metabolic and reproductive functions play a decisive role in the seasonal influence of metal levels on the tissues of these organisms.

Given the deposivoric diet of the organisms, the sedimentary bed would appear to be the principal source of metal accumulation.

The following classification illustrates the bioaccumulation of metals:

heavy accumulation	Zn	*Nereis* > *Scrobicularia*
noticeable accumulation	Cu > Ni = Cr	*Scrobicularia* > *Nereis*
lesser accumulation	Pb > Mn	*Scrobicularia* = *Nereis*
light accumulation	V > Fe	*Scrobicularia* > *Nereis*

The possibility that these metals may be carried into the estuary by sewage water from the cities of Rabat and Salé should certainly not be ignored.

Introduction

Focussing principally on the sediment of the Bou Regreg Estuary slikkes, this study deals with metal accumulation in two species of intertidal benthic animals: *Nereis diversicolor* (O. F. Müller) (Polychaete Annelidan) and *Scrobicularia plana* (Da Costa) (bivalve Mollusc). Of the 52 species counted along the estuary (Cheggour, 1988), the two above-mentioned species were selected as metal bioindicators because they best fulfilled the taxonomic, biogeographical and physiological requirements of the study, notably: widespread presence and a marked ability to accumulate metals.

Materials and methods

Surface sediment and fauna sampling was done at low and spring tide. The sampling surface was 0.25 m^2. The *Scrobicularia plana* and *Nereis diversicolor* were cleaned on the spot with a 1 mm sieve before being sorted and stocked in water from the sampling site. After an 18 hour purging period, they were thoroughly washed in distilled water. The fleshy parts of the *Scrobicularia plana* and the individual *Nereis diversicolor* were drained and dried at 80 °C until a constant weight (48 hrs) was obtained, in order to calculate their biomass. Samples were then finely ground and calcined.

The individual samples of *Scrobicularia plana* were divided into three groups:

group 1:
young animals Y 4 mm ≤ size < 10 mm
group 2:
average animals A 10 mm ≤ size ≤ 20 mm
group 3:
large animals L 20 mm < size

Sediment samples were heat dried at 40 °C. The silt-argillaceous content was separated by washing, by means of a 50 μm sieve, and then finely ground for chemical analysis.

Metal analyses of organic and sedimentary samples were carried out using an atomic absorption/flame spectrophotometer (Philips PYE UNICAM SP 9) after triacid induced mineralization (HF, $HClO_4$, HNO_3) at 150 °C. The resulting residue was mixed with 5 ml of 12 N, HCl and increased to a volume of 100 ml by the addition of 18 MΩ water. The solution was then filtered through a 0.45 μm pore size Whatman filter. The margins of error in metal concentrations are calculated as follows: Fe, Mn = 6%; Zn, Cr, Ni, Cu = 10% and Pb = 15%.

The mineralogical analysis procedure is described in Holtzapffel (1985). The analyses were carried out using a RIGAKU D MAX 1 A diffractometer. The fire loss was determined through differential weighing before and after calcination for four hours in a furnace at a temperature of 1000 °C.

Description of the area studied and its climatic conditions

The Bou Regreg Estuary is situated on the Atlantic coast of Morocco (Fig. 1) between the cities of Rabat and Salé (pop. ≈ 500000), which dump waste directly into the estuary or into the

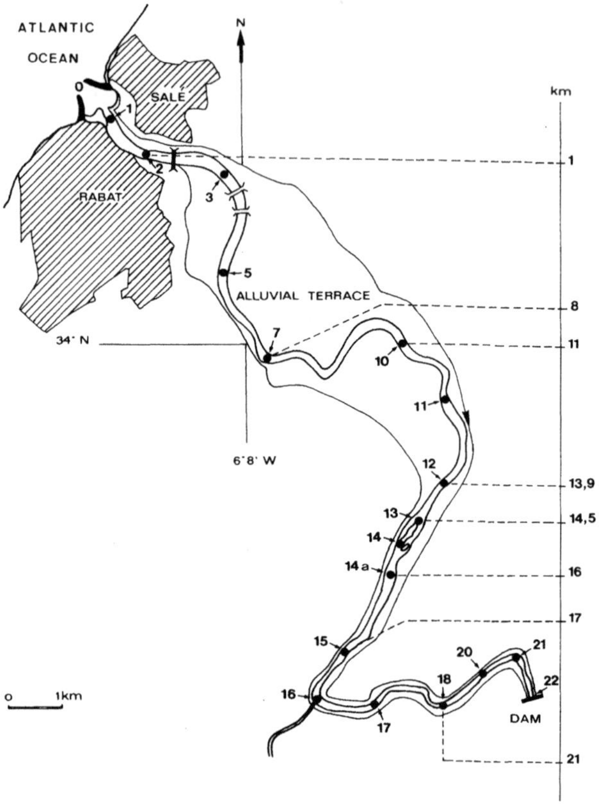

Fig. 1. Position of the stations along the Bou Regreg Estuary.

ocean on both sides of the mouth. The construction of a dam 22 km upstream changed the natural conditions of this area described by Elkaïm (1972). The virtual absence of river flow and the disappearance of flooding make this estuary a striking example of tide-governed hydrodynamics (Texier *et al.*, 1986).

The climate is classed as sub-humid Mediterranean and is characterized by a relatively long dry season followed by a cold season with irregular rainfall. Sampling took place during 1986, a rather wet year (566 mm). Rainfall generally occurs between November and April and peaks during the cold months of January and February (0 °C to 5 °C). The average summer temperature between May and October is 25 °C.

Results

Hydrological conditions

In the mouth (station 1), the water is maritime: pH = 8.1, salinity = 32.5. Among the other stations where sampling took place, four preferential sampling sites were selected for their hydrological and sedimentological representativeness (Gillet, 1986; Mayif, 1987; Cheggour, 1988). Station 2 situated 1 km from the mouth is the most maritime of the four. The clear waters are heavily salinated (30‰) and well oxygenated (saturation index: 60%). The pH of the water is slightly alkaline (7.6). Stations 7 and 10, situated respectively at 8 and 11 km from the mouth in the lower estuary, are characterized by urban pollution. The average salinity remains high (20‰). The water is turbid and the oxygen level is low (35%). The pH is slightly acidic (6.9). Station 15, 17 km into the upper estuary, is the most continental. The water is transparent, well oxygenated (80%) and still relatively salinated (10‰). The average annual pH is 8.0.

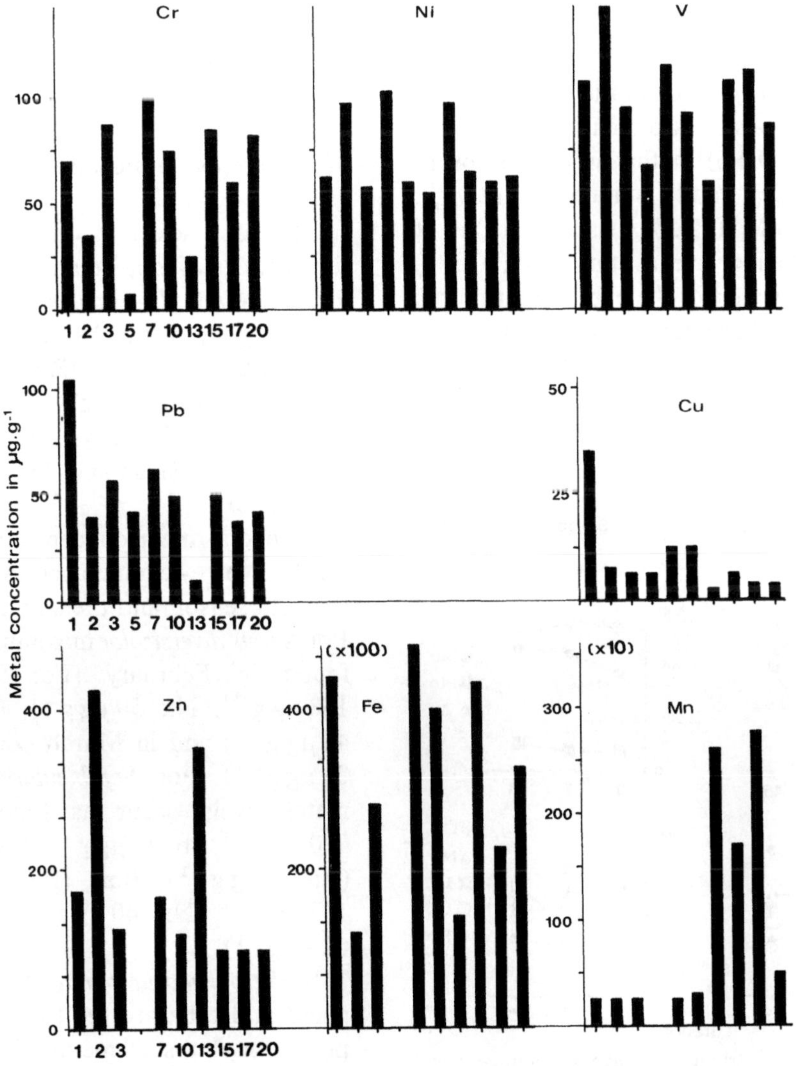

Fig. 2. Longitudinal distribution of metal elements in fine (<50 μm) sediment from slikkes.

Sedimentary environment

The sediment of the slikkes of stations 2, 10 and 15 contains a high percentage of arenite, *i.e.* particles measuring between 50 μm and 2 mm, compared to station 7 where fine matter dominates (70% lutite, *i.e.* particles inferior to 50 μm). Powder analysis indicates the presence of quartz, calcite, traces of dolomite and sylvite at every station. Feldspars are found only in the lower estuary, and sulphates were limited to stations 7, 10 and 15. The presence of sulfide at station 7 attests the reducing nature of the sediment. Argils were composed of illite + interstratified (60–70%), smectite (20–30%) and kaolinite (10%) (Texier *et al.*, 1989).

The longitudinal variation of the fire loss in the fine deposit reveals a significant quantity of organic matter (O.M.) in the urban dumps at station 7 (28.4%) and station 2 (18.0%). Upstream, the O.M. (15.5% at station 15) is of humic origin. The low level at station 10 (10.1%) reflects the importance of the ebb current in this sector. The percentages are two to four times as high in all stations in the summer and the post-summer period preceding the first rainfalls. Organic pollution results in a major precipitation

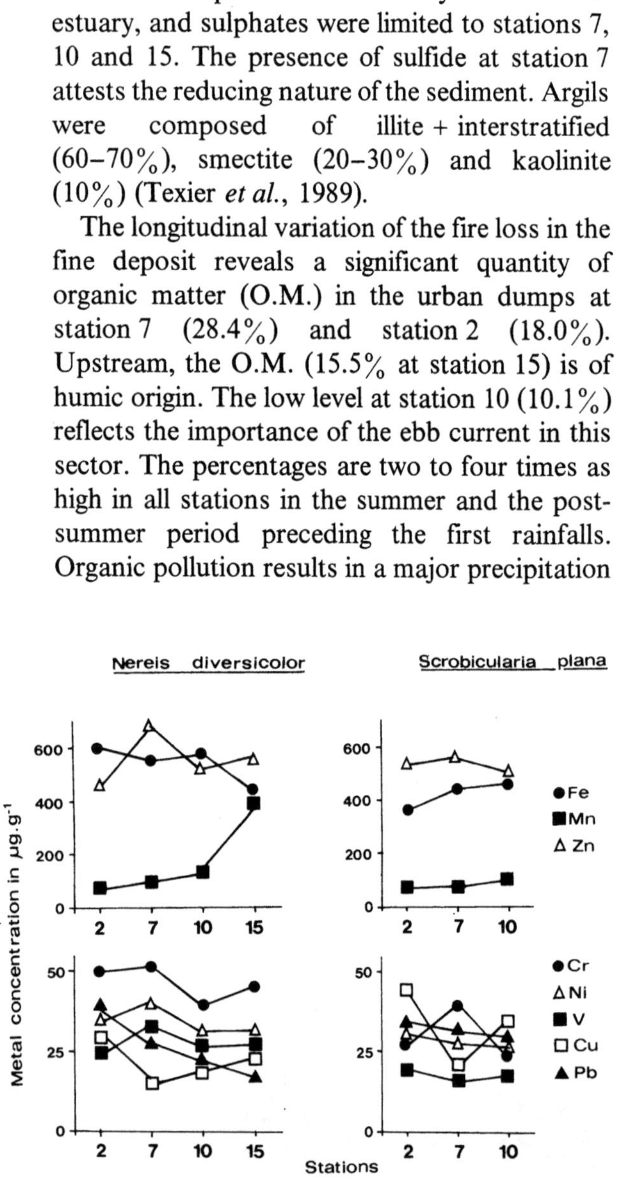

Fig. 3. Longitudinal distribution of average annual metal concentrations in *Nereis diversicolor* and *Scrobicularia plana* tissues (dry weight)

of carbonates in the built-up sludge between stations 2 and 7.

Metals in sediment

The highest concentrations of copper and lead are found at the mouth of the estuary (Cu: 36 $\mu g\ g^{-1}$, Pb: 105 $\mu g\ g^{-1}$) (Fig. 2). Heavy concentrations of zinc are found at stations 2 (420 $\mu g\ g^{-1}$) and 13 (360 $\mu g\ g^{-1}$). Iron (12000 to 47800 $\mu g\ g^{-1}$), vanadium (60 to 142 $\mu g\ g^{-1}$), nickel (53 to 102 $\mu g\ g^{-1}$) and chrome (9 to 101 $\mu g\ g^{-1}$) are found in varying concentrations along the length of the estuary. Manganese is abundant in the upper estuary (2500 $\mu g\ g^{-1}$).

Metals present in organisms

In the four selected stations (Fig. 3), average tissue levels of Zn, Fe and Mn are notably high for the *Nereis diversicolor* (Zn: 555 $\mu g\ g^{-1}$, Fe: 5430 $\mu g\ g^{-1}$, Mn: 191 $\mu g\ g^{-1}$) and for the *Scrobicularia plana* (Zn: 563 $\mu g\ g^{-1}$, Fe: 427 $\mu g\ g^{-1}$, Mn: 88 $\mu g\ g^{-1}$). The average levels of other metals vary from 18 $\mu g\ g^{-1}$ (V in *Scrobicularia plana*) and 46 $\mu g\ g^{-1}$ (Cr in *Nereis diversicolor*).

In general, metal concentrations are higher during the two key climatic periods: the wet period of the cold season and the summer season (Fig. 4). For *Nereis diversicolor* maximum levels are mainly found in February (Fe: 9224 $\mu g\ g^{-1}$, Mn: 158 $\mu g\ g^{-1}$, Ni: 59 $\mu g\ g^{-1}$, Cu: 53 $\mu g\ g^{-1}$, V: 44 $\mu g\ g^{-1}$) and in March (Zn: 654 $\mu g\ g^{-1}$, Cr: 83 $\mu g\ g^{-1}$). For *Scrobicularia plana* maximum metal levels occur as follows: March (Zn: 970 $\mu g\ g^{-1}$, Pb: 45 $\mu g\ g^{-1}$, Cu: 40 $\mu g\ g^{-1}$), April (V: 25 $\mu g\ g^{-1}$), June (Fe: 6267 $\mu g\ g^{-1}$, Mn: 197 $\mu g\ g^{-1}$, Ni: 40 $\mu g\ g^{-1}$) and August (Cr: 62 $\mu g\ g^{-1}$).

In *Scrobicularia plana* (Fig. 5), the digestive gland and tube are the two main accumulation points for almost all metals, Fe: 3100 $\mu g\ g^{-1}$, Zn: 670 $\mu g\ g^{-1}$, Pb: 73 $\mu g\ g^{-1}$, Mn: 71 $\mu g\ g^{-1}$, Cu: 40 $\mu g\ g^{-1}$, Ni: 36 $\mu g\ g^{-1}$, V: 25 $\mu g\ g^{-1}$. The gills

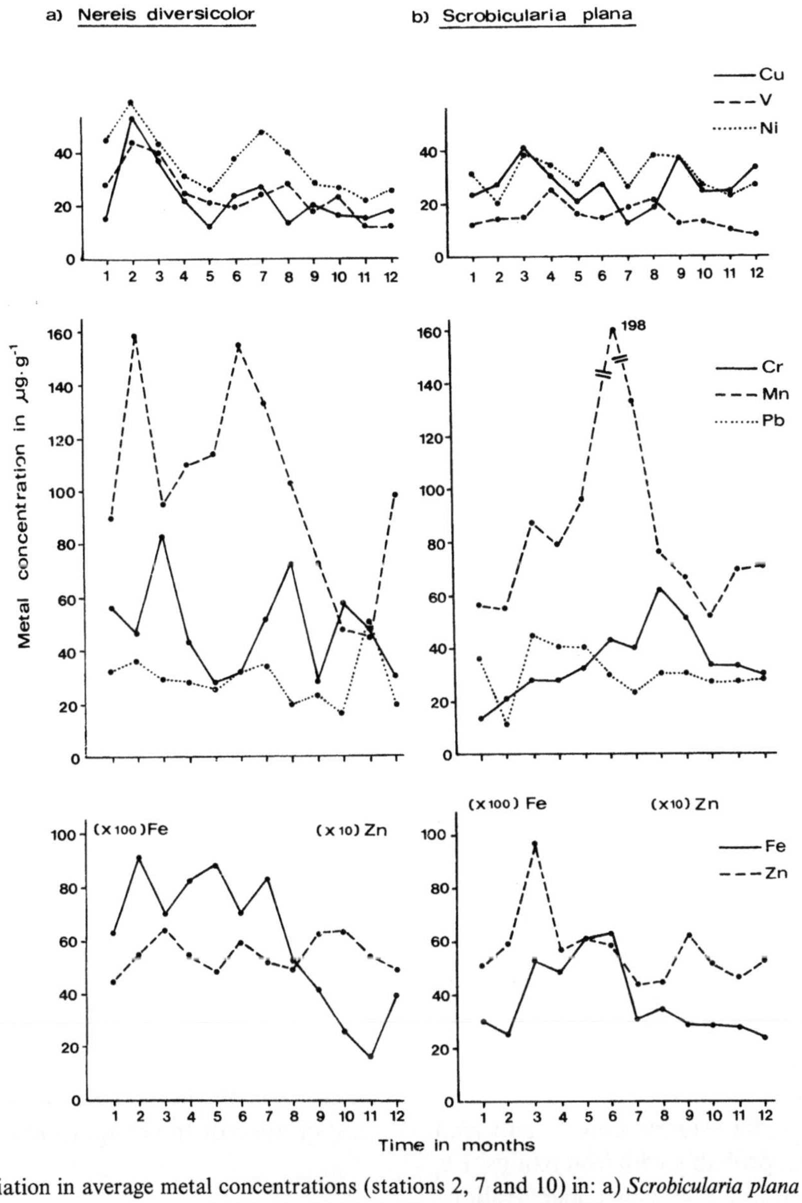

Fig. 4. Monthly variation in average metal concentrations (stations 2, 7 and 10) in: a) *Scrobicularia plana* tissues (dry weight); b) *Nereis diversicolor* tissues (dry weight).

(Zn ≫ Pb > Cu > Ni > V) and the mantle (Fe ≫ Mn > Cr) are also involved in the bioaccumulation of the metals. The quantities of Zn and Cu are higher in older animals (Zn: 530 $\mu g\ g^{-1}$, Cu: 33 $\mu g\ g^{-1}$) (Fig. 6). On the other hand, the concentrations of Pb (58 $\mu g\ g^{-1}$), Cr (48 $\mu g\ g^{-1}$), Ni (62 $\mu g\ g^{-1}$) et V (26 $\mu g\ g^{-1}$), are higher in younger animals with a more active metabolism.

Discussion

Metals in sediment

The average concentration of heavy metals in surface sediment is comparable to concentrations found in other estuaries except for the excessive zinc concentration in the lower estuary (Martin & Meybeck, 1979). The percentage of zinc in the

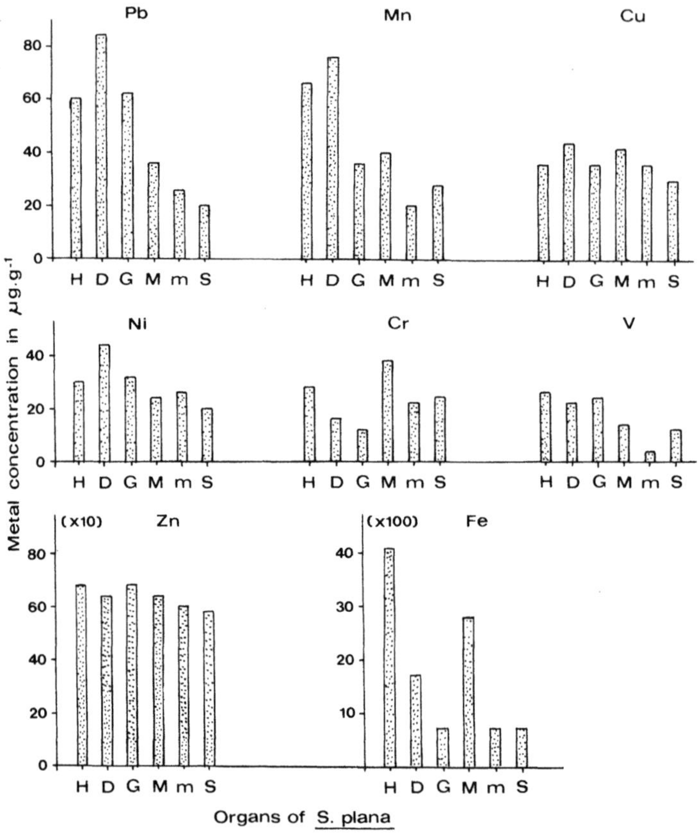

Fig. 5. Distribution of metals in the organs of the *Scrobicularia plana*: H: Hepatopancreas, D: Digestive tube, G: Gills, M: Mantle, m: Muscles, S: Siphons.

built-up sludge (station 2) is a sign of a more varied, largely inorganic, complexation. The abundance of chlorides, sylvite and carnalite present in this zone (Elkaïm, 1972; Mayif, 1987) could play an important role.

In the mouth of the estuary, copper and lead tend to precipitate, probably with iron oxides. Fe, V, Ni and Cr are found in varying concentrations along the length of the estuary. The chemical forms of these precipitations vary from one hydrological zone to the next. Nickel and vanadium, most often associated with organic matter, can precipitate with manganese oxides and ferro-humic complexes in contact with salinated waters. These two metals are the most labile and ubiquitous.

Manganese can be considered as a sign of fluviatile deposits. It precipitates upon contact with the water of the upper estuary (alkaline pH, salinity) to form insoluble oxides. It is the only metal to increase in concentration during the period of winter rainfall.

Concentrations of Zn, Pb, Cr and Cu reach their peak in the summer period, indicating the importance of the evaporation.

Metals present in organisms

The concentrations of copper found in the estuary are comparable to those found by Gillet (1987), but Gillet's average levels of zinc are much lower. This is most probably due to the choice of stations. Three out of four sampling sites used in this study are in the contaminated zones, while in the case of Gillet, this is true of only one out of two.

The longitudinal variation of metals, particu-

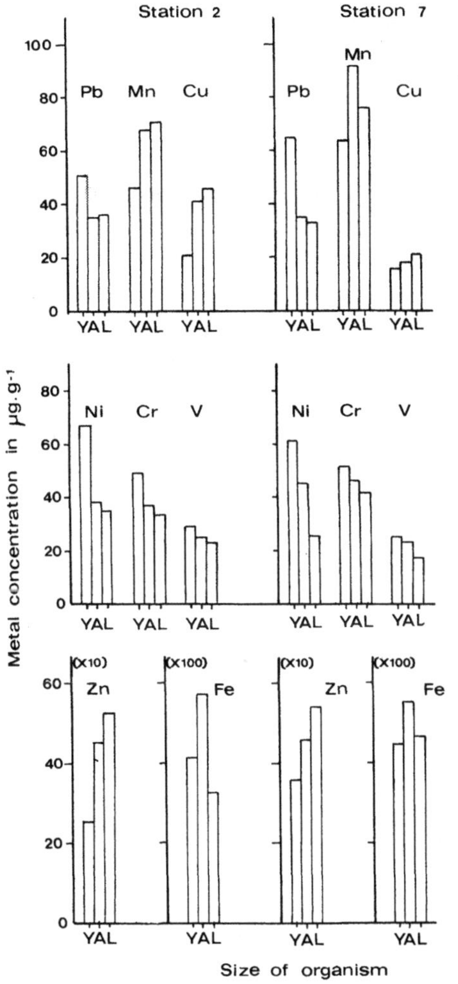

Fig. 6. Variation in metal concentrations according to the size of individual *Scrobicularia plana* animals. Y: Young, A: Average, L: Large.

larly significant in the case of lead and manganese, is comparable to the variation of sediments for *Nereis diversicolor*. For *Scrobicularia plana*, this bioavailability shows certain variations, except in the case of manganese.

Interactions between metals are an effect of stimulation (Luoma, 1983) or of antagonism (Rada, 1985). One can observe a strong correlation between the amount of iron and manganese in the tissues of the two species (Table 1). This can be mainly attributed to synergic coprecipitation in the form of hydrated manganese dioxide or hydrated ferric oxides. In deeper sediment, precipitates are often reduced to more soluble

Table 1. Correlational matrix of metals.

	Cu	Zn	Pb	Ni	Cr	Mn
Cu						
Zn	★★★					
Pb						
Ni	○○○		★★★			
Cr			★	★★		
Mn		★	–	–		
Fe		★★			★	★★★ ○○○○
V			★	★★	★★	

	N. diversicolor		S. plana	
	Positive	Negative	Positive	Negative
highly significant :	○○○	⊖⊖⊖	★★★	– – –
very significant :	○○	⊖⊖	★★	– –
significant :	○	⊖	★	–

manganous and ferrous forms. This results in higher concentrations of Mn and Fe in interstitial waters (Bryan, 1976). Upstream, a significant correlation exists between Ni and Cu in *Nereis diversicolor* and between both Ni-Pb and Cu-Zn in *Scrobicularia plana*. The same holds true for sediment concentrations. Berthet (1986) has also shown that the bioaccumulation of Zn in mussels is affected by a Cu contamination. Negative correlations are obtained between Zn and Mn ($r = -0.66$) and between Zn and Fe ($r = -0.66$) in *Nereis diversicolor*. This can be attributed to the different longitudinal variations of these metals in the estuary.

Major seasonal variations in metal concentrations occur in the two species. They can be attributed to either biotic or abiotic factors. Decreases in salinity and temperature in the winter and post-winter period appear to be linked to a bioaccumulation of metals in the two species. These findings corroborate those of other researchers (Bryan & Hummerstone, 1971; Bryan *et al.*, 1983; Rada, 1985). Nevertheless high concentrations of manganese in the *Nereis diversicolor* population can be found in areas of low salinity because of a possible relationship between manganese absorption and calcium metabolism (Bryan, 1976). A high upstream accumulation of Cu, Pb and Ni goes hand in hand with an alkaline pH, even though slightly acidic conditions in the center of the estuary (station 7) lead to the absorption of Zn in the two species. Baudin (1981) has shown that physico-chemical factors and a stable zinc concentration modify the zinc 65 concentration in several species of the estuary.

Physiological processes linked to metabolic functions, especially reproduction – gametogenesis and emission of eggs, described by Cheggour (1988) – seem to play a key role in determining seasonal variations in the level of metals present in the tissue of the two species. With few exceptions, the results of this study follow the reproductive cycle of *Nereis diversicolor*. In spring and autumn, a drop in metal levels is observed throughout the mating period. This observation was particularly true for the lower estuary, especially as regards the levels of Zn, Fe, Mn, Ni and Pb. In winter and summer, during the gametogenetic phase (Gillet, 1986), a period during which cells are metabolically active, more elevated levels of metal integration are observed. Well-developed gonads seem to be particularly prone to metal retention. In its reproductive cycle *Scrobicularia plana* is subject to a certain environmental influence. At station 7 the mating periods extended from January (22% of 6 mm individuals) to April (28%) and from September (15%) to December (19%). At station 2, mating took place from February (20%) to May (18%) and from October (30%) to December (27%). Kourradi (1987) obtained slightly different results at other stations. Generally speaking, with the exception of Pb, Cu and V the levels of metal concentrations follow the reproductive cycle of *Scrobicularia plana*. A drop in metals levels in March and April coincides with the first mating period. At the end of summer, when a second egg-laying period begins, another drop in metal concentration can be observed. This laying period is followed by autumn mating characterized by low metal levels. At the beginning of winter and especially at the beginning of summer, two missing periods, metal concentrations clearly peak.

Variations in metal concentrations according to tissue and size of the Scrobicularia plana

The hepatopancreas, the organ which serves to regulate and recycle food, and the digestive tube, which serves to transport food particles, are the most affected of the organs, as a direct result of their function. These findings tally with those of Bryan *et al.* (1980) obtained in United Kingdom estuaries. The accumulation of certain metals in the gills of the *Scrobicularia plana* can in part be explained by the fact that it can feed as a filtering animal, particularly at high tide (Guelorget & Mayère-Mazoyer, 1983). Muscles (adductors and feet) also contain concentrations of all the metals

Table 2. Organisms/sediment correlational matrix (symbols are the same as Table 1).

N. diversicolor

	Cu	Zn	Pb	Ni	Cr	Mn	Fe	V
Cu		○		○○○				
Zn								○
Pb	○○○		○○		○			
Ni							○○	
Cr	○		○	○○○	○○○			
Mn	⊖	⊖	⊖⊖⊖		⊖⊖	○○		
Fe								
V				○○	○○			

S. plana

	Cu	Zn	Pb	Ni	Cr	Mn	Fe	V
Cu				★			— —	— —
Zn				★★			— —	—
Pb			★	★★				
Ni		—		★	★			
Cr						—		★
Mn	— — —					★★★		—
Fe		★						
V								★★

studied with the exception of vanadium. Their fibrilla structure, which is linked to biochemical processes during muscle activity, is conducive to metal accumulation. Siphons, while not completely free of metal concentrations, are the least affected parts.

The results of the analysis of metal bioaccumulation according to size confirm previously mentioned metal interaction tendencies. It is possible that elimination of Zn and Cu is a very slow process or that these metals are stored in a non-toxic form in particular organs (the hepatopancreas and the digestive tube). Zatta (1985) has demonstrated that Zn^{2+} has 4 to 5 fixation sites on the hemocyanin. Little variation can be noticed in the levels of manganese and iron. Natural concentrations of iron and its role in numerous enzymatic activities in osmoregulation should be taken into consideration. In conclusion, there is no definitive rule concerning concentration related to size, as has been observed in other invertebrate organisms.

Correlation between organisms and sediments

The correlation between the levels of manganese and lead in *Nereis diversicolor* and *Scrobicularia plana* tissues and the levels of these same elements in sediment samples is significant (Table 2). These metals are principally transferred through the food supply, upstream for manganese and downstream for lead. Chrome accumulation in *Nereis diversicolor* and nickel and vanadium accumulation in *Scrobicularia plana* are also largely sedimentary in origin. The intake of metallic elements varies according to the type of sediment. The differences of availability can be attributed to variations in the strength of bonds formed with sedimentary particles. The misleading correlation of iron may be due to a high level of iron in the sediment bed as compared to the level in the organisms sampled, and a naturally high level of iron in the tissues of the species. The correlation of copper is not significant. It should be remembered that the precipitation of this element is principally confined to the mouth of the estuary. The levels of copper in the digestive gland and the gills are identical in *Scrobicularia plana*. This observation leads to the supposition that a certain amount of copper is taken in through interstitial water.

In Table 3, the sediment/species ratios (R) are calculated according to the average annual level of each element. For all the elements analysed with the exception of zinc, metal levels in sediments are higher than those in organisms. This leads us to postulate a heavy intake of zinc, which is then stored in tissues, and/or a very low level of excretion, further reinforced in the case of

Table 3. Sediment/species ratio – calculated with anuual average of each metal element.

N. diversicolor

Station	Zn	Cu	Ni	Cr	Pb	Mn	V	Fe
2	0.58	2.70	2.70	2.53	3.68	5.28	6.15	14.47
7	0.43	3.42	2.65	2.46	4.24	4.25	4.96	16.05
10	0.53	2.45	3.17	3.15	3.82	4.09	5.80	15.08
15	0.26	2.43	3.35	2.40	4.48	5.27	6.19	22.80

S. plana

Station	Zn	Cu	Ni	Cr	Pb	Mn	V	Fe
2	0.50	1.66	3.10	3.90	4.09	5.84	6.72	23.99
7	0.51	3.12	3.42	3.14	3.34	5.36	10.11	20.98
10	0.49	1.51	3.30	3.60	3.20	4.50	7.76	18.01

Scrobicularia plana by the influence of size. Given these findings, the possibility of high levels of water contamination in this climatic zone should be taken into consideration.

Conclusion

The level of bioaccumulation varies for each metal. Zinc is the element with the highest degree of accumulation in the two species. Three different categories of concentration can be distinguished using sediment/species ratios (R): $R < 1$ (zinc); $1 \leq R \leq 5$ (copper, nickel, chrome, lead, manganese); $R > 5$ (vanadium, iron). These findings are summarized in the following qualitative distribution schema of bioaccumulation:

heavy accumulation	noticeable accumulation
Zn	Cu > Ni = Cr
Nereis > *Scrobicularia*	*Scrobicularia* > *Nereis*
lesser accumulation	light accumulation
Pb > Mn	V > Fe
Scrobicularia = *Nereis*	*Scrobicularia* > *Nereis*

Even though iron and, to a lesser extent, manganese ratios are lower, iron and manganese oxides and hydroxides play a far from insignificant role as vectors for sediment-based metals entering organisms. Inorganic complexes (Zn and Cu) appear to be more retained than organic complexes (V). Nonetheless, within the framework of this initial general study, it has proved difficult to determine the influence of the various types of sedimentary beds encountered.

The role played by interstitial waters in the accumulation of Zn and Cu cannot be ignored. For *Scrobicularia plana* the digestive gland and tube are the organs with the highest metal concentrations. Seasonal variations in the levels of metal concentration are significant. They correspond to different periods in the reproductive cycle. Levels of metal concentration drop during the spawning seasons and rise during gametogenesis.

Acknowledgements

The authors would like to thank J. C. Massio who gave the inspiration for this research project, J. Louail who offered the service of his laboratory for the geochemical and mineralogical analyses and A. Boussaïd, A. Rada and D. Titoulet who provided precious assistance.

References

Baudin, J. P., 1981. Etude expérimentale de la contamination d'un sédiment et d'une annélide polychète (*Nereis diversicolor*, Müller) par le zinc 65. Radioprotection, 16/3: 175–183.

Berthet, B., 1986. Biological indicators of metal contamination in coastal areas: field and experimental studies of the limits of their use. In J. N. Lester, R. Perry & R. M. Sterrit., Chemicals in the environment, int. conf. chaired by Pr. R. Perry, Lisbon: 171–178.

Bryan, G. W., 1976. Heavy metal contamination in the sea. In R. Johnston., Marine pollution, Acad. press., London: 186–302.

Bryan, G. W. & L. G. Hummerstone, 1971. Adaptation of the polychaete *Nereis diversicolor* to estuarine sediments containing high concentrations of heavy metals. 1: General observations and adaptation to copper. J. mar. biol. Ass. U.K. 51: 845–863.

Bryan, G. W., W. J. Langston & L. G. Hummerstone, 1980. The use of biological indicators of heavy metal contamination in estuaries. J. mar. biol. Ass. U.K. Occ. publ. 1: 73 p.

Bryan, G. W., W. J. Langston, L. G. Hummerstone, G. R. Burt & Y. R. Ho, 1983. An assessment of the gastropod *Littorina Littorea*, as an indicator of heavy metal contamination in United Kingdom estuaries. J. mar. biol. Ass. U.K. 63: 327–345.

Cheggour, M., 1988. Contribution à l'étude de l'estuaire du Bou Regreg (côte atlantique marocaine). Conditions écologiques globales, étude de la contamination métallique. Thèse 3ème cycle, Ecole Normale Supérieure Takaddoum, Rabat, 337 p.

Elkaïm, B., 1972. Contribution à l'étude d'un estuaire atlantique marocain: l'estuaire du Bou Regreg. Bull. Soc. Sci. nat. Phys. Maroc: 131–339.

Gillet, P., 1986. Contribution à l'étude écologique des annélides polychètes de l'estuaire du Bou Regreg (Maroc). Thèse de l'Université, Aix-Marseille III, 187 p.

Gillet, P., 1987. Bioaccumulation du cuivre et du zinc chez *Nereis diversicolor* (Annélide Polychète) de l'estuaire du Bou Regreg (Maroc). Cah. Biol. mar. 28: 339–350.

Guelorget, D. & C. Mayère-Mazoyer, 1983. Croissance, biomasse et reproduction de *Scrobicularia plana* dans une

lagune méditerranéenne: l'étang de Prévost à Palavas, Hérault (France). Vie marine 3: 13–22.
Holtzapffel, T., 1985. Les minéraux argileux. Préparation, analyse diffractométrique, détermination. Soc. géol. Nord, Lille, 12, 136 p.
Kourradi, R., 1987. Dynamique de la population de *Scrobicularia plana* (Da Costa) d'un estuaire marocain: le Bou Regreg. Thèse 3ème cycle, Université Mohamed V, Rabat, 193 p.
Luoma, S. N., 1983. Bioavailability of trace metals to aquatic organisms – a review. Sci. total env. 28: 1–22.
Martin, J. M. & M. Meybeck, 1979. Elemental mass-balance of material carried by major world rivers. Mar. Chem. 7: 173–206.
Mayif, M. O. A., 1987. un modèle estuarien limite: le Bou Regreg (Maroc). Contribution à l'étude hydrodynamique, hydrologique, géochimique et minéralogique. Thèse 3ème cycle, Ecole Normale Supérieure Souissi, Rabat, 136 p.
Rada, A., 1985. Ecotoxicologie des métaux lourds. Mise au point bibliographique. Contamination métallique de l'oued Tensift (région de Marrakech, Maroc). Thèse 3ème cycle, Université Cadi Ayyad, Marrakech, 113 p.
Texier, H., M. O. A. Mayif, A. Nasseh & M. Teffal, 1986. Contribution à l'étude hydrologique d'un modèle estuarien limite: le Bou Regreg (Maroc). In Fac. sci. Marrakech, 1er colloque franco-marocain sur l'eau, Marrakech, 1: 56.
Texier, H., G. Moguedet, T. Holtzapffel & M. O. A. Mayif, 1989. Origines minéralogiques des sédiments de l'estuaire du Bou Regreg (Maroc). In Cent. intern. form. éch. géol., Colloque de géologie franco-marocain, Paris, 16: 193.
Zatta, P., 1985. Interaction between Zn^{2+}, Co^{2+}, Mn^{2+} with hemocyanin from *Carcinus moenas*. Cah. Biol. mar., 26: 241–249.

Received 25 August 1989; in revised form 5 March 1990; accepted 4 May 1990

Hydrobiologia **207**: 221–226, 1990.
D. J. Bonin & H. L. Golterman (eds), Fluxes Between Trophic Levels and Through the Water–Sediment Interface.

Growth rate of four freshwater algae in relation to light and temperature

Alain Dauta,[1], Jean Devaux,[2], Françoise Piquemal[1] & Lhoussaine Boumnich[2]
[1] *Laboratoire d'Hydrobiologie, Université Paul Sabatier, 118, route de Narbonne, 31062 Toulouse Cedex, France;* [2] *Laboratoire d'Hydrobiologie, Université Clermont II, B.P. 45, 63170 Aubière, France*

Key words: phytoplankton, growth rate, model

Abstract

Four algae of freshwater phytoplankton were studied in monospecific culture: *Chlorella vulgaris, Fragilaria crotonensis, Staurastrum pingue* and *Synechocystis minima*. Experiments were performed to determine the growth rate over a wide range of light intensities (5–800 μE m^{-2} s^{-1}, 15/9 light/dark photoperiod) and temperatures (10–35 °C). The results provide a set of parameters (particularly the maximal growth rate associated to optimal conditions of light and temperature) for a three-equation model used to described the growth rate response of a non-nutrient-limited culture.

Introduction

Computer simulation models can now be used to predict plankton population dynamics and patterns of productivity (Di Toro *et al.*, 1974; Thomann *et al.*, 1975; Chen & Orlob, 1975; Dauta, 1986) and we still believe, as does Sommer (1989), that it is possible to understand the mechanisms driving species replacement. Beyond their structural properties and concepts, interacting with experimental results from the field and laboratory, such models require more and more process analysis. The success of a model is related to the accuracy of the original basic assumptions; we thus need to refine our knowledge of the mechanisms classically considered to be determinant factors in plankton physiology and ecology. In the belief that an experimental effort would serve both theoretical and practical objectives, and moreover improve the precision and reality of the models, we carried out a major research project on the ecophysiology of algae (Guerri *et al.*, 1981; Dauta, 1982a, b). Growth can be reduced from maximum by suboptimal temperature, light or nutrient conditions: in the present study, we examine the growth response of four algal species to physical control by light and temperature.

Material and methods

Two species were isolated from the eutrophic Lake Aydat (France): *Fragilaria crotonensis* Kitton and *Staurastrum pingue* Teilling. *Chlorella vulgaris* Beiger n° 369 and *Synechocystis minima* Bourr. n° 153 were provided by the collection of the Museum National d'Histoire Naturelle (Paris, France).

The cultures were grown in batches in mineral medium (Dauta, 1982a) with air bubbling (0.5 l air l^{-1} min^{-1}) under controlled irradiance in temperature-programmable chambers, with a 15/9 light/dark photoperiod. The light was provided by 400 watt Phyto-Claude halogen lamps, which give a spectrum close to that of daylight. The incident light intensity was measured with a Li-Cor 185-A quantum sensor. A stock culture was maintained in exponential growth. In the first stage, a sample of this concentrated culture was placed in the dark 24 hours before the experiment to induce synchronization and to avoid preadaptation to light. For the experiment, the sample was diluted to avoid self-shading (since the cell density was low, the incident light intensity was almost equal to the intensity experienced by the algae in the flasks) and divided among different flasks (about 10). Simultaneously, for a given temperature, the flasks were exposed to irradiance levels ranging from 5 to 800 $\mu\mathrm{E\,m^{-2}\,s^{-1}}$. For each temperature condition (10, 15, 20, 25, 30, 35 °C), the duration of the experiment was 24 hours. Population density was measured after that time (t = 1 day) and growth defined from the relation:

$$\mu = t^{-1} \cdot \ln(N_t/N_0) \quad \text{unit (day}^{-1}) \qquad (1)$$

where N_0 is the biomass existing at the beginning of the experiment and N_t the biomass after the incubation time.

The cell densities were followed by absorbance (O.D.) readings at 750 nm (5 cm cuvette).

Results

Figures 1–4 show, for each species, the observed growth rate versus the various light-temperature

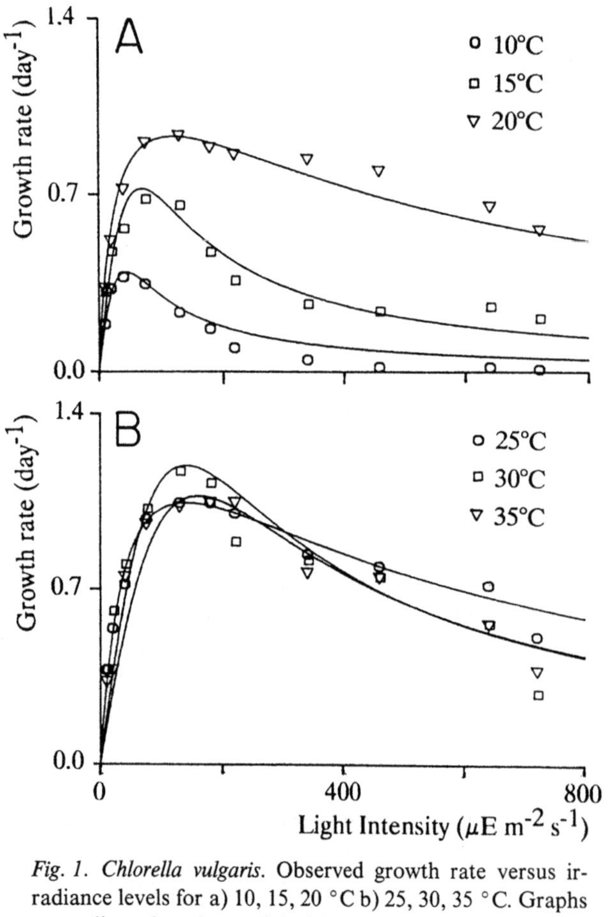

Fig. 1. Chlorella vulgaris. Observed growth rate versus irradiance levels for a) 10, 15, 20 °C b) 25, 30, 35 °C. Graphs adjusted to the model of Peeters & Eilers (1978).

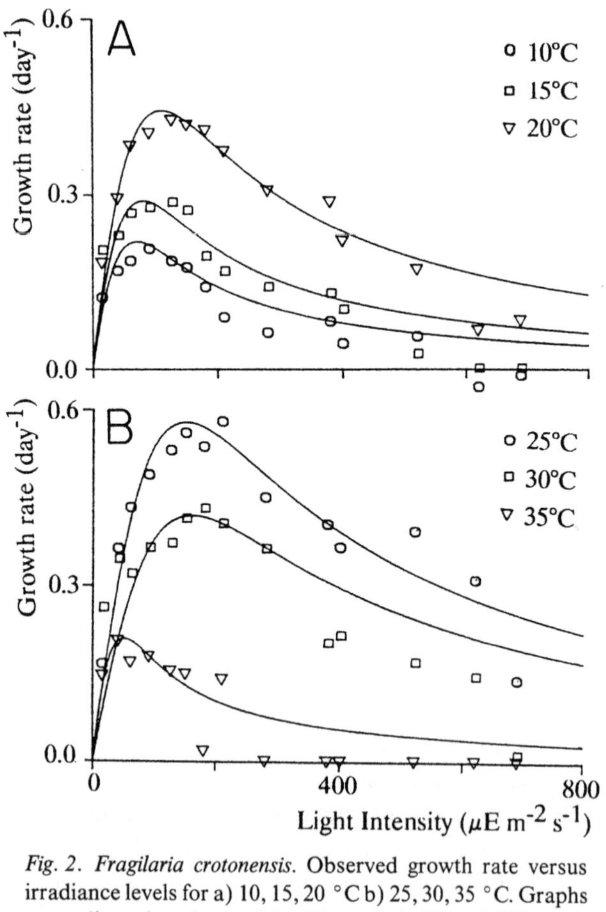

Fig. 2. Fragilaria crotonensis. Observed growth rate versus irradiance levels for a) 10, 15, 20 °C b) 25, 30, 35 °C. Graphs adjusted to the model of Peeters & Eilers (1978).

conditions. These plots show the interspecific differences in sensivity to temperature and irradiance. For each temperature condition, it is possible to determine a maximal growth rate (μmax_T) and the associated optimal light intensity ($Iopt_T$). At higher temperature, the algae generally tolerate much higher light intensities than at lower temperature. Two algae (*Chlorella* and *Synechocystis*) have a very similar response to the temperature (growth rate and Q_{10}), but *Synechocystis* needs lower light intensities. When temperature is higher than the optimal level of irradiance, Iopt, *Fragilaria* and *Staurastrum* present i) a notable drop in growth rate, ii) a decrease in optimal light intensity. The variations of μmax and Iopt are reported in Figs 5–6, which summarize the different behaviour of the four algae.

Two models may used to reproduce the

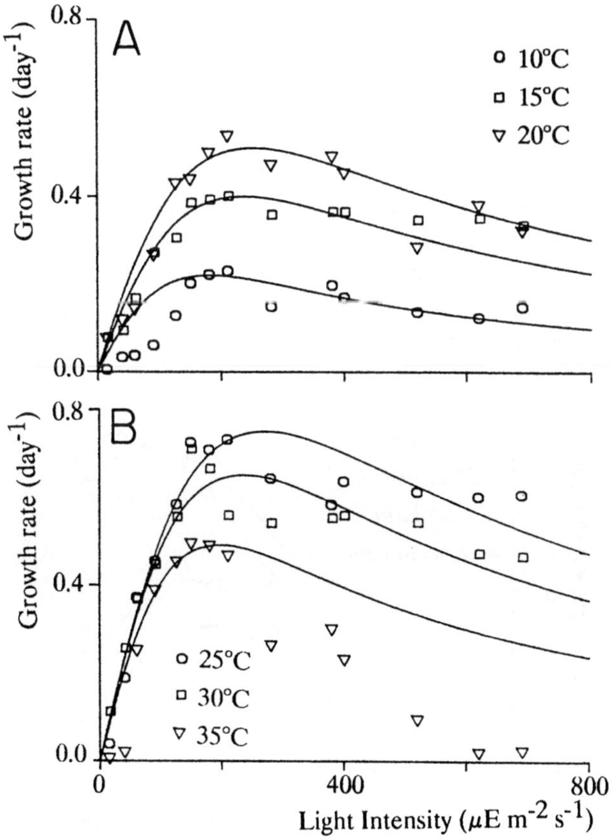

Fig. 3. Staurastrum pingue. Observed growth rate versus irradiance levels for a) 10, 15, 20 °C b) 25, 30, 35 °C. Graphs adjusted to the model of Peeters & Eilers (1978).

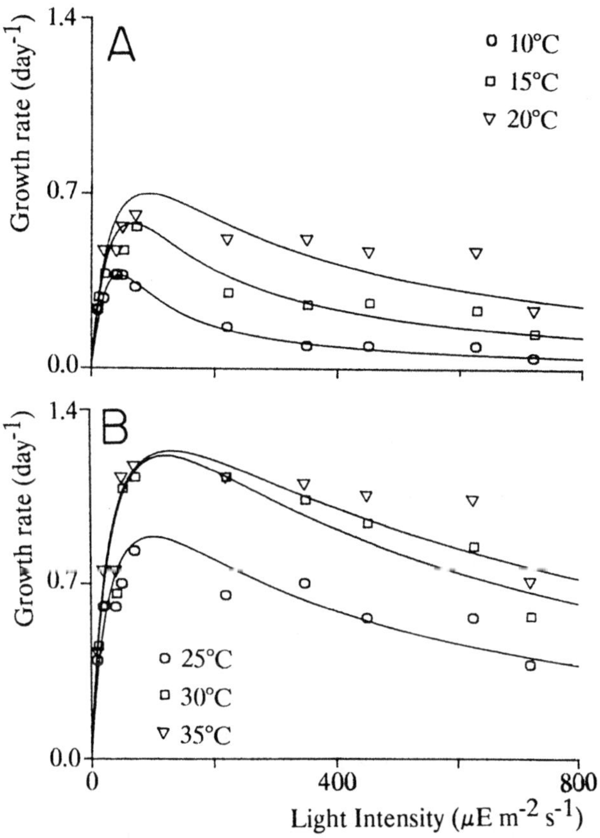

Fig. 4. Synechocystis minima. Observed growth rate versus irradiance levels for a) 10, 15, 20 °C b) 25, 30, 35 °C. Graphs adjusted to the model of Peeters & Eilers (1978).

variation of growth rate (μ) versus light intensity (I) with particular attention to growth inhibition at high irradiance levels:

$$\mu = \mu max_T \cdot Iopt_T^{-1} \cdot e(1 - I/Iopt_T) \quad \text{(Steele, 1962)} \qquad (2)$$

$$\mu = max_T \cdot 2 \cdot (1 + \beta) \cdot I_r / (I_r^2 + 2 \cdot \beta \cdot I_r + 1) \quad \text{(Peeters \& Eilers, 1978)} \qquad (3)$$

where μmax_T = maximal growth rate at temperature T, $Iopt_T$ = Optimal irradiance at temperature T, $I_r = I/Iopt_T$ and β = attenuation coefficient.

Equation (3) was found to give the best adjustment to the values observed, and for each experiment, the parameters μmax_T, $Iopt_T$, β were calculated by the least squares method.

Table 1. Values of the main parameters related to growth of *Chlorella vulgaris*, *Fragilaria crotonensis*, *Staurastrum pingue* and *Synechocystis minima*.

Species	μmax day^{-1}	Topt °C	Iopt μE m^{-2}s^{-1}	Q_{10}
Chlorella vulgaris	1.30	30	140	2.16
Fragilaria crotonensis	0.58	25	150	3.85
Staurastrum pingue	0.77	27	270	4.14
Synechocystis minima	1.32	32	125	2.59

Temperature-dependent relative growth (μmax_T) and saturating irradiance ($Iopt_T$) may be represented by a skewed normal distribution (Lehman *et al.*, 1975), as e.g. μmax_T:

$$\mu max_T = \mu max \cdot \{-2.3 \times [(T - Topt)/(Tsup - Topt)]^2\}, \text{ for } T > Topt \quad (6)$$

$$\mu max_T = \mu max \cdot \{-2.3 \times [(T - Topt)/(Tinf - Topt)]^2\}, \text{ for } T < Topt. \quad (6')$$

Growth rate (μmax_T) is highest at the optimal temperature (*Topt*) and declines as temperature deviates from the optimum toward either higher or lower limits. *Tsup* and *Tinf* are limits where $\mu max_T = 10\%$ of the maximal growth rate under optimal conditions.

Table 1 presents the main parameters and the Q_{10} values of the maximal growth rate in the range 10–20 °C. The Q_{10} values are calculated from extrapolated values of the growth model (Fig. 5).

From this set of parameters, the growth rate ($\mu_{I,T}$) is calculated, using eq. 3, from the results for μmax_T and $Iopt_T$ obtained from eq. 6.

Discussion

This study adds results to existing data on algal growth constants related to temperature (Cloern, 1976; Foy *et al.*, 1976; Ahlgren, 1978; Konopka & Brock, 1978; Goldman, 1979; Morgan & Kalf, 1979; Straskraba & Dvorakova, 1980; Zevenboom, 1980; Dauta, 1982a; Feuillade & Feuillade, 1987; Reynolds, 1988) with the aim of obtaining a homogeneous set of parameters.

Some species are well known, as is the case of *Chlorella vulgaris*, but we present here a wide range of responses for a clone: the maximal growth rate (1.3 d^{-1}) is lower than the *Chlorella* clone generally studied (close to 2 d^{-1}, Tamya *et al.*, 1953). The values obtained for *Fragilaria crotonensis* are quite similar to those of Hartig & Wallen (1986) and complete their study. No data were available for *Staurastrum pingue* and *Synechocystis minima*.

The variation of maximal growth rate and saturating irradiance (Fig. 5, 6) are in accordance with the experimental observations of Post *et al.* (1985) and Reynolds (1989), and indicate the interdependence between growth rate, irradiance and temperature.

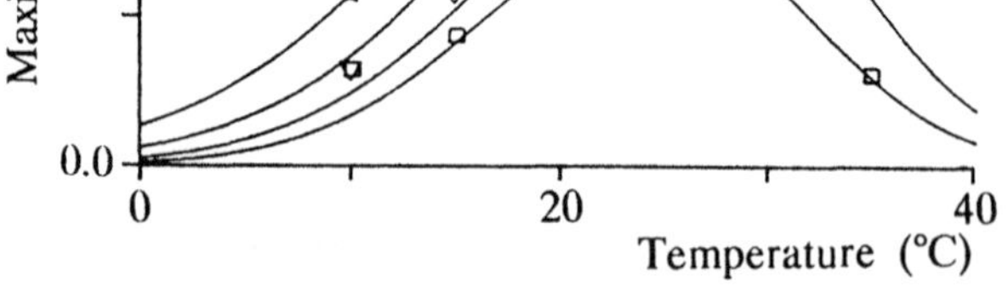

Fig. 5. Variation of maximal growth rate (μmax_T) versus temperature for *Chlorella vulgaris*, *Fragilaria crotonensis*, *Staurastrum pingue* and *Synechocystis minima*.

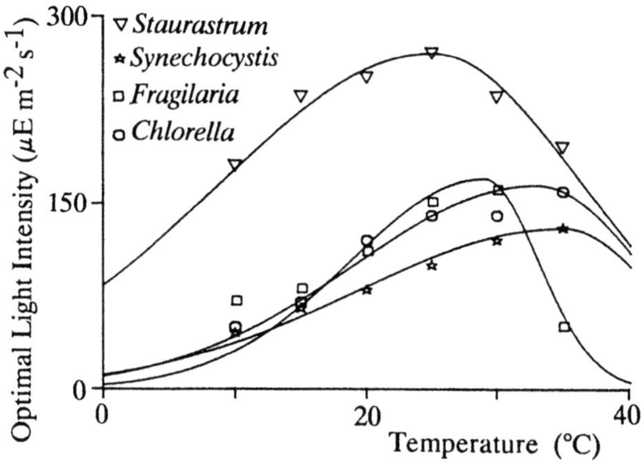

Fig. 6. Variation of optimal light intensity ($Iopt_T$) versus temperature for *Chlorella vulgaris*, *Fragilaria crotonensis*, *Staurastrum pingue* and *Synechocystis minima*.

The results show the temperature-dependent shift of Iopt, particularly for *Fragilaria crotonensis* and *Staurastrum pingue*. Two processes interact: i) an increase of light saturated growth with temperature, ii) a temperature-dependent enhancement of light inhibition of growth at high irradiance, very effective when the temperature is higher than *Topt*. Still not completely understood (Harris, 1978), the mechanism of photoinhibition appears to be purely photochemical at very high irradiance (Jones & Kok, 1966) with a temperature-sensitive effect (Satoh, 1970).

Beyond all the subtle mechanisms of algal ecophysiology, through the growth model proposed, growth becomes a global response involving rate of photosynthesis, assimilation of photosynthetate into proteins and new cell material, a daily result of anabolic and catabolic processes with various specific Q_{10} values (Reynolds, 1989) and a significant relationship with Surface/Volume ratio (Reynolds, 1984).

To establish general models of algal growth for use in long-term prediction, it is necessary to express the temperature effect in quantitative terms (Ahlgren, 1987). Early models usually treated phytoplankton as a single unit, without any regard for interspecies differences. Simulation of seasonal periodicity and changes in species composition serves both theoretical and practical objectives. Such models require complete homogeneous data for a number of algal species: evidence for the usefulness of the study of experimental ecophysiology of phytoplankton.

Acknowledgements

This research programme was partially financed by the PIREN 'Cycle Biogéochimique du phosphore'.

References

Ahlgren G., 1978. Growth of *Oscillatoria agardhii* in chemostat culture. 2. Dependance of growth constants on temperature. Mitt. Int. Ver. Limnol. 21 : 88–102.

Ahlgren G., 1987. Temperature functions in biology and their application to algal growth constants. Oikos 49: 177–190.

Chen C. W. & G. T. Orlob, 1975. Ecological simulation for aquatic environments. In B. C. Patten (ed.), Systems analysis and simulations in ecology. 3, Acad. Press, N.Y.: 475–528.

Cloern J. E., 1976. Population dynamics of *Cryptomonas ovata*: a laboratory, field and computer simulation study. Ph. D., Washington State University, 102 p.

Dauta A., 1982a. Conditions de développement du phytoplancton. Etude comparative de huit espèces en culture. I. Détermination des paramètres de croissance en fonction de la lumière et de la température, Annls Limnol. 18 : 217–262.

Dauta A., 1982b. Conditions de développement du phytoplancton. Etude comparative de huit espèces en culture. II. Rôle des nutriments: assimilation et stockage. Annls Limnol. 18 : 263–292.

Dauta A., 1986. Modélisation du développement du phytoplancton dans une rivière canalisée eutrophe: le Lot (France). Annls Limnol. 21: 199–132.

Di Toro D. M., D. J. O'Connor & R. V. Thomann, 1971. A dynamic model of phytoplankton populations in the Sacramento-San Joaquim delta. In R. F. Gould (ed.), Non equilibrium systems in natural water chemistry, Advances in Chemistry, Ser. 106, Am. Chem. Soc., Washington: 131–180.

Feuillade J. B. & M. Feuillade, 1987. Modelling steady-state growth and photosynthesis rates of *Oscillatoria rubescens* continuous cultures in relation to temperature and irradiance. J. Plankton Res. 9: 445–457.

Foy R. H., C. E. Gibson & R. V. Smith, 1976. The influence of daylength, light intensity and temperature on the growth rates of planktonic blue-green algae. Br. Phycol. J. 11: 151-163.

Goldman J. C., 1979. Physiological processes, nutrient availability, and the concept of relative growth rate in marine phytoplankton ecology. In P. G. Falkowski (ed.), Primary productivity in the sea, New-York, Plenum: 179–194.

Guerri M. M., Dauta A. & Brunel L., 1981. Interaction de la lumière et de la température sur le taux de croissance de *Scenedesmus crassus*. Annls Limnol. 17: 97–104.

Harris G. P., 1978. Photosynthesis, productivity and growth: the physiological ecology of phytoplankton. Ergebn. Limnol. 10: 1–163.

Hartig J. H. & D. G. Wallen, 1986. The influence of light and temperature on growth and photosynthesis of *Fragilaria crotonensis* Kitton. J. Freshwat. Ecol. 3: 371–382.

Jones L. W. & B. Kok, 1966. Photoinhibition of chloroplasts reactions. I. Kinetics and action spectra. Plant Physiol. 41: 1037–1043.

Konopka A. E. & T. D. Brock., 1978. Effect of temperature on blue-green algae (Cyanobacteria) in lake Mendota. Appl. Envir. Microbiol. 36: 572–576.

Lehman J. T., D. B. Botkin & G. E. Likens, 1975. The assumptions and rationales of a computer model of phytoplankton population dynamics. Limnol. Oceanogr. 20: 343–364.

Morgan K. C. & J. Kalff., 1979. Effect of light and temperature interactions on growth of *Cryptomonas erosa* (Cryptophyceae). J. Phycol. 15: 127–134.

Peeters J. C. & P. Eilers, 1978. The relationship between light intensity and photosynthesis: a simple mathematical model. Hydrobiol. Bull. 12: 134–136.

Post A. F., R. de Witt & L. R. Mur, 1985. Interactions between temperature and light intensity on growth and photosynthesis of the cyanobacterium *Oscillatoria agardhii*. J. Plankton Res. 7: 487–495.

Reynolds C. S., 1984. The ecology of freshwater phytoplankton, Cambridge University Press, London, 384 p.

Reynolds C. S., 1988. The concept of biological succession applied to seasonal periodicity of freshwater phytoplankton. Verh. Int. Ver. Limnol. 23: 683–691.

Reynolds C. S., 1989. Physical determinants of phytoplankton succession. In U. Sommer (ed.), Plankton Ecology. Springer Verlag. Berlin: 9–56.

Satoh K., 1970. Mechanism of photoinactivation in photosynthetic systems. I. The dark reaction in photoinactivation. PI. Cell Physiol., Tokyo 11: 15–27.

Sommers U., 1989. The role of competition for resources in phytoplankton succession. In U. Sommer (ed.), Plankton Ecology. Springer Verlag, Berlin: 57–106.

Steele J. H., 1962. Environmental control of photosynthesis in the sea. Limnol. Oceanogr.7: 137–150.

Straskraba M. & M. Dvorakova, 1980. Application to the continuous culture concepts to mathematical modeling of natural phytoplankton populations. Continuous Cultivation of Microorganisms, Proc. 7th Symp. Prague, 1978: 409–416.

Tamya H., T. Iwamura, K. Shibata, E. Hase & T. Nihei, 1953. Correlation between photosynthesis and light independant metabolism in the growth of *Chlorella*. Biochim. Biophys. Acta 12: 24–40.

Thomann R. V., D. M. Di Toro, D. R. P. Winfield & D. J. O'Connor, 1975. Mathematical modelling of phytoplankton in lake Ontario. I-Model development and verification. U.S. Envir. Prot. Agency, Corvallis, Oregon, 176 p.

Zevenboom W., 1980. Growth and nutrient uptake kinetics of *Oscillatoria agardhii*. A comparative investigation of continuous cultures and natural populations of a cyanobacterium. Ph. D., Univ. Amsterdam, 178 p.

Received 28 June 1989; in revised form 30 May 1990; accepted 30 June 1990

Hydrobiologia **207**: 227–232, 1990.
D. J. Bonin & H. L. Golterman (eds), Fluxes Between Trophic Levels and Through the Water–Sediment Interface.

Dynamics of carbon assimilation by St Lawrence estuary phytoplankton at the end of summer

Chantal Descolas-Gros[1] & Michel R. Fontugne[2]
[1] *Laboratoire Arago 66 650 Banyuls sur mer, France;* [2] *Centre des Faibles Radioactivités Laboratoire mixte CNRS-CEA 91 190 Gif sur Yvette, France*

Key words: estuary, carbon, carboxylases, $\partial^{13}C$, production, phytoplankton

Abstract

This study presents data of *in situ* measurements of inorganic carbon assimilation by phytoplankton communities of the St Lawrence estuary during the end of summer 1982. We used carboxylase activity measurements (ribulose-1,5-bisphosphate carboxylase, β carboxylases) and the $^{13}C/^{12}C$ ratio of phytoplankton organic carbon, expressed as $\partial^{13}C$, to study patterns of assimilation. Upper estuary phytoplankton communities showed a smaller turn-over rate in carbon assimilation than lower estuary phytoplankton communities. Carbon assimilation was limited by light intensity in the upper estuary and by CO_2 availability in the lower estuary. In the St Lawrence estuary, stable carbon isotope ratios of phytoplankton organic carbon seemed to be controlled by inorganic carbon availability rather than by phytoplankton metabolism.

Introduction

In estuarine ecosystems phytoplankton communities are submitted to high frequency variations of physical factors (Legendre, 1981; Legendre & Demers, 1984; Levasseur *et al.*, 1984; Demers *et al.*, 1986). Horizontal, vertical and temporal gradients induce changes in phytoplankton carbon metabolic patterns (Harding & Jones, 1988). In such environments, incubation techniques are not suitable to measure phytoplankton carbon assimilation (Grobbelaar, 1985; Collos *et al.*, 1989). Instantaneous measurements are more appropriate to study *in situ* processes of carbon assimilation. In fact, enzymatic measurements allow us to determine the adaptation time scale of biological systems induced by abiotic factors.

The pattern of carbon fixation in phytoplankton was hypothesized by Kremer & Berks (1978) and Appleby *et al.* (1980). In diatoms, Mortain-Bertrand *et al.* (1987, 1988a) and Mortain-Bertrand (1988) have shown that carbon assimilation occurs mainly through the Calvin-Benson cycle via the enzyme ribulose-1,5-bisphosphate carboxylase (Rubisco). Inorganic carbon fixation by β carboxylation is superimposed on this cycle. The substrate of β carboxylations comes either from an intermediate of the Calvin cycle (3-PGA) in light or from a polysaccharide in darkness. The β carboxylase might be phosphoenolpyruvate carboxylase, phosphoenolpyrvate carboxykinase or pyruvate carboxylase depending on the species. The same basic carbon fixation pathway has also been demonstrated by Owttrim & Colman (1988) in a cyanobacterium. Carbon assimilation patterns in natural phytoplankton communities may be obtained by the measurements of carboxylase activity (Rubisco, β carboxylases) and by the $\partial^{13}C$ of organic carbon in phytoplankton tissue.

Rubisco involved in the Calvin-Benson cycle, discriminates against the ^{13}C. When this enzyme functions, the value of $^{13}C/^{12}C$ ratio of organic carbon is smaller (more negative), see O'Leary (1981). However, in aquatic plants and phytoplankton, this ratio is also affected by the $^{13}C/^{12}C$ ratio of inorganic carbon, by CO_2 diffusion and by CO_2 concentration mechanisms Raven *et al.* (1987). Phytoplankton $\partial^{13}C$ values reflect both carboxylation and processes occurring prior to carboxylation (Descolas-Gros & Fontugne, 1990). The simultaneous use of carboxylase activity measurements and of $\partial^{13}C$ in phytoplankton organic carbon provides information on carbon fixation and related *in situ* processes. The aim of this paper is to study the variation scales of physiological adjustements in estuarine phytoplankton, using carboxylase activity and $\partial^{13}C$ measurements.

Materials and methods

Sampling took place in the upper and lower part of the St Lawrence estuary (Quebec) during a cruise in August and September 1982 (Fig. 1). The upper estuary is turbid with low phytoplankton biomass in contrast to the lower estuary which is characterized by higher salinities and higher biomass; El Sabh (1979). All stations were sampled during a 30 hour period with a 6 hour average frequency (see Demers *et al.*, 1985).

Carboxylase activity measurements were performed on board the RV 'Dawson' according to Descolas-Gros & Fontugne (1985, 1988). Activities (expressed in nmol CO_2 fixed per hour and per liter of water: nmol $CO_2\ h^{-1}\ l^{-1}$) were determined by measuring the incorporation of radioactive bicarbonate into stable products. Ribulose-1,5 bisphosphate carboxylase (Rubisco), phosphoenolpyruvate carboxylase (PEPC) and phosphoenolpyruvate carboxykinase (PEPCK) activities were measured on the same crude extract after filtration of 3 liters of seawater on a GF/C filter. Such activity measurements reflect the enzymatic pool of phytoplanktonic cells when sampling occurred. We used the βC/C3 (%) ratio which was the ratio of β carboxylase activities versus Rubisco activity × 100 in order to get an integrated view of the carbon pathway.

The $^{13}C/^{12}C$ ratio of organic carbon was measured after filtration onto precleaned GF/C glass-filters decarbonated with dilute 0.1 N HCl. The filters were dried at 60 °C and stored in the dark at 4 °C before combustion in oxygen. The

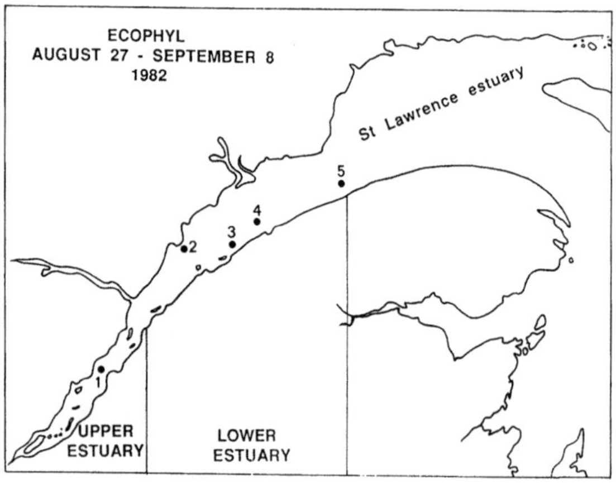

Fig. 1. Location of stations sampled in the St Lawrence estuary.

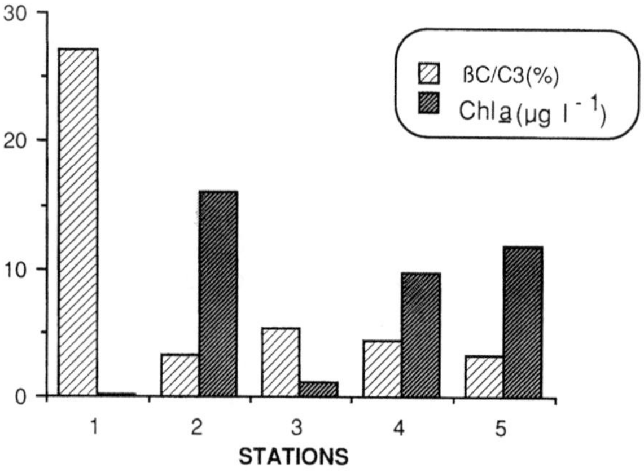

Fig. 2. Biomass (Chlorophyll *a* in $\mu g\ l^{-1}$) and βC/C3 (%) ratio at station 1 in the upper estuary and at stations 2, 3, 4, 5 in the lower estuary. βC/C3 (%) ratio was the ratio of β carboxylase activities (PEPCK) versus Rubisco activity × 100.

CO_2 gas obtained was analysed using a V.G. 'Micromass 602D' mass spectrometer and results expressed as $\partial^{13}C = ((^{13}C/^{12}C \text{ sample}: {}^{13}C/^{12}C \text{ standard}) - 1) \times 1000$. The standard was limestone from the Peedee formation, South Carolina (PDB).

Chlorophyll *a* concentrations were determined on acetone extracted Whatman GF/C filters (250 ml filtrations) using the fluorometric method of Yentsch & Menzel (1963).

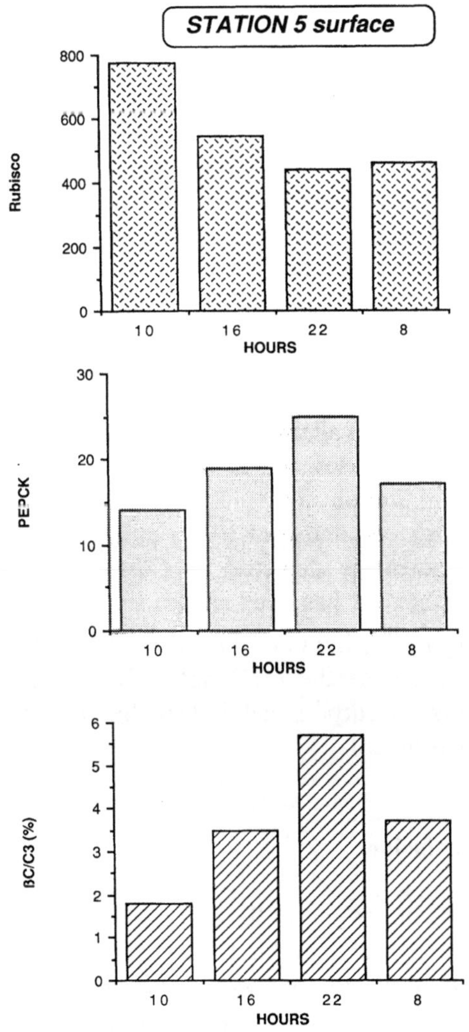

Fig. 3. Nycthemeral variations of Rubisco activity (nmol $CO_2\ h^{-1} l^{-1}$), β carboxylase activity (PEPCK) and $\beta C/C3$ (%) ratio on surface at station 5 (lower estuary).

Results

Carboxylase activities

In all samples (station × sampling date) two enzyme activities were detected. These activities correspond to ribulose-1,5-bisphosphate carboxylase (Rubisco), and only one β carboxylase: phosphoenolpyruvate carboxykinase (PEPCK). Rubisco activity was always higher than β carboxylase activity. For all samples these two activities were positively correlated with phytoplankton biomass (Chl *a*): $r = 0.90$; $n = 66$; $P < 0.1\%$; $r = 0.77$, $n = 66$, $P < 0.1\%$ for Rubisco and PEPCK, respectively).

Station 1 in the upper estuary clearly differed from the four other stations (Fig. 2) when considering average enzymatic activities for each station. Station 1 was characterized by low Chl *a* biomass and greater contribution of PEPCK to carbon fixation, revealed by a $\beta C/C3$ (%) ratio of 30%. In contrast, the $\beta C/C3$ (%) ratio remained below 10% for the stations of the lower estuary.

Temporal variations were apparent at Station 5 (Fig. 3). Rubisco activity was higher at midday and lower at night. The PEPCK activity followed the inverse variation. The highest values of the $\beta C/C3$ ratio never reached 10%. However such temporal variations were smaller than those

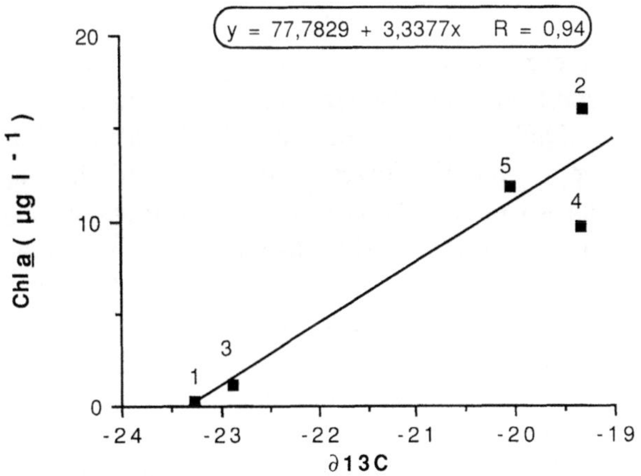

Fig. 4. Variation of $\partial^{13}C$ ($^{13}C/^{12}C$ ratio of organic carbon) and Chl *a* in the five stations of St Lawrence estuary (average values).

found between the upper estuary and the lower estuary.

Stable carbon isotope ration ($\partial^{13}C$)

Isotopic ratios ranged between $-19.01‰$ and $-23.92‰$. There were no significant differences between station 1 (upper estuary) and the other stations (lower estuary). However, there was an inverse relationship between the $^{13}C/^{12}C$ ratio and phytoplanktonic biomass (Chl *a*). The low (more negative) $\partial^{13}C$ values occurred in stations characterized by low phytoplanktonic biomass (Fig. 4).

Discussion

Carboxylase activities

In the St Lawrence estuary, phytoplankton communities are principally composed of diatoms (stations 2, 4, 5) and flagellates (Stations 1, 3) (pers. com. by Berard-Therriault). In these phytoplankton cells PEP carboxykinase was the dominant β carboxylase (Holdsworth & Bruck, 1977; Descolas-Gros & Fontugne, 1988). The correlation between Rubisco activity and biomass parameters has already been found in other environments, such as Portugese coast (Descolas-Gros & Fontugne, 1985) and Mediterranean sea (Descolas-Gros & Fontugne, 1988), indicating a good physiological state of the communities.

Differences between the $\beta C/C3$ ratios of the upper and the lower estuary reflected different turn-over of carbon fixation. In the upper estuary, the low light intensities available for phytoplankton cells, due to turbidity, restrained Rubisco from functioning and phytoplankton cells increased fixation by β carboxylation. The same phenomena may be observed in the open sea with increasing depth (Descolas-Gros & Fontugne, 1985) or in cultures, with decreasing light intensity (Mortain-Bertrand *et al.*, 1988b).

We observed the same tendency (increase of β carboxylase activity associated with decreases in Rubisco activity) between day and night. A strong interaction between light intensity and Rubisco activity has already been reported (Cu *et al.*, 1984; Duke *et al.*, 1986; Mortain-Bertrand *et al.*, 1988a; Descolas-Gros & Fontugne, 1990). However, during the present study the variation scale was smaller than between the two parts of the estuary at the same sampling hour.

Stable carbon isotope ratio ($\partial^{13}C$)

The $^{13}C/^{12}C$ ratio of organic carbon in the St Lawrence estuary was inversely correlated with phytoplanktonic biomass (Chl *a*). The isotopic ratio was smaller (more negative) when biomass was low. This pattern was observed in all stations; $r = 0.57$, $n = 67$, $P < 0.1\%$ for all samples. Regarding carboxylase activity results, the lowest values of the $^{13}C/^{12}C$ ratio should be found in the maritime estuary at station 2, 4, 5 where diatoms were numerous with high Rubisco activity. Therefore, the metabolic effect (Rubisco activity and associated discrimination against ^{13}C) was counterbalanced by other mechanisms. The most probable mechanism was CO_2 limitation. CO_2 availability decreases when salinity increases, it is low at station 5 (S ‰: 29.19) and higher at Station 1 (S ‰: 24.85). In environments with high cell densities (as in cultures), the cell environment is depleted in CO_2. In such an environment it has been shown that Rubisco no longer discriminates between ^{12}C and ^{13}C. In such a case the $\partial^{13}C$ not only reflects phytoplankton carbon metabolism but also the CO_2 environment.

Conclusion

In the upper part of the St Lawrence estuary light intensity seems to limit carbon fixation of phytoplanktonic communities, whereas in the lower part of the estuary carbon fixation of phytoplanktonic communities is controlled by CO_2 availability. In addition, our measurements showed that greater variations in the carbon assimilation are associated with the spatial scale rather than the temporal one.

Obviously, these mechanisms of CO_2 limitation and their subsequent impact on the $^{13}C/^{12}C$ ratio of phytoplankton should be considered in food chain studies where $\partial^{13}C$ is used as tracer of carbon origin (Fry & Sherr, 1984). The simultaneous use of a stable carbon isotopic ratio and carboxylase activities allows for the determination of such a limitation.

Acknowledgements

We are especially grateful to S. Demers, L. Legendre and J-C Therriault for their scientific collaboration and for enabling our participation in the cruise aboard the R/V Dawson. This study and participation of C. D-G in the cruise was financed by France-Québec and France-Canada scientific exchange programs. M.R.F.'s research was supported by CEA and CNRS. We thank A. Gremare for his comments on an earlier draft of the manuscript.

References

Appleby, G., J. Colbeck, E. S. Holdsworth & H. Wadman, 1980. β-carboxylation enzymes in marine phytoplankton and isolation and purification of pyruvate carboxylase from *Amphidinium carterae* (Dinophyceae). J. Phycol. 16: 290–295.

Collos, Y., S. Y. Maestrini & J. M. Robert, 1989. Nocturnal synthesis and diurnal degradation of phytoplankton biomass in surface waters. Mar. Biol. 101: 457–462.

Cu, J. J. V., L. H. Allen & G. Bowes, 1984. Dark/Light modulation of ribulose bisphosphate carboxylase activity in plants from different photosynthetic categories. Plant Physiol. 76: 843–845.

Demers, S., L. Legendre & J-C. Therriault, 1986. Phytoplankton responses to vertical tidal mixing. Lectures notes on Coastal and Estuarine Studies. In J. Bowman, M. Yentsch & W. T. Peterson (eds), Tidal mixing and Plankton dynamics. Springer-Verlag. Berlin: 1–40.

Demers, S., J. C. Therriault, L. Legendre & J. Neveux, 1985. An *in vivo* fluorescence method for the continuous *in situ* estimation of phytoplankton photosynthetic characteristics. Mar. Ecol. Prog. Ser. 27: 21–27.

Descolas-Gros C. & M. R. Fontugne, 1985. Carbon fixation in marine phytoplankton: carboxylase activities and stable carbon isotope ratios; physiological and paleoclimatological aspects. Mar. Biol. 87: 1–6.

Descolas-Gros C. & M. R. Fontugne, 1988. Carboxylase activities and carbon isotope ratios of Mediterranean phytoplankton. In H. J. Minas & P. Nival (eds), Océanographie pélagique méditerranéenne. Oceanol. Acta N° SP 9: 245–250.

Descolas-Gros C. & M. R. Fontugne, 1990. Stable carbon isotope fractionation by marine phytoplankton during photosynthesis. Plant Cell Envir. 13: 207–218.

Duke, C. S., B. E. Lapointe & J. Ramus, 1986. Effects of light on growth, RuBPCase activity and chemical composition of Ulva species (Chlorophyta). J. Phycol. 22: 362–370.

El-Sabh, M. I., 1979. L'océanographie de l'estuaire du Saint-Laurent. Nat. can. 106: 1–276.

Fry, B. & E. B. Sherr, 1984. $\partial^{13}C$ measurements as indicators of carbon flow in marine and freshwater ecosystems. Contr. mar. Sci. 27: 13–47.

Grobbelaar, J. U., 1985. Phytoplankton productivity in turbid waters. J. Plankton Res. 7: 653–663

Harding, L. W Jr & T. W. Jones, 1988. Carbon metabolism in estuarine phytoplankton: regulation of photosynthetic fixation patterns. Bio. Oceanogr. 5: 183–214.

Holdsworth, E. S. & K. Bruck, 1977. Enzymes concerning with β-carboxylation in marine phytoplankter purification and properties of phospho-enolpyruvate carboxykinase. Arch. Biochem. Biophys. 182: 87–94.

Kremer, B. P. & R. Berks, 1978. Photosynthesis and carbon metabolism in marine and freshwater diatoms. Z. Pflanzenphysiol. 87: 149–165.

Legendre, L., 1981. Hydrodynamic control of marine phytoplankton production: the paradox of stability. In J. C. J. Nihoul (ed.) Ecohydrodynamics, Elsevier, Amsterdam: 191–207.

Legendre, L. & S. Demers, 1984. Towards dynamic biological oceanography and limnology. Can. J. Fish. aquat. Sci. 41: 2–19.

Levasseur, M., J-C. Therriault & L. Legendre, 1984. Hierarchical control of phytoplankton succession by physical factors. Mar. Ecol. Prog. Ser. 19: 211–222.

Mortain-Bertrand, A., 1988. Photosynthetic metabolism of an antarctic diatom and its physiological responses to fluctuations in light. Polar Biol. 9: 53–60.

Mortain-Bertrand, A., C. Descolas-Gros & H. Jupin, 1987. Short term ^{14}C incorporation in *Skeletonema costatum* (Greville) Cleve (Bacillariophyceae) as a function of light regime. Phycologia 26: 262–269.

Mortain-Bertrand, A., C. Descolas-Gros & H. Jupin, 1988a. Pathway of dark inorganic carbon fixation in two species of diatoms: influence of light regime and regulator factors on diel variations. J. Plankton Res. 10: 199–217.

Mortain-Bertrand, A., C. Descolas-Gros & H. Jupin, 1988b. Growth, photosynthesis and carbon metabolism in the temperate marine diatom *Skeletonema costatum* adapted to low temperature and low photon flux density. Mar. Biol. 100: 135–141.

O'Leary, M. H., 1981. Carbon isotope fractionation in plants. Phytochemistry 20: 553–567.

Owttrim, G. W. & B. Colman, 1988. Phosphoenolpyruvate carboxylase mediated carbon flow in cyanobacterium. Biochem. Cell Biol. 66: 93–99.

Raven, J. A., J. J. Mac Farlane & H. Griffith, 1987. The application of carbon discrimination techniques. Br. ecol. Soc. spec. Publ. 5: 129–152.

Yentsch, C. S. & D. W. Menzel, 1963. A method for the determination of phytoplankton chlorophyll and phaeophytin by fluorescence. Deep-Sea Res. 10: 221–231.

Received 12 July 1989; in revised form 21 December 1989; accepted 7 March 1990.

Hydrobiologia **207**: 233–240, 1990.
D. J. Bonin & H. L. Golterman (eds), Fluxes Between Trophic Levels and Through the Water–Sediment Interface.

Alkaline phosphatase activity fluctuations and associated factors in a eutrophic lake dominated by *Oscillatoria rubescens*

Jacques Feuillade, Mauricette Feuillade & Paul Blanc
Institut de Limnologie, I.N.R.A., B.P. 511, F-74203 Thonon cedex, France

Key words: phosphorus, alkaline phosphatase activity, phytoplankton, Cyanobacteria, nitrogen, correspondence analysis

Abstract

Alkaline phosphatase activity (APA) was investigated monthly for 4 years in the eutrophic Lake Nantua, a lake colonized by a large population of Cyanobacteria. Total enzymatic activity as well as specific activities (the ratios between APA and biomass expressed as dry weight, chlorophyll *a*, cell phosphorus) varied strongly but they followed a similar pattern during each year. The data were processed using correspondence analysis. Specific APA was never related to depth but highest activities were always associated with low particulate phosphorus and nitrogen, low dissolved inorganic phosphorus (DIP) concentrations, low chlorophyll *a* to filament number ratio and zero nitrate in the waters, indicating P and N limiting conditions. However a high N/P ratio, close to Redfield optimum also occurred at these conditions. Low activities were associated only with high chlorophyll *a* to filament number ratio.

The results suggest that, during summer P-depletion and as long as the N/P ratio is close or above an optimum value, the DIP enzymatically regenerated from DOP pool by phosphatases could temporarily contribute to the algal phosphorus supply.

Introduction

In many lakes, phosphorus is the nutritional factor that controls the growth of phytoplankton. Although Lake Nantua is eutrophic, in the summer, in epilimnic waters, dissolved inorganic phosphorus (DIP), cellular phosphorus contents and N/P ratios, were characteristic of phosphorus limitation. This limitation was particularly drastic after the sewage diversion in 1972–73 (Feuillade *et al.*, 1985; Feuillade, 1988).

Although DIP is the preferred form of phosphorus directly taken up by plankton, many organisms can utilize organically bound phosphorus. This ability depends on the presence of phosphatases which are classified either as acid or alkaline. Because the conditions favouring their activity are more frequent in lakes, the second are more often measured. These phosphatases regenerate phosphorus by hydrolysis of mainly phosphomonoesters, which are produced by decomposition and excretion (see Bonin & Maestrini (1981) and Jansson *et al.* (1988) for a review). Inorganic polyphosphates are also hydrolysed by these enzymes.

The level of APA increases usually when DIP becomes limiting, so the use of the specific activity of alkaline phosphatase has been suggested as an indicator of DIP-limited algal growth (Kuenzler & Perras, 1965). However, contradictory observations have been reported; Huber & Hamel (1985) found that APA was independent of the external concentration of DIP in lake water. Hino (1988) observed a decline of APA in summer when DIP was not detected and he suggested that this was due to the succession of dominant species of phytoplankton. According to several authors (Wynne, 1981; Gage & Gorham, 1985: Huber & Hamel, 1985) APA levels were related to intracellular phosphorus content, the 'phosphorus cell quota', indicating that cellular rather than external phosphorus controlled phosphatase activity.

A new interest in alkaline phosphatases has resulted from recent studies. Currie & Kalff (1984a, 1984b) and Currie *et al.* (1986) suggested that in phosphorus deficient lakes, the bacterioplankton is responsible for most orthophosphate uptake while the phytoplankton obtains most of its phosphorus from excreted organic P compounds. However, according to Chróst (1986), the bacteria hydrolyzed phosphorus from dissolved organic phosphorus (DOP) to utilize only the organic part as a source of carbon, the remaining phosphorus becoming available to phytoplankton. Due to this new interest and because a study related to Lake Nantua, 16 years after sewage diversion, was planned in the forthcoming years, we reevaluated our 1971–1974 alkaline phosphatase data.

The study lake

In the eutrophic Lake Nantua (France) *O. rubescens* represented from 70 to 90% of the total phytoplanktonic biovolume; its population from 50 to 4000 filaments (mean length = 100 μm) per milliliter restricted the Secchi transparency to 3.5 m in winter and ≤ 1 m in summer. The lake stratified each year from April with a defined thermocline during summer and autumn (stagnation period), constituting thus an ideal habitat for *O. rubescens*.

Methods

APA was measured monthly, during 4 years, concomitantly with several physico-chemical and biological variables. The samples were collected at fixed points above the maximum depth. Only measurements from the epilimnion (depths: 0, 2.5, 5 and 7.5 m) will be considered in this paper.

Alkaline phosphatase activity (APA) was assayed by a modification of the spectrophotometric method of Fitzgerald & Nelson (1966) and determined on the seston collected on Whatman GF/C glass fiber filter. After washing twice with phosphorus free medium the particles were resuspended in 2 ml of Tris buffer at pH 8.5 containing 1 mg ml^{-1} of p-nitrophenyl phosphate and was incubated at 37 °C for 1 h. The reaction was stopped by adding 10 ml of 0.2 N NaOH. Then, after centrifugation, the absorbance of the supernatant was read at 410 nm. A blank without seston was substracted. The specific activities were expressed as units of enzyme per mg DW seston, μg Chl *a*, or μg P_{part}. One unit was defined as the amount of enzyme which liberates 10 μg nitrophenol h^{-1} (a practical biomedical laboratory unit).

Although in Nantua Lake waters, pH varied from 7.6 to 8.4 during this 4 year study, APA were always assayed at pH 8.5. It was verified that results were not significantly different for pH between 7.5 and 8.5. By using the same pH and high temperature throughout the year, we obtained comparable and maximized potential activities.

Analytical methods related to the following variables were described in Feuillade *et al.* (1985): dissolved inorganic phosphorus (DIP), dissolved phosphorus (DP), dissolved organic phosphorus (DOP), particulate phosphorus (P_{part}), dissolved nitrate (N-NO_3), particulate nitrogen (N_{part}), chlorophyll *a* (chl *a*) and dry weight of the seston (DW). Results are the averages of triplicate measurements for DW, Chl *a*, P_{part}, N_{part} and APA, and of duplicate measurements for the others variables.

The factorial analysis of correspondence was applied according to Lebart & Fénelon (1971). This form of data processing groups together variables having the same variation patterns (same 'profiles' in the original theory). Each quantitative variable retained was distributed among 4 equilibrated classes with only the exception of $N\text{-}NO_3$ whose zero values constituted a few larger class. Three qualitative variables were considered: the 4 sampling depths, the 4 years of study and the 12 months.

Results

In winter, phosphorus in the water column from 0 to 7.5 m was found mainly in the DIP fraction (Fig. 1). Maximal concentrations (from 30 to 70 $\mu g\ P\ l^{-1}$) were observed during February. From April onwards, phosphorus was transferred from DIP to the particulate form (seston). The highest concentrations of particulate phosphorus (from 50 to 70 $\mu g\ P\ l^{-1}$) were observed during spring and were concomitant with the highest concentrations of chlorophyll *a*. The largest sestonic biomass was found throughout the stagnation period with two maxima; the first in spring (April–May), the second in summer (July or September). Therefore, in summer and autumn, the seston cellular phosphorus content decreased from $> 1\%$ to $\approx 0.5\%$ of the DW (Fig. 2b), indicating a deficiency (Feuillade *et al.*, 1985). This deficiency worsened in 1973, in agreement with a decrease in DIP that can be attributed to the sewage diversion. In 1971 and 1972, from May to November, DOP usually constituted the dominant form of dissolved phosphorus (Fig. 1). Values of DOP varied from 0 to 37 $\mu g\ P\ l^{-1}$ while from 1973, they fluctuated between 0 and 10 $\mu g\ P\ l^{-1}$. This decrease may be attributed to the reduction of allochtonous inputs and/or to an increase of uptake.

Phosphatase activities were measured throughout the 4 years. They showed some large seasonal fluctuations (Fig. 2): from May to October they were from 4 to 25 times higher than in winter and early spring. At this period DIP was low, sometimes undetectable. Increase of phosphatases activities cannot be attributed to biomass increase only since not relationship was found between these 2 parameters. However, the specific activities expressed per unit of DW, Chl *a* or particular phosphorus (P_{part}), were also higher (Fig. 2). Maximum levels of specific APA occurred when the contents of P_{part} in the seston were below 0.5% of the DW. Specific activities presented the same fluctuations patterns for samples collected at the surface, 2.5, 5, or 7.5 m.

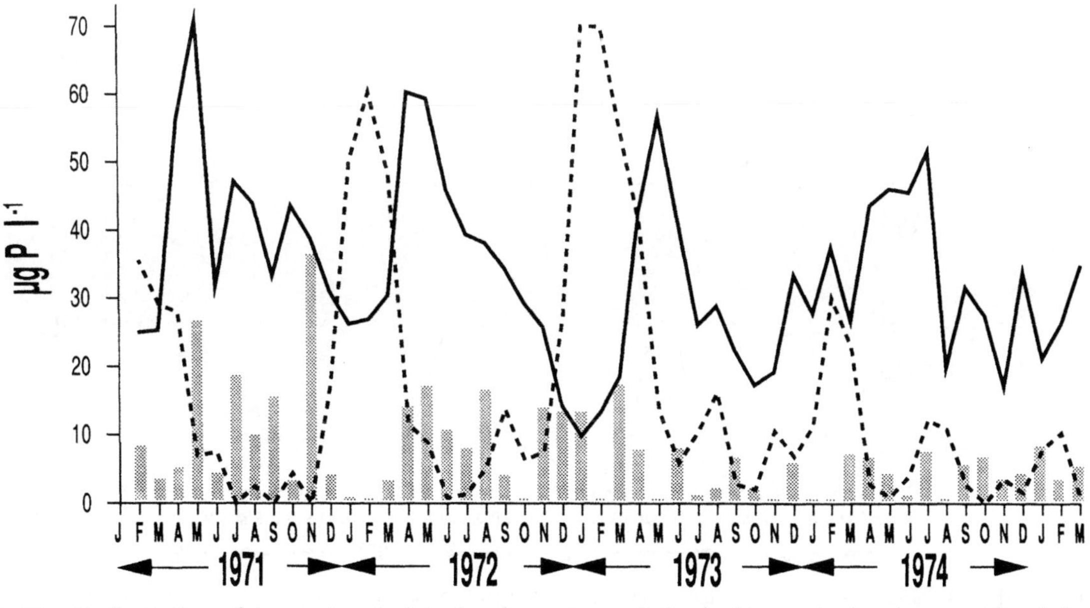

Fig. 1. Monthly fluctuations of: (———) particulate phosphorus, (-----) dissolved inorganic phosphorus and (bars) dissolved organic phosphorus. Integrated values, calculated in $\mu g\ P\ l^{-1}$, for the water column (0–7.5 m).

The relationships between specific APA and dissolved inorganic phosphorus (Fig. 3a), and between specific APA and particulate phosphorus (Fig. 3b), indicated that the specific APA increased only when both phosphorus forms were low; but such a condition appeared not to be sufficient as many low phosphorus measurements did not correspond to high APA. It was not possible to adjust the values to the classical hyperbolic equation with a significative r^2 (determination

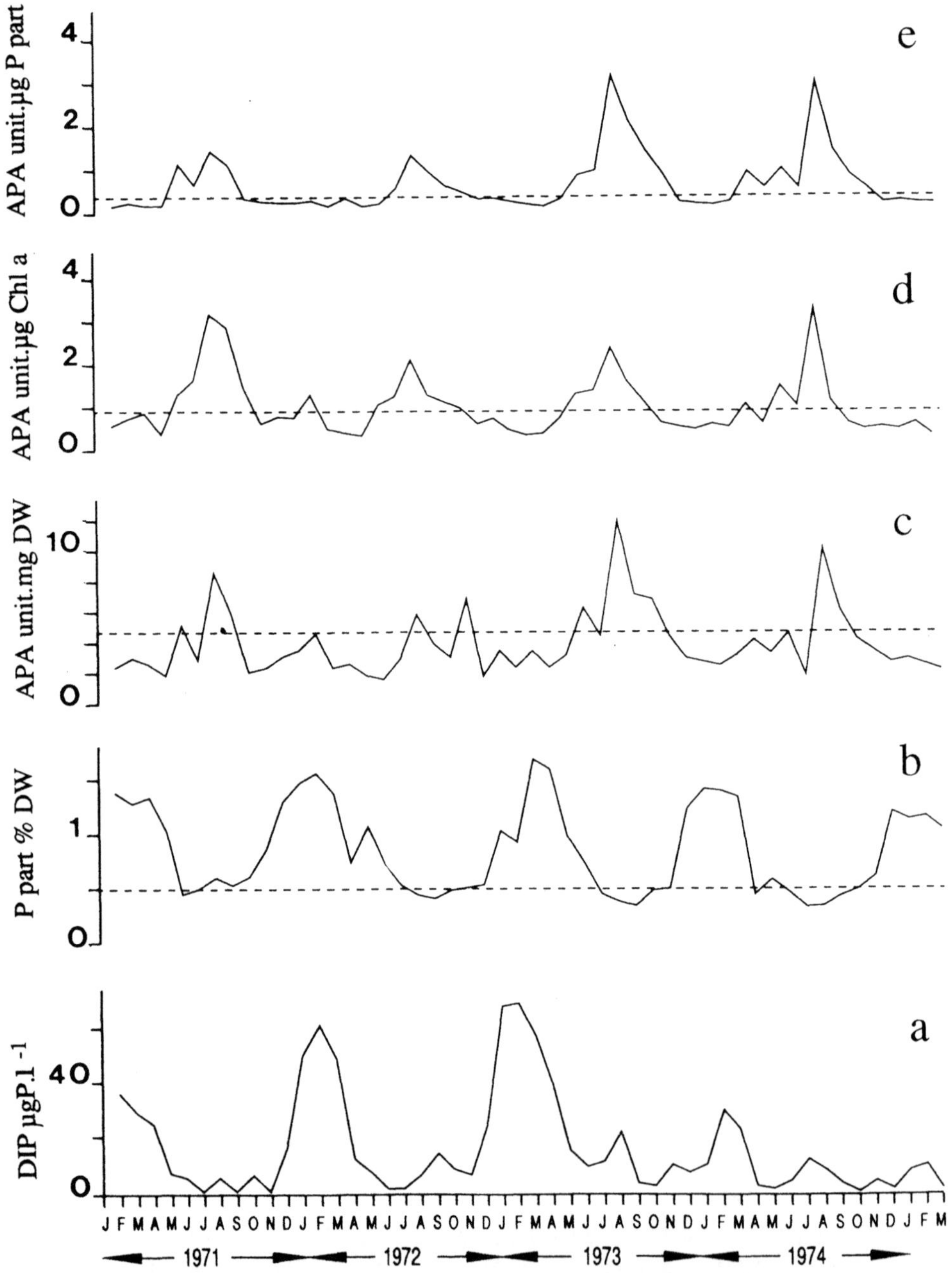

Fig. 2. Monthly variations at 2.5 m depth of: (a) dissolved inorganic phosphorus, in μg P l^{-1}, (b) seston particulate phosphorus (P_{part}) content, in% DW, (c, d, e) seston specific alkaline phosphatase activity in Units mg^{-1} DW (c), Units μg^{-1} chl *a* (d) and Units μg^{-1} P_{part} (e). Dotted lines are reference levels to make easier the lecture of the curves.

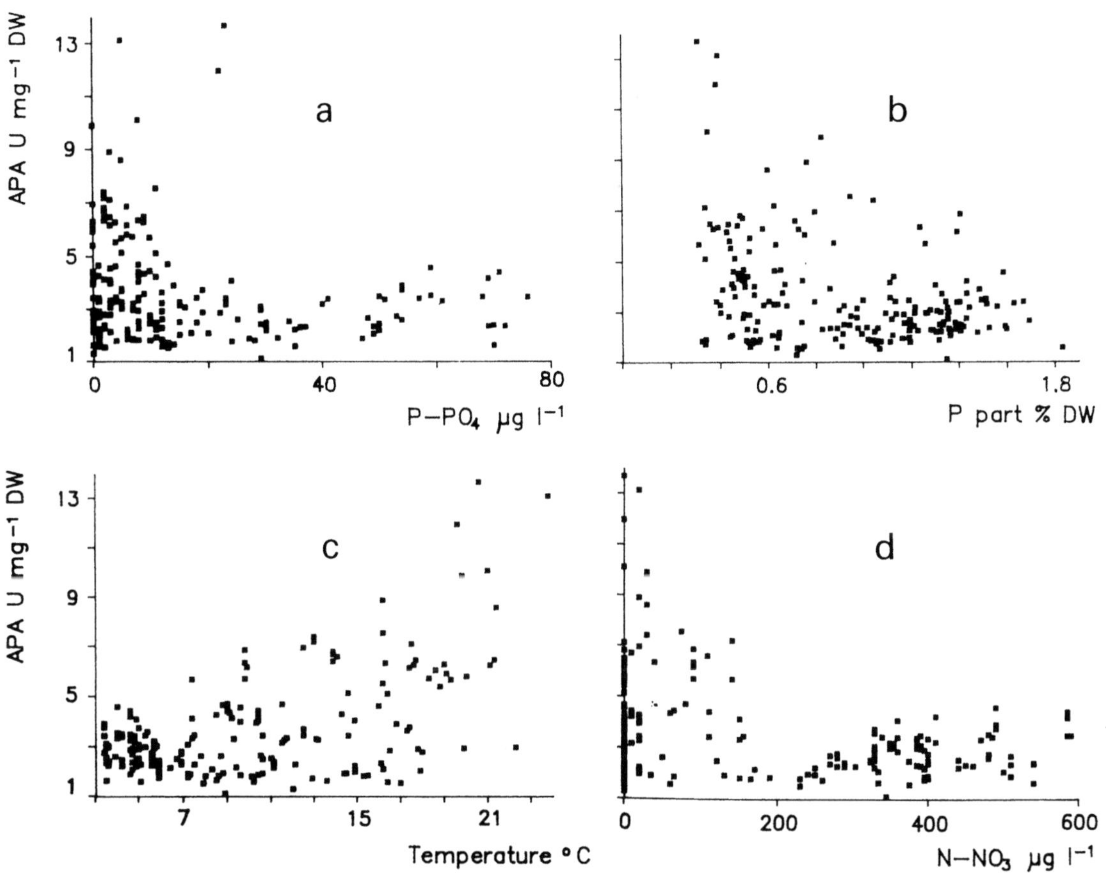

Fig. 3. Relationship between specific APA (Units mg^{-1} DW) and: (a) inorganic phosphorus, (b) particulate phosphorus, (c) water temperature and (d) nitrate nitrogen, for the water column 0–7.5 m in Lake Nantua.

coefficient). Simple linear correlation coefficients were also unsatisfactory. The best relationship was found with temperature, APA increasing with this variable (Fig. 3c). Because the determination coefficient between APA and nitrate nitrogen in the water was significant the relationship was examined (Fig. 3d). When N-NO_3 reached 200 $\mu g\, l^{-1}$, APA was minimum (≤ 2 U mg^{-1} DW), above 200 $\mu g\, l^{-1}$, APA was always less than 5 Units. Below this N-NO_3 value APA could rise to its maximum but some low values were also observed.

Factorial analysis of correspondence indicated a negligible weight of years and depths, these variables were therefore discarded (Fig. 4). Two sets of variables classes are opposed. Firstly, a set referred as 'A', grouping together around the highest APA values (the months from June to September possibly October), then the lowest values for P_{part}, N_{part}, N-NO_3 and chlorophyll *a* to filament ratio with the highest values for temperature and cellular N/P ratio. Secondly, a set referred as 'B', grouping together: the months from January to March, then the highest values for P_{part}, DIP and N-NO_3 with the lowest values for temperature and N/P ratio. It should be noted that set B is not associated with low APA but is opposed to the highest APA. These two sets are separated by ordinates on the 1st axis while the lowest APA class is exactly located at the point of equilibrium; low APA is neither associated with nor opposed to a particular environmental or cellular class of variable.

Along the 1st axis, many classes are distributed in increasing or decreasing orders: P_{part}, N_{part}, N-NO_3, P-PO_4, N/P and temperature; the 1st

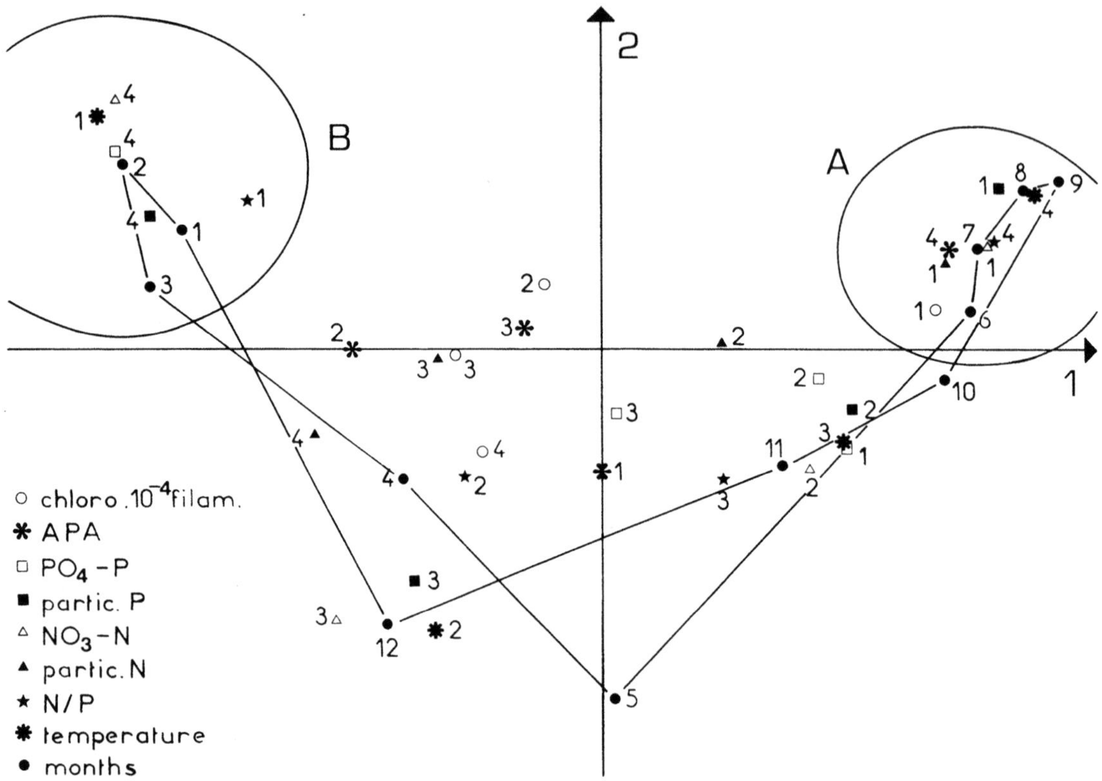

Fig. 4. First and second axes of the analysis of correspondence. Symbols are numbered from 1 to 12 for the 12 months and from 1 to 4 for the 4 others variables classes. The lines joining together the months plots, point out the graph structure. A and B are the two most separated sets of variables classes.

axis represents mostly the annual evolution of the lake from homogeneity to stratification. Months (qualitative variables) are clearly distributed along this first axis. January, February and March, are associated with group B classes; while months from June to October are associated with group A classes. April and May are transitory state months (towards the period of stratification); November and December are also transitory months but in reverse direction, (towards the period of mixing).

Classes are not so clearly distributed along the 2nd axis. This axis differentiates mostly between well established stable conditions: mixing and stratification, and transitory conditions. APA classes are found along this axis from low to high activities.

The close association of higher APA with July appears contradictory with results in Fig. 2 (c, d, e). The explanation is that the whole set of variables most frequently associated with higher APA is also strongly associated with July. If July was characterized by low APA, it was also characterized by low $N\text{-}NO_3$ and concomitant growth of many nitrogen fixing *Aphanizomenon flos-aquae* (Feuillade *et al.*, 1985).

Discussion

Alkaline phosphatase activity of the seston was essentially due to phytoplankton of which *Oscillatoria* was the major constituent. Other observations having showed that bacteria accounted for less than 1% of the seston biovolume (Feuillade *et al.*, 1988).

Although the lacustrine phosphorus decrease from 1973, correspondence analysis showed no difference between the four annual variation patterns of APA. Consequently, no different relationships between APA and the studied variables were involved before and after the sewage diversion.

Specific APA increased only when the DIP reached critically low concentrations, $< 10\ \mu g\ P\ l^{-1}$ (Fig. 2). The decrease of DIP was not immediately accompanied by changes in enzyme activity, probably due to the presence of phosphorus compounds stored in the phytoplankton cells (surplus P). According to Chróst & Overbeck (1987), it is only when the surplus P drops below 10% of the seston particulate P, that the sestonic fraction produces APA. For this reason, the enhancement of specific phosphatase activity was related to particulate phosphorus content. The relationship between APA and P_{part} was an inverse non-linear function. Maximal APA corresponded to a particulate phosphorus content of $\leq 0.5\%$ of the dry weight (Fig. 2), such values indicate phosphorus deficiency level (Healey & Hendzel, 1980; Feuillade *et al.*, 1985). In the correspondence analysis, the lowest class of DIP was found near group A (high APA associated classes) but not in it; whereas low P_{part} was in this group and very close to high APA; this confirms that APA variations were more directly (and inversely) dependent on internal than on external phosphorus.

To explain the recovery of high APA in August, we suggest that *O. rubescens* used the organic nitrogen excreted by the *Aphanizomenon* (Feuillade *et al.*, 1988) which was also responsible for the production of new alkaline phosphatases.

The relationship between APA and nitrogen status of the cells appears in Fig. 4. High APA was associated with both low N_{part} and P_{part} (because both nutrients were simultaneously depleted by phytoplankton). But is was also associated with the highest cellular N/P ratio class, indicating that APA developed only if sufficient nitrogen remained in the cells to meet an optimum N/P ratio close or above the Redfield ratio (16:1).

In the summer, after correction for lake temperature as in Chróst & Overbeck (1987), the APA potential activities from 1 mg DW seston, measured in Lake Nantua waters, could liberate from 20 up to 120 nmoles $PO_4^{3-}\ h^{-1}$. Then, hydrolysis of DOP by alkaline phosphatase could satisfy a significant part of the sestonic phosphate demand during the period of DIP starvation. However, a high APA will be nutritionally effective, only if the enzymatically hydrolysable substrate is present. The entire DOP fraction in the water cannot be hydrolysed by alkaline phosphatase; a large part of the DOP in natural waters is composed of a high molecular weight fraction resistant to alkaline phosphatase. However, the organic phosphorus in phytoplankton is easily hydrolysable (Hino, 1988); phosphomonoesters are one of the important constituents of living cells and the hot water-extractable algal phosphorus is also easily hydrolysed by alkaline phosphatase (Wynne & Berman, 1980). Since a large phytoplankton standing crop was always present in Lake Nantua, higher concentrations of organic phosphorus compounds might therefore occur in the form of easily hydrolysable DOP released from algal cells after lysis or by excretion. After the sewage diversion from 1972, the DIP supply decreased and the APA increased (Fig. 2). So, the rate of hydrolysis was probably greater than the DOP supply, explaining that DOP was below $10\ \mu g\ P\ l^{-1}$ and sometimes undetectable (Fig. 1).

If bacteria are more effective in taking up both organic P (Berman, 1985; 1988) and orthophosphates (Currie & Kalff, 1984b; Berman, 1985; Currie *et al.*, 1986), an explanation for the competitive success of *O. rubescens* in Lake Nantua must be found. This species still flourishes 10 years after sewage diversion, although the phosphorus deficiency has become increasingly severe. Among the possibilities proposed by Berman (1985), the control of the bacterial standing crop by active grazing can apply to Lake Nantua which has many Rotifers, all the more because *Oscillatoria* is very rarely consumed.

Acknowledgements

We acknowledge E. Menthon and our colleagues of the 'Institut de Limnologie' for their technical assistance and cooperation to the present research and the anonymous reviewers for improving the manuscript.

References

Berman, T., 1985. Uptake of ^{32}P orthophosphate by algae and bacteria in Lake Kinneret. J. Plankton Res. 7: 71–84.

Berman, T., 1988. Differential uptake of orthophosphate and organic phosphorus substrates by bacteria and algae in Lake Kinneret. J. Plankton Res. 10: 1239–1249.

Bonin, D. J. & S. Y. Maestrini, 1981. Importance of organic nutrients for phytoplankton growth in natural environments: implications for algal species succession. In T. Platt, (ed.). Physiological bases of phytoplankton ecology. Can. Bull. Fish. aquat. Sci. 210: 279–911.

Chróst, R. J., 1986. Algal-bacterial metabolic coupling in the carbon and phosphorus cycle in lakes. In F. Megusar & M. Gantar (eds), Perspectives in microbial activity. Slovene Soc. Microbiol., Ljubljana: 360–366.

Chróst, R. J. & J. Overbeck, 1987. Kinetics of alkaline phosphatase activity and phosphorus availability for phytoplankton and bacterioplankton in Lake Plußsee (North German eutrophic lake). Microb. Ecol. 13: 229–248.

Currie, D. J. & J. Kalff, 1984a. The relative importance of bacterioplankton and phytoplankton in phosphorus uptake in freshwater. Limnol. Oceanogr. 29: 311–321.

Currie, D. J. & J. Kalff, 1984b. A comparison of the abilities of freshwater algae and bacteria to acquire and retain phosphorus. Limnol. Oceanogr. 29: 298–310.

Currie, D. J., E. Bentzen & J. Kalff, 1986. Does algal-bacteria phosphorus partitioning vary among lakes? A comparative study of orthophosphate uptake and alkaline phosphatase activity in freshwater. Can. J. Fish. aquat. Sci. 43: 311–318.

Feuillade, J., 1988. Nutrient status and response of an eutrophicated lake to reduced external nutrient loading. In G. Balvay (ed.), Eutrophication and lake restoration. Water quality and biological impacts. Thonon-les-Bains. 99–111.

Feuillade, J., G. Balvay, P. Blanc, M. Feuillade, A. Orand, J. Pelletier & F. Chahuneau, 1985. Caractérisation et essais de restauration d'un écosystème dégradé: le lac de Nantua. Feuillade, J. (ed.), I.N.R.A., Versailles. 165 pp.

Feuillade, M., J. Bohatier, G. Bourdier, P. Dufour, J. Feuillade & H. Krupka, 1988. Amino acid uptake by a natural population of *Oscillatoria rubescens* in relation to uptake by bacterioplankton. Arch. Hydrobiol. 113: 345–358.

Fitzgerald, G. P. & T. C. Nelson, 1966. Extractive and enzymatic analyses for limiting or surplus phosphorus in algae. J. Phycol. 2: 32–7.

Gage, M. A. & E. Gorham, 1985. Alkaline phosphatase activity and cellular phosphorus as an index of the phosphorus status of phytoplankton in Minnesota lakes. Freshwat. Biol. 15: 227–233.

Healey, F. P. & L. L. Hendzel, 1980. Physiological indicators of nutrient deficiency in lake phytoplankton. Can. J. Fish. aquat. Sci. 37: 442–453.

Hino, S., 1988. Fluctuation of algal alkaline phosphatase activity and the possible mechanisms of hydrolysis of dissolved organic phosphorus in Lake Barato. Hydrobiologia 157: 77–84.

Huber, A. L. & K. S. Hamel, 1985. Phosphatase activities in relation to phosphorus nutrition in *Nodularia spumigena* (Cyanobacteriaceae). I. Field studies. Hydrobiologia 123: 145–152.

Jansson, M., H. Olsson & K. Pettersson, 1988. Phosphatases; origin, characteristics and function in lakes. Hydrobiologia 170: 157–175.

Kuenzler, E. J. & J. P. Perras, 1965. Phosphatases of marine algae. Biol. Bull. 128: 271–284.

Lebart, L. & J. P. Fénelon, 1971. Statistique et informatique appliquées. Dunod, Paris. 426 pp.

Wynne, D., 1981. Phosphorus, phosphatases and the *Peridinium* bloom in Lake Kinneret. Verh. int. Ver. Limnol. 21: 523–527.

Wynne, D. & T. Berman, 1980. Hot water extractable phosphorus: an indicator of nutritional status of *Peridinium cinctum* (Dinophyceae) from Lake Kinneret? J. Phycol. 16: 40–46.

Received 28 June 1989; in revised form 10 January 1990; accepted 23 February 1990

Hydrobiologia **207**: 241–249, 1990.
D. J. Bonin & H. L. Golterman (eds), Fluxes Between Trophic Levels and Through the Water–Sediment Interface.

Zooplankton feeding on seston in the Rhone River plume area (NW Mediterranean Sea) in May 1988

Raymond Gaudy, Marc Pagano & François Lochet
Centre d'Océanologie de Marseille, CNRS, URA 41, Faculté des Sciences de Luminy, case 901, 13288 Marseille Cedex 9, France

Key words: River plume, zooplankton, particle spectrum, grazing

Abstract

In the Rhone delta area the seston concentration was maximum in the plume waters where small-sized particles dominated, while marine waters were characterized by large particles. The food conditions were optimal for grazers in the interface layer, where the high chlorophyll concentration contrasted with the low values found under the plume.

The mesozooplankton specific composition did not show marked difference in and outside the plume. No special pattern for particle size selection by grazers appeared in the plume, contrary to marine waters, where largest particles were the most intensively grazed. The calculations of specific ingestion and filtration rates show that the plume waters (particularly at the salinity interface) were more favourable to zooplankton feeding than the marine ones (outside and under the plume). Nevertheless, taking into account the high level of seston biomass in the whole area studied, the daily grazing pressure of mesozooplankton was very low.

Introduction

In a marine environment, discontinuous hydrological zones such as fronts are known to be of biological interest (Le Fèvre, 1986). High phytoplankton biomasses, which sometimes lead to dystrophy, *e.g.* red tides (Margalef *et al.*, 1979), are often found in these zones. Favourable trophic situations can occur, leading to high productivity of pelagic ecosystems (Frontier, 1986; Le Fèvre, 1986).

In the case of river plume fronts, only few studies deal with biological productivity (Parsons *et al.*, 1969a; Stepien *et al.*, 1981; Harrison *et al.*, 1983; Cadée, 1984; Skreslet, 1986) and phytoplankton-zooplankton relationships (Parsons *et al.*, 1969b; Chervin *et al.*, 1981; Tester & Turner, 1988; Turner & Tester, 1989). In the Rhone delta area, previous works deal with phytoplankton and zooplankton distribution (Blanc *et al.*, 1969), seston size distribution (Loeillet & Leveau, 1985) and bacterial production (Kirchman *et al.*, 1989).

In this paper, we describe the qualitative (particle size spectrum) and quantitative (particle volume, carbon and chlorophyll contents) aspects of food environment for zooplankton inside and outside the Rhone River plume, and we examine their incidence on grazing.

Material and methods

The sampling stations (Fig. 1) were chosen according to the plume boundary position. Vertical

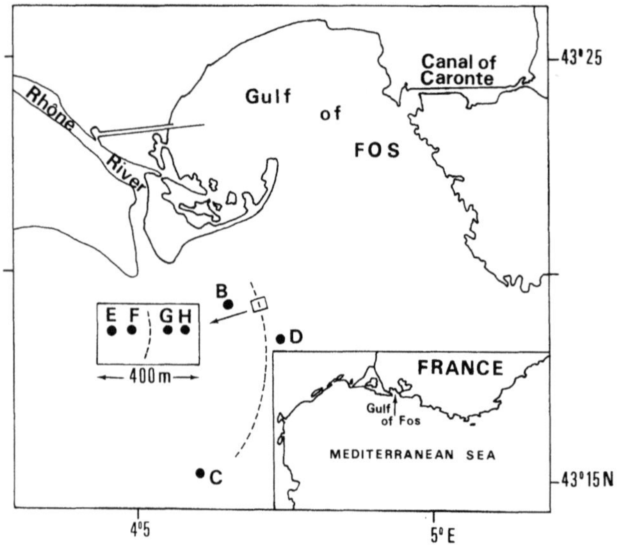

Fig. 1. Study site and position of stations. The horizontal border of the plume extension is indicated by a dashed line.

sampling and grazing experiments were carried out in a marine station (D) and in Rhone proximal (B) and distal (C) plume stations. In addition, surface water samples were collected along a transect across the front (B, E, F, G, H).

Salinity (derived from conductivity), temperature and depth were measured with a CTD probe. Water was sampled using a $10\ \mathrm{l\ min^{-1}}$ pump for upper levels (0–10 m) or a Niskin bottle for deepest layers (more than 10 m). Size distribution of seston particles was analyzed using a Coulter Counter TA II equipped with 100 and 220 μm aperture tubes. Chlorophyll (fluorimetric method: Holm-Hansen *et al.*, 1965) and carbon (CHN autoanalyser) analyses were performed on the same samples. Zooplankton was collected at 0 and 10 m in the marine station (D) and at three depth strata: surface, interface (*i.e.* the vertical limit of the plume) and deep layer in the plume stations (B, C) with a $700\ \mathrm{l\ min^{-1}}$ centrifugal pump discharging water into a 200 μm mesh plankton net for population analysis and biomass measurements (carbon analysis). Additional living material for experiments was collected by WP2 horizontal hauls (200 μm mesh size).

Grazing experiments were started about three hours after collection. Total zooplankton from each layer was gently cleared of large detritus and gelatinous organisms and incubated for 24 h in 300 ml flasks filled with *in situ* water from the corresponding layer, *i.e.* same water as analyzed for seston carbon and chlorophyll. Three experimental bottles (with animals at a density of $0.5–1.5\ \mathrm{mgC\ l^{-1}}$) and two controls (same water without animals) were prepared for each layer. Settlement of seston was minimized by manual agitation of the bottles every two hours. The experiments were carried out in the dark, at the temperature of collection. Ingestion rate was calculated for the various particle-size classes from the differences in particle volume (Coulter Counter measurements) between experimental and control bottles at the end of the experiment. For a given size channel, it was counted as 0 if particle concentration was higher in the experimental bottle than in the control (Tackx & Polk, 1986). This phenomenon which occurred particularly in the two smallest channels could be due to bacterial development, cell breakage by zooplankton (Harbison & McAlister, 1980) and egestion products, and results in probable underestimation of the true total ingestion values. However, several studies show that electronic counting of particles gives results comparable to other methods (Hargis, 1977; Kiørboe *et al.*, 1985). Rates were expressed in particle volume ingested per day and per zooplankton carbon weight (CHN analysis). Ingestion was also estimated in chlorophyll and carbon units, assuming that the consumed chlorophyll- or carbon-containing particles were in proportion to their availability in seston. Clearance rates (in $\mathrm{ml\ \mu gC^{-1}\ d^{-1}}$) were calculated according to the formula:

$$F = [(\ln C_1 - \ln C_0) * 24/(t_1 - t_0)] * V/B$$

where C_1 and C_0 are the total particle concentrations (in $\mu\mathrm{l\ l^{-1}}$) at final (t_1) and initial (t_0) times (in hours), V is the experimental vessel volume (in ml) and B the zooplankton biomass (in μgC). As for ingestion rates, C_0 was estimated from the concentration noted in control vessels at the end of the experiments.

Results

Temperature and salinity

The plume was characterized by an abrupt salinity gradient from the surface to about 1 to 2 m depth (Fig. 2). However, outside the plume (station D), a slight salinity gradient was still observed, due to the influence of the Durance River flowing into the northern part of the Gulf of Fos through the canal of Caronte. A clear temperature gradient was not observed in the different stations.

Size distribution of seston particles (Fig. 3)

At the marine station (D) a unimodal size spectrum (maximal frequency between 14 and 23 μm) was observed from the surface to 5 m, while below this layer, the particle spectrum was characterized by a higher frequency of larger particles (>40 μm). Inside the plume (stations B and C), the particle size distribution differed markedly with depth. At the proximal station (B), the particle size peak in surface waters was observed for fine particles (about 6 μm). On the other hand, the spectrum in interface waters (1 m) was very similar to that found in surface waters outside the

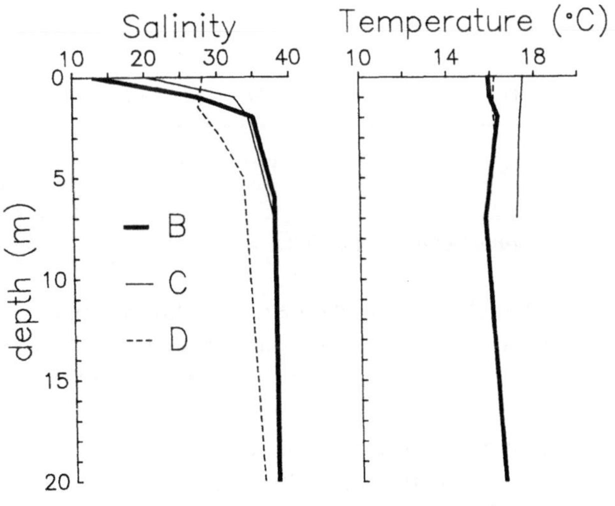

Fig. 2. Vertical profiles of temperature (°C) and salinity at the proximal (B) and distal (C) plume stations and in marine reference stations (D) (see Fig. 1).

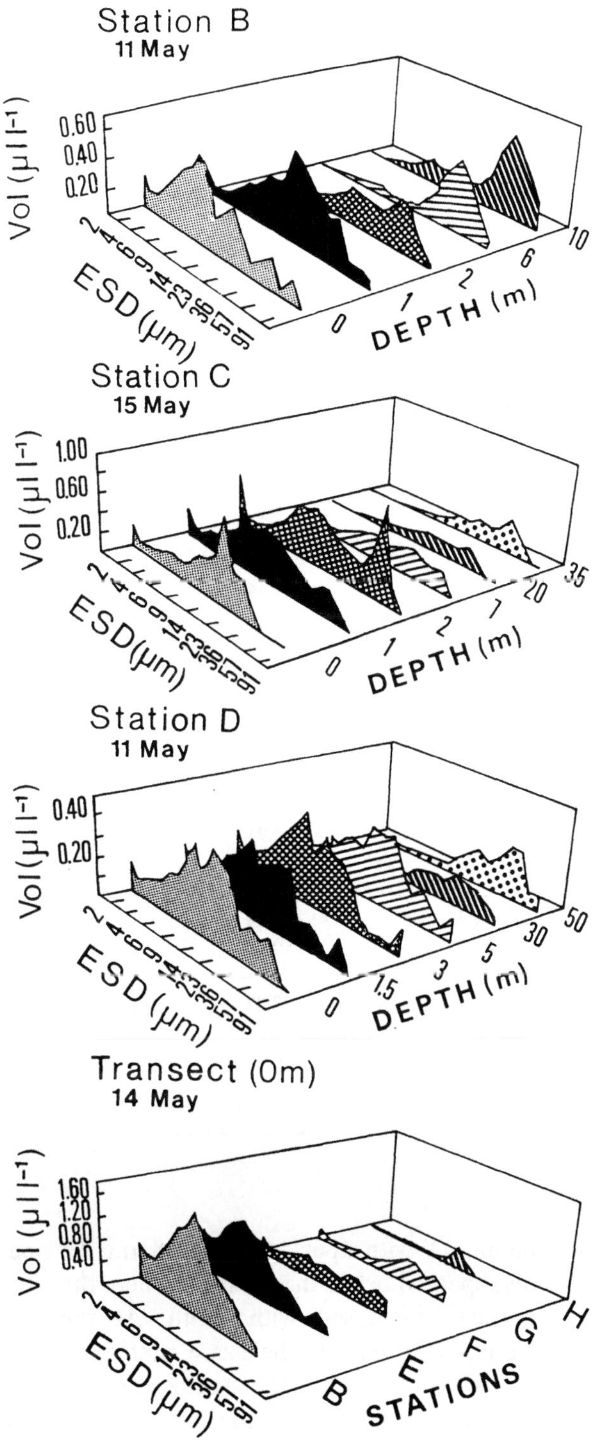

Fig. 3. Sestonic particle-size (equivalent spherical diameter, ESD, in μm) distribution per volume ($\mu l\, l^{-1}$) at different depths at proximal (st. B) and distal (C) plume stations, outside the plume area (D) and in surface waters from a transect across the plume limit (B, E, F, G, H).

Table 1. Percentage abundance of the main taxa and total abundance and biomass of zooplankton collected in pump samples at different depths (0, 2, 6, 10 m) from plume (B, C) and marine (D) stations.

Taxa	Station B			Station C			Station D	
	0 m	2 m	6 m	0 m	2 m	6 m	0 m	10 m
COPEPODS								
Copepod *nauplii*	2.08	0.23	0.00	0.24	0.39	0.73	0.31	0.00
Acartia spp.	3.13	0.86	1.77	2.06	0.69	0.49	2.96	1.36
Paracalanus spp.	21.61	22.08	43.66	8.58	16.59	5.15	59.44	31.88
Centropages typicus	11.20	7.65	7.14	0.18	0.03	2.43	9.05	3.54
Candacia sp.	0.78	0.41	1.77	0.20	0.15	1.75	0.16	0.82
Oithona helgolandica	8.59	7.01	11.56	3.69	1.75	4.42	1.40	5.18
Others	3.12	2.19	5.54	2.13	2.11	1.65	0.31	6.81
Total	50.52	40.42	71.43	17.08	21.72	16.00	73.63	49.59
OTHER HOLOPLANKTONIC TAXA								
Evadne spinifera	0.26	0.18	0.00	1.55	1.93	0.10	0.16	0.27
Limacina sp.	3.65	23.90	8.11	13.82	11.76	14.89	1.25	13.62
Siphonophores	36.46	27.77	14.45	64.82	62.44	65.06	22.46	33.24
Others	1.04	2.09	0.72	0.25	0.60	0.87	0.47	0.27
Total	41.41	53.94	23.27	80.45	76.74	80.92	24.34	47.41
MEROPLANKTON								
Balanoid larvae	3.91	1.55	0.08	0.56	0.51	0.05	0.16	0.27
Echinoderm larvae	1.04	0.77	0.40	0.60	0.36	0.19	0.31	0.27
Fish eggs and larvae	1.04	1.09	0.32	0.07	0.09	0.10	0.00	0.00
Bivalve larvae	2.08	2.00	3.45	1.00	0.30	1.31	1.09	0.54
Others	0.00	0.18	1.04	0.25	0.27	0.83	0.31	1.90
Total	8.07	5.60	5.30	2.47	1.54	2.48	1.87	2.99
Abundance (ind m^{-3})	70	330	389	809	304	184	962	551
Biomass (mgC m^{-3})	0.089	0.366	0.699	0.386	0.186	0.147	1.642	0.949

plume (D) with a peak of about 18 μm. This peak disappeared below the interface where the size of particles increased with depth. At the distal station (C), almost the same pattern was observed, but the 6 μm surface peak was replaced by a maximum between 8 and 23 μm. The surface transect across the frontal limit (stations B, E, F, G, H) showed an increase of the size spectrum from low-salinity to marine waters, similar to the vertical profile from surface to the salinity interface inside the plume.

Zooplankton population

The relative abundance of the main zooplankton taxa and their densities are presented in Table 1. There appeared to be no barrier effect at the front on the distribution of species: all taxa, which were dominated by copepods and siphonophores, were present in the three stations. The total number and biomass of zooplankton were slightly higher outside (D) than inside the plume (B, C).

Table 2. Seston volume (VOL), carbon (C) and chlorophyll (Chl *a*) concentrations measured in *in situ* samples corresponding to the same water as that used in grazing experiments.

Area	Station	Depth m	VOL $\mu l\ l^{-1}$	C $\mu g\ l^{-1}$	Chl *a* $\mu g\ l^{-1}$	C/Chl *a*	C/VOL	Chl *a*/VOL
Plume proximal	B	0	6.90	864	2.67	323.6	125	0.387
	B	2	5.12	644	4.60	140.0	126	0.998
	B	6	2.99	340	0.75	453.3	114	0.251
Plume distal	C	0	4.24	463	2.05	225.9	109	0.484
	C	2	6.37	876	7.64	114.7	137	1.199
	C	7	2.88	226	0.73	309.6	78	0.253
Marine	D	0	3.60	622	3.37	184.6	173	0.936
	D	5	1.52	340	2.96	114.9	90	1.947
	D	30	1.88	270	0.25	1080.0	180	0.133

Zooplankton grazing in relation to particle distribution (Fig. 4)

In marine surface water (station D), there was a wide range of ingested particle sizes. Selectivity (Ivlev' electivity index) was rather low but tended to increase positively with increasing particle size. This tendency was observed in the water layers below the plume (B, C, at 10 m). In plume surface and interface layers, no special pattern for size selectivity was observed, ingestion being more or less dependent on particle size peaks.

Ingestion and clearance rates in relation to food abundance

The total seston volumes and the carbon and chlorophyll concentrations, which represent the food availability to zooplankton, and the corresponding daily food rations and filtration rates are shown in Tables 2 and 3. In the plume, particle volumes were generally more abundant than outside the plume, maximal value being found near the river mouth (B), due to the inflow of detrital material from land runoff. The chlorophyll

Table 3. Grazing experiments on total mesozooplankton: Specific ingestion and clearance rates and grazing pressure estimated on the basis of clearance rates (Col. D), zooplankton biomass (Table 1) and *in situ* chlorophyll (Table 2). Col. A expressed in $10^3\ \mu m^3\ \mu gC^{-1}\ h^{-1}$, Col. B expressed in $\mu gC\ \mu gC^{-1}\ d^{-1}$, Col. C expressed in μgChl *a* $mgC^{-1}\ d^{-1}$, Col. D expressed in ml $\mu gC^{-1}\ d^{-1}$, Col. E expressed in μgChl *a* $m^{-3}\ d^{-1}$, Col. F expressed in % Chl *a* d^{-1}. ND = no data.

Area	Station	Depth (m)	Ingestion rates			Clearance rate	Grazing pressure	
			A	B	C	D	E	F
Plume proximal	B	0	82.0	0.246	0.761	0.331	0.079	0.0029
	B	2	109.1	0.329	2.351	0.417	0.703	0.0153
	B	6	147.8	0.364	0.802	2.038	1.068	0.1424
Plume distal	C	0	122.8	0.307	1.358	1.029	0.814	0.0397
	C	2	48.5	0.160	1.396	0.207	0.294	0.0039
	C	7	140.4	0.264	0.852	2.165	0.232	0.0318
Marine	D	0	29.0	0.120	0.652	0.182	1.006	0.0299
	D	5	ND	ND	ND	ND	ND	ND
	D	30	ND	ND	ND	ND	ND	ND

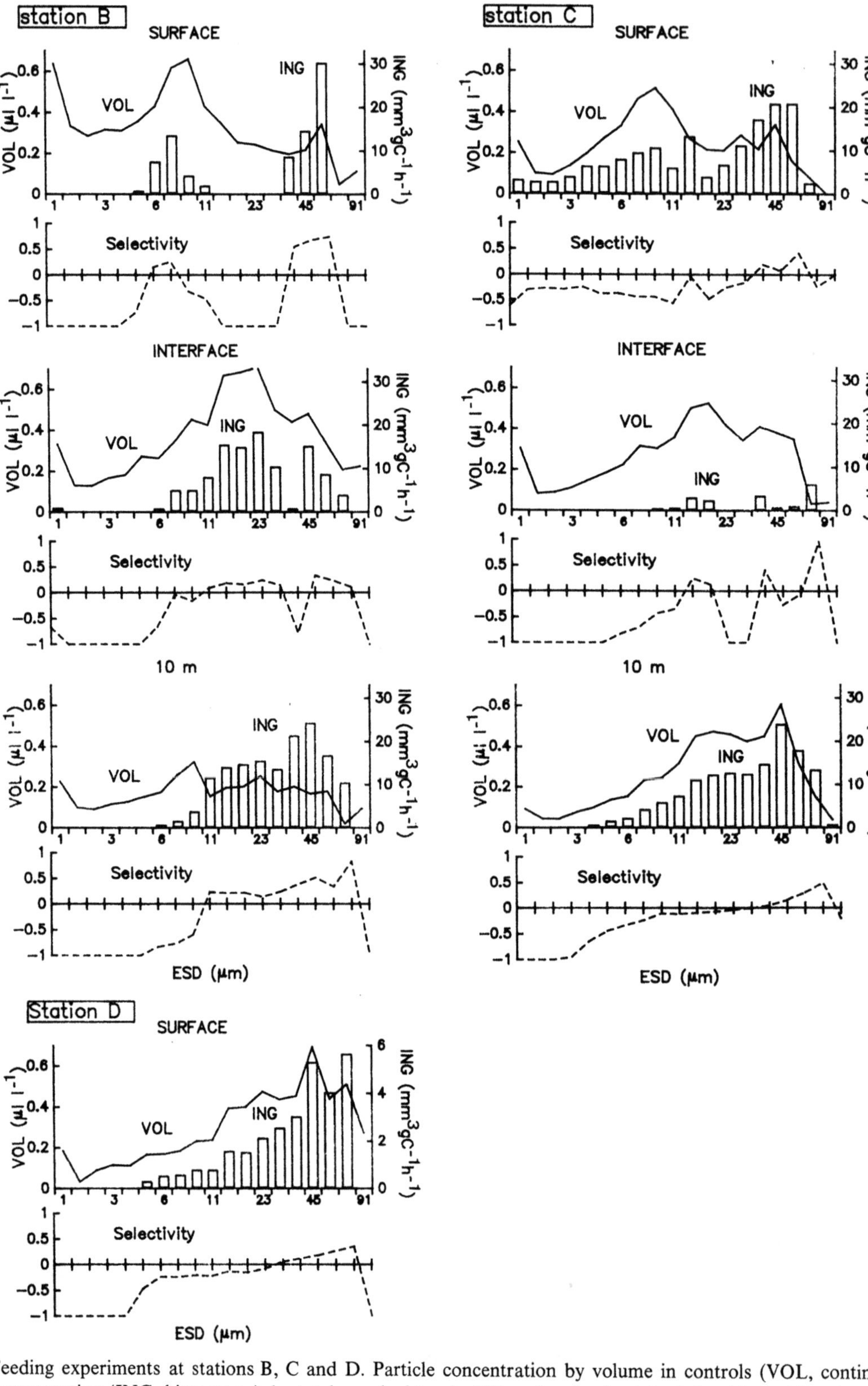

Fig. 4. Feeding experiments at stations B, C and D. Particle concentration by volume in controls (VOL, continuous lines), particle consumption (ING, histograms), by total zooplankton (> 200 μm) and selectivity index (dashed lines), related to the particle-size (equivalent spherical diameter, ESD, μm).

concentration of marine waters under the plume was very low compared with the value found in the marine station (D). In the plume, the chlorophyll values displayed a marked maximum at the interface layer.

The ingestion rates were higher inside and under the plume than in the marine station, particularly if they were expressed on volume basis (Table 3). The clearance rate of zooplankton from marine waters appeared to differ: under the plume (B, C, at 10 m), they were considerably higher than in the marine area (D), while the corresponding daily chlorophyll food intake was similar. The daily grazing pressure of the total zooplankton population upon the chlorophyll stock was very low inside and outside the plume, with a maximum percentage of utilization lower than 0.15% d^{-1}.

Discussion

The analysis of the zooplankton composition shows no marked differences in the faunistic assemblage of the different water masses associated with the plume, contrary to some other river plume areas (*e.g.* the Fraser River, Mackas & Louttit, 1988). The bulk of the species were euryhaline coastal forms such as *Paracalanus* and *Acartia*, which are common inhabitants of temperate neritic assemblages (Gaudy, 1985). Siphonophores (mainly calycophores) constituted the main carnivorous component of the mesozooplankton community. As they are markedly vertically migrating animals, they probably contributed significantly to the export outside the plume of the organic matter synthetized by the secondary production.

Size distribution analyses of sestonic particles have allowed characterization of water masses (Sheldon *et al.*, 1972; Kranck, 1980; Loeillet & Leveau, 1985; Kahru *et al.*, 1986). Our results also led to the identification of water masses of very different origins. The unimodal spectrum of upper layer waters (0–5 m) outside the Rhone front area is characteristic of a blooming phytoplankton development (Kranck, 1980; Tackx *et al.*, 1989), confirmed in this case by microscopic examination (Velasquez & Cruzado, 1989). This kind of spectrum was also found at the plume interface layer. This probably resulted from the transport of the surrounding surface waters by a convergence effect, as shown by water movement studies carried out in the same area (Forget, pers. comm.).

In the plume, the chlorophyll concentration at the interface layer was higher than at the surface. This could be attributed to both the passive accumulation of algal material due to convergence, and the increase in primary production enhanced by the richness in nutrients (Leveau *et al.*, 1990). Thus, the interface layer appeared to be particularly favourable to algal-food intake by filter feeders. Under the plume, while the particle concentration was similar to that found in the marine area, its nutritional values was very poor (high C/Chl *a* ratio; see Table 2). This results from the low primary production (Videau & Leveau, 1990) due to poor light penetration (high turbidity), and from the low level of nutrients under the plume (Leveau *et al.*, 1990).

Several variations observed in the feeding activities of total zooplankton from the different water masses cannot be ascribed to differences in its specific composition (see above). In the plume, the higher food intake, compared with that obtained in marine waters, can be related to differences in seston concentration according to the classic direct relationship between food abundance and food intake under nonsaturated conditions, which is also found in other plume areas (*e.g.* Parson *et al.*, 1969b; Chervin *et al.*, 1981; Turner & Tester, 1989). The chlorophyll intake found in the marine waters under the plume, despite the low available chlorophyll at this level, was comparable with that found in the richer marine area. This resulted from an increase of the filtering activity under the plume, which could be interpreted as a classic compensation mechanism to obtain a sufficient food intake in low trophic conditions. This pattern is illustrated by the strong relationship observed between filtration rates and logarithm of food concentration calculated from the data of Tables 2 and 3: correlation coefficients are respectively -0.93, -0.94 and -0.79 (d.f. = 6) for food concentration expressed as chlorophyll, carbon and volumes.

No particular pattern for food size selection was found inside the plume, despite the large amounts of food available. This is in contradiction with the optimal foraging theory (Pyke *et al.*, 1977; Cowles, 1979) of increasing selectivity with increasing food levels. However, our results are consistent with those of Turner & Tester (1989), who also found no size selectivity for several copepods in the Mississippi River plume. Nevertheless, on the basis of the present results, it is difficult to discuss further the problem of particle selection which appears to depend on a wide variety of mechanisms reviewed by Price (1988).

Expressed as percentage body-carbon, the daily food intake ranged between 12 and 36% (Table 3, col. B), which are in the same order of magnitude as reported by Parson *et al.* (1969b) for different copepod species in the Fraser River plume, and by Tester & Turner (1988) for *Centropages velificatus* (maximal values) and *Acartia tonsa* (average value) in the Mississippi River plume. In most cases, if assimilation rate is considered, food intake could hardly compensate metabolic expenditure (11 to 45% of body carbon per day; unpublished personal data for parallel experiments). Some caution may be necessary in extrapolating the values obtained in bottle experiments to those actually found in the field. Moreover, the electronic counting method probably underestimated the food intake levels (see methods). Nevertheless, our results indicate that in May, the daily ingestion rate was maximum for the mesozooplankton inhabiting the plume environment, but, when considering relative energy acquisition by food, secondary production must have been relatively low in the area studied even in the Rhone River plume. This result seemingly conflicts with other findings concerning front environments, where biomass and production of zooplankton are generally enhanced (Smith *et al.*, 1986). In the case of river plumes, Chervin *et al.* (1981) also showed an increase of secondary production in the Hudson River plume. However, in the Fraser River plume, Parsons *et al.* (1969b) observed that the zooplankton biomass depended on the flood level of the river: when flooding was low, the biomass markedly exceeded that of the marine environment (April), but in May, when flooding was at maximum level, the zooplankton was diffused in the surrounding waters.

The negligible grazing pressure from mesozooplankton observed in May 1988 in the Rhone River plume area seems to be fairly characteristic of most rich coastal areas (Deason, 1980; Stearns *et al.*, 1987).

In conclusion, although no enhancement of zooplankton biomass was observed in the Rhone River plume waters in May 1988, this zone appeared to be the most favourable for zooplankton algal food acquisition and filtering activity. The reasons for this discrepancy could be investigated in future studies, paying special attention to the export of transformed animal organic matter (predation, diffusion by current, vertical migration, etc.).

Acknowledgements

This work was carried out during the PANACHE 1 cruise, as part of the French FRONTAL program (INSU, TOAE).

References

Blanc, F., M. Leveau & K. H. Szekielda, 1969. Effets eutrophiques au débouché d'un grand fleuve (Grand Rhône). Mar. Biol. 3: 233–242.

Cadée, G. C., 1984. Particulate and dissolved organic carbon and chlorophyll *a* in the Zaire River, estuary and plume. Neth. J. Sea Res. 17: 426–440.

Chervin, M. B., T. C. Malone & P. J. Neale, 1981. Interactions between suspended organic matter and copepod grazing in the plume of the Hudson River. Estuar. coast. Shelf Sci. 13: 169–183.

Cowles, T. J., 1979. The feeding response of copepods from the Peru upwelling system: Food size selection. J. mar. Res. 37: 601–622.

Deason, E. E., 1980. Grazing of *Acartia hudsonica* (*A. clausi*) on *Skeletonema costatum* in Narragansett Bay (USA): influence of food concentration and temperature. Mar. Biol. 60: 101–113.

Frontier, S., 1986. Studying fronts as contact ecosystems. In

J. C. J. Nihoul (ed.), Marine interfaces echohydrodynamics. Elsevier Oceanogr. Ser. 42: 55–66.

Gaudy, R., 1985. Features and peculiarities of zooplankton communities from the western Mediterranean. In M. Moraitou Apostolopoulou & V. Kiortsis (eds), Mediterranean Marine Ecosystems. Plenum Press, New York: 279–301.

Harbison, G. R. & V. L. McAlister, 1981. Fact and artifact in copepod feeding experiments. Limnol. Oceanogr. 25: 971–981.

Hargis, J. R., 1977. Comparison of techniques for the measurement of zooplankton filtration rates. Limnol. Oceanogr. 22: 942–945.

Harrison, P. J., J. D. Fulton, F. J. R. Taylor & T. R. Parsons, 1983. Review of the biological oceanography of the Strait of Georgia: Pelagic environment. Can. J. Fish. aquat. Sci. 40: 1064–1094.

Holm-Hansen, O., C. J. Lorenzen, P. W. Holmes & J. D. H. Strickland, 1965. Fluorimetric determination of chlorophyll. J. Cons. int. Explor. Mer. 30: 3–15.

Kahru, M., S. Nommaan, M. Simm & K. Vilbaste, 1986. Plankton distributions and processes in the Baltic boundary zones. In J. C. J. Nihoul (ed.), Marine interfaces echohydrodynamics. Elsevier Oceanogr. Ser. 42: 273–274.

Kiørboe, T., F. Møhlenberg & H. U. Riisgard, 1985. *In situ* feeding rates of planktonic copepods: a comparison of four methods. J. exp. mar. Biol. Ecol. 88: 67–81.

Kirchman, D., Y. Soto, F. Van Wambeke & M. Bianchi, 1989. Bacterial production in the Rhone River plume: effect of mixing on relationships among microbial assemblages. Mar. Ecol. Prog. Ser. 53: 267–275.

Kranck, K., 1980. Variability of particulate matters in a small coastal inlet. Can. J. Fish. aquat. Sci. 37: 1209–1215.

Le Fèvre, J., 1986. Aspects of the biology of frontal systems. Adv. mar. Biol. 23: 163–299.

Leveau, M., F. Lochet, M. Goutx & F. Blanc, 1990. Effects of a plume front on the distribution of inorganic and organic matter off the Rhone River. Hydrobiologia, this volume.

Loeillet, C. & M. Leveau, 1985. Influence des apports rhodaniens sur la structure du système particulaire dans le Golfe du Lion. C. r. Acad. Sci., Paris 301, Ser. II, 6: 397–402.

Mackas, D. L. & G. C. Louttit, 1988. Aggregation of the copepod *Neocalanus plumchrus* at the margin of the Fraser River plume in the Strait of Georgia. Bull. mar. Sci. 43: 810–824.

Margalef, R., M. Estrada & D. Blasco, 1979. Functional morphology of organisms involved in red tides, as adapted to decaying turbulence. In D. L. Taylor & H. H. Seliger (eds), Toxic dinoflagellates blooms. Elsevier/North Holland, Amsterdam: 89–94.

Parsons, T. R., K. Stephens & R. J. Lebrasseur, 1969a. Productions studies in the strait of Georgia. I. Primary production under the Fraser River plume, February to May, 1967. J. exp. mar. Biol. Ecol. 3: 27–38.

Parsons, T. R., R. J. Lebrasseur, J. D. Fulton & O. D. Kennedy, 1969b. Productions studies in the Strait of Georgia. II. Secondary production under the Fraser River plume, February to May, 1967. J. exp. mar. Biol. Ecol. 3: 39–50.

Price, H. J., 1988. Feeding mechanisms in marine and freshwater zooplankton. Bull. mar. Sci. 43: 327–343.

Pyke, G. H., H. R. Pulliam & E. L. Charnov, 1977. Optimal foraging: a selective review of theory and tests. Quart. Rev. Biol. 52: 137–154.

Sheldon, R. W., A. Prakash & W. H. Sutcliffe Jr., 1972. The size distribution of particles in the Ocean. Limnol. Oceanogr. 17: 327–340.

Skreslet, S., 1986. Effects of freshwater inflow on the distribution, composition and production of plankton in the Dutch coastal waters of the North Sea. In S. Skreslet (ed.) The role of freshwater outflow in coastal marine ecosystems. NATO Adv. Stud. Inst. Ser. G, Ecol. Sci. 7: 241–250.

Smith, S. L., B. H. Jones, L. P. Atkinson & K. H. Brink, 1986. Zooplankton in the upwelling fronts off Pt. Conception, California. In J. C. J. Nihoul (ed.), Marine interfaces ecohydrodynamics. Elsevier Oceanogr. Ser. 42: 195–214.

Stearns, D. E., W. Litaker & G. Rosenberg, 1987. Impacts of zooplankton grazing and excretion on short-interval fluctuations in chlorophyll *a* and nitrogen concentrations in a well-mixed estuary. Estuar. coast. Shelf Sci. 24: 305–325.

Stepien, J. C., T. C. Malone & M. B. Chervin, 1981. Copepod communities in the estuary and coastal plume of the Hudson River. Estuar. coast. Shelf. Sci. 13: 185–195.

Tackx, M. L. M. & P. Polk, 1986. Effect of incubation time and concentration of animals in grazing experiments using a narrow size range of particles. Syllogeus 58: 604–609.

Tackx, M. L. M., C. Bakker, J. W. Francke & M. Vink, 1989. Size and phytoplankton selection by Oosterschelde zooplankton. Neth. J. Sea Res. 23: 35–43.

Tester, P. A. & J. T. Turner, 1988. Comparative carbon-specific ingestion rates of phytoplankton by *Acartia tonsa*, *Centropages velificatus* and *Eucalanus pileatus* grazing on natural phytoplankton assemblages in the plume of the Mississippi River (Northern Gulf of Mexico continental shelf). In G. A. Boxshall & H. K. Schminke (eds), The biology of copepods, Hydrobiologia/Dev. Hydrobiol. 167/168: 211–217.

Turner, J. T. & P. A. Tester, 1989. Zooplankton feeding ecology: non selective grazing by the copepods *Acartia tonsa* Dana, *Centropages velificatus* De Oliveira and *Eucalanus pileatus* Giesbrecht in the plume of the Mississippi River. J. exp. mar. Biol. Ecol. 126: 21–43.

Velasquez, Z. R. & A. Cruzado, 1989. Project FRONTAL Cruise PANACHE I (May 1988). Results of nutrients and phytoplankton analysis II. Phytoplankton. Informes Techn. CEAB Ser I. 23b: 20 pp.

Videau, C. & M. Leveau, 1990. Biomasse et production phytoplanctoniques dans le front du panache rhodanien. Situation printanière. C. r. Acad. Sci., Paris, in press.

Received 31 July 1989; in revised form 1 February 1990; accepted 4 April 1990

Hydrobiologia **207**: 251–258, 1990.
D. J. Bonin & H. L. Golterman (eds), Fluxes Between Trophic Levels and Through the Water–Sediment Interface.

The effects of bioturbation on particle redistribution in Mediterranean coastal sediment. Preliminary results

Magali Gerino
Centre d'Océanologie de Marseille, U.R.A. 41, Station Marine d'Endoume, Rue de la Batterie des Lions, 13007 Marseille, France

Key words: Bioturbation, Flux, Sediment, Macrobenthos

Abstract

In order to quantify bioturbation processes in a coastal Mediterranean ecosystem, experiments were performed to determine sediment mixing rates resulting from macrobenthos activity. Particle flux was measured *in situ* for 22 days using luminophores, which are colored sediment particles with sizes ranging from 10 to 200 μm.

In sediment depths from 0–5 cm, particle mixing was intensive due to high macrobenthos abundance. A small quantity of luminophores was transported down to a depth of 14 $\pm$ 2 cm, where the macrofauna was represented principally by Polychetes. In a control experimental structure – without benthic fauna – no transfer of luminophores into the sediment was recorded.

Sediment particle mixing measured in the ecosystem studied is intensive, and is the result of high macrobenthos activity. Different mixing modes occur with scales and rates depending on the organisms present. The luminophore profile resulting from bioturbation processes is explained by an intensive bioadvection sediment mixing added to a biodiffusive mixing with an order of magnitude of 10^{-6} cm^2 s^{-1}. Tracer accumulations between 1 and 2 $\pm$ 1 cm and between 4 and 5 $\pm$ 1 cm are attributed to bioadvection activity of two or more distinct populations. Studies over a larger time scale have been undertaken to monitor developments in the observed subsurface maxima.

Introduction

Bioturbation is a process of sediment mixing that results from macrofauna burrowing, feeding and reworking at the water-sediment interface. The macrofauna's ability to mix sediment alters the structure and distribution of the primary sediment. Moreover, knowledge of particle movement is essential for understanding the movement of fecal pellets and organic matter across the sediment-water interface.

In a particular sedimentary deposit each species present induces a particular particle transport mechanism depending on its feeding mode and behavior. In the literature, different modes of bioturbation have been described. Random burrowing by abundant organisms leads to an exchange between adjacent parcels of sediment at all depths within the mixing zone (Benninger *et al.*, 1979). These random sediment particle fluxes are usually modelled by analogy with diffusive processes (Goldberg & Koide, 1962; Guinasso & Schink, 1975; De Master & Cochran, 1982). Other organisms such as surface deposit feeders or head-down deposit feeders, that ingest sediment at depth and excrete it at the surface, may create intensive vertical particle fluxes in the sediment (Robbins *et al.*, 1979; Fisher *et al.*, 1980; Rice, 1986; Bianchi, 1988). These organisms (mainly Polychetes) are called 'conveyor-belt feeders' (Rhoads, 1974) and generate bioadvective processes in sediment.

Bioturbation studies require the use of sediment particle tracers to mark particle migrations under macrofauna action. Interpretation of vertical distributions of natural or artificial tracers such as radionuclides or microtecktites has been applied widely to determine biological mixing rates in sample cores (Turekian *et al.*, 1978; Benninger *et al.*, 1979; De Master & Cochran, 1982; Imboden & Stiller, 1982; Mauviel *et al.*, 1982; Aller & Demaster, 1984; Cochran, 1985; De Master *et al.*, 1985; Kershaw, 1985; Yokoyama *et al.*, 1985; Rice, 1986). Because of the lack of complementary biological analysis in most research, it is difficult to come to any definite conclusions on the relationships between organisms and sediment. Some laboratory experiments have accordingly been carried out to measure sediment reworking by monospecific or natural populations (Robbins *et al.*, 1979; Guinasso & Schink, 1975; Aller & Cochran, 1976; Rice, 1986; Mahaut & Graf, 1987; Bianchi, 1988). This research must be extended to *in situ* experiments to assess the bioturbation activity of whole benthic assemblages. In order to estimate the time scale of bioturbation processes in a coastal environment, and their effects on particle flux, a series of experiments was conducted over a period of three months. In this paper, we present the preliminary results of the first group of samplings.

Experimental design

The experimental site

The study site was in the Gulf of Fos (Carteau area), located 12 miles east of Marseilles near the mouth of the river Rhone (Fig. 1). The station at 5 m depth presented a muddy sand sediment occupied by a characteristic 'muddy sand of sheltered areas' assemblage (Pérès & Picard, 1964; Stora, 1982).

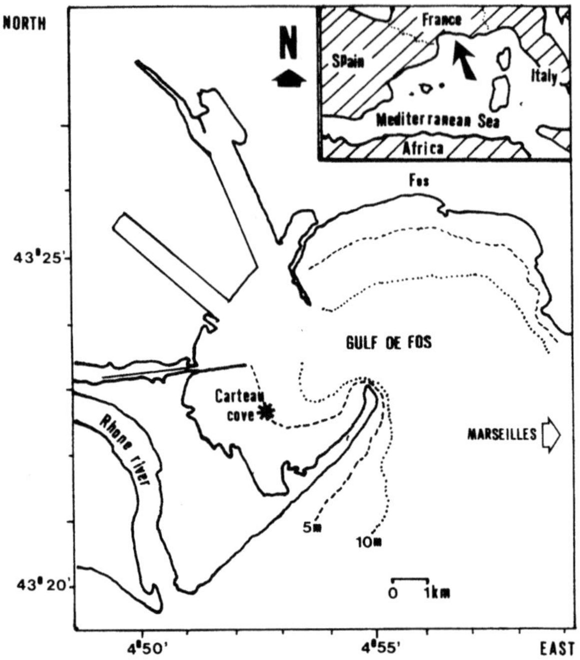

Fig. 1. The gulf of Fos with the experimental station.

Materials and methods

The method for measuring particles sediment flux is based on the utilisation of tracers, the luminophores (Mahaut & Graf, 1987), which are colored sediment particles with diameters ranging from 10 to 200 microns. The luminophores become luminous under microscope U.V. light and can be easily identified in sediment samples in this condition.

Three open cores (diameter = 11 cm and length = 30 cm) were driven into the sediment to a depth of 25 cm to isolate three portions of biotopes with the *in situ* fauna. To characterize the action of macrobenthos on the bioturbation process, one control core with defaunated sediment was installed under the same experimental conditions as the others. Defaunation was ob-

Table 1. Number of individuals per sediment volume (mean ± CL) in the three open cores as a function of the sediment depth.

Number of individuals per cm³						
Depth	Polychetes	Bivalvia	Crustacea	Gastropoda	Sipunculidea	Total
0–1	4.76E − 02	5.29E − 02	4.92E − 01	2.01E − 01	0	7.94E − 01 ± 1.70E + 00
1–2	3.17E − 02	5.82E − 02	1.11E − 01	1.59E − 02	0	2.17E − 01 ± 7.12E − 01
2–3	7.94E − 02	1.06E − 02	0	0	0	8.99E − 02 ± 1.27E − 01
3–4	1.16E − 01	0	0	0	5.29E − 03	1.22E − 01 ± 2.58E − 01
4–5	1.59E − 01	1.06E − 02	0	0	0	1.69E − 01 ± 1.88E − 01
5–6	1.22E − 01	0	0	0	0	1.22E − 01 ± 1.86E − 01
6–7	1.80E − 01	0	0	0	0	1.80E − 01 ± 1.59E − 01
7–8	1.32E − 01	0	0	0	0	1.32E − 01 ± 2.17E − 01
8–9	1.27E − 01	0	0	0	0	1.27E − 01 ± 1.81E − 01
9–10	6.88E − 02	0	0	0	0	6.88E − 02 ± 1.27E − 01
10–12	2.65E − 02	0	0	0	0	2.65E − 02 ± 8.20E − 02
12–14	5.29E − 03	0	0	0	0	5.29E − 03 ± 2.28E − 02
14–16	5.29E − 03	0	0	0	0	5.29E − 03 ± 2.28E − 02
16–18	0	0	0	0	0	0
Total	1.10E + 00 ± 1.60E + 00	1.32E − 01 ± 2.94E − 01	6.03E − 01 ± 1.60E + 00	2.17E − 01 ± 4.64E − 01	5.29E − 03 ± 2.01E − 02	2.06E + 00 ± 3.98E + 00

tained by freezing the sediment previously collected from the experimental site. To prevent macrofaunal settlement during the experiment, the control defaunated sediment was contaminated with a solution of tetraethyl lead at a concentration of 0.5 g kg^{-1}. At the beginning of the experiment, in December 1988, divers introduced luminophores in the 4 open cores. In each core, a little box containing 2 g of luminophores was carefully opened in the over-lying water. Cores were immediately closed to let the bulk of luminophores deposit; after 2 hours all the cores were opened. The 4 cores were collected 22 days later. At the laboratory, the cores were first frozen, then cut into a series of horizontal sections at 1 centimeter intervals. For each section, the macrofauna were extracted from sediment using a sieve with a 0.5 mm diagonal mesh and each species was counted. The remaining sediment fractions were dried at 70 °C for 48 hours and carefully mixed to homogenize sediment and luminophores. The luminophores were identified and counted in 2 subsamples of 0.25 g for each sediment fraction of the 4 cores.

Results

Values of macrobenthos abundance (mean ± CL) present at different depths in the three normal cores are given in Table 1. (The confidence limits,

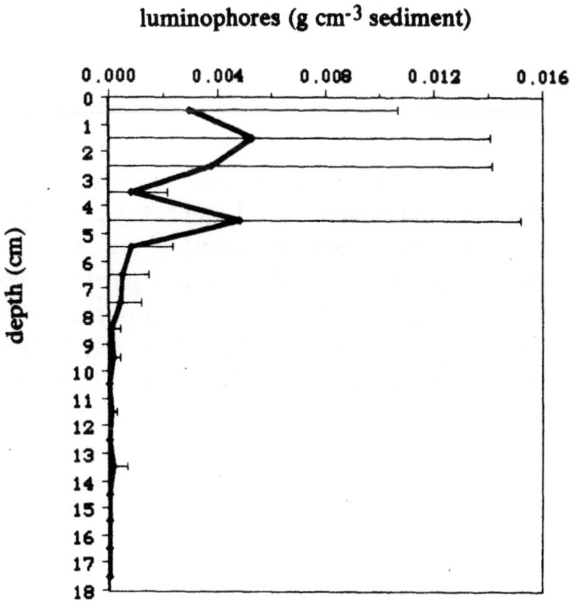

Fig. 2. Luminophore (mean ± CL) partition in the three open cores with macroinfauna.

CL, of means are calculated using the statistical t test and are equal to $t_{0.05}$ SE with n = 3, and SE the standard error.) In the control core, macrobenthos was poorly represented by Gastropoda (number of individuals cm^{-3} = 4.7), and only in the top (0–1 cm) of the sediment; however no infaunal populations were encountered. In the three other cores with macrobenthos, most of the classes, Bivalvia, Gastropoda, Crustacea, Echinodermata and Sipunculidea were present only in the upper 5 ± 1 cm of sediment and could be herbivores, deposit feeders or suspension feeders. Polychetes dominated the community and are essentially deposit feeders. They penetrated down to 16 ± 2 cm in the sediment with most occurring between 6 and 7 cm. Total macroinfauna abundance decreased with sediment depth.

Vertical distribution of luminophores in the 3 cores with macrobenthos, after 22 days, is shown in Fig. 2. Luminophore data (mean ± CL) are calculated from the 3 cores and expressed in luminophore weight (g) per volume of sediment (cm^3). In the control core, all the luminophores were still near the surface of sediment at the end of the experiment. The total quantity of tracer in all cores at the end of the experiment represented 80% of the initial quantity incorporated in the cores. The loss of luminophores is attributed to under-water manipulation or to resuspension in the course of the experiment. In the three cores with macrobenthos, after 22 days, luminophores were present in the sediment to a depth of 14 ± 2 cm. The concentration of luminophores has maxima between 1 ± 1 and 5 ± 1 cm and decreased with sediment depth from 5 ± 1 to 14 ± 2 cm. Because all the tracer particles were placed at the interface at the start of the experiment, it was possible to determine particle fluxes for the duration of the experiment. Fluxes were calculated at each level, based on the weight of luminophores present both at a given level and in the layers below. Sediment fluxes (mean ± Cl) from the three cores with macroinfauna, are expressed in g sed cm^{-2} for 22 days. These fluxes represent the quantities of sediment which has passed through a particular depth layer in the sediment (Fig. 3). Vertical particle sediment fluxes exhibit high intensity in the first 5 ± 1 cm of sediment, and decrease rapidly in the sediment below.

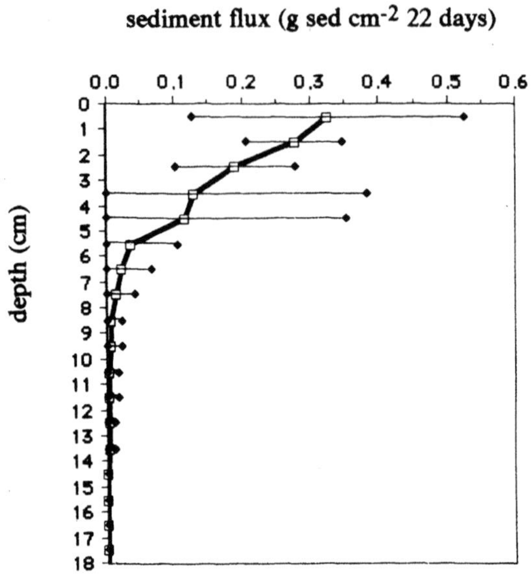

Fig. 3. Sediment particle fluxes (mean ± CL) as a function of the sediment depth.

Discussion

In the control core without macrobenthos, tracer movements in the sediment were controlled by hydrodynamic processes. The non-penetration of luminophores in the absence of macroinfauna indicates that tracer migration measured in normal cores resulted from macrobenthic activity. Macroinfauna activity in the sediment generates sediment particle fluxes. Luminophores were transported *via* bioturbation to a depth of 14 ± 2 cm, which corresponds to the penetration of macrobenthos in the cores. Tracer quantities are maximal between 0 and 5 ± 1 cm and are associated with high macrofaunal abundance. Furthermore, luminophore accumulations between 1 and 2 ± 1 cm and between 4 and 5 ± 1 cm indicate fluxes from the interface to these preferential depths, despite a small proportion of luminophores transported down to the 5 to 14 ± 2 cm layer. These fluxes can be attributed to the activity of Polychetes which are the dominant macrofauna at these depths.

In the literature, sediment particle fluxes due to bioturbation are usually attributted to biodiffusive mixing and bioadvection (Guinasso & Schink, 1975; Benninger *et al.*, 1979; Matissof, 1982; Rice, 1986; Sharma *et al.*, 1987; Boudreau, 1986).

The random redistribution of solid particles in the sediment is described mathematically by a biodiffusive mixing model introduced by Goldberg & Koide (1962). This model has been applied widely to radionuclide profiles in deep sediments (Guinasso & Schink, 1975; Turekian *et al.*, 1978; De Master & Cochran, 1982; Imboden & Stiller, 1982; Mauviel *et al.*, 1982; Aller & De Master, 1984; Cochran, 1985; De Master *et al.*, 1985; Kershaw, 1985; Yokoyama *et al.*, 1985) and in shallow sediments (Benninger *et al.*, 1979). It describes multidirectional particle fluxes resulting from suspension feeder and deposit feeder burrowing activity in sediments. In the differential equation of this model, the particle biodiffusion coefficient DB is treated as an eddy diffusion coefficient and represents the rate of biodiffusive mixing. The fact that luminophores remain at the top of the sediment in the control core indicates that sediment accumulation can be considered as negligible over the experimental period, and the non-transient character of the tracer particles makes it possible to use Cochran's reduced equation (1985) of Fick's second law of diffusion:

$$\frac{\partial A}{\partial t} = DB \frac{\partial^2 A}{\partial z^2} \qquad (1)$$

where

A = concentration of luminophores ($g\ cm^{-3}$)
DB = particle mixing coefficient ($cm^2\ y^{-1}$)
z = depth in the sediment column (cm)

DB is constant on the mixing zone from 0 to 14 cm.

The boundary conditions for this equation are defined as follows:

$$A = A_0 \text{ at } z = 0$$
$$A \to 0 \text{ as } z \to \infty$$

The solution of the differential equation (1) is, with A_0 the tracer concentration at $z = 0$ (Cochran, 1985):

$$A = A_0 \exp\left(-z^2/4DBt\right) \qquad (2)$$

Determination of the *DB* value is achieved by adjustment of theoretical distributions from equation (2) with different values of *DB* to the measured tracer profile. The statistical method to fit the model consists in the least squares procedure that minimizes the sum of the squares of the difference between measured and calculated values of tracer concentrations (Lynch & Officer, 1984). The biodiffusive mixing rate is estimated from the luminophore profile to be $6.97\ 10^{-6}\ cm^2\ s^{-1}$ in a mixed layer of 14 + 2 cm. Figure 4 represents the theoretical distribution of tracer calculated with this biodiffusive coefficient, and the measured luminophore profile. Matissof (1982) reported different *DB* values collected in the literature from radionucleide profiles in littoral bioturbated sediments, and estimated a biodiffusive mixing coefficient in shallow water with an order of magnitude of $\approx 10^{-6}\ cm^2\ s^{-1}$. The present *DB* value is considered as consistent with

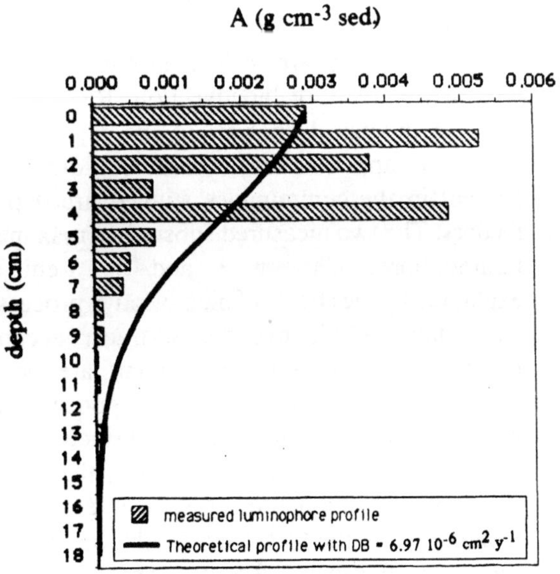

Fig. 4. Theoretical partition of tracer calculated with DB = $6.97\ 10^{-6}\ cm^2\ s^{-1}$, and the measured luminophore profile.

those in the literature. However, it appears that theoretical distribution does not fit the observed luminophore profile closely, in particular with respect to the 2 subsurface maxima. Although random mixing occurs generally in inhabited sediment, the different modes of bioturbation are not necessarily perfectly random. Research in marine (Aller, 1982; Rice, 1986; Garner *et al.*, 1987: Sharma *et al.*, 1987; Smith *et al.*, 1987; Bianchi, 1988) and lake environments (Robbins *et al.*, 1979; Fisher *et al.*, 1980) has shown the action of organisms such as Polychetes, Oligochaetes and Crustacea on non-random bioadvective processes. The processes entail intensive vertical particle fluxes in the sediment. The head-down deposit feeders that ingest the sediment at depth and excrete it at the surface are called 'conveyor belt feeders' by Rhoads (1974), and constitute a major component of bioadvective sediment mixing. Inversely, surface deposit feeders such as Sipunculidea excrete material at depth in the sediment (Smith *et al.*, 1987). Some other organisms, such as Crustacea, may influence particle redistribution by digging large burrows which are subsequently backfilled with surface sediment (Gardner *et al.*, 1987). It is clear that this type of mixing and 'conveyor-belt' phenomena create unidirectional particle flux in the sediment on a larger scale than the diffusive process. These types of non-diffusive biogenic mixing are called non-local mixing (Boudreau, 1986) or bioadvection (Rice, 1986).

From biological analysis of the experimental cores, it appears that numerous organisms present in the communities studied are deposit feeders. The two measured subsurface maxima of luminophores between 1 and 5 ± 1 cm are explained by the effects of bioadvective processes. According to this hypothesis, tracer accumulations, at the end of the experiment, are located in the egestion zone of surface deposit feeder populations or result from subsurface deposit feeding modes. In the former case, the feeding zone of the population responsible must be located below the tracer accumulation. Most of the conveyor-belt feeders are Polychetes (Rhoads, 1974). The dominant species of Polychetes in the cores are represented by *Mediomastus californiensis* (Capitellidae), and *Paraonidae lyra* (Paraonidae) and are subsurface deposit feeders and surface deposit feeders (Fauchald & Jumard, 1979). The possible involvement of crustacean burrowers in the formation of the upper tracer accumulation cannot be discounted, as they occur at the corresponding depths. The great diversity of macrofauna in the first 5 ± 1 cm of sediment in the cores renders it difficult to identify the organisms responsible for these events. In any case, the presence of two distinct luminophore maxima indicates the contribution of two different populations in the bioadvective processes. The relatively large distance covered in short periods of time by the recorded flux, and the quantity of sediment particles moving *via* bioadvection revealed the importance of these processes in the sediment studied.

In order to elaborate an accurate model of the observed sediment mixing, we must use a bioadvective model equation including a bioadvective term to account for non-random mixing in addition to the biodiffusive term. Models using transport analogies other than biodiffusion have been proposed in the literature, but these models are neither numerous nor widely applied (Fisher *et al.*, 1980; Rice, 1986; Gardner *et al.*, 1987; Smith *et al.*, 1987). Moreover, these models have been designed to describe the contribution of single conveyor-belt or crustacean populations in laboratory or *in situ* conditions. In this paper, tracer profile results from bioturbation activity of a whole benthic assemblage and two or more distinct populations contributed to bioadvective particle transports. The different bioadvective models proposed in the literature are limited to single population contribution and must be improved before being applied to whole benthic assemblage with more than one distinct process of non-diffusive sediment transport.

This research needs to be extended to provide further information on sediment transport in shallow benthic assemblage. This information might be used to produce a more realistic model to include the diverse bioadvective activities of a benthic assemblage.

Conclusion

Sediment particle fluxes measured in a littoral Mediterranean ecosystem are intensive, and are the result of a high rate of sediment mixing by macrobenthos activity.

Different mixing modes occur with scales and rates depending on the organisms present. In the environment studied, bioturbation processes result from intensive bioadvection sediment mixing added to a biodiffusive mixing with an order of magnitude of $10^{-6}\ cm^2\ s^{-1}$. Tracer accumulations between 1 and 5 $\pm$ 1 cm are attributed to bioadvection activity of two or more distinct populations. Studies over a larger time scale have been undertaken to monitor vertical fluctuations of the subsurface maxima observed.

Further research needs to be carried out to identify the organisms responsible for the bioturbation process. The intensity of the measured particle flux indicates that bioturbation generates major changes in the initial structure and composition of the sediment-water interface. There is a rapid turn-over of the superficial sediment which increases the potential for exchanges of matter through this interface.

References

Aller, R. C., 1982. The effects of macrobenthos on chemical properties of marine sediment and overlying water. In: P. L. Mc Call & M. J. S. Tevesz (eds), Animal-sediment relations. Plenum Press; New York: 53–102.

Aller, R. C. & J. K. Cochran, 1976. $^{234}Th/^{238}U$ disequilibrium in nearshore sediment: particle reworking and diagenetic time scales. Earth Planet. Sci. Lett. 29: 37–50.

Aller, R. C. & D. J. De Master, 1984. Estimates of particle flux and reworking at the deep-sea floor using $^{234}Th/^{238}U$ disequilibrium. Earth Planet. Sci. Lett. 63: 308–318.

Benninger, L. K., R. C. Aller, J. K. Cochran & K. K. Turekian, 1979. Effects of biological sediment mixing on the ^{210}Pb chronology and trace metal distribution in a Long Island Sound sediment core. Earth Planet. Sci. Lett. 43: 241–259.

Bianchi, T. S., 1988. Feeding ecology of subsurface deposit-feeder *Leitoscoloplos fragilis* Verrill. I. Mechanisms affecting particle availability on intertidal sandflat. J. exp. mar. Biol. Ecol. 115: 79–97.

Boudreau, B. P., 1986. Mathematics of tracer mixing in sediments: I. Spatially-dependent, diffusive mixing. II. Nonlocal mixing and biological conveyor-belt phenomena. Am. J. Sci. 286: 199–238.

Cochran, J. K., 1985. Particle mixing rates in sediments of the eastern equatorial Pacific. Evidence from ^{210}Pb, $^{239,240}Pu$, and ^{137}Cs distribution at MANOP sites. Geochim. Cosmochim. Acta 49: 1195–1210.

De Master, D. J. & J. K. Cochran, 1982. Particle mixing rates in deep-sea sediments determined from ^{210}Pb and ^{32}Si profiles. Earth Planet. Sci. Lett. 61: 257–271.

De Master, D. J., B. A. Mc Kee, C. A. Nittrouer, D. C. Brewster & P. E. Biscaye, 1985. Rates of sediment reworking at the HEBBLE site based on measurements of ^{234}Th, ^{137}Cs and ^{210}Pb. Mar. Geol. 66: 133–148.

Fauchald, K. & P. A. Jumars, 1979. The diet of worms: a study of polychaete feeding guilds. Oceanogr. Mar. Biol. Ann. Rev. 17: 193–283.

Fisher, J. B., P. L. Mc Call & J. A. Robbins, 1980. Vertical mixing of lake sediments by tubificid oligochaetes. J. geophys. Res. 85: 3997–4006.

Gardner, L. R., P. Sharma & W. S. Moore, 1987. A regeneration model for the effect of bioturbation by fiddler crabs on ^{210}Pb profiles in salt marsh sediments. J. envir. Radioactivity 5: 25–36.

Goldberg, E. D. & M. Koide, 1962. Geochronological studies of deep-sea sediments by the ionium-thorium method. Geochim. Cosmochim. Acta 26: 417–450.

Guinasso, N. L. & D. R. Schink, 1975. Quantitative estimates of biological mixing rates in abyssal sediments. J. geophys. Res. 80: 3032–3043.

Imboden, D. M. & M. Stiller, 1982. The influence of radon diffusion on the ^{210}Pb distribution in sediments. J. geophys. Res. 87: 557–565.

Kershaw, P. J., 1985. ^{14}C and ^{210}Pb in NE Atlantic sediments: Evidence of biological reworking in the context of radioactive waste disposal. J. envir. Radioactivity 2: 115–134.

Lynch, D. R. & C. B. Officer, 1984. Nonlinear parameter estimation for sediment cores. Mar. Geol. 44: 203–225.

Mahaut, M. L. & G. Graf, 1987. A luminophore tracer technique for bioturbation studies. Oceanol. Acta. 10: 323–328.

Matissof, G., 1982. Mathematical models of bioturbation. In: P. L. Mc Call & M. J. S. Tevesz (eds), Animal-sediment relations. Plenum Press, New York: 289–330.

Mauviel, A., H. V. Nguyen, R. Chesselet, M. Sibuet, Y. Yokoyama & G. Auffret, 1982. Etude des variations des taux de bioturbations par la spectrométrie gamma non destructive (GeHP), dans trois zones sédimentaires de l'Atlantique Nord, à 2000 m et à plus de 4000 m de profondeur. Bull. Inst. Géol. d'Aquitaine, Bordeaux. 31: 257–275.

Pérès, J. M. & J. Picard, 1964. Nouveau manuel de bionomie benthique de la mer Méditerranée. Rec. Trav. Stn mar. Endoume 81 (47): 137 p.

Rhoads, D. C., 1974. Organism-sediment relations on the muddy sea floor. Oceanogr. Mar. Biol. Ann. Rev. 12: 263–300.

Rice, D. L., 1986. Early diagenesis in bioadvective sediments: Relationships between the diagenesis of beryllium-7, sediment reworking rates, and the abundance of conveyor-belt deposit-feeders. J. mar. Res. 44: 149–184.

Robbin, J. A., P. L. Mc Call, J. B. Fisher & J. R. Krezoski, 1979. Effect of deposit feeders on migration of ^{137}Cs in lake sediments. Earth Planet. Sci. Lett. 42: 277–287.

Sharma, P., L. R. Gardner, W. S. Moore & M. S. Bollinger, 1987. Sedimentation and bioturbation in a salt marsh as revealed by ^{210}Pb, ^{137}Cs, and ^{7}Be Studies. Limnol. Oceanogr. 32: 313–326.

Smith, J. N., B. P. Boudreau & V. Noshkin, 1987. Plutonium and ^{210}Pb distributions in northeast Atlantic sediments: subsurface anomalies caused by non-local mixing. Earth Planet. Sci. Lett. 81: 15–28.

Stora, G., 1982. Recherches de bionomie descriptive et experimentale dans quelques biotopes littoraux soumis à des variations naturelles ou artificielles des conditions du milieu. Thèse Doct. ès Sci. Univ. Aix-Marseille II, 430 pp.

Turekian, K. K., J. K. Cochran & D. J. De Masters, 1978. Bioturbation in deep sea deposits. Rates and consequences. Oceanus 21: 34–41.

Yokoyama, Y., H. V. Nguyen, C. E. Lambert & R. Chesselet, 1985. Etude de la bioturbation dans les sédiments superficiels de la Mediterranée occidentale par le traceur ^{210}Pb. Oceanol. Acta 8: 285–291.

Received 23 August 1989; in revised form 5 March 1990; accepted 15 June 1990

Hydrobiologia **207**: 259–267, 1990.
D. J. Bonin & H. L. Golterman (eds), Fluxes Between Trophic Levels and Through the Water–Sediment Interface.

The effects of selective fish predation on the horizontal distribution of pelagic Cladocera in a southern French reservoir

Didier Pont* & Jamila Amrani
*Laboratoire d'Hydrobiologie, Université de Provence, Marseille, France; *Present address: Université Claude Bernard (Lyon), CNRS URA 367, Laboratoire d'Ecologie des Systèmes fluviaux, 1 Rue Parmentier, 13200 Arles, France*

Key words: Cladocera, Reservoir, Spatial Distribution, Fish predation, *Alburnus alburnus*

Abstract

A four-year study showed a clear seasonal succession of species within the cladoceran community of the large, oligotrophic Sainte-Croix reservoir (S.E. France). *Diaphanosoma brachyurum* and *Ceriodaphnia pulchella* were strictly limited to the warm stratified period (July to October), whereas *Bosmina longirostris* and *Bosmina coregoni* were dominant during spring and autumn. *Daphnia longispina* was the only species to occur throughout the year with higher densities in spring.

In spring and late autumn, the discharge of the inflowing river Verdon was high and the abundance of all species showed a gradient over the whole lake with lower densities close to the inflow. During the stratified period, water inflow was very low and species showed different patterns. Densities of the small form *Ceriodaphnia pulchella* were similar all along the long axis of the lake, whereas *Daphnia* densities were significantly higher near the dam. The distribution pattern of *Diaphanosoma*, an intermediate-sized species, showed similar trends to that of *Daphnia*. The only planktivorous fish in the pelagic zone, the bleak (*Alburnus alburnus*), fed mostly on large-bodied species (> 1.0 mm) and was more abundant close to the inflow current. A comparison between the length frequency distributions of cladocera upstream and downstream provided a clear demonstration of the effects of size-selective predation on prey populations. Finally, the interactions between spatial heterogeneity and long-term development of the zooplankton community and the indirect effects of predation are discussed.

Introduction

The spatial distribution of aquatic organisms and its practical impacts on population sampling is well documented (Cochran, 1977; Frontier, 1983). Its importance for ecological theories, such as the regulation of community diversity or population stability have also been widely studied (Steele, 1974; Wiens, 1976).

Most investigations on distribution patterns of zooplankton populations have been focused both on the spatial scales of maximum heterogeneity and on the factors involved (Wiebe, 1970; Steele, 1978; Malone & Mac Queen, 1983; Butorina, 1986; Lenz *et al.*, 1986; Pont, 1986; Pinel-Alloul *et al.*, 1988).

In the pelagic zone of lakes, horizontal distribution patterns of zooplankton populations had been attributed to a wide variety of physical and biotic factors, including wind, physical turbulence and water discharge (Georges & Edwards, 1976; Patalas, 1981), thermal structure of the system

(Patalas, 1969), non-homogeneity in food distribution (Dagg, 1977; De Nie *et al.*, 1980), behavior and predation (Siebeck, 1980; Jakobsen & Johnsen, 1987).

The horizontal distribution of plankton populations in reservoirs is strongly influenced by the discharge of inflowing rivers (Balvay, 1985). Densities of primary and secondary producers may be higher in downstream areas because of turbidity and low transparency of water in the upstream area (Gannon & Stemberger, 1981; Armengol, 1984).

Predation is recognised as an important factor in regulation of the composition and vertical distribution of zooplankton (Brooks & Dodson, 1965; Nilsson & Pejler, 1973; Zaret & Suffern, 1976), but its influence on the horizontal distribution of zooplankton in reservoirs has not previously been clearly demonstrated.

The aim of this study was i) to describe the spatial pattern of zooplankton distribution in the pelagic zone of the large recently-constructed Ste-Croix reservoir, ii) to identify the main factors that govern horizontal distribution patterns of dominant species along the long axis of the reservoir, with special attention to predation.

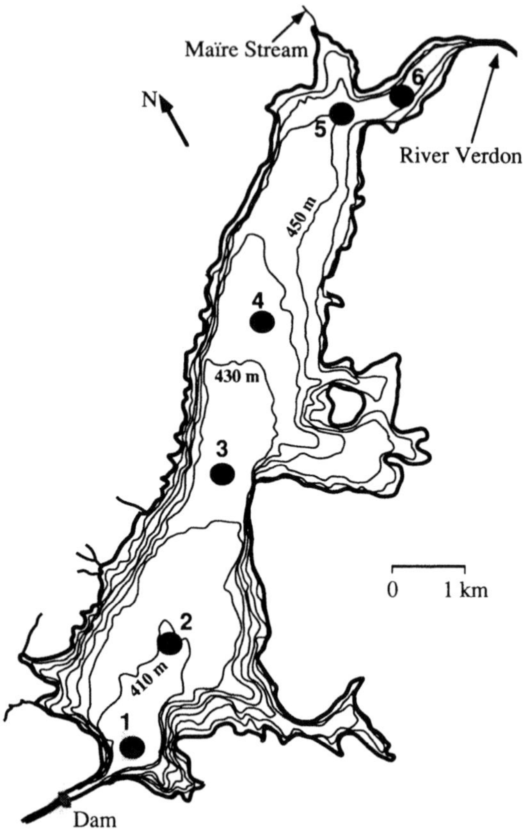

Fig. 1. Bathymetric map of Sainte-Croix reservoir (contours expressed in meters above sea level) showing the six sampling stations.

Study area

The Sainte-Croix man-made reservoir is situated in southern France. It was constructed as one of 5 hydro-electric stations on the river Verdon and was filled in 1974. Its catchment area covers a total area of 1 591 km^2. The impoundment has a surface area of 21.82 km^2 and a volume of 0.763 km^3. Its shape is elongated, 12 km in length and 2 km in width (Fig. 1). Depth increases from upstream (less than 10 m) to downstream (maximum depth 83 m), so a hypolimnion is present only downstream. The average turnover time is quite long (280 days). Maximum water level fluctuations generally occur in winter (up to 16 m).

The water inflow from the river verdon (Fig. 2a) is not very high, with an annual mean value of 18.9 $m^3\ s^{-1}$. High discharges occur in spring (up to a monthly average of 46.8 $m^3\ s^{-1}$) and November (up to 33.9 $m^3\ s^{-1}$). In spring, the influence of the river Verdon was clearly distinguishable as far as station 5, based on ionic composition measurements.

The lake is a warm monomictic waterbody. The annual heat budget varied between 65 000 and 75 000 cal m^{-2} (Grégoire, 1981). Because of the warm climatic conditions, the lake was stratified during 4 to 5 months (Fig. 2b). The temperature of the surface pelagic waters increased quickly from mid-May, reached a maximum of 25 °C at the end of July and remained over 20 °C until mid-October. The thermocline depth was 10 to 13 m throughout most of the summer, while the bottom water temperature reached 9 °C. In winter, the water temperature never fell below 6 °C.

According to nutrient concentration criteria, the lake can be classified as oligotrophic to meso-

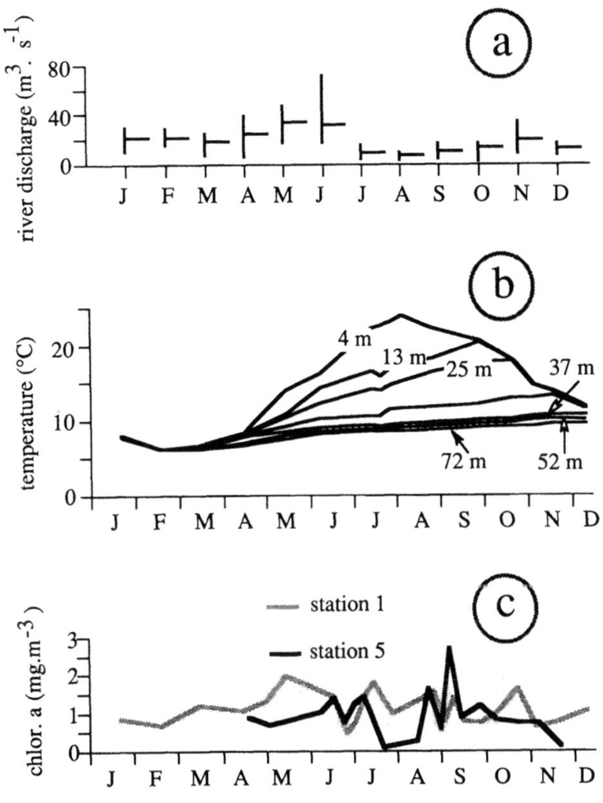

Fig. 2. Annual variations of: a) average and extreme monthly values of River Verdon discharge over a five-year period (1982 to 1986); b) temperatures recorded at seven depths, from the surface (4 meter deep) to the vicinity of the bottom (72 m deep); c) chlorophyll *a* concentrations at stations 1 and 5 (from El Hafa, 1987).

trophic with annual average concentrations of P_{tot} and N_{tot} of 19.7 and 236 $mg \cdot m^{-3}$ respectively (El Hafa, 1987, but $P\text{–}PO_4$ concentrations were always lower than 1 $mg \cdot m^{-3}$. Consequently, chlorophyll *a* values varied between 0.5 and 3 $mg \cdot m^{-3}$ (Fig. 2c).

Materials and methods

Zooplankton was sampled at 6 stations located along the long axis of the reservoir (Fig. 1), twice a month, in 1986. Stations 1, 3, 5 and 6 were sampled only 7 times. Vertical tows were taken with a 12.3 cm diameter, 1.50 m-long plankton net (mesh size 50 μm). A flowmeter was placed at the mouth of the net, so filtrated water volume was estimated at each tow. Four vertical tows were carried out at each station. The total number of vertical tows was 430. All samples were preserved in a 4% formalin solution and examined under a stereoscopic microscope. The lengths of at least 400 individuals of each of the main species collected in station 1 and 5 were measured.

Results

Seasonal succession

The macrozooplankton community has been studied by different authors since the lake was filled (Grégoire, 1981; Bressac & Champeau, 1985).

During the last 6 years, the annual cycle of zooplankton each year was similar (Pont *et al.*, 1989). Cladoceran species dominated: representing more than 81% of total organisms. High densities occurred at the same period but the mean annual abundance of *Daphnia longispina* has declined continuously. The general features of phytoplankton dynamics (El Hafa, 1987) and cladoceran seasonal succession (Fig. 3) in the Sainte-Croix Reservoir were close to the PEG seasonal succession model (Summer *et al.*, 1986).

In 1986, winter densities were low, being mostly composed of *Bosmina longirostris*, *D. longispina* and *Asplanchna priodonta*. In spring, three successive peaks of *Asplanchna priodonta* in April, *B. longirostris* in May and last of all *D. longispina* were observed. Densities of the latter large-bodied species immediately decreased at the end of June, while the *B. longirostris* population declined. Typical summer species, *Diaphanosoma brachyurum* and *Ceriodaphnia pulchella*, appeared at the end of July and were dominant in September. These two species disappeared from the lake two months later. The automn overturn period was characterized by the abundance of *Bosmina coregoni*. Throughout the warm period, *Macrocylops albidus* densities slowly increased.

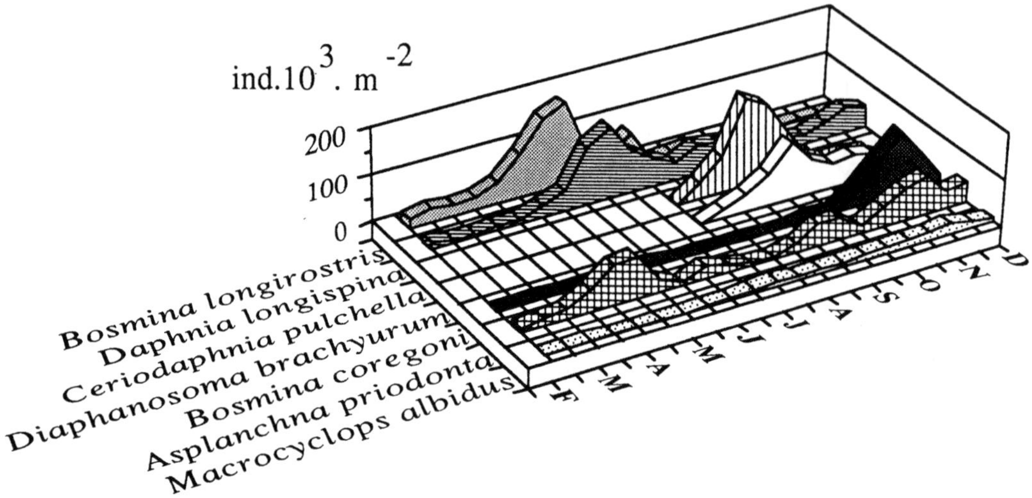

Fig. 3. Seasonal variations in the abundance (ind. $10^3\ m^{-2}$) of seven macrozooplankton species over one year (1986) in Sainte-Croix Reservoir.

Spatial patterns of total zooplankton and cladoceran species

Comparison of mean annual densities of total macrozooplankton at each station (Table 1 showed that there was no significant difference between stations 1 to 4 (Wilcoxon signed rank test); values ranged from 278 to 337.10^3 ind · m^{-2}. But total densities at station 5 (53.10^3 ind · m^{-2}) and station 6 (2.10^3 ind · m^{-2}) differed significantly from other sites and from each other. The two *Bosmina spp.*, which developed high densities in spring and late autumn when water discharge was high, showed similar patterns: numbers differed significantly only between stations 4 and 5 ($p < 0.01$). The summer species of *D. brachyurum* and *C. pulchella* showed lower densities in station 6 only. However, *C. pulchella* numbers were higher in the central part (stations 3 to 5: 74.10^3 ind · m^{-2}) than downstream

Table 1. Comparison of mean annual densities (ind. $103\ m^{-2}$) of total macrozooplankton and cladoceran species at each sampling station using Wilcoxon signed rank statistical test. ns: non significant, *: $p < 0.05$, **: $p < 0.01$; $n = 14$.

Stations	1		2		3		4		5		6
Total zooplankton	309		326		337		278		178		75
		ns		ns		ns		**		**	
D. longispina	70		87		63		37		14		9
		ns		ns		*		**		ns	
D. brachyurum	73		53		58		55		36		13
		*		ns		ns		ns		*	
B. coregoni	35		28		20		23		7		2
		ns		ns		ns		**		ns	
B. longirostris	37		55		54		46		22		10
		ns		ns		ns		**		ns	
C. pulchella	39		49		88		70		64		32
		ns		*		ns		ns		**	

(station 1 and 2: 44.10^3 ind · m^{-2}), whereas *D. brachyurum* densities reached maximum values at station 1 (73.10^3 ind · m^{-2}, against 50.10^3 ind · m^{-2} at stations 2–5). *D. longispina* appeared to be the only species with high densities only in the deeper downstream part (stations 1 and 2: 78.10^3 ind · m^{-2}) as well as a typical gradient pattern and a stable decrease towards the inflow.

The spatial distribution of total macrozooplankton (Fig. 4) changed throughout the year. In spring (high water inflow), total densities increased only in stations 1 to 3 or 4. In summer, total zooplankton densities were high throughout the lake with the exception of station 6. During the autumn, overall densities were high, but those in station 5 tended to be lower than in stations 3 and 4.

In spring, both *B. longirostris* and *D. longispina* populations increased only in the downstream and central parts of the lake (stations 1 to 4), with a more restricted development area for *D. longispina* (Fig. 5). A very similar pattern appeared for *B. coregoni* in November. During these two periods, the inflow of Verdon water seemed to be the controlling factor affecting all cladocerans of all sizes.

C. pulchella densities dropped in August in all six stations. In contrast, *D. longispina* did not show any significant increase in density in the three upstream stations 4, 5 and 6. Densities of *D. brachyurum*, a typical summer species, increased in stations 1 to 5 but higher values are noted only in the downstream zone (stations 1 to 3).

From July to October, water discharge and scarcity of food could not account for such dissimilarities between species patterns. The input of Verdon water (Fig. 2a) was very low (3.0 to 9.4 $m^3 \cdot s^{-1}$) and chlorophyll *a* concentrations (Fig. 2c) in station 5 (1.52 mg · m^{-3}) were slightly but significantly higher ($p < 0.05$) than in station 1 (1.16 mg m^{-3}).

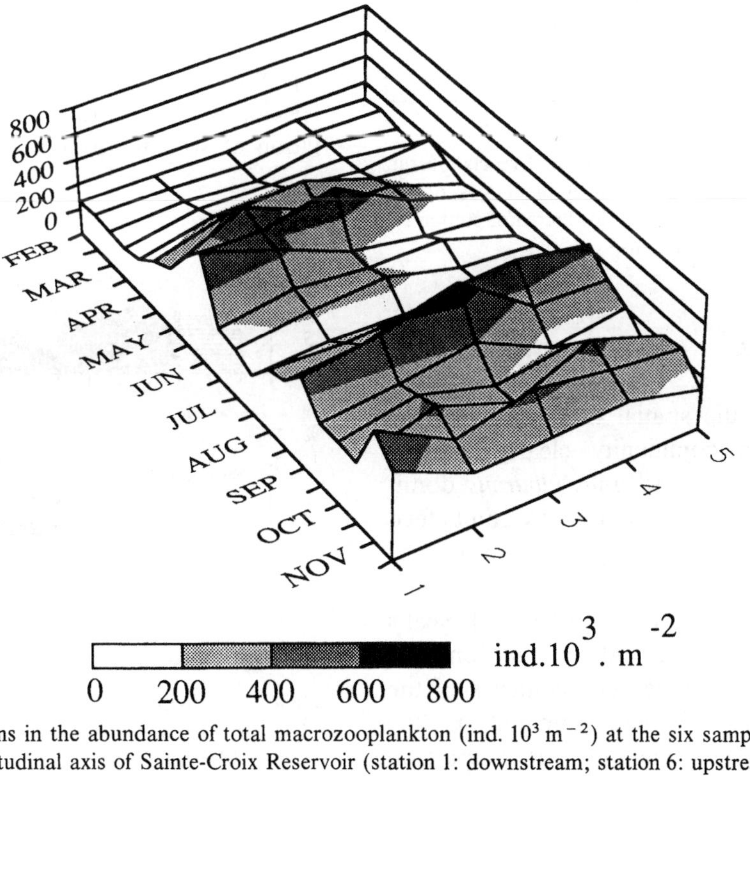

Fig. 4. Annual variations in the abundance of total macrozooplankton (ind. 10^3 m^{-2}) at the six sampling stations along the longitudinal axis of Sainte-Croix Reservoir (station 1: downstream; station 6: upstream).

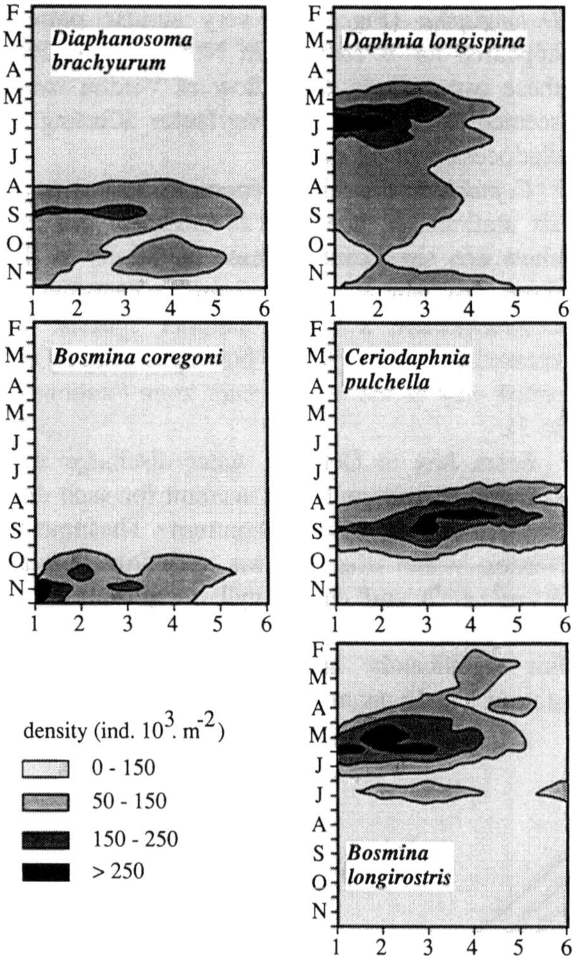

Fig. 5. Spatio-temporal distribution of the five cladoceran species (ind. 10^3 m^{-2}) along the longitudinal axis of the Sainte-Croix Reservoir (station 1: downstream; station 6: upstream).

Predation pressures

Recent work on the spatial distribution of the Sainte-Croix fish community clearly demonstrated that the bleak, *Alburnus alburnus* dominates the pelagic zone and that bleak adults feed mostly on zooplankton during the warm months (Chappaz *et al.*, 1987).

In May 1987, simultaneous samples of bleaks (using a 10 mm-mesh net) and zooplankton were taken in the pelagic zone. The contents of the anterior gut of ten fish were examined and the lengths of the prey were measured for comparison with the length frequency distribution of zooplankton in the epilimnion (Fig. 6). *Daphnia longispina* was the dominant species in the lake during this time and represented 81% of the organisms. *Bosmina longirostris* (6%), the two cyclopoids *Cyclops vicinus vicinus* and *Macrocyclops albidus* (9%) and the rotifer *Asplanchna priodonta* (4%) were also present. *D. longispina* (90% of total number of preyed organisms) and cyclopids (mainly *Cyclops vicinus vicinus*) were ingested by adult bleak (11.5 to 14 cm length). More than 98% of the ingested prey organisms were longer than one millimeter, whereas small-sized forms (less than one millimeter) were dominant in the lake (60%).

The horizontal distribution of fish populations was studied during two intensive surveys in 1987 and 1988 by means of an echosounding system (Guillard pers. com.). Density estimates of pelagic fish in the upstream zone (stations 4 to 6), were four times higher than in the central zone (stations 3 and 4) and ten times higher than in the downstream zone. This indicated that predation pressure was higher in the upstream part of the reservoir and especially on large zooplankters.

The importance of predation as a factor affecting the horizontal distribution of cladoceran species was demonstrated by the seasonal variation of size distribution of the five cladoceran

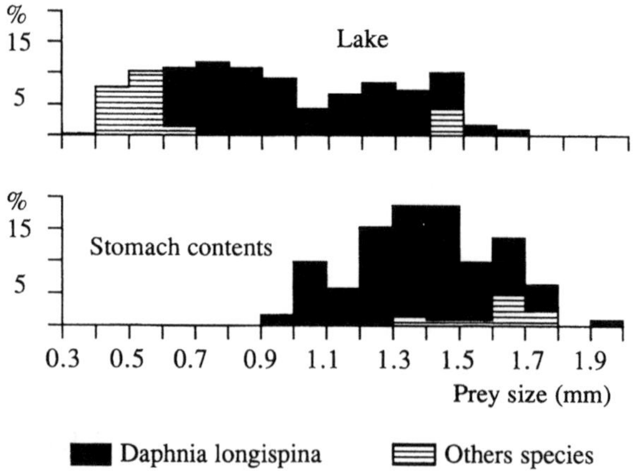

Fig. 6. Size distribution of *D. longispina* from gut contents of bleak ($N = 10$) and in the lake collected simultaneously in May 1987 (in % of total number of prey).

species at stations 1 (downstream area) and 5 (upstream area) throughout the year (Fig. 7).

In winter, no differences were apparent between stations 1 and 5. From spring to autumn, the length frequency distributions of small species, *Ceriodaphnia pulchella* (up to 0.9 mm long), *Bosmina longiristris* (up to 0.8 mm long), *Bosmina coregoni* (up to 0.8 mm long), were quite similar at both ends of the lake. In contrast, significant differences in the length distributions of *D. longispina* (up to 2.0 mm long) were recorded between the two stations in spring and especially in summer (*i.e.* the active feeding season for fish). In the upstream area, large organisms were less represented than near the dam, and there was a clear displacement towards a distribution dominated by small-sized individuals. In summer, 62% of the population of *D. longispina* at station 1 was longer than 1 mm, compared to less than 32% at station 2. In autumn, the two length distributions became more similar.

A similar trend was noted for *D. brachyurum* (up to 1.2 mm long) in autumn, with a decrease in relative abundance of the largest individuals (more than one millimeter long).

Discussion

Results presented indicate that spatial distribution of the cladoceran community along the longi-

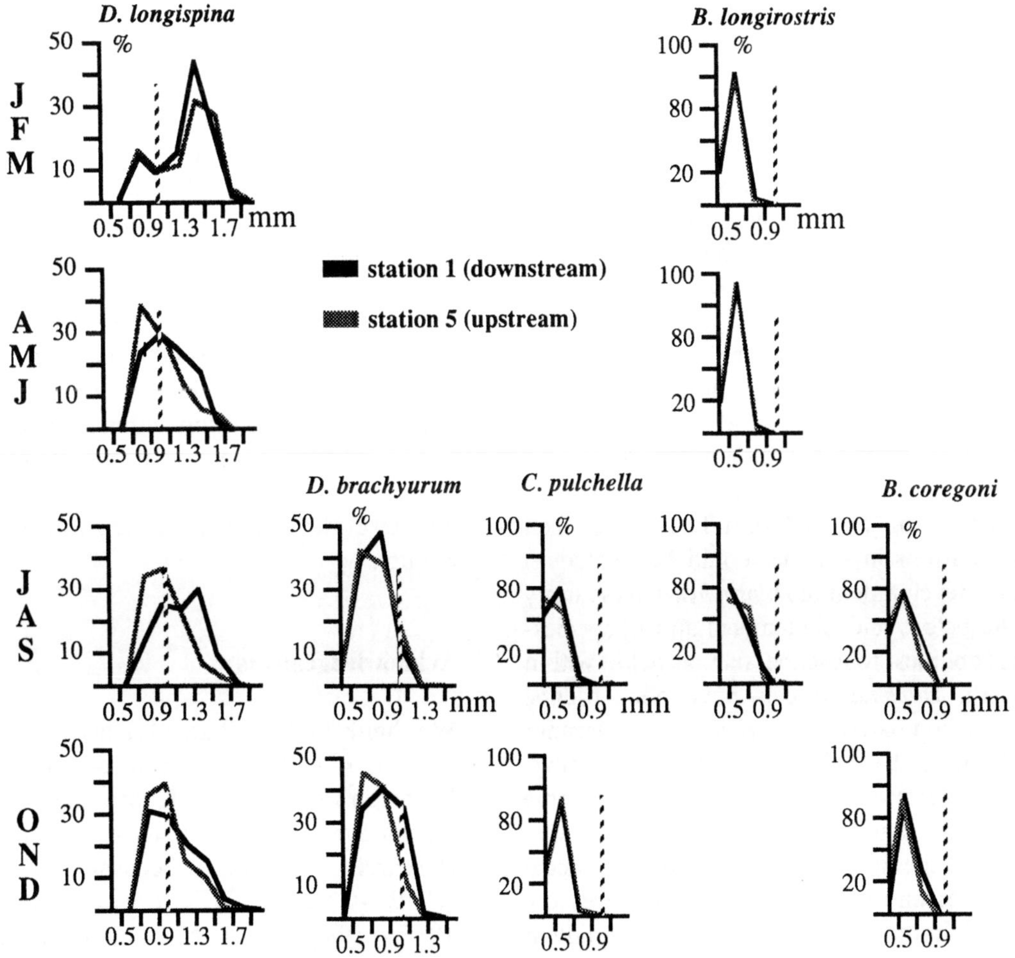

Fig. 7. Graphic comparison of mean size distributions of *D. longispina* in the downstream (1) and upstream (5) stations at each season (in% of total number of prey).

tudinal axis of the Sainte-Croix Reservoir varied both during the year and from one species to another. In spring and fall, when the inflow from the Verdon River was high, cladoceran densities were higher downstream, near the dam (*Daphnia longispina, Bosmina longirostris, Bosmina coregoni*). But in summer, only the largest species of *D. longispina* did not develop high densities in the upstream part of the lake. In contrast, the small summer species *Ceriodaphnia pulchella* achieved high densities along the whole axis in August. The bleak fed mostly on large-bodied forms (more than one mm) and was preferentially abundant upstream. This abundance of bleak in the upstream part of the lake is probably related to the fact that this fish was concentrated during the cold season in the Verdon River where it feeds mostly on macro-invertebrates (Chappaz *et al.*, 1987). Phytoplankton biomass was almost similar at the two extreme ends of the reservoir during summer and fall.

The comparison between the length frequencies of Cladocera in stations at the two opposite ends of the lake clearly demonstrates the consequence of size-selective predation on the horizontal distribution of the large-sized prey *D. longispina*. The preferential development of *D. longispina* in the downstream part of the lake can be considered as the consequence of low predation rates in this area.

Length frequency distribution of *Diaphanosoma brachyurum* was quite similar in summer upstream and downstream. The lack of difference between the two stations in summer could be related to i) the late development of *D. brachyurum* (scarcity of adults before mid-September) and ii) species-specific abilities to escape fish attacks within Cladocera. *Daphnia* species have poor escape abilities as compared to *D. brachyurum* (Drenner & McComas, 1980). Length frequency distribution of *D. brachyurum* at stations 1 and 5 became different only in autumn. During this period, discharge from the river increased and both physical disturbances and differential predation pressure could explain the *D. brachyurum* density gradient from upstream to downstream.

In contrast, no differences were noted during the year between length frequency distribution of small-sized forms upstream and downstream (*Bosmina longirostris, Bosmina coregoni, Ceriodaphnia pulchella*).

The examination of the long-term changes in the zooplankton community (Pont *et al.*, 1989) shows that, three years after the lake was filled, *Daphnia longispina* dominated throughout the stratified period, whereas *Diaphanosoma brachyurum* was absent. The replacement of *Daphnia* by *Diaphanosoma* in late summer occurred between 1978 and 1982, when the bleak population was intensively increasing (Chappaz, 1986).

This can be explained by the instability of the predator-prey system (Murdoch & Bence, 1987) because i) predator generation time is much longer than that of the prey, ii) the predator is able to change its diet when large zooplankton prey become less abundant. The classical density-dependence mechanisms do not stabilize the predatory-prey system and the main planktonic prey tend to disappear.

Nowadays, the effect of fish predation is considered to be one of the main factors explaining the dramatic drop in zooplankton density in late spring and the dominance of small species in late summer (Summer *et al.*, 1986). The present results show that fish predation is also an important factor in accounting for the horizontal distribution of zooplankton within the pelagic area of lakes, and that there is a strong interrelationship between present spatial heterogeneity and the long-term development of the zooplankton community.

Acknowledgements

We thank Pr Eric Pattee for his criticisms. This work is a part of the Sainte-Croix Ecological Research Project of the Hydrobiological Laboratory of Provence University and of the National Research Project 'Ecosystèmes lacustres (CNRS-GRECO 71)'. It was supported from 1985 to 1987 by the C.N.R.S., the PIREN and by regional funds.

References

Armengol, J., 1984. Ciclo anual y heterogeneidad espacial en el zooplancton de una cadena de embalses del rio Guadiana. Oecologia aquatica 7: 43–72.

Balvay, G., 1985. Réseau trophique dans les retenues. In D. Gerdeaux & R. Billard (eds), Gestion piscicole des lacs et retenues artificielles. INRA, Paris: 29–66.

Bressac, Y. & A. Champeau, 1985. Contribution à l'étude du réseau trophique des lacs de barrage: le zooplancton des retenues de Serre-Ponçon et Sainte-Croix. In: Gerdeaux, D. & Billard, R. (eds), Gestion piscicole des lacs et retenues artificielles. INRA, Paris: 67–77.

Brooks, J. L. & S. I. Dodson, 1965. Predation, body size and composition of plankton. Science 150: 28–35.

Butorina, L. G., 1986. On the problem of aggregations of planktonic crustaceans *Polyphemus peliculus* (L.), Cladocera. Arch. Hydrobiol. 105: 355–386.

Chappaz, R., 1986. Etude piscicole de la retenue de Sainte-Croix-Fontaine l'Evêque. Thèse de 3ème cycle. Université de Provence Aix-Marseille I, 183 pp.

Chappaz, R., G. Brun & G. Olivari, 1987. Mise en évidence de différences de régime alimentaire dans une population d'ablettes *Alburnus alburnus* (L.) dans la lac de Sainte-Croix. Conséquences sur la croissance et la fécondité. Ann. Limnol. 23: 245–252.

Cochran, W. G., 1977. Sampling techniques. Wiley, N.Y., 428 pp.

Dagg, M., 1977. Some effects of patchy food environments on copepods. Limnol. Oceanogr. 22: 99–107.

De Nie, H. W., H. J. Bromley & J. Vijverberg, 1980. Distribution patterns of zooplankton in Tjeukemeer, The Netherlands. J. Plankton Res. 2: 317–334.

Drenner R. W. & S. R. McComas, 1980. The roles of zooplankton escape ability and fish size selectivity in the selective feeding and impact of planktivorous fish. In W. C. Kerfoot (ed.), Evolution and ecology of zooplankton communities. The University Press of New England, Hanover (N.H.), Lond.: 587–593.

El Hafa, M., 1987. Bilan des apports de sels nutritifs au lac de Sainte Croix (Provence) et leur utilisation par les phytoplancton dans la retenue. Thèse de 3ème cycle Université de Provence Aix-Marseille I, 131 pp.

Frontier, S., 1983. Stratégies d'échantillonnage en écologie. Collection d'Ecologie, Masson, Paris, 494 pp.

Gannon J. E. & R. S. Stemberger, 1981. Impact of influent river on the distribution of zooplankton in lake Skadar. In S. Karaman & A. M. Beeton (ed.), The biota and limnology of lake Skadar. Titograd: 202–218.

Georges, D. G. & R. W. Edwards, 1976. The effect of wind on the distribution of chlorophyll *a* and crustacean zooplankton in a shallow eutrophic reservoir. J. appl. Ecol. 13: 667–690.

Grégoire, A., 1981. Contribution à l'étude hydrobiologique d'une rivière aménagée: le Verdon (Alpes de Haute Provence). Les lacs de barrage et les tronçons de cours d'eau à débit régulé. Cah. Lab. Hydrobiol. Montereau 13: 1–172.

Jakobsen, P. J. & G. H. Jonhsen G. H., 1987. The influence of predation on horizontal distribution of zooplankton species. Freshwat. Biol. 17: 501–507.

Lenz, P. H., S. D. Cooper, J. M. Melack & D. W. Winkler, 1986. Spatial and temporal patterns of three levels in a saline lake. J. Plankton Res. 8: 1054–1064.

Malone, B. J. & D. J. MacQueen, 1983. Horizontal patchiness in zooplankton populations in two Ontario kettle lakes. Hydrobiologia 99: 101–124.

Murdoch, W. W. & J. Bence, 1987. General predators and unstable prey populations. In W. C. Kerfoot & A. Sih (eds), Predation. Direct and indirect impacts on aquatic communities. The University Press of New England, Hanover (N.H.), Lond.: 17–30.

Nilsson, N. A. & B. Pejler, 1973. On the relation between fish fauna and zooplankton composition in North Swedish lakes. Inst. Freshwater Res. Drottningholm Rep. 53: 51–77.

Patalas, K., 1969. Composition and horizontal distribution of crustacean plankton in Lake Ontario. J. Fish. Res. Bd Can. 26: 2135–2164.

Patalas, K., 1981. Spatial structure of the crustacean planktonic community in the lake Winnipeg, Canada. Verh. int. Ver. Limnol. 21: 305–311.

Pinel-Alloul, B., J. A. Downing, M. Perusse & G. Codin-Blumer, 1988. Spatial heterogeneity in freshwater zooplankton: systematic variation with body-size, depth and sampling scale. Ecology 69: 1393–1400.

Pont, D., 1986. Structure spatiale d'une population du cyclopide *Acanthocyclops robustus* dans une rizière de Camargue (France). Acta Oecol., Oecol. gen. 7: 289–302.

Pont, D., R. Chappaz, G. Brun & A. Champeau, 1989. Interactions zooplancton-poissons dans une retenue oligotrophe de mise en eau récente (Ste-Croix, Provence, France). Rev. Sci. Eau 2: 777–792.

Siebeck, H. O., 1980. Optical orientation of pelagic crustaceans and its consequences in the pelagic and littoral zones. In W. C. Kerfoot (ed.), Evolution and ecology of zooplankton communities. The University Press of New England, Hanover (N.H.), Lond.: 28–38.

Steele, J. H., 1974. Spatial heterogeneity and population stability. Nature 248: 83–85.

Steele, J. H., 1978. Some comments on plankton patches. In J. H. Steele (ed.), Spatial pattern in plankton communities, Plenum Press, N.Y.: 1–20.

Summer, U., Z. M. Gliwicz, W. Lampert & A. Duncan, 1986. The PEG-model of seasonal succession of planktonic events in freshwaters. Arch. Hydrobiol. 106: 433–471.

Wiebe, P. H., 1970. Small-scale spatial distribution in oceanic zooplankton. Limnol. Oceanogr., 15: 205–217.

Wiens, J. A., 1976. Populations responses to patchy environments. Annu. Rev. Ecol. Syst. 7: 81–120.

Zaret, T. M. & J. S. Suffern, 1976. Vertical migration in zooplankton as a predator avoidance mechanism. Limnol. Oceanogr. 21: 804–813.

Received 12 August 1989; in revised form 4 January 1990; accepted 19 March 1990

Hydrobiologia **207**: 269–277, 1990.
D. J. Bonin & H. L. Golterman (eds), Fluxes Between Trophic Levels and Through the Water–Sediment Interface.

Organic matter distribution in burrows of the thalassinid crustacean *Callichirus laurae*, Gulf of Aqaba (Red Sea)

Jean de Vaugelas[1] & Roselyne Buscail[2]
[1] *Groupe de Recherches Marines – Université de Nice – Parc Valrose – 06034 Nice – France et Mission Océanographique Française au Moyen-Orient;* [2] *Laboratoire de Sédimentologie et Géochimie Marines, U.A. 715 C.N.R.S. – Université de Perpignan, Avenue de Villeneuve – 66025 Perpignan – France*

Key words: organic carbon, humic substances, bioturbation, burrows, *Callichirus laurae*, Red Sea, sampling

Abstract

Analysis of sediments in and around the large and deep burrows made by the callianassid shrimp *Callichirus laurae*, Jordan Gulf of Aqaba (Red Sea), showed that organic carbon (OC), humic matter (HM), humic (HA) and fulvic acids (FA) are redistributed both quantitatively and qualitatively by sediment reworking activities:

1. OC and humic matter (HM) concentrations are *ca.* 30 times higher in the stomach content than in surface sand, proving a selective food uptake.
2. a 11 to 17 times increase in OC and HM is observed in the mucus-rich burrow wall when compared to ambient sediment.

Among humic substances, FA seem preferentially absorbed by the shrimp, while HA are concentrated in the fecal pellets.

The spatial distribution of organic matter in the form of a 3D network is discussed in relation with sampling procedures. Due to inadequate sampling gears, organic rich networks such as those of thalassinid shrimps are rarely considered in conventional sampling strategies, leading to an underestimation of the sediment organic content.

Introduction

Callianassid shrimps are burrowing crustaceans commonly found in fine sand of intertidal and shallow subtidal zones. The bioturbation for which they are responsible has long been recognized, as the quantity of sand processed per unit time can be large (Mac Ginitie, 1934; Suchanek, 1983; Vaugelas & Saint Laurent, 1984). Dense monospecific populations are a characteristic feature of many lagoons and embayments, giving the bottom a bumpy aspect due to numerous volcanoes of excavated sediment. Permanent burrows of larger species (adult size up to 15 cm long from rostrum to telson) commonly extend deeper than 2 m in the sand, resulting in a complete alteration of the sediment stratigraphy (Risk *et al.*, 1978; Suchanek, 1983, Tudhope & Scoffin, 1984).

It has been frequently mentioned that thalassinid species consolidate their galleries with fine particles cemented by a mucus (Dworschak, 1983: 39 for a review). However, except for a short mention in Thompson's (1972) work on

Upogebia pugettensis, no detailed study on the distribution of organic matter has been done so far on thalassinid burrows (Alongi, 1989). The burrow wall is a key micro-environment for biogeochemical processes, as it is a place of transient exchange of various solutes between the ambient sediment and the burrow water (Aller *et al.*, 1983; Waslenschuk *et al.*, 1983). It supports higher levels of bacteria than ambient sediment (Branch & Pringle, 1987; Dobbs & Guckert, 1988) and it was demonstrated that metals and radionuclides are enhanced (Abu-Hilal *et al.*, 1988; Aller *et al.*, 1983; Pemberton *et al.*, 1976; McMurtry *et al.*, 1986; Whitehead *et al.*, 1988). During sediment diagenesis, cementation processes can harden the burrow wall and callianassid burrows are common ichnofacies in the fossil record (Weimer & Hoyt, 1964; Curran, 1984).

The present work analyzes the effects of sediment reworking by *Callichirus laurae* Saint Laurent 1984 on the distribution of organic carbon and humic substances in sediments, burrows and shrimp digestive organs. Humic substances are complex biogeopolymers constituting the bulk of alkaline extractable organic matter in soils and marine sediments, and their reactivity with trace elements is well established (Rashid, 1985). The statistically correlated distributions of uranium, thorium and various trace elements with humic substances in *C. laurae* burrows were analyzed in separate papers (Abu-Hilal *et al.*, 1988; Whitehead *et al.*, 1988).

Description of the site studied

The study was carried out in the northern part of the Gulf of Aqaba (29° 25′ N; 34° 58′ E). Climate, hydrography, circulation pattern, seasonality and nutrients from the area are summarized in Reiss & Hottinger (1984). The study site was located in the bay bordering the Marine Science Station of Aqaba, Jordan (Fig. 1).

Sediment

An extensive *Halophila stipulacea* seagrass bed covers up to 80% of the soft-bottom, from 1.5 to 35 m depth (Hulings, 1979) (see Fig. 1 – Detail).

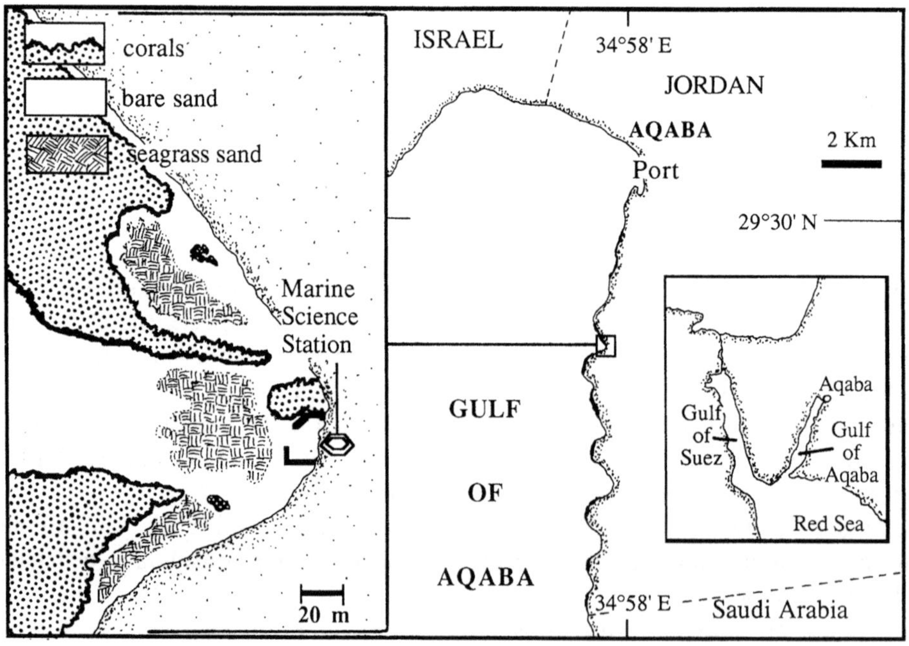

Fig. 1. Location of the study area in the Jordan Gulf of Aqaba (Red Sea).

In areas adjacent to corals, a 5 to 10 m wide stretch of bare sand separates the seagrass bed from the corals. *C. laurae* burrows appear as conspicuous mounds at the surface of both types of sands. Seagrass and bare sand are both terrigenous fine sand facies (Gabrié & Montaggioni, 1982). Representative granulometric data for the surface layer are mean diameter (Md) = 125 to 200 μm, Q_1 (25%) = 75 μm, Q_2 (75%) = 200 to 275 μm, Sorting (Trask) = 1.63 to 1.91 (well sorted), 20 to 30% carbonates.

Callichirus laurae – bioturbation

The average number of burrows similar in size to the one displayed on Fig. 2 is 0.2 to 0.3 m^{-2}, with a single shrimp per burrow (Vaugelas & Saint Laurent, 1984). The main functional aspects of burrow architecture are as follows:

a) The sediment sorted for food by the shrimp comes from the surface layer. After the careful removal of a part of the organic elements it contains through sifting, the finer fraction (particle diameter < 2 mm) of the sand is rejected from the burrow, where it accumulates at the exit opening in the form of a large mound (20 cm high; 60 cm in diameter).

b) The coarser fraction of the sorted sand (> 2 mm) cannot be pumped through the restricted aperture of the exit shaft (Vaugelas, 1984). It is then black-filled by the shrimp in adjacent tunnels branching off the main gallery. These coarse skeletal elements (Md = 420 μm, Q_1 (25%) = 175 μm, Q_2 (75%) = 2000 μm, So = 1.5, carbonates = 15%) are mainly gravel, bits of coral, mollusc shells, large foraminifers and *C. laurae* chelipeds of exuviae. Pockets full of seagrass leaves and accumulations of *C. laurae* fecal pellets are occasionally stacked in between other skeletal debris, but were not considered in this study.

c) In open sections of the 2.5 cm internal diameter tunnels where the shrimp can freely circulate, *C. laurae* burrow wall consists of a smooth, 1 to 2 mm thick slimy inner lining spreaded at the surface of a *ca.* 5–15 mm thick dark, cohesive matrix, whose outer layer is irregular (see Fig. 2 – Detail). The matrix is made of particles finer than ambient sediment and richer in carbonates (Md = 100 to 150 μm; Q_1 (25%) = 60 to 70 μm, Q_2 (75%) = 200 to 500 μm, So = 1.3 to 1.5, very well sorted; carbonates = 30 to 60%). Mucus is evident as a sticky coating at the inner surface of freshly made or maintained walls. The biochemical properties of *C. laurae* mucus were not analysed but we assume that it is mainly muccopolysaccharides (MPS), as it was demonstrated for the thalassinid crustacean *Upogebia pugettensis* (Thompson, 1972).

Materials and methods

Sample collection

a) Sediment and burrows (Fig. 2): surface sand (0–2 cm: sample 1), ambient sediment (– 10 cm: sample 2) and rejected sand (sample 4) were collected with a 7 cm diameter, 50 cm long hand corer at a water depth of 10 m. Cohesive sections of burrow walls (sample 3) and skeletal elements stored in back-filled tunnels (sample 5), coming from different parts of the network, were separated from the surrounding sediment using a sediment sucker and collected in plastic bags. Resin casts of shrimp burrows were made according to the method of Shinn (1968). A complete cast, amounting to *ca.* 15 Kg of polyester resin, is shown on Fig. 2.

b) *C. laurae*: immediately following collection with the 'weighted lines' (Vaugelas, 1985), specimens of adults *C. laurae* were measured, weighed to the nearest mg and dissected to remove the complete digestive tract, before storage in a deep-freezer. For simplification purposes the foregut is termed 'stomach' while the 'intestine' corresponds to the posterior midgut and the hindgut (*i.e.* excluding the digestive glands). The organic carbon content of stomach (Table 1: sample A) and intestine (Table 1: sample B) of 6 specimens (4 in seagrass sand and 2 in bare sand) were analyzed following the procedure of Strickland & Parsons

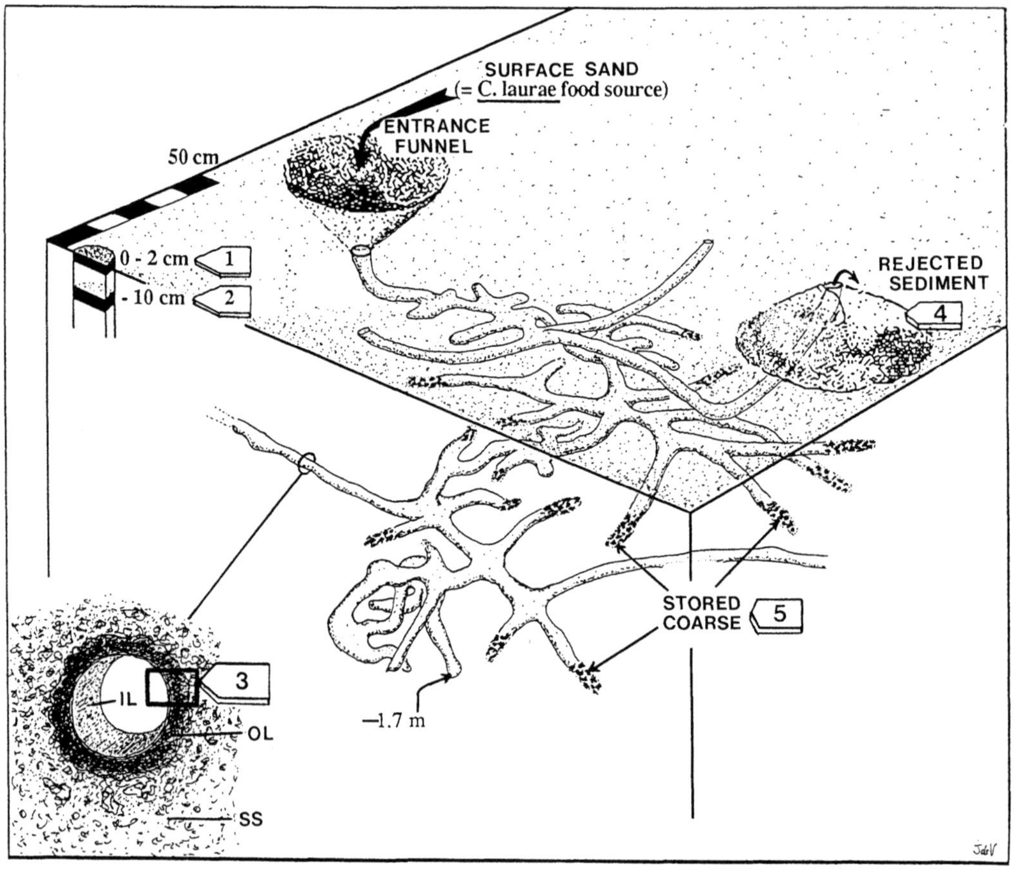

Fig. 2. Callichirus laurae burrow: line-drawing based on a resin cast (3/4 top view). Legend: IL = smooth, slimy, inner lining; OL = outer lining = matrix; SS = surrounding sediment; 1, 2, 3, 4, 5 = samples.

Table 1. Distribution of organic carbon humic matter, fulvic acids (FA), humic acids (HA) and the FA/HA ratio in: – Sand and burrows of *Callichirus laurae* (samples 1, 2, 3, 4, 5); – *Callichirus laurae* stomach and intestine contents (samples A, B, C).

No.	Sample	Organic Carbon ($mg\ C\ g^{-1}$)	Humic Matter ($mg\ C\ g^{-1}$)	Fulvic Acids ($mg\ C\ g^{-1}$)	Humic Acids ($mg\ C\ g^{-1}$)	FA/HA
Sediment and burrow						
1	Surface sand (0–2 cm)	2.50 ± 0.70 (6)*	0.70 ± 0.37 (6)	0.32 ± 0.13 (6)	0.37 ± 0.24 (6)	0.86
2	Ambient sed. (– 10 cm)	0.75 ± 0.15 (6)	0.36 ± 0.19 (6)	0.10 ± 0.01 (6)	0.25 ± 0.18 (6)	0.40
3	Burrow lining	11.25 ± 4.35 (8)	2.16 ± 1.03 (8)	0.96 ± 0.43 (8)	1.20 ± 0.60 (8)	0.80
4	Rejected sand	0.95 ± 0.25 (4)	0.20 ± 0.07 (4)	0.16 ± 0.04 (4)	0.04 ± 0.03 (4)	4.00
5	Stored coarse	2.00 ± 0.50 (2)	1.02 ± 0.25 (2)	0.17 ± 0.04 (4)	0.85 ± 0.21 (2)	0.20
Callichirus laurae						
A	Stomach	67.8 ± 19.8 (6)				
B	Intestine/feces	47.2 ± 13.5 (6)				
C	Intestine/feces		13.5 ± 3.5 (3**)	1.96 ± 0.29 (3**)	11.84 ± 4.02 (3**)	0.16

*: ± standard deviation (no. of samples).

**: Triplicate measurements on a pool of 16 intestine contents (see Materials and Methods).

(1972). Taking 8 other specimens collected in the bare sand area and 8 in the seagrass sand, 16 intestine contents were pooled in order to provide a sufficient quantity (*ca.* 6 g w.w.) to perform the humic acid analysis (Table 1: sample C). We could not pool enough material to perform the humic acid analysis on stomach contents.

Analysis

a) Geochemical analysis: organic carbon (OC) content of the dry sediment was obtained by combustion in a LECO analyzer (after acidification of the samples with H_3PO_4 for OC). Total humic matter (HM) of sediments, burrows and intestine contents was extracted with alkalin solvent (0.1 N NaOH/1% Na_4 P_2O_7, pH 13) (Debyser & Gadel, 1983); humic acids (HA) were precipitated after acidification to pH 2 and separated from the fulvic acids (FA). Dissolved organic carbon was measured on aliquot fractions of total humic and fulvic acids solutions (Coulomat 702 from Strohlein Instruments). Humic acids were dialysed and purified on H^+ ion-exchange DOWEX resin.

b) Sedimentological analysis: Grain size distribution of sediments, burrow walls and coarse elements stored in *C. laurae* burrow, was carried out using an AFNOR sieve column ($r = \sqrt[10]{10}$). Cumulative grain-size curves were used to determine the mean diameter (Md), quartiles 25% (Q_1) and 75% (Q_2) and Trask sorting index (So) (Martin-Buscail *et al.*, 1981).

Unless otherwise specified, all weights are expressed as dry weight.

Results and discussion

Surface sand and ambient sediment

The organic carbon distribution was found similar to the figure observed in a previous study (Vaugelas & Naïm, 1982) – *i.e.* 2.5 ± 0.7 mg C g^{-1} at the surface and a decrease with increasing depth of burial (Table 1). Humic matter (HM) distribution was found parallel to that of OC.

Feeding processes

C. laurae feeding activities resulted in a quantitative and qualitative alteration of OC and HM distribution.

C. laurae is a micro-detritus feeder who ingests only the organic material gleaned at the surface of sifted sand grains, ingesting only a very small fraction of mineral particles (Vaugelas & Saint Laurent, 1984: 151). It was therefore not surprising to measure high enrichment factors for organic carbon and humic matter between surface sand (taken as the reference sample in Fig. 3A) and the shrimp's digestive organs (Table 1 and Fig. 3A). The quantitative results also confirmed the possibility of finding the rejected sand depleted of its organic content when compared to surface sand (Fig. 3A). Similarly, a loss of half the nitrogen content was observed by Suchanek *et al.* (1986: 148) for *Callianassa* sp. from Enewetak lagoon, between the intake pits (0.04%) and the mounds of rejected sediment (0.02%).

What is the fate of organic elements concentrated through *C. laurae* feeding processes? A part of it is used for nutrition, *i.e.* accounting for the *ca.* 30% loss of OC between stomach and intestine (Table 1). What remains is excreted in the form of fecal rods 5–8 mm long and 1–2 mm in diameter. The mean value obtained for *C. laurae* intestinal OC (47.2 ± 13.5 mg C g^{-1}) was found similar to the 45 mg C g^{-1} reported for *Callianassa major* fecal pellets (Frankenberg & Smith, 1967: 448). In macruran decapods most of the digestion and assimilation occurs in the digestive glands opening at the junction between the lower foregut and the upper midgut (Dall & Moriarty, 1983). We therefore assumed in this study that absorption of nutritive elements was negligible in the portion of the intestine from which we took samples (*i.e.* lower midgut + hindgut) and that this intestine content globally reflected what was to be excreted. As a consequence, the 'intestine' samples were referred to as 'intestine-

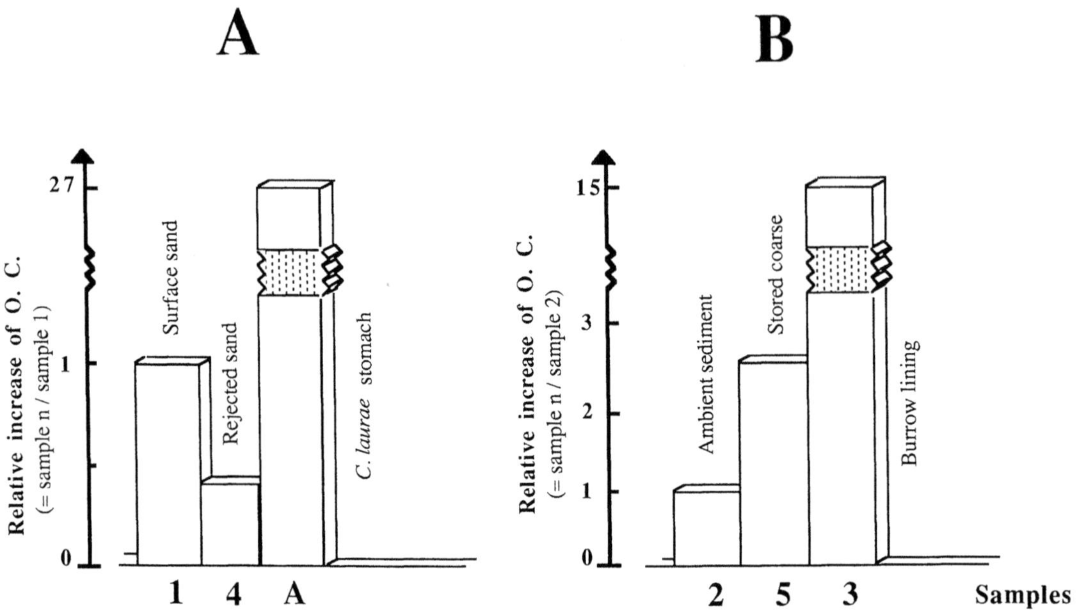

Fig. 3. Organic carbon quantitative distribution through *Callichirus laurae* feeding (3A) and burrowing (3B) processes.

feces' in the following discussion as well as in Table 1 and Fig. 4.

Due to their size, most fecal pellets were found stored in back-filled tunnels with the coarser fraction ≥ 2 mm and only a minor fraction was rejected from the burrow. This means that *C. laurae* burrow may act as a sink for the organically rich feces which are not pumped back to the surface, where their assimilable elements would be used by various detritus feeders (Frankenberg *et al.*, 1967).

The relative contribution of HA and FA to the pool of humic substances was analyzed in order to provide some information on the qualitative modifications of organic matter due to feeding processes. Table 1 and Fig. 4 clearly show a high concentration of HA and a low level of FA in the intestine-feces content, resulting in a very low

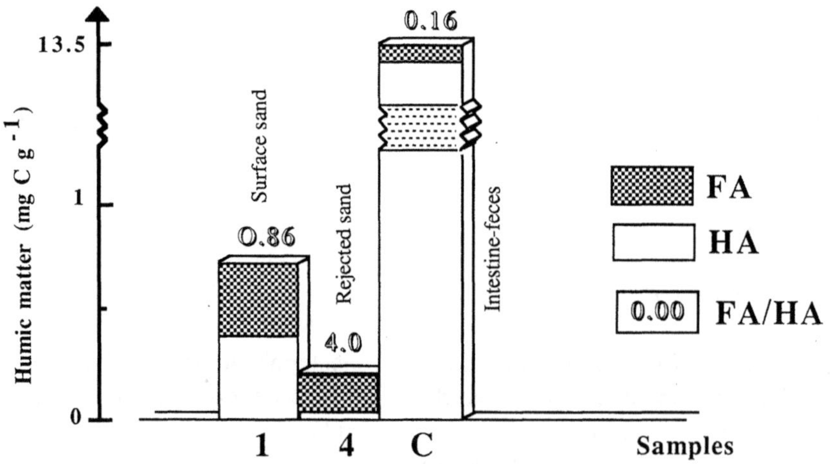

Fig. 4. Humic matter qualitative distribution through *Callichirus laurae* feeding processes.

FA/HA ratio (0.16). This contrasts with a FA/HA ratio close to 1 in surface sand and a strong dominance of FA in the rejected sand (FA/HA = 4). *C. laurae* seems to selectively concentrate and excret HA. According to Rashid (1985) HA are high molecular weight compounds, highly aromatized and thought to be the condensation products of fulvic acids. FA are low molecular weight, weakly aromatized biogeopolymers with a preponderance of aliphatic side chains. FA isotopic concentrations, particulary those of C^{12}, are closer than those in humic acids to the values for phytoplankton (Nissenbaum, 1974). Organic compounds such as aliphatic chains, simple carbohydrates, sugars, organic acids, and some amino acids are easy to breakdown and digest during feeding processes and are therefore the most readily available type of organic matter. HA represents the refractory part of the organic matter, therefore constituting up to 90% of the humic substances in the intestine-feces.

However, in the absence of experimental evidence for the selective assimilation of fulvics, a physical mechanism of separating fulvic rich material from humic rich material, perhaps via grain size selection and/or differences in molecular size, may account for a part of the results obtained.

Organic carbon quantitative distribution through burrowing activities

The distribution of organic substances in *C. laurae* burrows was found to be related to the tri-dimensional distribution of biogenic substrata such as burrow walls and organic accumulations in back-filled tunnels.

The burrow wall was enriched *ca.* 15 times in OC when compared to ambient sediment (taken as the reference sample in Fig. 3B) and the organic carbon content of skeletal elements stored in back-filled tunnels (phanerogams debris and *C. laurae* fecal pellets not included) was 2–3 times higher than in ambient sediment (Table 1 and Fig. 3B).

We found it difficult to compare out data with previous works. Despite the fact that many studies mention the incorporation of mucus in thalassinid burrow walls (Dworschak, 1983: 39), only Thompson (1972) actually measured the organic matter content of *Upogebia pugettensis* burrow walls with respect to levels in the surrounding sediment. In *U. pugettensis* burrows, the statistically significant increase between ambient sediment and the burrow wall was of only 1.7 to 1.8 times. This relatively low increase could account for the fact that in this species there are hardly any differences in granulometry and mineralogy between ambient sediment and the burrow wall and probably less mucus incorporated. Aller and Yingst (1978), studying the tubicolous Annelid *Amphitrite ornata*, found a 3–4 times increase in the burrow wall. *A. ornata* makes a lining whose structure is very similar to that of *C. laurae* – *i.e.* coating sand grains finer than ambient sediment with a polysaccharide mucus and successively packing small parcels of sediment into the wall, thus achieving a smooth inner surface through compaction. A similar process of construction was recently described for *Callianassa trilobata* (Dobbs & Guckert, 1988).

Cementation processes during sediment diagenesis are frequent wherever high concentrations of organic matter are associated with a porous environment (Rashid, 1985). Callianassid tunnels are conspicuous trace fossils in many sedimentary deposits of the world (Weimer & Hoyt, 1964; Curran, 1984). It is therefore legitimate to consider *C. laurae* burrows (and probably many other thalassinid species as well) as sinks for organic elements incorporated in the lining. In addition, it is frequent to find fossilized thalassinid pellets in consolidated sediments (Moore, 1932), demonstrating that the feces were not consumed and remained intact when stored at depth in back-filled tunnels.

With its mucus rich cemented walls and the storage of fecal pellets and organic debris, a *C. laurae* burrow can be assimilated to an extensive network of concentrated organic matter branching out within a relatively poor ambient sediment. This conclusion, while trite, should have implications for the sampling strategy used when studying bioturbated sediments.

Most currently available benthic samplers such as box corers and grabs do not properly take burrows into account, except when networks are both small and superficial (Romero-Wetzel, 1989). Large, deep networks, such as those of thalassinid shrimps and enteropneusts, start deeper than the average penetration depth of this type of equipment (30–50 cm deep in mud, 10–20 cm in sand). Only the long corers such as vibro-corers or Kullenberg types may eventually sample portions of the tunnels. However, the probability of getting a clear idea of the extent of the network is poor. Risk *et al.* (1978), using computer simulation, demonstrated that the number of 3 cm diameter galleries intersected by a 2 m long, 10 cm diameter corer was only 1.20 in a 9 *Axius serratus* m^{-2} population and 2.13 with 18 *A. serratus* m^{-2}, whose large burrows may extend deeper than 3 m (Pemberton *et al.*, 1976). Considering that most open tunnels will be destroyed beyond recognition during sampling (Tudhope & Scoffin, 1984: 1093) and that only predetermined levels will be sub-sampled in these long cores, the probability of obtaining a clear image of a network of tunnels is very low. The main consequence of this point, so rarely considered in conventional benthic studied that it needs to be stressed, may be a misinterpretation of the local distribution of organic matter and related elements, both quantitatively and qualitatively.

Large burrows significantly increase the surface of oxidized substratum (Ott *et al.*, 1976; Abu-Hilal *et al.*, 1988). However, studies on benthic respiration and fluxes between trophic levels rarely consider the significant extension of water-sediment interface in heavily burrowed sediment. This may influence the pattern of distribution of both oxygen and nutrients. Thus, low rates of nutrient releases, a characteristic of many tropical sediment, may be due to the activity of a highly productive benthos, including heterotrophic bacterial assemblages, associated with water-sediment interfaces extending far below the first 5 cm considered in most benthic studies (Alongi, 1989). The relation between deep bioturbation and nutrient fluxes needs to be specifically addressed in future benthic studies. This will give its full meaning to the emerging recognition that callianassid shrimps' bioturbated areas are 'major functional zones' in many tropical lagoons (Alongi, 1989).

Acknowledgements

The senior author's work was carried out with the financial support of the French Ministry of Foreign Affairs, through the French Oceanographic Mission in the Middle East, and with the logistic support of the Marine Science Station of Aqaba (Universities of Jordan and Yarmouk). Thanks are due to Captain Abu Taleb for assistance in the field and to Dr. A. Abu-Hilal for offering laboratory space and administrative aid. Dr. R. Buscail's work was carried out with the financial support of the CNRS Associated Laboratory UA 715, whose help is gratefully acknowledged.

The authors are much indebted to Dr. P. Dworschak (University of Vienna, Austria) for fruitful discussions on callianassid mud-shrimp's ecology and careful reviewing of the manuscript.

References

Abu-Hilal, A., M. Badran & J. de Vaugelas, 1988. Redistribution of trace elements in *Callichirus laurae* burrows (Crustacea: Thalassinidea), Jordan Gulf of Aqaba (Red Sea). Mar. envir. Res. 25: 233–248.

Aller, R. C. & J. Y. Yingst, 1978. Biogeochemistry of tube-dwellings: a study of the sedentary polychaete *Amphitrite ornata* (Leidy). J. mar. Res. 36: 201–254.

Aller, R. C., J. Y. Yingst & W. J. Ullman, 1983. Comparative biogeochemistry of water in intertidal *Onuphis* (polychaeta) and *Upogebia* (crustacea) burrows: temporal patterns and causes. J. mar. Res. 41: 571–604.

Alongi, D. M., 1989. The role of soft-bottom benthic communities in tropical mangrove and coral reef ecosystems. Reviews in Aquatic Sciences, 1: 243–280.

Branch, G. M. & A. Pringle, 1987. The impact of the sand prawn *Callianassa kraussi* Stebbing on sediment turnover and on bacteria, meiofauna, and benthic microflora. J. exp. mar. Biol. Ecol. 107: 219–235.

Curran, H. A., 1984. Ichnology of pleistocene carbonates on San Salvador, Bahamas. J. Paleontol. 58: 312–321.

Dall, W. & D. J. W. Moriarty, 1983. Functional aspects of nutrition and digestion. In L. H. Mantel (ed.), The Biology of Crustacea. Vol. 5: Internal anatomy and physiological regulation. Academic Press, New York: 215–261.

Debyser, Y. & F. Gadel, 1983. Géochimie des kérogènes. In CNRS (ed.), Géochimie organique des sédiments marins d'ORGON à MISEDOR. Editions du CNRS, Paris: 421–450.

Dobbs, F. C. & J. B. Guckert, 1988. *Callianassa trilobata* (Crustacea: Thalassinidea) influences abundance of meiofauna and biomass, composition, and physiologic state of microbial communities within its burrows. Mar. Ecol. Prog. Ser. 45: 69–79.

Dworschak, P. C., 1983. The biology of *Upogebia pusilla* (Petagna) (Decapoda, Thalassinidea). I. The burrows. P.S.Z.N.I.: Marine Ecology 4: 19–43.

Frankenberg, D., S. L. Coles & R. E. Johannes, 1967. The potential trophic significance of *Callianassa major* fecal pellets. Limnol. Oceanogr. 12: 113–120.

Frankenberg, D. & K. L. Smith Jr., 1967. Coprophagy in marine animals. Limnol. Oceanogr. 12: 443–450.

Gabrié, C. & L. F. Montaggioni, 1982. Sedimentary facies from modern coral reefs, Jordan Gulf of Aqaba, Red Sea. Coral Reefs 1: 115–124.

Hulings, N. C., 1979. The ecology, biometry and biomass of the seagrass *Halophila stipulacea* along the Jordanian coast of the Gulf of Aqaba. Bot. Mar. 22: 425–430.

Mac Ginitie, G. E., 1934. The natural history of *Callianassa californiensis* Dana. Am. Midl. Nat. 15: 116–117.

Martin-Buscail, R. T., F. Gadel & J. P. Baruseau, 1981. Holocene evolution of the Canet-St Nazaire lagoon (Golfe du Lion, France) as determined from a study of sediment properties. Sedimentology 28: 823–836.

McMurtry, G. M., R. C. Schneider, P. L. Colin, R. W. Buddemeier & T. H. Suchanek, 1986. Vertical distribution of fallout radionuclides in Enewetak Atoll lagoon sediments: effects of burial and bioturbation on the radionuclide inventory. Bull. mar. Sci. 38: 35–55.

Moore, H. B., 1932. The faecal pellets of the Anomura. Proc. roy. Soc. Edinburgh 52: 296–308.

Nissenbaum, A., 1974. The organic geochemistry of marine and terrestrial humic substances: implications of carbon and hydrogen isotopes studies. In B. Tissot & F. Bienner (eds), Advances in organic geochemistry 1973. Technip, Paris: 39–52.

Ott J. A., B. Fuchs, R. Fuchs & A. Malasek, 1976. Observations on the biology of *Callianassa stebbingi* Borrodaille and *Upogebia litoralis* Risso and their effect upon the sediment. Senckenbergiana marit 8: 61–79.

Pemberton, G. S., M. J. Risk & D. E. Buckley, 1976. Supershrimp: deep bioturbation in the Strait of Canso, Nova Scotia. Science 192: 790–791.

Rashid, M. A., 1985. Geochemistry of marine humic compounds. Springer-Verlag, Berlin, 291 pp.

Reiss, Z. & L. Hottinger, 1984. The Gulf of Aqaba. Ecological micropaleontology. Springer-Verlag, Berlin, 354 pp.

Risk, M. J., R. D. Venter, S. G. Pemberton & D. E. Buckley, 1978. Computer simulation and sedimentological implications of burrowing by *Axius serratus*. Can. J. Earth Sci. 15: 1370–1374.

Romero-Wetzel, M. B., 1989. Branched burrow-systems of the enteropneust *Stereobalanus canadensis* (Spengel) in deep-sea sediments of the Vöring-Plateau, Norwegian Sea. Sarsia 74: 85–89.

Shinn, E. A., 1968. Burrowing in recent lime sediments of Florida and the Bahamas. J. Paleontol. 42: 879–894.

Strickland, J. D. H. & T. R. Parsons, 1972. A practical handbook of seawater analysis. Bull. Fish. Res. Bd Can. 167: 311 pp.

Suchanek, T. H., 1983. Control of seagrass communities and sediment distribution by *Callianassa* (*Crustacea*, *Thalassinidea*) bioturbation. J. mar. Res. 41: 281–298.

Suchanek, T. H., P. L. Colin, G. M. McMurtry & C. S. Suchanek, 1986. Bioturbation and redistribution of sediment radionuclides in Enewetak Atoll lagoon by callianassid shrimp: biological aspects. Bull. mar. Sci. 387: 144–154.

Thompson, R. K., 1972. Functional morphology of the hindgut gland of *Upogebia pugettensis* (Crustacea, Thalassinidea) and its role in burrow construction. PhD, University of California, Berkeley, 202 pp.

Tudhope, A. W. & T. P. Scoffin, 1984. The effects of *Callianassa* bioturbation on the preservation of carbonate grains in Davies Reef lagoon. Great Barrier Reef, Australia. J. sedim. Petrol. 54: 1091–1096.

Vaugelas, J. de, 1984. Preliminary observations on two types of callianassid (*Crustacea*, *Thalassinidea*) burrows, Gulf of Aqaba, Red Sea. Proc. Symp. Coral Reef Environ. Red Sea, Jeddah, Jan. 1984: 520–539.

Vaugelas, J. de, 1985. A new technique for collecting large-sized callianassid mud-shrimp (*Crustacea*, *Decapoda*, *Thalassinidea*). Crustaceana 49: 105–109.

Vaugelas, J. de & O. Naïm, 1982. Organic matter distribution in marine sediments of the Jordanian Gulf of Aqaba. Proc. 4th Int. Coral Reef Symp., Manila 1981 1: 405–410.

Vaugelas, J. de & M. de Saint Laurent, 1984. Premières données sur l'écologie de *Callichirus laurae* (de Saint Laurent), sp. nov., (*Crustacea*, *Thalassinidea*): son action bioturbatrice sur les formations sédimentaires du Golfe d'Aqaba (Mer Rouge). C.r. Acad. Sci., Paris t. 298, ser. III, 6: 147–152.

Waslenchuk, D. G., E. A. Matson, R. N. Zajac, F. C. Dobbs & J. M Tramontano, 1983. Geochemistry of burrow waters vented by a bioturbating shrimp in Bermudian sediments. Mar. Biol. 72: 219–225.

Weimer, R. J. & J. H. Hoyt, 1964. Burrows of *Callianassa major* Say, geologic indicators of littoral and shallow neritic environments. J. Paleontol. 38: 761–767.

Whitehead, N., J. de Vaugelas, P. Parsi & M. C. Navarro, 1988. Preliminary study of Uranium and Thorium redistribution in *Callichirus laurae* (Crustacea: Thalassinidea) burrows and adjacent sediments, Jordan Gulf of Aqaba (Red Sea). Oceanol. Acta 11: 259–266.

Received 26 July 1989; in revised form 22 January 1990; accepted 25 February 1990

Hydrobiologia **207**: 279–286, 1990.
D. J. Bonin & H. L. Golterman (eds), Fluxes Between Trophic Levels and Through the Water–Sediment Interface.

Seasonal variation in the interactions between piscivorous fish, planktivorous fish and zooplankton in a shallow eutrophic lake

Jacobus Vijverberg[1], Maarten Boersma[1], Wim L.T. van Densen[1,2], Wim Hoogenboezem[1], Eddy H.R.R. Lammens[1] & Wolf M. Mooij[1]
[1] *Limnological Institute, Tjeukemeer Laboratory, De Akkers 47, 8536 VD Oosterzee, The Netherlands*;
[2] *Department of Fish Culture and Fisheries, Agricultural University, Wageningen, The Netherlands*

Key words: food web, trophic interactions, production, consumption

Abstract

The interactions between the higher trophic levels in a shallow eutrophic lake were studied during the course of a year. Three fish species determined the main pathways of organic matter flow within the system: the predominantly planktivorous bream (*Abramis brama*), the obligate planktivorous smelt (*Osmerus eperlanus*), and the piscivorous pikeperch (*Stizostedion lucioperca*). Of the thirteen common zooplankton taxa *Daphnia hyalina* and cyclopoid copepods were utilized most by the planktivorous fish, while the large production of small cladocerans is almost left unutilized.

The seasonal variations of production and consumption are large. This is mainly affected by seasonal variation of the water temperature. The production of O + smelt is efficiently utilized by the pikeperch. Being the most important zooplankton consumer, as well as the most important prey group, O + fish plays a key role in the Tjeukemeer food web.

Introduction

Food web theory has centered around the idea of connectance, i.e. the fraction of total number of possible trophic connections actually observed (Rejmánek & Starý, 1979). The interaction strength proposed by May (1973) also focusses on this idea. The quantities of connectance and interaction strength are easily extracted from published studies which describe food webs in natural systems (see e.g. Cohen, 1978; Pimm, 1982). In a recent evaluation Paine (1988) criticized this static approach. He has not challenged the food web theory itself, but the way natural communities have been analyzed using simple equations for connectance and interaction strength. According to Paine food web studies should focus on the dynamic, rather than on the static properties of the ecosystem, thus taking into account the spatial and temporal variations within age classes of species.

In this study we investigate the food web of lake Tjeukemeer, using Paine's approach. In previous studies on this lake Vijverberg & van Densen (1984), Lammens *et al.* (1985) reported on the year to year changes in the zooplankton species- and size composition, as well as on changes in the condition and growth of bream and pikeperch because of large variations in the stocks of O + fish. Production of O + fish may differ by two orders of a magnitude from year to year. In the years 1976–1983 it ranged from 0.12 dry wt m^{-2} yr^{-1} (6 kg fresh wt ha^{-1} yr^{-1}) to 11.7 g dry wt m^{-2} yr^{-1} (585 kg fresh wt ha^{-1} yr^{-1}), with a geometric mean production of 2.6 g dry wt m^{-2} yr^{-1} (130 kg fresh wt ha^{-1} yr^{-1}).

We will focus on the seasonal changes of trophic interactions between piscivorous and planktivorous fish on the one hand and between planktivorous fish and zooplankton on the other. We select the year 1978 for the data analysis, because O + fish production in this year (3.0 g dry wt m^{-2}) is close to the geometric mean value.

Study area

Tjeukemeer, situated in the North of the Netherlands, is a shallow (mean depth = 1.5 m), eutrophic freshwater lake and has a surface area of 2150 ha. The lake is a part of an interconnected system of water bodies, which act as a reservoir for the surrounding polders. This system receives water from the near by IJsselmeer (120000 ha) from May to September and polder-water which is rich in humic compounds in winter. Secchi-disk depth is usually 20–35 cm and the algal biomass is dominated by the blue-green *Oscillatoria agardhii*. The littoral zone of the lake is poorly developed, representing only 1% of the lake's surface area.

Methods

Zooplankton

Zooplankton was sampled with a Friedinger closing sampler at five fixed stations at fortnightly intervals in May–September and at monthly intervals during the rest of the year. Although rotifers are often very abundant during spring, only copepods and cladocerans are important as fish food, and hence only microcrustacean zooplankton are considered. For each zooplankton species the population density and the size distribution were assessed (see Vijverberg, 1977; Vijverberg & Richter, 1982a). Mean monthly biomass was calculated using the appropriate length-weight relationship (Table 1).

Table 1. Length-weight relations used in this study, using ln W = b ln L + ln a. W is dry weight in g, and L is length in mm.

Species / age group	b	ln a	Source
Zooplankton			
Daphnia hyalina	2.52	- 11.356	Bottrell *et al.*, 1976
Bosmina spp.	2.53	- 11.104	Bottrell *et al.*, 1976
Chydorus sphaericus	4.64	- 8.543	Vijverberg & Frank, 1976
Cyclopoid copepods	2.40	- 11.862	Bottrell *et al.*, 1976
O + fish			
smelt	3.2	- 14.251	Mooij own obs.
pikeperch	3.2	- 14.182	Mooij own obs.
perch	3.2	- 13.675	Mooij own obs.
≥I + fish			
bream			
sub adult	3.0	- 12.687	Mooij own obs.
spring	3.06	- 13.043	Lammens, 1982
autumn	3.10	- 13.177	Lammens, 1982
pikeperch			
spring male	3.51	- 14.885	van Densen & Vijverberg, 1982
spring female	3.23	- 13.057	van Densen & Vijverberg, 1982
autumn male	3.10	- 12.140	van Densen & Vijverberg, 1982
autumn female	2.94	- 11.176	van Densen & Vijverberg, 1982

Monthly production was calculated using P/B ratios, which were assessed under experimental conditions in earlier studies (Vijverberg, 1980; Vijverberg & Richter, 1982a, 1982b).

Fish

Young of the year fish (O +) were sampled using hoop nets from May to the first half of June, a 3 m trawl in the second half of June and a 5 m beam trawl from July onwards (van Densen, 1985). Older fish were caught using gill nets and a 5 m beam trawl. The number and size distribution of each species in the catch were noted. Absolute densities were estimated from the catches with the 5 m beam trawl, using a catching efficiency of 50% (van Densen, own obs.). Since the net efficiency for larger pikeperch is not known, pikeperch densities were estimated using the numbers of O + pikeperch in the previous years, combined with an instantaneous mortality rate per day (M) of 0.00055 d^{-1}, which is a conservative estimate based on Pauly's emperical formula for the natural mortality rate in fish stocks as a function of growth parameters and water temperature (Pauly, 1980). Mean monthly biomass (B) was assessed using Ricker's (1946) formula:

$$B = \frac{N_O \cdot W_O \cdot (e^{(G-M)\cdot t} - 1)}{(G-M)\cdot t} \qquad (1)$$

where N_O is the density and W_O is the mean dry weight at the start of the month; the weight was computed from the size distribution using the appropriate length-weight relationship, thus taking into account seasonal changes in condition (Table 1), G is the instantaneous growth rate per day, $t = 30$ days, and M is the instantaneous mortality rate per day; M was estimated from a negative exponential curve relating numbers with time. Monthly production (P) was calculated using:

$$P = G \cdot B \cdot t \qquad (2)$$

Gut contents were analyzed microscopically. The number and size distribution of each species in the gut were noted (Lammens *et al.*, 1985; van Densen, 1985). Fish consumption was estimated using conversion factors between fish production and consumption taken from literature. Although many factors may influence this conversion (Brett & Groves, 1979; Soofiania & Hawkins, 1985), we did not take into account the effects of these factors, because they are very difficult to estimate quantitatively under field conditions. For O + fish we used a conversion of 25%, which is the average of the conversion efficiency for herbivorous and carnivorous young fish reported in literature (Brett & Groves, 1979). Consumption rates for ≥I+ piscivorous fish were calculated using a conversion efficiency of 20% (Popova, 1978) and for ≥I+ planktivorous fish of 12.5% (O'Grady & Spillett, 1985).

Results

Zooplankton

About 98% of the zooplankton densities was represented by *Daphnia hyalina*, *Bosmina coregoni*, *Bosmina longirostris*, *Chydorus sphaericus*, and cyclopoid copepods. *Leptodora kindtii* was the largest cladoceran present, but because of its very low densities (<1 ind. l^{-1}) its biomass was very low. The two *Bosmina* species (*B. longirostris* and *B. coregoni*) were lumped.

Zooplankton biomass showed a distinct seasonal pattern (Fig. 1a). During winter the zooplankton biomass was virtually zero. The cyclopoid copepod peak in May was followed by a peak of *Daphnia hyalina* in June. In July the small zooplankton species (*Bosmina* spp. and *Chydorus sphaericus*) exhibited their peak biomass. In autumn the large zooplankton, *D. hyalina* and cyclopoid copepods, increased again, but their biomass maxima were lower than in the preceding spring and early summer. Zooplankton production (Fig. 1b) increased steadily to reach a peak in July and thereafter decreased again. The relative importance of the species changed during the year. Production of the small species was more important in July, but

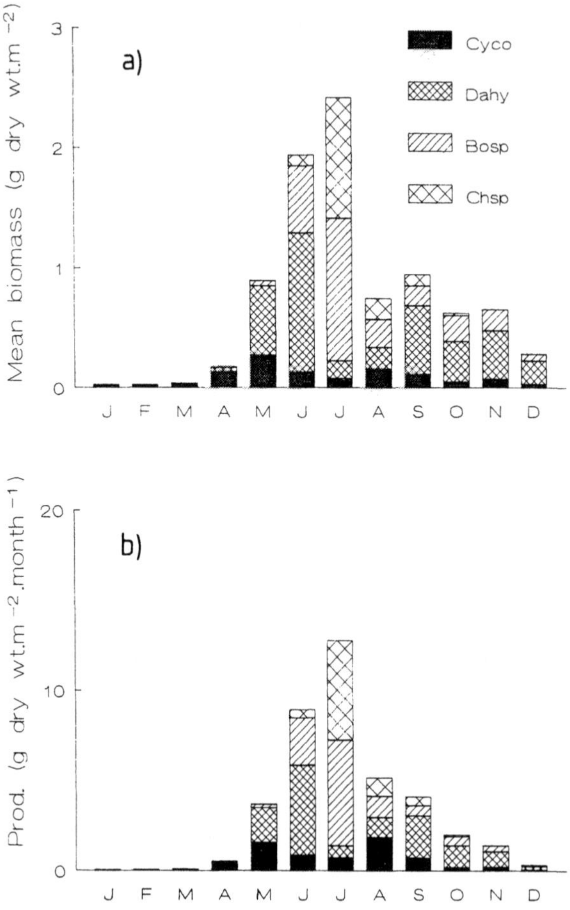

Fig. 1. Seasonal variation in zooplankton biomass and production. a. Mean monthly biomass (g dry wt m^{-2}), b. Monthly production (g dry wt m^{-2} $month^{-1}$). Dahy = *Daphnia hyalina*, Cyco = Cyclopoid copepods, Bosp = *Bosmina* spp., and Chsp = *Chydorus sphaericus*.

in spring, early summer and in autumn the large species contributed the most to the production.

Fish

The open-water fish community of the lake consists of eight species: bream (*Abramis brama*), white bream (*Blicca björkna*), roach (*Rutilus rutilus*), smelt (*Osmerus eperlanus*), pikeperch (*Stizostedion lucioperca*), perch (*Perca fluviatilis*), ruffe (*Gymnocephalus cernua*), and eel (*Anguilla anguilla*). The main pathways of organic matter flow within the community are determined by

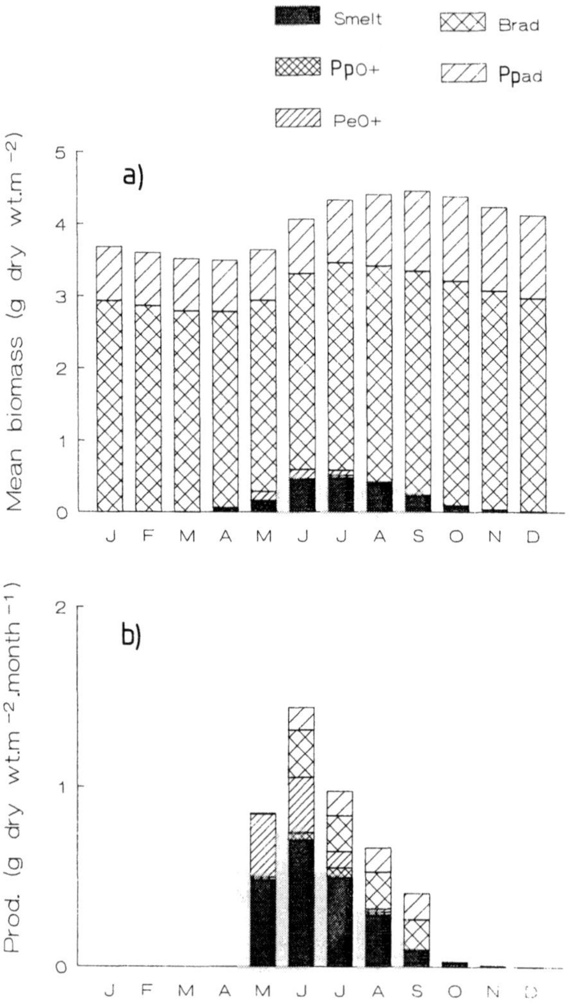

Fig. 2. Seasonal variation in fish biomass and production. a. Mean monthly biomass (g dry wt m^{-2}), b. Monthly production (g dry wt m^{-2} $month^{-1}$). SmO + = O + smelt, PpO + = O + pikeperch, PeO + = O + perch, Brad = ≥I + bream, and Ppad = ≥ I + pikeperch.

three groups; ≥I + bream, ≥I + pikeperch, and O + fish.

The bream population is the largest and the most stable component in the community (Fig. 2a). Its stability is based on a large number (15) of year classes. In 1978 approximately one quarter of the standing stock biomass was composed of III + (year class 1975) individuals, with individual lengths ranging from ca. 17 cm in spring to 22 cm in autumn. Production started in June and was relatively constant during the growing season (Fig. 2b).

Pikeperch is the main predator in Tjeukemeer and ≥I+ pikeperch is strictly piscivorous. In 1978 the standing stock biomass was composed mainly of two year classes: II+ (1976), and III+ (1975), with mean lengths of ca. 30 and 40 cm at the start of the season. Both the standing stock biomass and the production of pikeperch were relatively low compared with the biomass and production of bream (Fig. 2a, b).

The O+ fish are the most unstable component in the system. Their first appearance in April–May, as a result of spawning, depend on the ambient temperature (Mooij & Van Tongeren, 1990). The rate of decrease in O+ fish biomass during August–December depends largely on the predation pressure exerted by the pikeperch (Fig. 2a). Smelt dominates the O+ fish group, comprising 80–90% of the biomass. Total recruitment of smelt can be influenced considerably by migration of larvae from IJsselmeer (van Densen & Vijverberg, 1982). In 1978 O+ perch and O+ pikeperch also contributed substantially to the total O+ fish biomass. The biomass of O+ fish is low as compared with the biomass of the ≥I+ bream. However, the production of O+ fish is much higher than the production of bream (Fig. 2a, b). For instance the O+ fish production in June was 70% of the total fish production, but its biomass was only 10% of the total fish biomass. This is caused by the much higher growth rates of the O+ fish compared with the growth rates of the ≥I+ bream, and ≥I+ pikeperch. O+ fish production was highest in June, decreasing thereafter until the end of the year.

Zooplankton production and consumption by planktivorous fish

From the gut content analysis of the planktivorous fish species it was observed that O+ fish, the main tertiary producer, predates mainly on the large zooplankton species (see also van Densen, 1985). Usually the large prey species make up more than 95% of the fish diet. Only when the biomass of the large species in the environment is very low (July) are the small species also eaten considerably (ca. 40%). ≥I+ bream also feeds size selectively on the large zooplankton species, but its feeding behaviour differs. Bream larger than 15 cm are filter feeders, whereas smaller bream are generally particulate feeders (Lammens, 1984). The size selection of the filter feeding bream depends primarily on the mesh size of the branchiospinal system of the fish, which increases with increasing fish size (Lammens & Hoogenboezem, in press). Moreover the mean proportion of benthic cladocerans and larval chironomids on biomass basis in the bream's diet is ca. 50%, because large bream regularly switch to benthic feeding. This different feeding behaviour resulted in a different consumption pattern of the zooplankton of the O+ fish and ≥I+ bream.

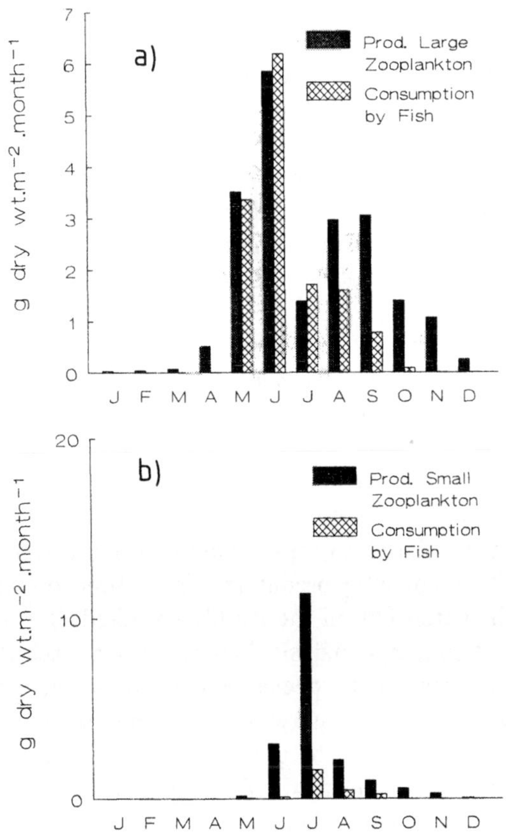

Fig. 3. Seasonal variation in the production of the zooplankton (g dry wt m^{-2} $month^{-1}$), and the total fish consumption (g dry wt m^{-2} $month^{-1}$). a. Production and consumption by fish of large zooplankton (*D. hyalina* and cyclopoid copepods), b. Production and consumption by fish of small zooplankton (*Bosmina* spp. and *Chydorus sphaericus*).

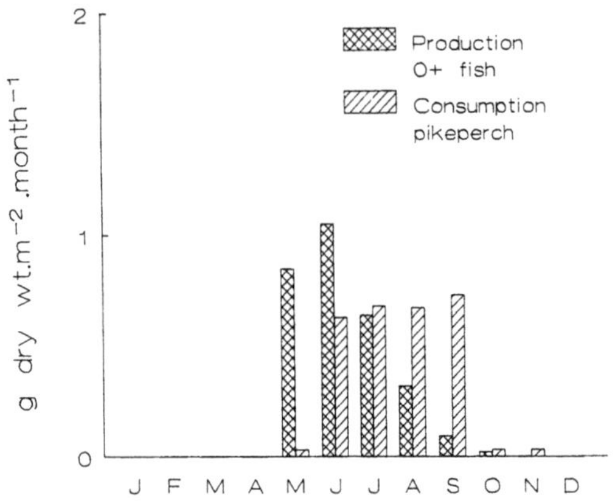

Fig. 4. Seasonal variation in the total production of O + fish (g dry wt m^{-2} $month^{-1}$) and the consumption by pikeperch (g dry wt m^{-2} $month^{-1}$).

In June O + fish only rarely fed on *D. hyalina*, whereas ≥I + bream fed almost exclusively on *D. hyalina* at that time.

During the growth season almost the entire production of the large zooplankton was eaten by the planktivorous fish (Fig. 3a), in contrast with a very small proportion of the small zooplankton production that was consumed (Fig. 3b).

Planktivorous fish production and consumption by piscivorous fish

During winter and spring, ≥I + pikeperch fed predominantly on I + fish, and occasionally also on II + fish. Larger prey cannot be handled by the 30–40 cm long predators. From June onwards pikeperch fed on the freshly recruited O + fish. Gut contents analysis does not indicate selection for certain fish species, and hence smelt which was most abundant was also eaten most often. In the first part of the growing season (May–June) pikeperch consumed only a small part of the O + fish production, later in the year, however, the consumption of pikeperch exceeded the production of O + fish (Fig. 4).

Discussion

On an annual basis planktivorous fish consumed 40% of the total zooplankton production. This may seem low, but the food chain efficiency is much higher if only the large zooplankton is taken into account; 70% of the large zooplankton production is consumed by planktivorous fish. Since ≥I + bream is not eaten by II + and III + pikeperch the food chain efficiency of pikeperch consumption is only moderate. But the efficiency is very high if only O + fish are taken into account, since the O + production eaten by pikeperch is ca. 95% (Fig. 5).

Smelt only rarely fed on *D. hyalina* in June, although the biomass of this prey is considerable at that time, and from mid June onwards no morphological limitations exist to prevent consumption of even the largest *D. hyalina* present. We cannot explain this, but it could well be related to the earlier feeding history, namely feeding on copepods (Furnass, 1979). This not feeding on *D. hyalina* by O + smelt in June is offset by the high predation pressure of ≥I + bream on *D. hyalina*; almost 100% of the bream diet consists of *D. hyalina*, which results in a consumption of almost the entire production of large zooplank-

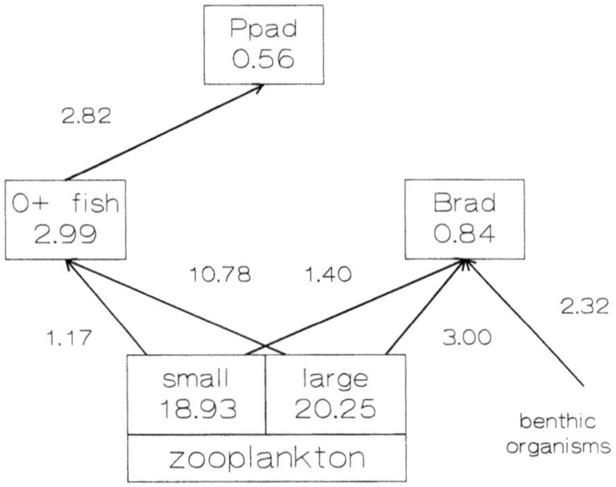

Fig. 5. The Tjeukemeer foodweb in 1978. Production (within squares), and consumption (next to arrows) of the most important groups (g dry wt m^{-2} $year^{-1}$). Ppad = ≥I + pikeperch, Brad = ≥I + bream, O + fish = Total of O + smelt, O + pikeperch, and O + perch.

ton in June. In July the O + fish exerts a very high predation pressure upon the large zooplankton, which wipes out almost all large zooplankton. The low biomass of large zooplankton in July causes both the O + fish and ≥I + bream to switch to feeding on small zooplankton which is consumed in substantial amounts at that time. From August onwards the production of the large zooplankton is higher than the consumption of the planktivorous fish, and this production is thought to be lost for the next trophic level. During 1978 the size of the O + fish stock was moderate, and consequently the predation pressure on the zooplankton was moderate too. In years when the O + fish stock is larger, the large zooplankton will be reduced to low levels during July–September, and a greater part of the small species will be utilized by the O + fish (van Densen, 1985). Hence, a low utilization of the small zooplankton species by fish is characteristic of years with low or moderate O + fish stocks. Most of this small zooplankton production is probably consumed by invertebrate predators (i.e. cyclopoid copepods, *Leptodora kindtii*, water mites). When high O + fish stocks reduce the biomass of large zooplankton, large bream is forced to switch to a benthic feeding behaviour, because of its filter feeding habit and hence its relatively low ability to feed on small zooplankton (Lammens, 1984). Since the biomass of benthic invertebrates is generally low in Tjeukemeer the condition of ≥I + bream will deteriorate (Lammens, 1982).

At first glance there is a poor agreement between the production of O + fish and its consumption by pikeperch. Pikeperch consumption is too low in the beginning of the year and too high at the end. However, this does not mean that the O + biomass produced in the beginning of the year is lost for consumption by pikeperch, since most of the production is accumulated in O + fish biomass. This O + fish biomass is still available for the pikeperch in the autumn when O + fish production is lower than the consumption by pikeperch. This conclusion is corroborated by the observation that O + fish biomass starts to decrease at the same time (July–August) that pikeperch consumption exceeds O + fish production.

It can be concluded that the seasonal variations in production and consumption are large, caused mainly by differences in the water temperature (range 1–23 °C). Both zooplankton (Vijverberg, 1980) and fish growth (Mooij & van Tongeren, 1990) are very much temperature related. Temperature also affects the spawning and recruitment cycle of the fish, and hence the time O + fish enters the system. Being the most important zooplankton consumer, as well as the most important prey group, O + fish plays a key role in the Tjeukemeer food web.

Acknowledgements

We thank Aafje Frank-Landman, Theo Frank, Bertus Lemsma, Peter MacGillavry, and Koos Swart for their assistance in the field and the laboratory, Anja van der Honig for the typing of the manuscript and Nico van Benthem for preparing the illustrations. We acknowledge Ramesh Gulati for his critical comments on different versions of the manuscript, and Onno van Tongeren for his interest and stimulating discussions.

References

Bottrell, H. H., A. Duncan, Z. M. Gliwicz, E. Grygierek, A. Herzig, A. Hillbricht-Ilkowska, H. Kurasawa, P. Larsson & T. Weglenska, 1976. A review of some problems in zooplankton production studies. Norw. J. Zool. 24: 419–456.

Brett, J. R. & T. D. D. Groves, 1979. Physiological energetics. In W. S. Hoar, D. J. Randall & J. R. Brett (eds), Fish Physiology. Vol. 8. Academic Press, New York: 279–352.

Cohen, J. E., 1978. Food webs and niche space. Princeton University Press, Princeton, New Jersey, USA, 250 pp.

Densen, W. L. T. van, 1985. Feeding behaviour of major O + fish species in a shallow, eutrophic lake (Tjeukemeer, The Netherlands). Z. angew. Ichthyologie 1: 49–70.

Densen, W. L. T. van & J. Vijverberg, 1982. The relation between O + fish density, zooplankton size and vulnerability of pikeperch *Stizostedion lucioperca*, to angling in the Frisian lakes. Hydrobiologia 95: 321–336.

Furnass, T. I., 1979. Laboratory experiments on prey selection by perch fry (*Perca fluviatilis*). Freshwat. Biol. 9: 33–43.

the fluorometric approach has been widely used. Other assays for the measurement of the activity of more specific phosphatases, such as 5′-nucleotidases, in aquatic environments have also been introduced (Karl & Craven, 1980; Ammerman & Azam, 1985).

In our early studies of phytoplanktonic P nutrition we used variations of the pNPP spectrophotometric assay (Berman, 1970; Wynne, 1977, 1981). We now describe a simple modification of the method which permits sensitive, rapid and easy quantification of organism associated alkaline or acid (AcP) phosphatase activities in natural waters. Instead of measuring the enzymatic activity of 'whole' water samples containing suspended organisms, we determine the activity of microplankton collected on filters of various porosity. This technique enables considerable concentration of the microorganisms responsible for the enzymatic activity. Moreover, by using filters of different pore size, enzymatic activity can be assigned to specific size fractions of the microplankton and the relative contributions of various sized organisms to the total, particulate phosphatase activity in the water may be determined. We note that Hino (1988) has previously briefly mentioned the determination of AP from algal samples collected on GF/C filters but without any elaboration of experimental detail.

Methods

In order to establish the protocol for the 'filter' assay we determined AP and AcP activities in axenic algal cultures which were harvested at exponential phase and transferred to SCM culture medium (Moss, 1972) without phosphorus for 3–4 days prior to analysis. Enzyme activities were also measured in natural water samples taken from lake Kinneret and a nearby water impoundment.

For 'filter' assays of AP and AcP activity a suitable volume of lake water (usually 200–400 ml) or algal culture (10–25 ml) was passed through a 47 mm Schleicher and Schull filter (0.2 μm and 0.45 μm for lake and culture samples respectively) which had been prewashed in boiling distilled water. For size fractionation experiments, Nitex netting (20 μm mesh), prerinsed Nuclepore polycarbonate filters (3 μm and 0.8 μm) and Schleicher and Schull filters (0.2 μm) were used. Maximum vacuum for filtration did not exceed 100 mm Hg. The filters were placed at the bottom of a glass beaker and covered with 2.6 ml of either 0.05 M Tris (pH 9), containing 1 mM Mg Cl_2, or 0.2 M acetate buffer (pH 5) for AP and AcP determinations respectively. Reactions for the spectrophotometric assay were initiated by addition of 0.1 ml 10 mM p-nitrophenyl phosphate (pNPP). After incubation for one hour at 37 °C, the reaction was terminated with 0.3 ml of 1 M NaOH and the supernatant of the reaction mixture poured into a small centrifuge tube. A brief centrifugation (3 min at 5000 rpm) was used to remove any cell debris or filter fibers, and the optical density of the supernatant at 410 nm was measured. The extinction coefficient for pNP was taken as 16200 (Garen & Leventhal, 1960).

The 'filter' spectrophotometric assay was compared to the method described previously by Berman (1970). For the latter, AP activities were determined in 1.3 ml samples of filtered and unfiltered Lake Kinneret water incubated with 1 ml of 1 mM pNPP, 0.7 ml of 1 M Tris buffer (pH 8.6) and 0.1 ml chloroform in test tubes at 37 °C for 2–4 days. AcP was similarly measured using 0.2 M acetate buffer (pH 5) with an addition of 0.3 ml lM NaOH at the end of the incubation.

The fluorometric AP assay was carried out as described by Pettersson (1980) using 4-methylumbelliferyl phosphate (MUP; Sigma Chemical Co.) as substrate. Filtrations were performed as outlined above for the spectrophotometric assay, except that only 10–20 ml of lake water was required. A reaction mixture containing 4 ml of 0.1 M Tris buffer, pH 8.3, 1 mM Mg Cl_2, and 0.5 ml 0.1 mM MUP was added to the filters, which were incubated at 37 °C for 1 hour. The reaction mixture was then filtered (0.45 μm) and the fluorescence read on a Turner Model 430 spectrofluorimeter (excitation: 360 nm; emission: 450 nm). Fluorescence units were converted to

Hydrobiologia **207**: 287–294, 1990.
D. J. Bonin & H. L. Golterman (eds), Fluxes Between Trophic Levels and Through the Water–Sediment Interface.

Phosphatases revisited: analysis of particle-associated enzyme activities in aquatic systems

Thomas Berman, David Wynne & Bina Kaplan
Kinneret Limnological Laboratory, Israel Oceanographic & Limnological Research, P.O. Box 345, Tiberias, 14102 Israel

Key words: Microplankton phosphatases, acid-alkaline, size fractionation

Abstract

A modified assay for alkaline or acid phosphatases associated with microorganisms in aquatic environments has been developed. This is based on collecting microplankton on filters and subsequently determining the enzyme activities spectrophotometrically or fluorometrically, using p-nitrophenyl- phosphate or 4-methylumbelliferyl phosphate, respectively as substrates. The assay is simple and rapid, and has the further advantage of permitting phosphatase activities to be assigned to specific size fractions of the natural microplankton. In samples taken from Lake Kinneret and a nearby reservoir, a consistently high proportion of the total alkaline or acid phosphatase activity was associated with the size fraction $<0.8\ \mu m$ $>0.2\ \mu m$ indicating the potentially high contribution of bacteria to these activities. This approach can also be used to examine the enzymatic potential of microplankton to release orthophosphate from other organo-phosphate substrates.

Introduction

Phosphatase activities in natural waters have been the focus of much research attention over the past two decades (Reichardt *et al.*, 1967; Berman, 1969; Perry, 1972; Hino, 1988). The original rationale for this interest was the assumption that the level of alkaline phosphatase (AP) activity measured in natural water samples could be used as an indicator of the nutritional status of phytoplankton in respect to phosphorus (Berman, 1970) and also as a measure of lake trophy (Chróst & Krumbeck, 1986). Although in many microorganisms, AP is induced in situations where orthophosphate availability is low, it has been difficult to define absolute levels of this activity which would be a characteristic indicator for phosphorus limitation or sufficiency in natural waters (but see Healey, 1978). Nevertheless, phosphatases continue to be measured and, when used with appropriate caution, can be useful indicators of phosphorus cycling and availability to microorganisms in aquatic ecosystems (Karl & Craven, 1980; Pettersson, 1980; Gage & Gorham, 1985; Huber & Kidby, 1985; Hino, 1988).

Generally, two analytical techniques have been used to determine phosphatase activities in natural waters. The classical spectrophotometric assay (Lindhardt & Walter, 1965) based on the enzyme mediated hydrolysis of p-nitrophenol phosphate (pNPP) has been supplemented by more sensitive fluorescent methods which employ substrates such as 3-O-methylfluorescein phosphate (Perry, 1972) or 4-methylumbelliferyl phosphate (Pettersson & Jansson, 1978). For studies in natural waters where enzyme activities are low,

Lammens, E. H. R. R., 1982. Growth, condition and gonad development of bream (*Abramis brama*) in relation to its feeding conditions in Tjeukemeer. Hydrobiologia 95: 311–321.

Lammens, E. H. R. R., 1984. A comparison between the feeding of white bream (*Blicca björkna*) and bream (*Abramis brama*). Verh. int. Ver. Limnol. 22: 886–890.

Lammens, E. H. R. R. & W. Hoogenboezem, in press. Diets and feeding behaviour. In I. J. Winfield & J. S. Nelson (eds), The biology of cyprinid fishes. Chapman & Hall Publ., London: **XXX–XXX**.

Lammens, E. H. R. R., H. W. de Nie, J. Vijverberg & W. L. T. van Densen, 1985. Resource partitioning and niche shifts of bream (*Abramis brama*) and eel (*Anguilla anguilla*) mediated by predation of smelt (*Osmerus eperlanus*) on *Daphnia hyalina*. Can. J. Fish. aquat. Sci. 42: 1342–1351.

May, R. M., 1973. Stability and complexity in model ecosystems. Princeton University Press, Princeton, New Jersey, USA, pp. 235.

Mooij, W. M. & O. F. R. van Tongeren, 1990. The growth of O+ roach (*Rutilus rutilus*) in relation to temperature and size in a shallow eutrophic lake: comparison of field and laboratory data. Can. J. Fish. aquat. Sci. 47: 960–967.

O'Grady, K. T. & P. B. Spillett, 1985. Gross nutrition and conversion efficiency of intensively and extensively reared carp (*Cyprinus carpio* L.). In C. B. Cowey, A. M. Mackie & J. G. Bell (eds), Nutrition and feeding in fish. Academic press, London: 269–280.

Paine, R. T., 1988. Food webs, road maps of interactions or grist for theoretical development? Ecology 69: 1648–1654.

Pauly, D., 1980. On the interrelationships between natural mortality, growth parameters, and mean environmental temperature in 175 fish stocks. J. Cons. perm. int. Explor. Mer. 39: 175–192.

Pimm, S. L., 1982. Food webs. Chapman & Hall, London, U.K., pp. 230.

Popova, O. A., 1978. The role of predaceous fish in ecosystems. In S. D. Gerking (ed.), Ecology of freshwater fish production. Blackwell Scientific Publ., London: 215–249.

Rejmánek, M. & P. Starý, 1979. Connectance in real biotic communities and critical values for stability in model ecosystems. Nature 280: 311–313.

Ricker, W. E., 1946. Production and utilization of fish populations. Ecol. Monogr. 16: 374–391.

Soofiani, N. M. & A. D. Hawkins, 1985. Field studies of energy budgets. In P. Tytler & P. Calow (eds), Fish energetics – New perspectives. Croom Helm, London: 283–307.

Vijverberg, J., 1977. Population structure, life histories and abundance of copepods in Tjeukemeer, The Netherlands. Freshwat. Biol. 7: 579–597.

Vijverberg, J., 1980. Effect of temperature in laboratory studies on development and growth of Cladocera and Copepoda from Tjeukemeer, The Netherlands. Freshwat. Biol. 10: 317–340.

Vijverberg, J. & Th. H. Frank, 1976. The chemical composition and energy contents of freshwater copepods and cladocerans in relation to their size. Freshwat. Biol. 6: 333–345.

Vijverberg, J. & A. F. Richter, 1982a. Population dynamics and production of *Daphnia hyalina* Leydig and *Daphnia cucullata* Sars in Tjeukemeer. Hydrobiologia 95: 235–259.

Vijverberg, J. & A. F. Richter, 1982b. Population dynamics and production of *Acanthocyclops robustus* (Sars) and *Mesocyclops leuckarti* (Claus) in Tjeukemeer. Hydrobiologia 95: 261–274.

Vijverberg, J. & W. L. T. van Densen, 1984. The role of the fish in the foodweb of Tjeukemeer, The Netherlands. Verh. int. Ver. Limnol. 22: 891–896.

Received 28 June 1989; in revised form 8 February 1990; accepted 25 February 1990 concentrations using a standard curve (0.4–40 nM) of 4-methylumbelliferone (MU; Sigma) and activity expressed as μmoles MU formed l^{-1} h^{-1}.

We also tested the ability of lake microplankton, immobilised on filters as above, to hydrolise organo-phosphorus substrates (D-glucose 6 phosphate, AMP, ATP, ribose 5-phosphate, O-phospho-DL-Tyrosine, O-phosphothreonine and phytic acid). In this case, the reaction mixture consisted of 100 μM P of substrate in 14.3 ml of 0.05 M Tris buffer (pH 9) containing 1 mM Mg Cl_2. After 3 hour incubation at 37 °C, the supernatant was decanted, centrifuged (3 min at 5000 rpm) to remove particles and the concentration of released orthophosphate measured (Strickland & Parsons, 1968). Orthophosphate in controls, taken at zero time, were subtracted from the experimental samples. All samples were run in triplicate.

Results and discussion

Reaction conditions for the modified 'Filter' assay

Comparison of various filters showed that the lowest reagent mixture blanks were obtained with 0.45 μm or 0.2 μm cellulose nitrate (Schleicher & Schull), or with polycarbonate (Nuclepore) filters rather than with precombusted GF/C or GF/F glass-fibre filters. Variability between replicates was low, coefficients of variation for a lake water sample were <8.9% and <9.5% (n = 10) for AP and AcP respectively. For routine assays, all samples were run in triplicate.

A range of substrate concentrations of pNPP between 0.03 to 3.3 mM gave optimal activity at 0.33 mM. The activities of both AP and AcP were linear for at least 12 hours under our conditions. Samples collected on filters could be stored at − 10 °C for at least two weeks without loss of enzymatic activity. We found that this assay worked well both with algal cultures (*Pediastrum*, *Chlorella* sp., *Chlamydesmonas* sp., *Scenedesmus* sp., *Selenastrum* sp., *Synedra* sp., *Cosmarium* sp) and with lake or reservoir water samples.

Table 1. Inhibition (%) of alkaline and acid phosphatase activities in Lake Kinneret water* by orthophosphate (P_i).

Added P_i (mM)	% Inhibition	
	AP	AcP
0.01	23.6	11.3
0.1	38.6	21.6
1.00	75.0	43.5
10.00	93.4	69.4

* Samples taken from 1 m depth on 30 May 1988 for AP and on 31 May 1988 for AcP at a central station A.

Both AP and AcP activities were inhibited by orthophosphate (Table 1) although the array of enzymes from lake microplankton populations appeared to be less sensitive than some specific phosphatases (Pettersson, 1980). AcP activities as determined with this assay were less sensitive to orthophosphate inhibition than AP. Phosphatases from the dinoflagellate, *Peridinium cinctum* were previously found to be insensitive to orthophosphate inhibition at environmentally relevant concentrations (Wynne, 1981).

Comparison of the 'Filter' and 'Test tube incubation' methods

A series of experiments was run with Lake Kinneret water to compare the modified 'filter' assay with the 'test tube incubation' method used in previous studies (Reichardt *et al.*, 1967; Berman, 1969). At the levels of AP and AcP activity normally found in this lake, 'test tube' experiments required a 2–4 day incubation period in order to achieve absorbancies which could be accurately and reproducibly determined (Berman, 1970). In the present experiments, we compared particle associated phosphatase activities measured by pNP release on filters with those determined by the 'test tube' method on unfiltered and 0.45 μm filtered water (Table 2). For the latter method, we assumed that the difference between the activities in unfiltered and filtered samples (dissolved enzyme activity) reflected the

Table 2. Phosphatase activities* in Lake Kinneret: comparison of 'Filter' method and 'Test tube incubation' method (Berman, 1970).

Sample	Filter method	'Test tube incubation'	
Date (1988)		Particle associated	Dissolved
Alkaline phosphatase			
12 April	4.3 (0.1)	2.5 (0.2)	1.1 (0.5)
1 May	5.9 (0.2)	1.8 (0.5)	1.6 (0.4)
25 May	5.4 (0.9)	2.2 (0.2)	1.6 (0.4)
6 July	8.2 (0.1)	5.8 (0.3)	b.d.
Acid phosphatase			
20 June	9.2 (0.1)	3.6 (0.3)	12.9 (1.0)
6 July	9.3 (0.1)	10.3 (0.4)	21.4 (0.9)

* Activities are given as μmoles pNP formed l^{-1} h^{-1}. All experiments were run in triplicate, with the standard deviation given in parentheses. b.d. Below detection limit.

particle associated phosphatases. With the exception of AcP in the July 6th sample, the 'filter' method always gave higher particle associated activities than the 'test tube' technique (Table 2). Note that the 'filter' method does not provide any measurements of 'free dissolved enzymes' (Reichardt *et al.*, 1967) which were relatively high in these samples.

Size partitioning of phosphatase activities

The present method permits the size fractionation of plankton on filters of different pore size and the subsequent determination of phosphatase activities. We emphasize that size fractionation does not give absolute size differentiation. Some larger organisms do pass through nominally smaller meshes: some smaller organisms are retained on larger mesh filters and others may disintegrate as a result of filtration. Nevertheless differential filtration does give a useful indication of the general size classes of microorganisms contributing to the observed activity. As an example of this application, we show the results of a series of experiments with water taken from Lake Kinneret and from a nearby water reservoir at Amiad (Fig. 1, Fig. 2, also Table 3).

When phosphatases from particulates of four size fractions (>20 μm; $>3<20$ μm; $>0.8<3$ μm and >0.2 <0.8 μm) were assayed, a substantial proportion of both AcP and AP activities were always associated with microorganisms in the smallest size range, 0.2–0.8 μm. This fraction contains mainly bacteria although some picocyanobacteria may also have been present. In contrast, most ($>80\%$) of the chlorophyll in these samples was found in particles >3 μm (see also Table 3). This result emphasises the potentially important contribution which may be made by bacteria to total phosphatase activities in aquatic ecosystems. Furthermore, this observation indicates the difficulty of attempting to normalize AP or AcP activities to phytoplankton biomass or chlorophyll. Other investigations (e.g. Francko, 1983) have previously shown that significant AP activity may be present in particles less than 1 μm. The pattern of activities associated with different size fractions will obviously depend on the seasonal changes of microplankton populations and the characteristics of the specific aquatic environment being examined.

Other consistent features of the samples taken from Lake Kinneret in late winter and spring were the relatively low AP and AcP activities associated with organisms in the size classes between

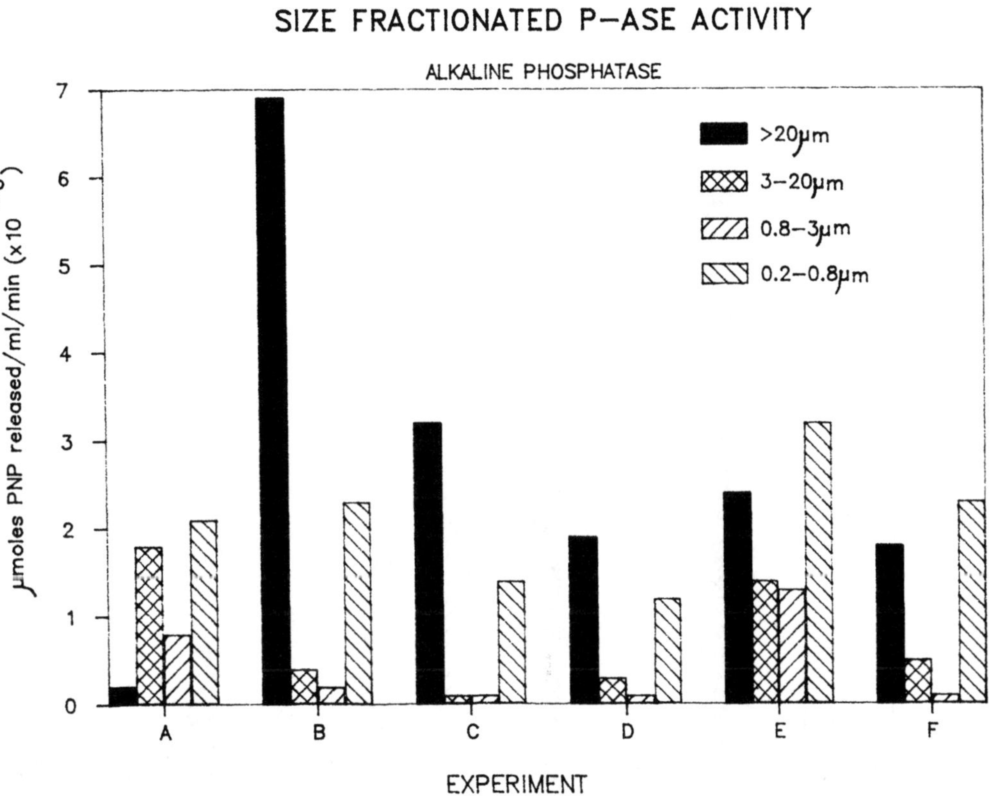

Fig. 1. Alkaline phosphatase activities (µmole PNP released ml^{-1} min^{-1} ($\times 10^{-6}$) associated with various size fractions of microplankton in natural water samples. A, B, D, E, and F: Lake Kinneret surface water, 12 January, 8 February, 21 February, 1 March. C. Amiad Reservoir 17 February 1988.

0.8 and 20 µm compared to those in the > 20 µm fraction. The high proportion of activity in the 20 µm fraction clearly reflects the dominance of large algae, *Melosira* sp. and *Peridinium gatunense* (formerly *P. cinctum*) which occurred at this season. A similar partitioning of AP and AcP activities in Lake Kinneret was again observed in the winter-spring of 1989 (data not shown). In contrast to this pattern samples taken in the fall or early winter showed relatively low activities in the largest size fraction, and relatively higher activities in the nanoplankton (3–20 µm) range.

In this experimental series, the size distribution of AcP activity (Fig. 2) was generally similar to that of AP (Fig. 1).

Comparison between fluorimetric and spectrophotometric methods

The results obtained with the spectrophotometric method with pNPP were compared to those with the fluorometric AP determination with MUP. From Table 3, it can be seen that although in two experiments (9 April, 21 May) the overall size distribution of activities were roughly similar, this was not always the case. In most instances, relatively high activities were determined for the smallest size fraction by both methods. For the larger fractions AP and AcP activities could be normalised to chlorophyll but this would obviously be incorrect for activities measured in the smallest size fraction (< 3 µm) which have very low chlorophyll concentrations. Absolute values of AP activity were very different for the two methods,

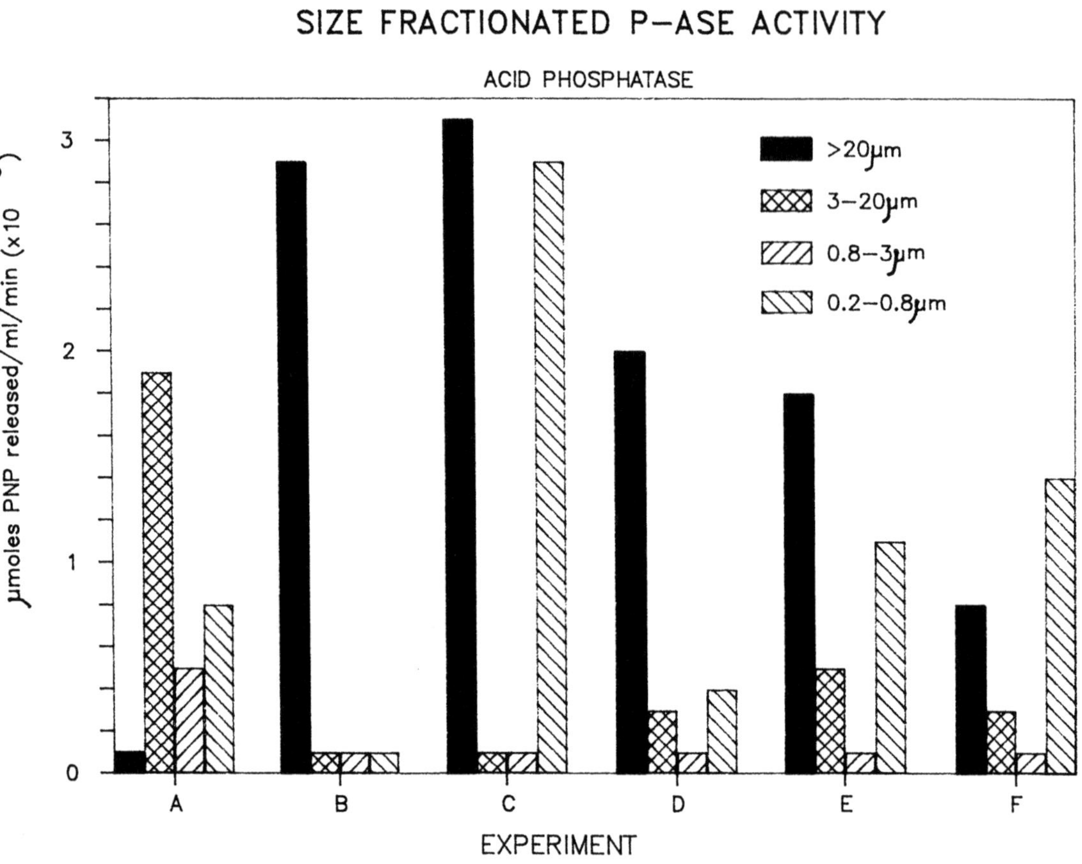

Fig. 2. Acid phosphatase activities associated with various size fractions of microplankton in natural water samples (Legend as in Fig. 1).

as might be expected because of the different enzyme affinities for the two substrates. The fluorometric method however had greater variability and probably could be improved. It should be emphasied that neither pNPP or MUP are substrates which are normally found in aquatic ecosystems and therefore the activities of microplankton AP or AcP as measured here may not reflect *in situ* activities with natural phosphate esters (see below).

Our results indicate that both spectrophotometric and fluorometric detection are suitable for AP determination using microorganisms immobilized on filters. Several differences should be pointed out: (a) absolute AP activity was higher and variability lower with pNPP than MUP (b) the fluorimetric method is more sensitive and required filtration of a much lower volume of water, and (c) addition of $MgCl_2$ was necessary for optimum AP activity in the fluorometric assay (data not shown), although in previous studies this has not been included (Pettersson, 1980).

Hydrolysis of P-esters

In view of the artificial nature of both pNPP and MUP as substrates we extended the simple method of immobilising microplankton on filters to check the potential of these organisms to release orthophosphate directly from a variety of organic phosphorus substrates which might be expected to be found in the dissolved organic phosphorus pool of natural waters. Hydrolysis of

Table 3. Chlorophyll concentrations (Chl) and comparison of spectrophotometric (S) and fluorometric (F) determinations of alkaline phosphatase in size fractionated surface water from Lake Kinneret.

Sample fraction	9 April 1989			21 May 1989		
	S	F	Chl	S	F	Chl
Unfiltered	0.93	0.14	2.3	1.09	0.02	3.2
3–20 μm	0.18	0.04	0.3	0.53	0.01	1.4
1–3 μm	0.07	0.04	0.4	0.03	0.005	0.4
0.2–1 μm	0.43	0.07	0.3	0.45	0.007	0.4
Sample fraction	23 May 1989			28 May 1989		
	S	F	Chl	S	F	Chl
Unfiltered	1.04	0.041	2.4	1.23	0.035	2.0
3–20 μm	0.55	0.056	1.3	0.69	0.068	1.1
0.8–3 μm	0.04	0.040	0.6	0.04	0.014	0.3
0.2–1 μm	0.41	0.027	0.2	0.40	0.014	0.1

Enzyme activities (S and F) given as μmoles pNP or MU formed $l^{-1} h^{-1}$. Chlorophyll (Chl) concentrations: $\mu g\, l^{-1}$. Averaged results of triplicate determinations.

the phosphate esters was found for all the substrates tested, with the exception of phytic acid (Table 4). Although this method is not as sensitive as those used by Karl & Craven (1980) or Ammerman & Azam (1985) it is simple and has the advantage of being applicable to a wider range of substrates. We have not yet investigated this method fully but it would appear to be useful in evaluating the potential for hydrolysis of natural organo-phosphate substrates by microplankton.

Table 4. Release of orthophosphate (μmoles $P\, l^{-1} h^{-1}$) from organophosphate substrates by Kinneret microorganisms*.

Substrate	Experiment date (1989)		
	11 June	15 June	15 July
AMP	17.4	38.0	9.2
ATP	–	70.0	10.2
Glucose 6-P	14.0	24.2	9.8
Ribose 5-P	11.6	26.2	11.8
O-P-Tyrosine	11.1	26.0	9.2
Threonine-P	9.6	14.8	9.8
Phytic acid	0	0	0

* Microplankton collected from 250 ml of near surface lake water on 0.45 μm filters – experimental conditions as described in methods.

Conclusions

Our experiments point out a basic problem associated with APA and AcPA measurements in natural waters; i.e. the use of 'unnatural' substrates (pNPP and MUP) may give highly biased results. Presumable a better approach would be to use as substrates those compounds which are more likely to be present in the dissolved organic phosphorus pool. We suggest that a comparative study using variants of the 'filter' method here would be helpful in assessing the applicability of using phosphatase activities to investigate phosphorus cycling or the phosphorus status of microorganisms in natural waters.

The modified 'filter' assay as described in this paper has several advantages over the previously used 'test tube' method. The technique allows rapid spectrophotometric or fluorometric measurement of AP and AcP activities associated with particulate fractions in natural water samples. Sufficient quantities of water can be easily filtered to insure adequate enzyme activity. Samples can be filtered and then stored at − 10 °C for at least two weeks until the assay is carried out. The most important advantage of the method is the possibility of determining directly the relative phosphatase activities associated with various size fractions of the plankton. The filter technique, however, does not permit estimation of free 'dissolved' phosphatases, which may sometimes contribute a considerable portion to total phosphatase activities in natural waters (Reichardt *et al.*, 1967).

Acknowledgements

This work was supported by a grant (No. WF12) to T. Berman from the Israel National Council for

Research and Development and by KFK, Karlsruhe, F. R. Germany, and by funding from the Israel Water Commissioner. A contribution of Israel Oceanographic & Limnological Research.

References

Ammerman, J. W. & F. Azam, 1985. Bacterial 5′-nucleotidase in aquatic ecosystems. A novel mechanism of phosphorus regeneration. Science 227: 1338–1340.

Berman, T., 1969. Phosphatase release of inorganic phosphorus in Lake Kinneret. Nature 224: 1231–1232.

Berman, T., 1970. Alkaline phosphatases and phosphorus availability in Lake Kinneret. Limnol. Oceanogr. 15: 663–674.

Chróst, R. J. & H. J. Krambeck, 1986. Fluorescence correction for measurements of enzyme activity in natural waters using methylumbelliferyl-substrates. Arch. Hydrobiol. 106: 91–117.

Francko, D. A., 1983. Size fractionation of alkaline phosphatase activity in lake water by membrane filtration. J. Freshwat. Ecol. 2: 305–309.

Gage, M. & E. G. Gorham, 1985. Alkaline phosphatase activity and cellular phosphorus as an index of the phosphorus status of phytoplankton in Minnesota lakes. Freshwat. Biol. 15: 227–333.

Garen, A. & C. Leventhal, 1960. A fine structure genetic and chemical study of the enzyme alkaline phosphatase of E. coli. Biochim. Biophys. Acta 38: 470–483.

Healey, F. P., 1978. Physiological indicators of nutrient deficiency in algae. Mitt. int. Ver. Limnol. 21: 34–41.

Hino, S., 1988. Fluctuation of algal alkaline phosphatase activity and the possible mechanisms of hydrolysis of dissolved organic phosphorus in Lake Barato. Hydrobiologia 157: 77–84.

Huber, A. L. & K. Kidby, 1985. An examination of the factors involved in determining phosphatase activity in estuarine waters. 2. Sampling procedures. Hydrobiologia 111: 13–19.

Karl, D. M. & D. B. Craven, 1980. Effects of alkaline phosphosphatase activity on nucleotide measurements in aquatic microbial communities. Appl. envir. Microbiol. 40: 549–561.

Lindhardt, K. & K. Walter, 1965. Phosphatases. In: H. U. Bergmeyer, (ed.), 779–785. Methods of Enzymatic Analysis, Verlag Chemie.

Moss, B., 1972. The influence of environmental factors on the distribution of freshwater algae: an experimental study. J. Ecol. 60: 917–932.

Perry, M. J., 1972. Alkaline phosphatase activity in subtropical Central North Pacific waters using a sensitive fluorimetric method. Mar. Biol. 15: 113–119.

Pettersson, K., 1980. Alkaline phosphatase activity and algal surplus phosphorus as phosphorus-deficiency indicator in Lake Erken. Arch. Hydrobiol. 89: 54–87.

Pettersson, K. & M. Jansson, 1978. Determination of phosphatase activity in lake water – a study of methods. Verh. int. Ver. Limnol. 20: 1226–1230.

Reichardt, W., J. Overbeck & L. Steubing, 1967. Free dissolved enzymes in lake waters. Nature Lond. 216: 1345–1347.

Strickland, J. P. H. & T. R. Parsons, 1968. A practical handbook of seawater analysis. Bull. Fish. Res. Bd Can. 167: 311 p.

Wynne, D., 1977. Alterations in activity of phosphatases during the *Peridinium* bloom in Lake Kinneret. Physiol. Pl. 40: 219–224.

Wynne, D., 1981. The role of phosphatases in the metabolism of *Peridinium cinctum* from Lake Kinneret. Hydrobiologia 83: 93–99.

Received 28 July 1989; in revised form 15 December 1989; accepted 25 February 1990

Hydrobiologia **207**: 295–301, 1990.
D. J. Bonin & H. L. Golterman (eds), Fluxes Between Trophic Levels and Through the Water–Sediment Interface.

Advantages of distinguishing the active fraction in bacterioplankton assemblages: some examples

Philippe Dufour [1], Jean-Pascal Torreton [2] & Michel Colon [3]
[1] *ORSTOM – Institut de Limnologie – B.P. 11 F – 74203 Thonon-les-Bains Cedex, France;* [2] *ORSTOM – Centre de Recherches Océanographiques – BP V18 – Abidjan – Côte d'Ivoire;* [3] *INRA – Institut de Limnologie – B.P. 11 F – 74203 Thonon-les-Bains Cedex, France*

Key words: bacterioplankton, abundance, dormancy, ETS activity, bacterial activity, methods

Abstract

The difficulty of distinguishing between active and dormant or dead bacterial cells is an important problem for the aquatic microbiologist.

Active cells can be detected under the microscope by the presence of an intact electron transport system able to reduce the colourless INT [2-(p-iodophenyl)-3-(p-nitrophenyl)-5-phenyl tetrazolium chloride] to an optically dense intracellular deposit.

An improvement of this method has been applied to Lake Geneva and to a fish pond in the Ivory Coast. The portion of INT-reducing bacterial cells ranged from 1 to 71%, depending on place, depth, season and time of the day. In all cases bacterial activity, determined by uptake of ^{3}H Thymidine or ^{14}C glucose, and frequency of dividing cells were better correlated with the number of INT-reducing cells than with the total number of cells. This means that counts of cells able to reduce INT have a better metabolic significance than total cell counts. Some examples are developed which show the advantages of applying this method in cases where it is useful to distinguish active cells in a bacterial assemblage.

Introduction

It has been recognized that only part of the bacterial aquatic flora is metabolically active at a given moment (Stevenson, 1978). Because of resting or dead cells, total abundance of bacteria does not reflect activity. Therefore, it may be pertinent to distinguish between active and inactive cells in a bacterial assemblage. This may further lead us to inquire what changes in a bacterium's environment could determine its activity, dormancy or death.

A method first proposed by Von Bielig *et al.* (1949) and further developed by Iturriaga & Rheinheimer (1975) allows the enumeration of respiring bacterial cells. Respiring cells are defined as being able to reduce colorless INT [2-(p-iodophenyl)-3-(p-nitrophenyl)-5-phenyl tetrazolium chloride] to INT formazan recognized as an opaque intracellular deposit. Zimmermann *et al.* (1978) improved this method by combining the recognition of individual cells with respiring activity and direct counts by fluorescence microscopy. We further improved this method and applied it to samples from three different trophic environments. The aim was to evaluate its interest and metabolic significance in conjunction with growth indicators such as thymidine incorporation or ^{14}C glucose uptake.

Material and methods

Water samples were collected from a central station in Lake Geneva at 12 levels from surface to bottom (300 m) in March and May 1989, and in a fish pond in Ivory Coast in March 1988 at a unique level (0.10 m) every two hours over 34 h.

Every sample was assayed for ^{3}H Thymidine (TdR) incorporation (Riemann *et al.*, 1982) and ^{14}C glucose assimilation (Wright & Hobbie, 1966) at *in situ* temperature immediately after collection. Total counts of bacterial cells were obtained after staining with Dapi (Porter & Feig, 1980) and frequency of dividing cells was also noted (Hagström *et al.*, 1979). Grazing on bacteria was estimated by comparing the change of bacterial abundance in undisturbed water and in water filtered by gravity onto 35 μm mesh size plankton net and 10 μm, 3 μm and 1 μm pore size nuclepore membranes. The percentage of respiring bacterial cells was determined by counting those showing an electron transport system (ETS) activity (Zimmermann *et al.*, 1978) by means of an improved procedure briefly described below.

Freshly collected samples were incubated with INT (Aldrich, 0.02% final concentration) and NaCN (0.2 mM) for one hour in the dark at *in situ* temperature. Formalin-fixed controls (2% final concentration) were included for each sample. Incubation was stopped by formalin and samples were then stored in the dark at 4 °C. Samples were further filtered onto 0.1 μm pore size cellulose membranes (Millipore VC). Air-dried filters were then cut into 6 wedges. One wedge was cleared by floating it on a drop of Cargille immersion oil (code 848, $n_D^{25\ °C} = 1.506$) on a microscopic slide, organisms up. Bacteria were then immediately stained by a drop of Dapi (Boehringer, 10^{-5} wt/vol) and a cover slip was added. The preparation was then examined under microscope. Each field was photographed, first under bright light transmitted through the membrane in order to detect intracellular granules of formazan (Fig. 1A), and secondly with a standard epifluorescence procedure for total cell detection (Fig. 1B). The procedure was repeated on other wedges, until at least 400 cells were photographed for each sample.

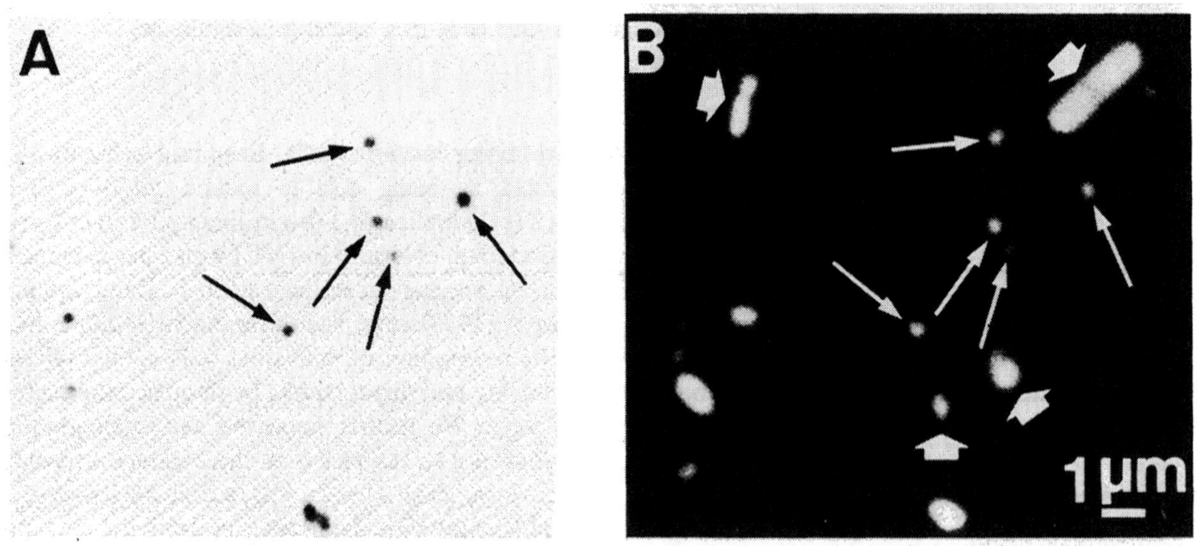

Fig. 1. Photomicrographs of bacterioplankton from Lake Geneva (9/6/89, surface, coastal zone). A: transmitted light; formazan cristals are accumulated in active cells. B: the same field under epifluorescent excitation shows all the cells. Large arrow: an inactive cell; fine arrow: an INT reducing cell. Note that even the smallest cells may present detectable formazan.

Results and discussion

Total bacterial abundance ranged from 0.1 to 9.4×10^6 cells ml^{-1}, in agreement with the trophic states of the different water bodies sampled. In Lake Geneva the average abundance in the whole water column (0–300 m) was 6×10^5 cells ml^{-1} in March and 9×10^5 cells ml^{-1} in May. Seasonal differences were more pronounced in the epilimnion (0–10 m) where 6×10^5 and 4.1×10^6 cells ml^{-1} were enumerated in March and May respectively. In the fish pond, abundance ranged from 5.6×10^6 at night to 9.4×10^6 cells ml^{-1} at sunset (average 7.1×10^6).

The percent of cells able to reduce INT ranged from 0.6 to 47% in Lake Geneva (average 10%) and from 28 to 71% in the tropical aquaculture pond (average 47%). These two values are included in the range reported in literature (Table 1). It may be interesting to check the significance of the large variations in the proportion of 'respiring' bacteria all through the water column for lake Geneva and during a diurnal cycle in the tropical aquaculture pond.

Spearman's rank correlation matrix (Table 2) shows that every index of activity (thymidine incorporation, glucose assimilation, abundance of dividing cells) was more closely related to INT-reducing cells than to total cell counts. As a matter of fact, in Lake Geneva in May, the profile of thymidine incorporation was more closely related to INT-reducing bacteria than to total bacteria (Fig. 2). In the epilimnion, the minimum of thymidine incorporation at depth 7.5 m was asso-

Table 1. Range of values for percent of INT-reducing bacteria in different water bodies.

Location	% active	References	Method
Kiel Fjord (FRG)	19–23	Iturriaga & Rheinheimer (1975)	Iturriaga & Rheinheimer (1975)
Westensee (lake, FRG)	23–36	Zimmermann *et al.* (1978)	Zimmermann *et al.* (1978)
Pond (FRG)	5– 9	Zimmermann *et al.* (1978)	
Baltic Sea (Kiel Firth, FRG)	12	Zimmermann *et al.* (1978)	
Baltic Sea (Kiel Bight, FRG)	7	Zimmermann *et al.* (1978)	
Eckernförde Bight, Baltic Sea setlting particles	8–97	Iturriaga (1979)	Iturriaga & Rheinheimer (1975)
Salt marsh creek (Calif., USA)			
surface layer	16	Harvey & Young (1980)	Zimmermann *et al.* (1978)
subsurface	5	Harvey & Young (1980)	cells >0.4 μm
Rice field, Camargue (Fr.)	5	Baldensperger (1981)	Zimmerman *et al.* (1978)
Salt marsh (Georgia, USA)			
Duplin river	31–36	Newell (1984)	Newell (1984)
Marsh creek	29–30	Newell (1984)	
Chesapeake Bay (Md, USA)			
1 m	25–94	Tabor & Neihof (1984)	Tabor & Neihof (1982)
8.5 m (near bottom)	29–62	Tabor & Neihof (1984)	
Lough Neagh (Ireland)	21–28	Quinn (1984)	Zimmermann *et al.* (1978)
Little Crooked lake (Ind., USA)	28–30	Lowell & Konopka (1985)	Zimmermann *et al.* (1978) cells >0.4 μm
Mc Loud lake (Florida, USA)	9	Dutton *et al.* (1985)	Bitton & Koopman (1982)
Wauberg lake (Florida, USA)	25	Dutton *et al.* (1985)	
Lake Geneva (Fr.)	1–47	This study	This study
Fish Pond (Ivory Coast)	28–71	This study	

Table 2. Spearman's rank correlation matrix between 3 measurements of bacterial activity and 2 abundance determinations (total and INT-reducing bacteria) in the three situations investigated. ns = not significant; * = significant $p < 0.05$; ** = $p < 0.01$; *** = $p < 0.001$.

	Total abundance	INT reducing bacteria	Sampling site
TdR incorporation	0.02 ns	0.64*	Lake Geneva (winter 89)
Glucose assimilation	o.30 ns	0.69*	Vertical profile (12 levels)
Abundance of dividing cells	0.71**	0.72**	Mesotrophic
TdR incorporation	0.67*	0.99***	Lake Geneva (spring 89)
Glucose assimilation	0.64*	0.79**	Vertical profile (12 levels)
Abundance of dividing cells	0.62*	0.71**	Mesotrophic
TdR incorporation	0.17 ns	0.58*	Fish pond (Ivory Coast)
Abundance of dividing cells	0.22 ns	0.61**	March 1988, 16 samples in 32 h Eutrophic

ciated with a minimum in INT-reducing bacteria (less than 3% of total counts), while total abundance was nearly constant from 0 to 10 m. This suggests that this minimum of thymidine incorporation rate per cell, an index of growth rate for the whole population, may be due to inadequate growth conditions for 97% of the total population. In March, again in Lake Geneva, total abundance was nearly constant in the whole water column (average 0.60×10^6 cell ml^{-1} in epilimnion, 0.62×10^6 cell ml^{-1} in hypolimnion), while thymidine incorporation rate decreased 12 times

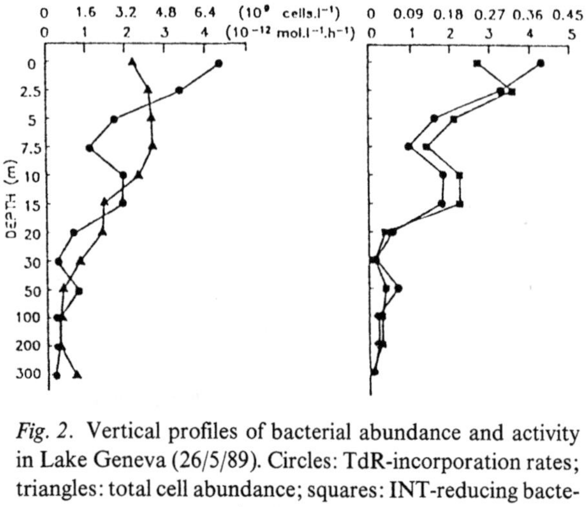

Fig. 2. Vertical profiles of bacterial abundance and activity in Lake Geneva (26/5/89). Circles: TdR-incorporation rates; triangles: total cell abundance; squares: INT-reducing bacteria abundance.

from epilimnion to hypolimnion. The enumeration of INT-reducing bacteria shows that the decrease in thymidine incorporation rate may be partly due to decreasing proportion of 'active' bacteria, from a 23% average for the epilimnion to a 9% average for the hypolimnion.

In the tropical fish pond, thymidine incorporation rate and abundance of dividing cells, both indices of bacterial cell production, increased by 1.4 and 1.2 respectively between sunrise and sunset, while the total bacterial abundance decreased 1.2 times during the same period (Table 3). The cells produced during the day may have disapeared by grazing, lysis and/or sedimentation. However, the active INT-reducing fraction

Table 3. Examples of bacteria-related parameters during a diurnal period in a fish pond in Ivory Coast (20/3/88).

	6 a.m.	8 p.m.
Total abundance 10^9 cells l^{-1}	7.6	6.4
INT reducing bacteria 10^9 cells l^{-1}	2.2	4.1
Dividing bacteria 10^9 cells l^{-1}	2.4	2.9
TdR incorporation rate 10^{-9} mol l^{-1} h^{-1}	1.7	2.5

rose from 29 to 64% of the total (Table 3). A constant or slightly increasing lysis of the cells (abundance is decreasing) is not compatible with the increase of the active fraction. Sedimentation is likely to be independent of the active proportion of the bacterial cells, but related only to the presence of sedimenting particles colonized by bacteria. Thus an increase of grazing on bacteria, acting probably equally on the growing and the inactive fraction, might be the major reason for bacterial losses. This is in agreement with the observed increase of thymidine incorporation per cell. An independent estimate of grazing by differential filtration confirmed that grazing was effective during the day and inexistent during the preceeding night.

The enumeration of INT-reducing bacteria allows also the more realistic estimation of generation time for bacteria. For example, in Lake Geneva, assuming exponential growth *in situ*, the experimentally determined conversion factor (according to Kirchman *et al.*, 1982) of 3.9×10^{18} cells per mol of incorporated thymidine allowed the estimation of a generation time of 146 h for the whole bacterial population of the surface sample in May (Fig. 2). Interpretation of INT reducing bacterial abundance suggests that there were, in that sample, only 7% of active cells with an average of 10 hours generation time.

An example of application in ecophysiology is given for Lake Geneva, where we sought to discover whether the bacterioplankton population in winter was mainly composed of strictly psychrophylic bacteria. A sample of surface water collected in February (6 °C *in situ* temperature), with only 9% of the total bacteria reducing INT, was incubated at 25 °C with 0.025% yeast extract (wt/vol, final concentration). Five hours later, the total number remained unchanged, but cell volumes were larger and 98% of the cells were found to be able to reduce INT. Cell-multiplication occurred later, after 9 h incubation. Direct examination of INT-reducing bacteria thus revealed that these bacteria are not strictly psychrophilic, but are able to grow at a higher temperature and are limited *in situ* by low temperature, and/or nutrient concentration.

This method could also be applied to check whether the lag phase before exponential growth in a water culture (Ammerman *et al.*, 1984) is due to 'awakening' of dormant bacteria and/or to the development of only a small part of the bacterial population. This question is important for the evaluation of the representativity of growing bacteria, since these cultures are often used to determine a conversion factor between thymidine or leucine incorporation and cell-production (Kirchman *et al.*, 1982).

These examples show the importance of distinguishing between active and dormant cells. Other methods have been used to make this distinction. Microautoradiography (Meyer-Reil, 1978) has been widely used, with an addition of significant amounts of substrate that may shift some bacteria from dormancy to activity, thus leading to an overestimation of *in situ* active bacteria. Moreover, it usually takes days before data can be provided. The colour emitted by cells stained with acridine orange is not directly related to cellular activity. (Jones, 1974; Quinn, 1984). The method proposed by Kogure *et al.* (1979), involving several hours incubations of bacteria with yeast extract and nalidixic acid, may also lead to overestimation, due to the substrate addition (Peele & Colwell, 1981), or to underestimation due to nalidixic acid resistant cells (Kogure *et al.*, 1979; Maki & Remsen, 1981). The fluorescein diacetate method proposed by Rotman & Papermaster (1966) has also been shown to be a method which underestimates active bacteria (Chrzanowski *et al.*, 1984; Quinn, 1984).

The INT-reducing bacterial count method implies neither substrate addition, nor long incubation times. The procedure of filter preparation used here is as simple as that described by Hobbie *et al.* (1977) for total cell count. It allows the enumeration of respiring and total cells on the same microscope field. When counting INT-respiring bacteria, errors due to subjectivity do not seem to be higher than those obtained by exclusive determination of the total number (Zimmermann *et al.*, 1978). Tetrazolium salt has shown no visible toxic effects during short incubations (Guha, 1957). Based on ETS activity, the reduction of INT is a component of virtually all bacterial types, including lithotrophs and methylotrophs (Smith & Hoare, 1977), facultative or obligate anaerobes (Packard *et al.*, 1983) and aerobic heterotrophs (Packard, 1985). The limit of the method is due to the size of formazan deposit into the cell. Cells with low activity and low-size deposits under detection may be considered as dormant or dead, leading to an underestimation of the active fraction. However, cells presenting such a low ETS activity may not be important for total community metabolism.

Conclusions

The above examples show that, with the use of this method of determination of active cells in conjonction with other specific methods, it may be possible to differenciate factors controlling bacterial production. Thus, in mesotrophic Lake Geneva at the time of experiment, bacteria seemed to be limited by starvation and/or low temperature, as attested by the low percentage of active cells, while in the eutrophic tropical fish pond, a high percentage of active cells suggests an important contribution to grazing. Therefore, we think that the improved INT method used here should be applied whenever the distinction between active and dormant or dead cells is necessary. It could help to improve knowledge of life strategies of bacteria, particularly in aquatic environments. This method is appropriate for a number of applications in ecophysiology and ecotoxicology.

Acknowledgements

This research was supported by ORSTOM and INRA (Paris, Fr.). N. Kaba's help with the English is gratefully acknowledged.

References

Ammerman, J. W., A. Hagström & F. Azam, 1984. Bacterioplankton growth in seawater: 1. growth kinetics and cellular characteristics in seawater cultures. Mar. Ecol. Prog. Ser. 18: 31–39.

Baldensperger, J. F., 1981. Short term variations of microbial and physicochemical parameters in submersion water over a rice field. Ann. Microbiol. 132 B: 101–122.

Bitton, G. & B. Koopman, 1982. Tetrazolium reduction-malachite green method for assessing the viability of filamentous bacteria in activated sludge. Appl. Envir. Microbiol. 43: 964–966.

Chrzanowski, T. H., R. D. Crotty, J. G. Hubbard & R. P. Welch, 1984. Applicability of the fluorescein diacetate method of detecting active bacteria in freshwater. Microb. Ecol. 10: 179–185.

Dutton, R. J., G. Bitton & B. Koopman, 1986. Application of a direct microscopic method of the determination of active bacteria in lakes. Wat. Res 20: 1461–1464.

Guha, A., 1957. A new technique for the viable count of bacteria. Nature 179: 1360.

Hagström, A., U. Larsson, P. Hörstedt & S. Normak, 1979. Frequency of dividing cells, a new approach to the determination of bacterial growth rates in aquatic environments. Appl. envir. Microbiol. 37: 805–812.

Harvey, R. W. & L. Y. Young, 1980. Enumeration of particle bound and unattached respiring bacteria in the salt marsh environment. Appl. envir. Microbiol. 40: 156–160.

Iturriaga, R., 1979. Bacterial activity related to sedimenting particulate matter. Mar. Biol. 55: 157–169.

Iturriaga, R. & G. Rheinheimer, 1975. Eine einfach Methode zur Auszählung von Bakterien mit aktivem Elektrontransportsystem in Wasser und Sediment-proben. Kiel. Meeresforsch. 31: 83–86.

Jones, J. G., 1974. Some observations on direct counts of freshwater bacteria obtained with a fluorescent microscope. Limnol. Oceanogr. 19: 540–543.

Kirchman, D., H. Ducklow & R. Mitchell, 1982. Estimates of bacterial growth from changes in uptake rates and biomass. Appl. envir. Microbiol. 44: 1296–1307.

Kogure, K., U. Simidu & N. Taga, 1979. A tentative direct microscopic method for counting living marine bacteria. Can. J. Microbiol. 25: 415–420.

Lowell, C. R. & A. Konopka, 1985. Seasonal bacterial production in a dimictic lake as measured by increases in cell numbers and thymidine incorporation. Appl. envir. Microbiol. 49: 492–500.

Maki, J. S. & C. C. Remsen, 1981. Comparison of two direct-count methods for determining metabolizing bacteria in freshwater. Appl. envir. Microbiol. 41: 1132–1138.

Meyer-Reil, L. A., 1978. Autoradiography and epifluorescence microscopy combined for the determination of number and spectrum of actively metabolizing bacteria in natural waters. Appl. envir. Microbiol. 36: 506–512.

Newell, S. Y., 1984. Modification of the gelatin-matrix

method for enumeration of respiring bacterial cells for use with salt-marsh water samples. Appl. envir. Microbiol. 47: 873–875.

Packard, T. T., 1985. Measurement of electron transport activity of microplankton. Adv. aquat. Microbiol. 3: 207–261.

Packard, T. T., P. C. Garfield & R. Martinez, 1983. Respiration and enzyme activity in aerobic and anaerobic cultures of the marine denitrifying bacterium, *Pseudomonas perfectomarinus*. Deep-Sea Res. 30: 227–243.

Peele, E. R. & R. R. Colwell, 1981. Application of a direct microscopic method for enumeration of substrate responsive marine bacteria. Can. J. Microbiol. 27: 1071–1075.

Porter, K. J. & Y. S. Feig, 1980. The use of DAPI for identifying and counting aquatic microflora. Limnol. Oceanogr. 25: 943–948.

Quinn, J. P., 1984. The modification and evaluation of some cytochemical techniques for the enumeration of metabolically active heterotrophic bacteria in the aquatic environment. J. appl. Bact. 57: 51–57.

Riemann, B., J. Fuhrman & F. Azam, 1982. Bacterial secondary production in freshwater measured by ^{3}H-thymidine incorporation method. Microb. Ecol. 8: 101–114.

Rotman, B. & B. W. Papermaster, 1966. Membrane properties of living mammalian cells as studied by enzymatic hydrolysis of fluorogenic esters. Proc. natl. Acad. Sci. U.S.A. 55: 134–141.

Smith, A. J. & D. S. Hoare, 1977. Specialist phototrophs, lithotrophs, and methylotrophs: a unity among a diversity of procaryotes. Bact. Rev. 41: 419–448.

Stevenson, L. H., 1978. A case for bacterial dormancy in aquatic systems. Microb. Ecol. 4: 127–133.

Tabor, P. S. & R. A. Neihof, 1982. Improved method for determination of respiring individual microorganisms in natural waters. Appl. envir. Microbiol. 43: 1249–1255.

Tabor, P. S. & R. A. Neihof, 1984. Direct determination of activities for microorganisms of Chesapeake bay populations. Appl. envir. Microbiol. 48: 1012–1019.

Von Bielig, H. G., G. A. Kausche & H. Haardick, 1949. Über den nachweis von reduktion in bakterien. Z. Naturforsch. 4b: 80–91.

Wright, R. T. & J. E. Hobbie, 1966. Use of glucose and acetate by bacteria and algae in aquatic ecosystems. Ecology 47: 447–464.

Zimmermann, R., R. Iturriaga & J. Becker-Birck, 1978. Simultaneous determination of the total number of aquatic bacteria and the number thereof involved in respiration. Appl. envir. Microbiol. 36: 926–935.

Received 31 August 1989; in revised form 19 February 1990; accepted 15 March 1990

Hydrobiologia **207**: 303–309, 1990.
D. J. Bonin & H. L. Golterman (eds), Fluxes Between Trophic Levels and Through the Water–Sediment Interface.

Some remarks on the presence of organic phosphates in sediments

Cornelis-Jan De Groot
Station Biologique de la Tour du Valat, Le Sambuc, 13200 Arles, France

Key words: Organic phosphate, fractionation, sediment

Abstract

This article describes a new method developed to assess the size and nature of the organic phosphate pool. Using sediment suspensions from the Rhone, Garonne and Po rivers, inorganic P compounds, Fe(OOH) and $CaCO_3$ were removed using mild extractants at sediment pH. The residual phosphate was then fractionated into an acid soluble organic phosphate pool and a residual organic phosphate pool by acid hydrolysis (0.5 M H^+). Both pools were quantitatively important, accounting for between 16 and 54% and 16 and 51% of total phosphate respectively. Acid hydrolysis was chosen since it yielded a distinct plateau, with high reproducibility, within 30 minutes.

This fractionation permits a further study of dynamics and bioavailability of sediment org-P, without interference of Fe(OOH) and $CaCO_3$.

In many studies in which changes in the organic pool were examined after extraction of inorganic phosphate, 0.5 M HCl was used to extract apatite bound phosphate. The results presented here show that this is likely to result in a considerable underestimation of the organic phosphate pool.

Introduction

Phosphate in sediments occurs in inorganic and organic forms. The inorganic part consists of iron and calcium bound phosphate, which will be abbreviated as Fe(OOH) $\approx$ P and $CaCO_3 \approx$ P in order not to enter the discussion of the exact chemical structure. The nature of the organic phosphates is less obvious; several components such as phytic acid, nucleic acids and humic substances have been suggested (Golterman, 1960; Krause, 1964; Herbes *et al.*, 1975, Ogram et al., 1978). Knowledge of the nature and size of the different sediment phosphate pools is essential for understanding fluxes between the pools and thus the P-cycle in lakes. To measure the different pools, 'functional' extractions are needed; i.e methods which extract phosphate of a chemically well defined pool (functional unit) consisting of a series of homologous compounds, by use of specific chemical features of this pool, without causing changes in the remaining phosphate pools.

Golterman & Booman (1988) presented a 'functional' extraction for inorganic phosphate fractions, by which method Fe(OOH) $\approx$ P is extracted with buffered Ca-NTA/dithionite and $CaCO_3 \approx$ P with Na-EDTA, both at a pH of approximately 8. The extractions are repeated until the $Fe(OOH) \approx P$ and $CaCO_3 \approx P$ pools are exhausted[1].

Instead of 'functional' extractions, less specific 'operational' extractions, yielding chemically poorly defined fractions, are normally used in studies of the different phosphate pools in sediments. Extractants like NaOH are used to extract $Fe(OOH) \approx P$, and H_2SO_4 or HCl to extract $CaCO_3 \approx P$ (Chang & Jackson, 1957), with or without preliminary use of a less aggressive extractant such as NH_4Cl (Hieltjes & Lijklema, 1980). De Groot & Golterman (1990) compared this 'functional' extraction scheme with the 'classical' ones using mixtures of model compounds: $Fe(OOH) \approx P$ and $CaCO_3 \approx P$, dead and live algal cells as organic phosphate and sediment suspensions. They showed that both a reprecipitation of extracted phosphate as $CaCO_3 \approx P$ and a hydrolysis of part of the organic phosphate occurred during the NaOH extraction, due to the high pH (13) and the long extraction period of 17 hrs. This caused an underestimation of the quantity of $Fe(OOH) \approx P$ and of the quantity of organic phosphate. Organic phosphate was also hydrolysed during the acid extraction (pH = 0.3) of $CaCO_3 \approx P$, causing a serious over-estimation of $CaCO_3 \approx P$ and an underestimation of RP[2].

[1] 1) $Fe(OOH) \approx P$:
The dithionite reduces $Fe(OOH) \approx P$ to Fe^{2+} dissolving the phosphate. The NTA prevents the Fe^{2+} from precipitating. The NTA is saturated with Ca^{2+}, which makes it inactive towards $CaCO_3 \approx P$. TRIS-buffer is added to keep the pH between ± 8.0 and 7.5, during the extraction.
2) $CaCO_3 \approx P$ extraction:
EDTA chelates calcium from the sediment, dissolving the calcium bound phosphate. This extraction is performed at pH = 7.8 – 8.0.

[2] Abbreviations

o-P	Dissolved ortho phosphate
inorg-P	Inorganic phosphate
org-P	Organic phosphate
RP	Residual phosphate
ASOP	Acid soluble organic phosphate
ROP	Residual organic phosphate
Tot-P	Total phosphate in sediment
$Fe(OOH) \approx P$	Iron bound inorganic phosphate
$CaCO_3 \approx P$	Calcium bound inorganic phosphate

Psenner *et al.* (1985) presented an extraction scheme to fractionate organic and inorganic phosphate compounds from sediments into ecologically relevant fractions, which was slightly modified by Psenner *et al.* (1988). Their extraction sequence was: 1) NH_4Cl 1.0 M, 2) dithionite 0.11 M buffered with $NaHCO_3$, 3) NaOH 1.0 M, 4) HCl 0.5 M, 5) final digestion. Although the introduction of a dithionite extraction step is an improvement, the extraction scheme has the same faults as the scheme of Hieltjes & Lijklema, since HCl is used to extract apatite-P and Fe(OOH) is not extracted.

In the past, two different approaches have been used to study organic phosphate pool(s) in sediments:
1) Studies in which changes in the organic pool were examined after extracting inorganic phosphate with insufficiently specific extractants such as HCl and NaOH (Boström, 1984, Boström *et al.*., 1985., Petterson *et al.*, 1988, Petterson & Istvanovics, 1988, Psenner *et al.*, 1985).
2) The organic pool was also studied under near natural conditions, at sediment pH, without fractionation of sediment P (Gächter *et al.*, 1988).

In this study we have tried to develop a method to distinguish between different organic fractions, which later can serve as a tool in the study of the dynamics and the bioavailability of organic phosphate in sediments.

To study the RP pool a number of different approaches can be chosen:
1) hydrolysis of org-P with phosphatases,
2) hydrolysis of org-P with H^+ or OH^-,
3) photo-chemical destruction of the C-O-P bond.
4) oxidation of the organic matter, thus liberating the org-P.

First, we prepared a sediment suspension containing only RP by removing the $Fe(OOH) \approx P$- and $CaCO_3 \approx P$ complexes from the sediment with Ca-NTA/dithionite and Na-EDTA. For practical reasons we used the approaches 2) and 4) and applied different extractants, both oxidizing- and hydrolysing. Acid hydrolysis was applied to a number of Rhone, Garonne and Po sediments.

To ensure that no $CaCO_3 \approx P$ remained in the sediment after Na-EDTA extraction, the extraction was tested on synthetic apatite as well as on ground phosphorite (a mineral rich in $CaCO_3 \approx P$).

Materials and methods

Chemical methods

All determinations were carried out according to the IBP Manual Nr 8 (Golterman *et al.* 1978); method nr 4.5.1 was used for Fe, with ascorbic acid as reductor; 5.6.2 for o-P; Tot-P was determined as o-P after destruction with 2.5 ml of 6 M H_2SO_4 + 1 ml of H_2O_2 (30%) at 135 °C during 3 hrs, followed by centrifugation and neutralization.

A suspension 'S' of all soil samples was prepared containing 1 g (d.w.) per 10 ml and sieved over 0.25 mm pore size. Tot-Fe was measured after destruction of 2 ml of 'S' with 2 ml of 6 M H_2SO_4 + 1 ml of H_2O_2 (30%) at 135 °C during 3 hours, followed by adequate dilution.

A Rhone sediment suspension 'O' containing only RP was prepared by extracting all inorg-P using Ca-NTA/dithionite and Na-EDTA. This suspension was used to test the different extracting agents.

$Fe(OOH) \approx P$ and $CaCO_3 \approx P$ were extracted as described by Golterman & Booman (1988) and De Groot & Golterman (1990). After the last Na-EDTA extraction, ASOP was extracted by shaking the residual sediment pellet intensively with 0.5 M H^+ (HCl or H_2SO_4) for 30 minutes, followed by centrifugation and determination of the o-P.

To aliquots of the suspension 'O' 25 ml of the following agents were added, respectively: H_2SO_4 0.25 M/H_2O_2 1%, HCl 0.5 M/H_2O_2 1%, NaOH 0.1 M/H_2O_2 1%, $KMnO_4$ 0.1 M, $KMnO_4$ 0.5 M and S_2O_8 0.1 M. After 17 hours, o-P was measured after centrifugation and neutralization.

To study the quantities of phosphate extracted with H_2O_2 3%, NaOH 0.1 M and H_2SO_4 0.25 M over time: at $t = 0$, the extracting agents were added to aliquots of suspension 'O', from which the inorg-P had already been extracted. The reaction mixtures were shaken continuously. From the reaction mixtures homogeneous samples were taken and neutralized immediately at $t = 30$, $t = 150$ and $t = 1140$ minutes. After centrifugation o-P and Fe concentrations were determined.

Samples of synthetic apatite (0.1 g) were extracted with 25 ml 0.05 M Na-EDTA (pH = 8.0).

After extraction of $Fe(OOH) \approx P$ and $CaCO_3 \approx P$ (Golterman & Booman, 1988) from samples of approximately 0.2 g of ground phosphorite, the acid extraction was applied to the residual.

Materials used

Suspensions ('S') of two Rhone and three Garonne sediment samples from different locations were used. Freeze-dried sediment samples of Po sediment and aliquots of each suspension were used for inorg-P, ASOP and ROP extractions. For a detailed description of the Po-samples see Barbanti & Sighinolfi (1988), for a detailed description of the Garonne sediments see Fabre (1991).

Results and Discussion

In order to make sure that after the extraction of $CaCO_3 \approx P$ with EDTA, no inorganic calcium bound phosphate was left, the Na-EDTA extraction was tested again. The results of the test on synthetic apatite and on phosphorite were satisfactory: 100.5% of the (0.1 g) synthetic apatite phosphate was recovered in one extraction step with 25 ml of 0.05 M NaEDTA. After three Na-EDTA extractions of the phosphorite, no phosphate was found in a subsequent acid extraction (Table 1). This shows that phosphate found in the acid extraction is not derived from $CaCO_3 \approx P$, since the phosphate salts present in the phosphorite may be considered to be the most inert that may occur in natural waters (eroded rock particles).

Table 1. Percentage of Tot-P extracted sequentially with Ca-NTA 0.02 M/dithionite 0.67%, Na-EDTA 0.05 M and HCl 0.5 M on phosphorite.

Ca-NTA dithionite	1st EDTA	2nd EDTA	3rd EDTA	HCl
0.7	80.2	18.4	0.5	0
0.4	72.2	25.5	1.5	0

Of the different extractions studied with the sediment suspension 'O' (NaOH, H_2SO_4, $Na_2S_2O_8$, $KMnO_4$, H_2O_2 and HCl), the acid hydrolysis yielded the highest results, with a high reproducibility (Tables 2 & 3). In a series of extractions, the acid extraction yielded 64.1% of the residual organic phosphate, with a standard deviation of 2.2 (Table 2).

The extraction of organic phosphate with 0.1 M NaOH, H_2O_2 3% and 0.25 M H_2SO_4 was followed over time. However, in neither the NaOH, nor the H_2O_2 extraction was a plateau reached. The acid extraction reached a plateau within 30 minutes (Fig. 1), but after 1 minute considerable quantities of o-P were already in solution. Interferences occurred neither with the phosphate nor iron determination.

Golterman (1988) defined criteria which should be met by extraction methods of sediment P. Of the extracting agents tested, only the acid hydrolysis (0.5 M H^+) approached these criteria.

Hydrolysis with 0.5 M H^+ was therefore applied to Garonne, Rhone and Po sediments (see table 4). The quantities of RP found varied from 35.7% in Garonne 1 to 80.4% in Rhone 2. The quantity of ASOP varied from 16.4% in Garonne 3 to 53.8% in Rhone 2. In all the Po samples analyzed, over 50% of total P was found to be RP. Considerable amounts of ASOP were found in all samples.

As 0.5 M HCl is generally used to extract apatite-P, a large part of the ASOP can be expected to be extracted with this fraction, thus systematically over-estimating the apatite-P fraction and underestimating the residual phosphate. Using 'operational' extractions, Boström *et al.*, (1985) found that the decrease in the

Table 2. Amounts of organic phosphate (in $\mu g\ g^{-1}$ and in % of Residual Phosphate) extracted from Camargue suspension 'O' with 0.5 M H^+ (HCl or H_2SO_4).

Extractant	$\mu g\ P \cdot g^{-1}$ (d.w.)	% of RP
H_2SO_4 0.25 M	396	62.3
H_2SO_4 0.25 M	406	63.8
H_2SO_4 0.25 M	406	63.8
HCl 0.5 M	421	66.2
HCl 0.5 M	406	63.8
HCl 0.5 M	403	63.3
HCl 0.5 M	384	60.4
HCl 0.5 M	412	64.8
HCl 0.5 M	434	68.2
Average	407	64.1
Stand dev	14	2.2

Table 3. Amounts of organic phosphate (in $\mu g\ g^{-1}$ and in % of Residual Phosphate) extracted from suspension 'O' with different extractants.

Extractant	$\mu g\ P \cdot g^{-1}$ (d.w.)	% of RP
H_2SO_4 0.25 M/H_2O_2 1%	434	68.2
H_2SO_4 0.25 M/H_2O_2 1%	359	56.5
HCl 0.5 M/H_2O_2 1%	415	65.2
NaOH 0.1 M/H_2O_2 1%	77.4	12.2
$KMnO_4$ 0.1 M	307	48.2
$KMnO_4$ 0.5 M	349	54.9
$KMnO_4$ 0.5 M	355	55.8
$K_2S_2O_8$ 0.1 M	22.5	3.5
$K_2S_2O_8$ 0.1 M	22.5	3.5

RP-pool accounted for the P released from the sediments of Lake Vallentunasjön (Sweden). As we have shown that this extraction is likely to underestimate the org-P pool, it is likely that the changes in the org-P pool were quantitatively different from those observed by the method used. Since the changes in the RP-pool were larger then the sum of the NaOH and HCl fractions, the authors' conclusions are not affected by the errors which might have occurred during the extraction with HCl.

In studies in which org-P is studied under near natural conditions, without fractionation of sediment P, shifts due to phosphate adsorption onto or desorption from iron(III)hydroxide or

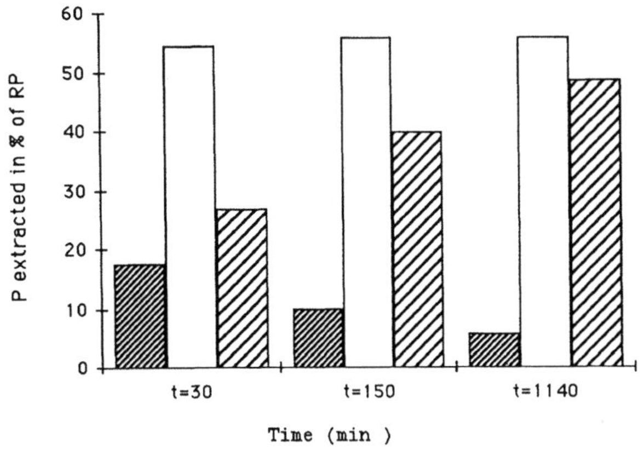

Fig. 1. Extraction of residual phosphate Time (min) from suspension 'O' (from which the inorg-P had been extracted previously) with NaOH 0.1 M (▨), H_2SO_4 0.25 M (□) and H_2O_2 1% (▨) followed over time.

$CaCO_3$ and/or precipitation or dissolution of calcium-phosphate compounds may occur.

Since, with the extraction of inorg-P, with Ca-NTA and Na-EDTA, some organic matter dissolves, the following organic phosphate fractions may be distinguished:

1) Iron-bound organic phosphate: phosphate bound to organic matter (probably humic substances) which is dissolved during the CaNTA/dithionite extractions.
2) Calcium-bound organic phosphate: phosphate bound to organic matter (probably humic substances) which is dissolved during the Na-EDTA extractions.
3) Acid Soluble Organic Phosphate (ASOP).
4) Residual organic phosphate (ROP).

To determine 1) and 2), destruction with H_2SO_4/H_2O_2 was performed on the Ca-NTA and Na-EDTA extracts, which did not usually yield more than 10% of the o-P extracted (unpublished data).

Although the acid extraction is an 'operational' extraction, it appears from the extraction characteristics that ASOP is a chemically well defined fraction, and we suggest the following possible chemical structures:

1) acid soluble organic compounds, which are dissolved during the resuspension of the pellet in the acid, and of which the C-O-P bond is being hydrolysed during the extraction, as well as during the o-P determination, which is performed under acid conditions (± 0.4 M H^+).
2) Ca(poly-valent cation)-organic phosphate complexes, which are not extracted by EDTA.
3) phosphate from living cells. Gächter *et al.* (1988) showed that bacterial biomass may contribute significantly to Tot-P, although it is difficult to judge the precision of the estimates made, since the influence of pH changes which occurred during the experiments on the adsorption of phosphate to Fe(OOH) were not discussed.

The quantity of calcium in the acid extract of two Rhone sediment samples was found to be 4.3 mg g^{-1} and 2.1 mg g^{-1}. Detailed studies should be performed to clarify the chemical structure and the ecological importance of the organic phosphate pools.

Little is known of the biological relevance of the organic phosphate fractions. From a series of Dutch lakes, De Graaf Bierbrauer-Würtz & Golterman, (1989) showed that there is a good correlation between the bio-availability of sediment phosphate and the total inorganic phosphate extracted with Ca-NTA and Na-EDTA. It should be noticed, however, that these results are obtained in short experiments, and it is possible that the org-P pool is also partly available on a larger time scale. Studies should also be carried out to assess the bioavailability of the organic phosphate fractions.

Conclusions

1) From these results it can be concluded that a systematic underestimation of the quantity of organic phosphate present in sediment has occurred in the past. This is because in many studies, using fractionation schemes in which 0.5 M HCl is used to extract calcium bound phosphate, a considerable part of the ASOP will erroneously be extracted as calcium bound phosphate. The quantity of ASOP extracted from the Garonne, Rhone and Po sediments may serve as an indication of the size of error which may occur.

2) In order to avoid erroneous interactions with Fe(OOH) $\approx$ P and CaCO3 $\approx$ P, the in-

Table 4. Amounts of ASOP extracted from some Rhone, Garonne and Po sediments in % of Tot-P.

	Tot inorg P	ASOP	Rest org-P	Tot org-P
Garonne sediments				
1)	64.4	19.6	16.1	35.7
	64.2	19.6	16.2	35.8
2)	59	22.4	18.6	41
	57.4	23.9	18.7	42.6
3)	62.6	16.8	21	37.8
	61.3	16.4	22.3	38.7
Rhone sediments				
1)	36.4	45.3	18.3	63.6
	36.8	46.8	16.4	63.2
2)	19.5	53.8	26.6	80.4
	19.5	53.8	26.6	80.4
Po sediments				
Sacca di Goro				
1)	30	18.9	51.1	70
	31.8	19.3	48.9	68.2
2)	46.8	21	32.2	53.2
	42.7	20.2	37.2	57.4
3)	46.9	19.8	33.3	53.1
	47.5	18.8	33.7	52.5
Suspended. sediment				
1)	41.9	22.6	35.5	58.1
	42.6	22.1	35.3	57.4
2)	45.8	21	33.2	54.2
	46.5	19.5	34	53.5
Bottom sediment				
1)	45	21.4	33.6	55
	45.4	19.5	35.1	54.6
2)	41.7	25	33.3	58.3
	44.8	28.1	27.1	55.2
3)	34.6	33.8	31.6	65.4
	34.6	31.9	33.5	65.4
Core Po delta				
1)	45.7	35.3	19	54.3
	42.9	38.1	19	57.1
2)	43.3	40.2	16.5	56.7
	47.1	33.4	19.5	52.9
3)	33.6	49	17.5	66.5
	34	49	17	66

organic phosphate pool (both inorganic phosphate and Fe(III)- and Ca-salts) should be removed by 'functional' extractions before org-P is studied.

3) An acid soluble organic phosphate (ASOP) fraction was found in sediments of the Rhone, the Garonne and the Po with 0.5 M H^+. The size of this pool varied between 16.4% and 53.8% of Tot-P. A plateau was reached within 30 minutes with a high reproducibility.

Acknowledgements

The author wishes to thank Dr. H. L. Golterman for useful discussions and for reading the manuscript, Dr. Andrea Barbanti with whom he analyzed the Po-sediment samples. Financial support was received from the C.E.E., the CNRS-PIREN and from the 'Ministère de l'Environnement SRETIE', France.

References

Barbanti, A & G. P. Sighinolfi, 1988. Sequential extraction of phosphate and heavy metals from sediments: methodological considerations. Envir. Tech. Letters, Vol 9: 127–134.

Boström, B., 1984. Potencial mobility in different types of lakes. Int. Revue ges. Hydrobiol. 69: 457–474.

Boström, B., I. Ahlgren & R. Bell, 1985. Internal nutrient loading in a eutrophic lake, reflected in seasonal variations of some sediment parameters. Verh. int. Ver. Limnol. 22: 3335–3339.

Chang, S. C. & M. L. Jackson, 1957. Fractionation of soil phosphorus. Soil Sci. 84: 133–144.

Fabre, A, 1991. Forms and evolution of inorganic phosphorus in exposed sediments of the river Garonne. Hydrobiologia submitted.

Gächter R, J. S. Meyer & A. Mares, 1988. Contribution of bacteria to release and fixation of phosphorus in lake sediments. Limnol. Oceanogr. 33: 1542–1558.

Golterman, H. L., 1960. Studies on the cycle of elements in fresh water. Acta Bot. Neerl. 9: 1–58.

Golterman, H. L., 1988. Reflections on fractionation and bioavailability of sediment bound phosphate. Arch. Hydrobiol., Beih. Ergebn. Limnol. 30: 1–4.

Golterman, H. L. & A. Booman, 1988. The sequential extraction of Ca- and Fe-bound phosphates. Verh. int. Ver. Limnol. 23: 904–909.

Golterman, H. L., R. S. Clymo & M. A. M. Ohnstad, 1978. Methods for Physical and Chemical Analysis of Freshwaters, IBP Manual No. 8 (2nd edn.). Blackwell sci. Publi., Oxford, 213 pp.

de Graaf Bierbrauer-Würtz, I. M. & H. L. Golterman, 1989. Fosfaat fracties in de bodem van een aantal Nederlandse plassen. (in Dutch, with English summary). H_2O. 13: 411–414.

de Groot, C. J. & H. L. Golterman, 1990. Sequential fractionation of sediment phosphate. Hydrobiologia 192: 143–148.

Herbes, S. E., H.E. Allen & K. H. Mancy, 1975. Enzymatic characterization of soluble organic phosphorus in lake water. Science 187: 432–434.

Hieltjes, A. H. M. & L. Lijklema, 1980. Fractionation of inorganic phosphates in calcareous sediments. J. Envir. Qual. 9: 405–407.

Krause, H. R., 1964. Zur Chemie und Biochemie der Zersetzung
von Süsswasserorganismen, unter besonderer Berucksichtigung des Abbaues der organische Phosphorkomponenten. Verh. int. Ver. Limnol. 15: 549–561.

Ogram A., G. S. Sayler, D. Gustin & R. J. Lewis, 1978. DNA adsorption to soils and sediments. Envir. Sci. Technol. 22: 982–984.

Petterson K. & V. Istvanovics, 1988. Sediment phosphorus in lake Balaton – forms and mobility. Arch. Hydrobiol., Beih. Ergebn. Limnol. 30: 25–41.

Petterson K., B. Boström & O. Jacobsen, 1988. Phosphorus in sediments – Speciation and analysis. Hydrobiologia 170: 91–101.

Psenner R., R. Pucsko & M. Sager, 1985. Die Fraktionierung organischer und anorganischer Phosphorverbindungen von Sedimenten. Versuch einer Definition ökologisch wichtiger Fraktionen. Arch Hydrobiol. Suppl. 70: 111–155.

Psenner R., B. Boström, M. Dinka, K. Petterson, R. Pucsko & M.Sager, 1988. Fractionation of phosphorus in suspended matter and sediment. Arch. Hydrobiol., Beih. Ergebn. Limnol. 30: 98–103.

Received 14 August 1989; in revised form 19 March 1990; accepted 25 June 1990

Hydrobiologia **207**: 311–318, 1990.
D. J. Bonin & H. L. Golterman (eds), Fluxes Between Trophic Levels and Through the Water–Sediment Interface.

Direct measurements of nanomolar nitrate uptake by the marine diatom *Phaeodactylum tricornutum* (Bohlin). Implications for studies of oligotrophic ecosystems

Patrick Raimbault, Gerd Slawyk & Valérie Gentilhomme
Centre d'Océanologie de Marseille, Faculté des Sciences de Luminy, case 901, 13288 Marseille cédex 9, France

Key words: nitrate uptake, nanomolar range, diatom, oligotrophy

Abstract

A new automated procedure for nanomolar nitrate analysis was applied to the study of uptake of low nitrate concentrations (< 100 ngat l^{-1}) by phytoplankton. The precision of this analytical method (± 3 ngat l^{-1}) made it possible to monitor the absorption of very low quantities of nitrate over short term periods by a low cell-density culture of the marine diatom *Phaeodactylum tricornutum*, where the levels of particulate nitrogen and chlorophyll were equivalent to those found in oligotrophic areas (0.5 μgat N l^{-1} and 0.4 μg l^{-1} respectively). By continuous monitoring of nitrate disappearance from the culture medium, we are able to describe accurately the transient uptake responses of the diatom after a spike addition of trace quantities of nitrate and thus to provide new information on the still largely unknown small-scale phenomenon of pulsed nitrate supply in the upper layer of stratified oceans and rapid uptake of these nitrate patches by phytoplankton.

The results show that a N-limited culture of *Phaeodactylum tricornutum* is immediately capable of taking up trace quantities of nitrate (< 100 ngat l^{-1}) at high rates (0.10–0.14 h^{-1}). These initial rates are one order of magnitude higher than the theoretical rates calculated from the Michaelis-Menten equation and are close to the level of V_{max} (0.15 h^{-1}) obtained when cells are exposed to saturating nitrate concentrations. This rapid initial uptake would be a considerable advantage in oligotrophic areas where nanomolar nitrate supply is thought to be episodic. The present results suggest that phytoplankton evolve adaptations to utilize the available nitrate at the spatial and temporal scales at which it occurs. On the other hand, we can consider this physiological adaptation as evidence of nitrate pulses in the field which would invalidate the steady-state approach to the oligotrophic ecosystems.

Introduction

Previous studies of short-term uptake have measured initial high nitrate uptake rates or 'surge uptake', following a sudden addition of nitrate which can exceed, in some cases, the maximum growth rate by a wide margin (Demanche *et al.*, 1979; Raimbault, 1984; Raimbault & Mingazzini, 1987). This behaviour has been interpreted as an ecological adaptation to fluctuating nutrient level in the environment such as nitrate pulses. However, these observations were obtained with high nitrate concentrations (> 1 μgat l^{-1}) which do not exist in the upper oligotrophic ocean. While short term input of nitrate following moderate winds have been recently revealed in the sea

(Klein & Coste, 1984; Eppley & Renger, 1988), measurements show very low values, less than 50–100 ngat l^{-1} (Garside, 1985; Eppley & Renger, 1986; 1988; Raimbault *et al.*, 1990). These mechanisms required that phytoplankton become rapidly capable of taking up nitrate at nanomolar levels. Consequently, an understanding of nitrate uptake mechanisms at low concentrations is clearly needed in order to determine how phytoplankton respond and adapt to periodic nitrate supplies and to evaluate the relative effectiveness of transient uptake of such trace patches of nitrate (<100 ngat l^{-1}). Such information is crucial if conclusions are to be reached concerning the possible mechanisms of phytoplankton survival in a poor and probably patchy environment.

A new automated procedure allows nitrate to be measured in the nanomolar range (Oudot & Montel, 1988; Raimbault *et al.*, 1990) and permits investigation of the responses of phytoplankton under 'simulated oligotrophic conditions'. The present experiments are designed to supply information on the capability of the diatom *Phaeodactylum tricornutum* to take up nitrate when levels in the medium are as low as encountered in the upper layer of the stratified ocean, *i.e.* <100 ngat l^{-1} (concentration which corresponds to the limit of detection of the conventional analytical method, Strickland & Parsons (1972). The procedure described here uses the conventional perturbation experiment (Caperon & Meyer, 1972) which simulates the meeting of N-limited phytoplankton with a nitrate pulse enrichment. Continuous recording of nitrate concentrations allows description of the short-term response of the algae following a very low nitrate supply and makes it possible to check the adequacy of the Michaelis-Menten equation generally applied to describe nutrient uptake kinetics by marine phytoplankton.

Materials and methods

Axenic stock cultures of the diatom *Phaeodactylum tricornutum* Bohlin were maintained on f/2 medium (Guillard & Ryther, 1962). The experimental cultures were grown in nitrogen-limited continuous cultures as described by Raimbault (1984). Nitrogen was supplied as $NaNO_3$ at a final concentration of 10 μgat l^{-1}. Dilution rate and experimental temperature were 1 day^{-1} and 20 °C respectively. Cultures were illuminated by cool white fluorescent lamps on a 12 h–12 h light-dark cycle. The irradiance measured with a Li-Cor quantameter was 135 $\mu E\ m^{-2}\ s^{-1}$. Steady-state was assumed when daily cell counts did not vary by $>10\%$ over five days. Cell density was monitored microscopically twice per day with a Neubauer hemacytometer. During the steady-state, the mean cell concentration was 10^5 cells ml^{-1}. The overflow was completely nitrate and nitrite depleted during the light period and the particulate nitrogen values were determined at steady-state from a mass balance where the disappearance of nitrate was assumed to be equal to the increase in particulate nitrogen in the culture (10 μgat N l^{-1}). The level of particulate nitrogen was determined with a CHN analyzer Leco before uptake experiments.

Nitrate uptake experiments

Firstly, a conventional perturbation experiment was performed as described in Raimbault (1984, 1986) where 10 μgat l^{-1} nitrate were added to one liter of the N-limited culture one hour after the beginning of the light period. Measurements of nitrate and nitrite were carried out on filtrates of subsamples, with a Technicon Autoanalyzer according to the method of Tréguer & Le Corre (1975), until nitrate was entirely exhausted from the medium. Continuous sampling and analyses provide a time series of disappearance of the limiting nutrient from the culture. Uptake rates are calculated from the concentration changes over short periods of time.

Secondly, two types of experiments were conducted to measure uptake at very low nitrate concentration. A trace quantity of nitrate (<100 ngat l^{-1}) was added to one liter of the experimental N-limited culture, named high cell-density culture (HCD culture) and to one liter of a low cell-density culture (LCD culture) obtained by dilution of 50 ml HCD culture in 950 ml nitrate-free medium (from filtered overflow). To ensure the validity of this dilution (by a factor 20), the particulate nitrogen and chlorophyll concentrations were measured before and after dilution and cell counts were made with a nageotte hematocytometer; the mean cell-density found after dilution was 5040 cell ml^{-1}. Nitrate addition was always performed during the first two hours after the onset of the light period. The concentrations of nitrate were monitored by a high sensitivity colorimetric method developped for the determination of nanomolar quantities of nitrate with a Technicon AutoAnalyzer (Oudot & Montel, 1988), applied here with some modifications (Raimbault *et al.*, 1990). The precision of this analytical method is ± 2.9 ngat l^{-1} nitrate (95% C.I.).

Continuous measurements of the consumption of trace quantities of nitrate were conducted according to the following procedure: The pen of the chart recorder was firstly adjusted to zero with the baseline control potentiometer of the colorimeter when high quality deionized water containing no more nitrate or nitrite than the Milli Q water was drawn off via the sampling tube. Then the sample needle surrounded by 2 layers of 10 μm mesh net was directly inserted in the reactor vessel. When the shift of the signal obtained on the recorder was no higher than the salt effect as referred to by Oudot & Montel (1988), culture medium was assumed to be nitrate depleted. After the few minutes required to obtain a baseline, nitrate was added to the culture medium. Both the preliminary sudden increase of nitrate concentration and the following decrease with time were recorded. Calibration with 2 or 3 standards prepared in deionized water was made just before the beginning of the experiment. Nitrate uptake was calculated from the differences over a short-term period (1–5 minutes) and reported as specific rates (h^{-1}) based upon the particulate nitrogen measured at the beginning of the experiment and then corrected from nitrate

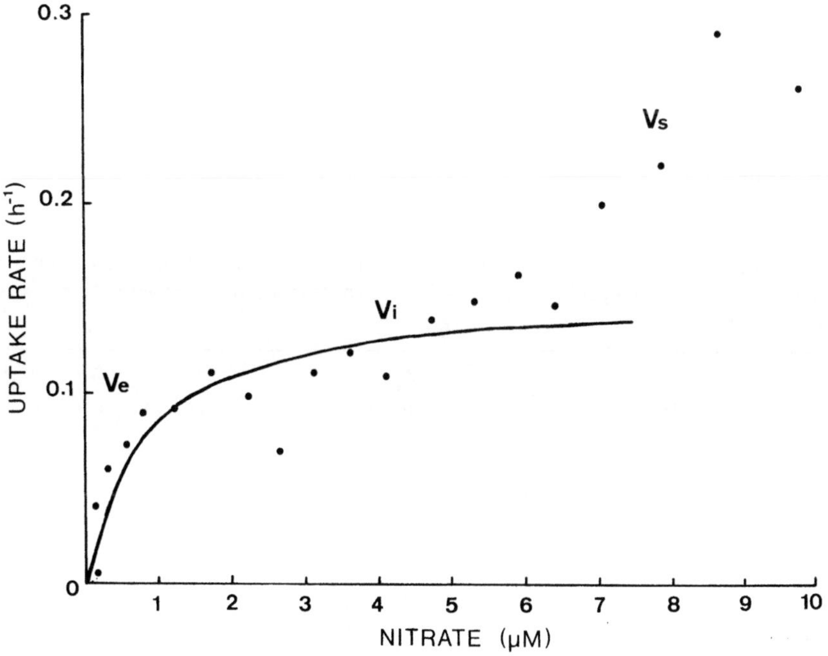

Fig. 1. Nitrate uptake rate as function of nitrate concentration for *Phaeodactylum tricornutum* in culture at high cell-density (10^5 cells ml^{-1}). The hyperbola curve is calculated with the constants of Michaelis-Menten equation according to Wilkinson (1961).

taken up. During these 'trace perturbations' nitrite was not simultaneously measured, but previous studies have shown that nitrite excretion by *P. tricornutum* was negligible when nitrate concentration was low (Raimbault, 1986).

Results

Figure 1 shows the typical relationship between nitrate uptake rates and nitrate (at high concentrations) in which three modes of uptake can be distinguished during this nitrate perturbation experiment corresponding to the type 2 defined by Collos (1983): *Vs*, also referred to as surge uptake (Conway *et al.*, 1976), in which nitrate uptake rate is initially high immediately after the nitrate addition, but decreases rapidly over a short-time period; *Vi*, during which uptake remains constant or slowly decreases over a long time period in spite of a considerable decrease in nitrate concentration, and *Ve* characterized by a rapid decrease of the uptake rate when external nitrate was almost exhausted from the medium (<1 μgat l^{-1}). Since the Michaelis-Menten expression

$$V = \frac{V_{max} \cdot C}{Ks + C},$$

where V is the uptake rate, V_{max} the maximum uptake rate, C the ambient nutrient concentration and Ks the half-saturation constant, has been used as a mathematical equivalence of uptake kinetic (Dugdale, 1967), we can calculate the affinity constant Ks and the maximum uptake rate V_{max} by a weighted non linear regression method (Wilkinson, 1961), excluding the initial transient phase Vs (Table 1). The value of V_{max} (0.15 h^{-1}) allows a transient growth rate four times higher than the steady-state growth rate. When the nitrate concentration (C) is much lower than the Ks value, the uptake rate is related to the nitrate concentration by the linear relationship $V = \alpha NO_3$ where α equals to the V_{max}/Ks ratio, *i.e.* 0.205 h^{-1} μM^{-1} (Table 1). Thus, the uptake rates calculated from this equation for nitrate concentrations lower than 100 ngat l^{-1} (<0.02 h^{-1}) are not high enough to sustain a growth rate of 1 day^{-1} (*i.e.* 0.04 h^{-1}) during the steady-state.

The patterns of nitrate absorption by the same high cell-density culture (HCD) after low nitrate additions (95 and 15 ngat l^{-1}) are shown in the Fig. 2 (experiments a and b). Nitrate is immediately absorbed without any time lag and entirely exhausted after 19 and 12 minutes respectively. In the preceeding experiment (HCD, 10 μgat l^{-1}), the 10 μgat l^{-1} nitrate was completely exhausted after 300 minutes. The disappearance of nitrate is not linear with time and consequently the uptake rate varies with time. The uptake rate given by the slope of the disappearance curves (Fig. 2), is high

Table 1. Half-saturation constant (Ks) and maximum velocity of uptake (V_{max}) calculated according to Wilkinson (1961) with data from Fig. 1 (perturbation with 10 μgat l^{-1}) and from Fig. 2 (perturbation with 95 ngat l^{-1}). α measured is the slope of the regression line from the Fig. 4a.

	Ks (μM)	V_{max} (h^{-1})	V_{max}/Ks ($h^{-1} \cdot \mu M^{-1}$)	α measured
Perturbation with 10 μgat l^{-1}				
1) $Ve + Vi$	0.73 ± 0.22	0.15 ± 0.01	0.205	–
2) $Ve + Vs$	1.78 ± 0.47	0.30 ± 0.02	0.167	–
Perturbation with 95 ngat l^{-1}	0.07 ± 0.01	0.15 ± 0.01	2.13	0.96
+ Vs*	0.19 ± 0.03	0.26 ± 0.01	1.36	0.96

Vs*: data from Fig. 1.

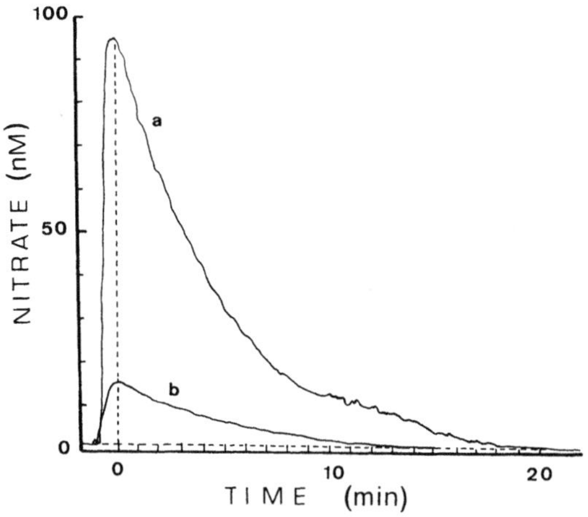

Fig. 2. Continuous nitrate disappearance traces obtained after addition of 95 ngat l^{-1} (a) and 15 ngat l^{-1} (b) nitrate to the same N-limited culture of *P. tricornutum* as in the Fig. 1.

during the first minutes and then declines until nitrate was entirely exhausted from the medium. During the experiment a (HCD, 95 ngat l^{-1}), the decrease of nitrate is rapid during the first 7 minutes where 75 ngat l^{-1} nitrate are absorbed. The last 20 ngat l^{-1} are consumed in 12 minutes, indicating a strong decrease in uptake velocity after the initial phase. In the same manner, the 15 ngat l^{-1} added in experiment b (HCD, 15 ngat l^{-1}) are entirely taken up after a relatively long time (12 minutes). In the experiment performed with the LCD culture, the 94 ngat l^{-1} nitrate added are entirely absorbed after 165 minutes, but the pattern of nitrate disappearance is quite similar to the preceeding (HCD, 95 ngat l^{-1}) experiments (Fig. 3). The nitrate uptake rate seems to be constant with time during an initially short period (30 minutes), and decreases slowly after this time lag. A dramatic change in the slope of the recorded trace, *i.e.* in the uptake rate, appears when nitrate concentration reaches about 20–25 ngat l^{-1}. A parallel experiment performed with discreet samples analyzed every 15 minutes after filtration on Whatman GF/F filters, shows the same nitrate changes with time (Fig. 3), except at the beginning because of an higher nitrate addition or because of contamination of the first samples. In all cases, the specific uptake rates calculated over every one minute for experiments with HCD samples or every 5 minutes for experiments with LCD samples are linearly related to the nitrate concentrations less than 100 ngat l^{-1} (Fig. 4a and b).

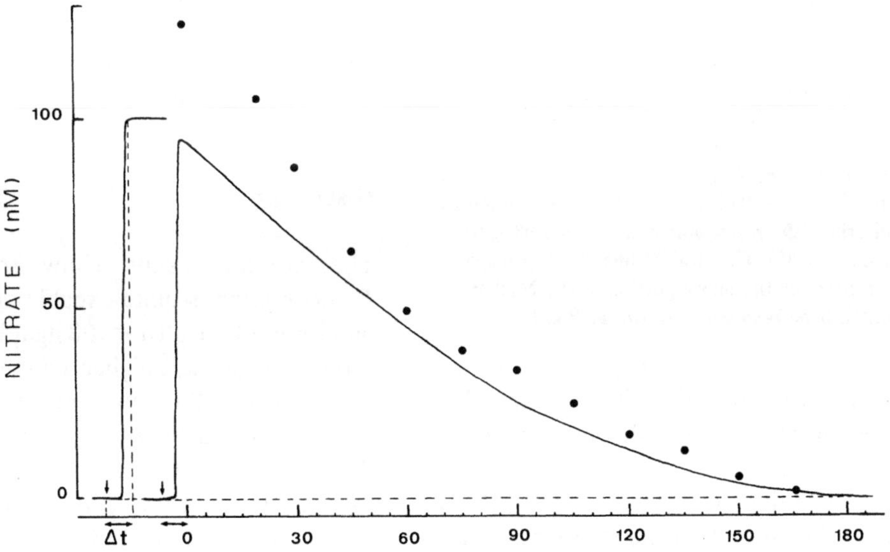

Fig. 3. Continuous nitrate disappearance traces obtained after addition of 94 ngat l^{-1} to a N-limited culture of *P. tricornutum* at low cell-density (5.10^3 cells ml^{-1}; particulate nitrogen = 0.5 μgat l^{-1}). A 100 ngat l^{-1} nitrate standard preceeds the experiment. Points indicate discreet samples performed on a replicate experiment.

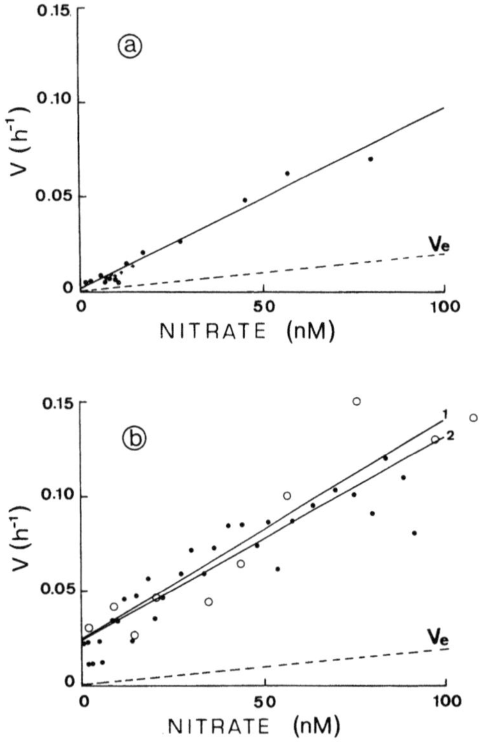

Fig. 4. Specific uptake rate (V in h^{-1}) versus nitrate concentration relationship obtained from the uptake experiments shown in Fig. 2 and 3. The line represents the least squares linear regression. Uptake rates are calculated during successive short time intervals and plotted against the average substrate concentrations during these intervals. a) Relationship obtained from the uptake experiment shown in Figure 2 with time interval of one minute; data for the two perturbations (15 and 100 nM) are pooled: $V = 0.96 \times 10^{-3}$ NO_3^- + 0.002, $r^2 = 0.95$. b) Relationship obtained from the experiments shown in Fig. 4 with time interval of 15 minutes for the discreet sampling (○, regression line 1: $V = 1.17 \times 10^{-3}$ NO_3^- + 0.021, $r^2 = 0.86$) and 5 minutes for the monitoring (●, regression line 2: $V = 1.08 \times 10^{-3}$ NO_3^- + 0.024, $r^2 = 0.85$). The straight line Ve ($V = 0.205 \times 10^{-3}$ NO_3^-) represents the linear portion of the Michaelis-Menten hyperbola curve shown in Fig. 1.

The two relations obtained with the LCD samples (Fig. 4b) are not significantly different with a confidence interval of 95% and are quite comparable to that found with the HCD (95 and 15 ngat l^{-1}) sampled (Fig. 4a). Based on the standard deviation of individual uptake rates from this line, the standard deviation for individual nitrate measurements for the 3 experiments ranged between 8 to 16 ngat l^{-1} (95% C.I.) which is a little higher than the expected precision of the analytical technique (see above). The linear relationship confirms the direct dependence of the uptake rate on low nitrate concentrations as described by the Michaelis-Menten equation, but the level of dependence experimentally observed and noted by the regression coefficient α is very different from the one obtained with the theoretical kinetic constants represented by the line Ve (HCD, 10 μgat l^{-1}) in Fig. 4a and 4b. In the same way the initial rates found with nitrate concentrations between 90–100 ngat l^{-1} (0.10 to 0.14 h^{-1}, Fig. 4a, b) are very close to the V_{max} value (0.15 h^{-1}) previously determined with nitrate concentrations between 5 and 8 μgat l^{-1} (Fig. 1). The regression lines obtained with the LCD samples have an intercept significantly different from 0 (Fig. 4a, b). Even at very low nitrate concentration (< 10 ngat l^{-1}), the uptake showed significant rates of 0.010–0.020 h^{-1}, while the phase Ve (from Fig. 1; HCD, 10 μgat l^{-1} experiment) gives a rate one order of magnitude lower (0.002 h^{-1}). These relationships indicate that an uptake rate of 0.04 h^{-1} can be sustained as long as the nitrate concentration is not lower than 40 ngat l^{-1} in the HCD culture and 10 ngat l^{-1} in the HCD cultures. This can explain the keeping of a high steady-state growth rate of 0.04 h^{-1} during the chemostat run while no nitrate was available in the outflow.

Discussion

The present results show that the diatom *P. tricornutum* is quite capable of taking up trace quantities of nitrate (< 100 ngat l^{-1}) at a rate one order of magnitude higher than the theoretical rate obtained for this concentration from a conventional perturbation experiment and closer to the V_{max} value. Our results are consistent with the hypothesis of McCarthy & Goldman (1979) based on ammonium uptake, that rapid uptake of inorganic nitrogen by marine phytoplankton is possible under regimes of heterogeneous nutrient supply in which the concentrations of ambient nutrients are very low. This confirms that the

Michaelis-Menten function used as an empirical description of phytoplankton uptake kinetics at saturating substrate levels is not a precise model of the underlaying system and cannot appropriately be used to describe nitrate uptake, and especially to calculate the *Ks* value. If the kinetic constants are recalculated as suggested by Conway *et al.* (1976) using only the externally controlled phase obtained here at low nitrate concentrations (< 100 ngat l^{-1}; data from Fig. 2), the estimation reveals a very low *Ks* with low confidence limits (Table 1) and a V_{max} value that does not differ from that previously found with high nitrate concentration (10 μgat l^{-1}; data from the Fig. 1). But the V_{max}/Ks ratio (2.1) is then twice as high as the experimentally measured α coefficient. If the calculation also includes the Vs phase (HCD, 10 μgat l^{-1}), as suggested by Collos (1983), the V_{max} value (0.26 h^{-1}) is not different from the surge uptake and the V_{max}/Ks ratio (1.36) is closer to the α coefficient found experimentally with low nutrient perturbation (Table 1). Therefore, nitrate uptake can be described by a Michaelis-Menten function considering the short-term uptake rates obtained at saturating nitrate to determine the true maximum uptake capacity and at very low nitrate concentrations to reveal the real affinity for nitrate. This also demonstrates the inadequacy of long term measurements of N-nutrient uptake which can lead to serious underestimates of productivity (Glibert & Goldman, 1981). But the question remains how the modulation of the short-term uptake capacity relates to the long term needs of the organisms. The present results suggest that phytoplankton evolve to utilize the injections of nitrate at the spatial and temporal scales at which they may occur in the oligotrophic areas. Data on nutrient pulse magnitude and frequency in the field are clearly needed in order to assess primary production on relevant scales. Although the existence and duration of microscale nutrient patches in the ocean have remained controversial during the last decade (Jackson, 1980; William & Muir, 1981), these data are now becoming available for nitrate (Klein & Coste, 1984; Eppley & Renger, 1988). Since the relationship between limiting nutrient concentration and phytoplankton growth rate is seen as a key physiological issue determining primary production in aquatic systems (Morel, 1987), we have now to rely on these field observations on algal cultures to gain insight into phytoplankton growth processes.

References

Caperon, J. & K. Meyer, 1972. Nitrogen-limited growth of marine phytoplankton. I. Changes in population characteristics with steady-state growth rate. Deep-Sea Res. 19: 601–618.

Collos, Y., 1983. Transient situations in nitrate assimilation by marine diatoms. 4. Non-linear phenomena and the estimation of the maximum uptake rate. J. Plankton Res. 5: 677–691.

Conway, H. L., P. J. Harrison & C. O. Davis, 1976. Marine diatoms grown in chemostat under silicate limitation. 2. Transient response of *Skeletonema costatum* and *Thalassiosira gravida* to a single addition of the limiting nutrient. Mar. Biol. 43: 33–43.

Demanche, J. L., H. C. Curl, D. W. Lundy & P. L. Donaghay, 1979. The rapid respond of the marine diatom *Skeletonema costatum* to changes in external and internal nutrient concentration. Mar. Biol. 53: 239–252.

Dugdale, R. C., 1967. Nutrient limitation in the sea: dynamics, identification and significance. Limnol. Oceanogr. 12: 685–595.

Eppley, R. W. & E. H. Renger, 1986. Nitrate-based primary production in nutrient-depleted surface water off California. Oceanogr. Trop. 21: 229–238.

Eppley, R. W. & E. H. Renger, 1988. Nanomolar increase in surface layer nitrate concentration following a small wind effect. Deep-Sea Res. 35: 1119–1125.

Garside, C., 1985. The vertical distribution of nitrate in open ocean surface water. Deep-Sea Res. 32: 723–732.

Glibert, P. M. & J. C. Goldman, 1981. Rapid ammonium uptake by phytoplankton. Mar. biol. Lett. 2: 25–31.

Guillard, R. R. L. & R. H. Ryther, 1962. Studies of marine plankton diatoms. I. *Cyclotella nana* (Hudstedt) and *Detonula confercea* (Cleve). Can. J. Microbiol. 8: 186–190.

Jackson, G. A., 1980. Phytoplankton growth and zooplankton grazing in oligotrophic ocean. Nature (London) 284: 439–441.

Klein, P. & B. Coste, 1984. Effects of wind stress variability on nutrient enrichment of the mixed layer. Deep-Sea res. 31: 21–37.

McCarthy, J. J. & J. C. Goldman, 1979. Nitrogenous nutrition of marine phytoplankton in nutrient depleted water. Science N.Y. 39: 670–672.

Morel, F. M. M., 1987. Kinetics of nutrient uptake and growth in phytoplankton. J. Phycol. 23: 137–150.

Oudot, C. & Y. Montel, 1988. A high sensitivity method for the determination of nanomolar concentrations of nitrate and nitrite in seawater with Technicon AutoAnalyzer II. Mar. Chem. 24: 239–252.

Raimbault, P., 1984. Influence of temperature on the transient response in nitrate uptake and reduction by four diatoms. J. exp. mar. Biol. Ecol. 84: 275–286.

Raimbault, P., 1986. Effect of temperature on nitrite excretion by three marine diatoms during nitrate uptake. Mar. Biol. 92: 149–155.

Raimbault, P. & M. Mingazzini, 1987. Diurnal variations of intracellular nitrate storage by marine diatoms: effects of nutrient state. J. exp. mar. Biol. EcoL. 112: 217–232.

Raimbault, P., G. Slawyk, B. Coste & J. Fry, 1990. Feasibility of using an automated colorimetric procedure for the determination of seawater nitrate in the 0 to 100 nM range: examples from filed and culture. Mar. Biol. 104: 347–351.

Strickland, J. D. H. & T. R. Parsons, 1972. A practical handbook of seawater analysis. Bull. Fish. Res. Bd Can. 167: 71–80.

Tréguer, P. & P. Le Corre, 1975. Manuel d'analyses des sels nutritifs dans l'eau de mer (Utilisation de l'AutoAnalyzer II Technicon). Laboratoire d'Océanographie chimique, Univ. de Bretagne Occidentale, Brest, 2e ed., 110 pp.

Wilkinson, G. N., 1961. Statistical estimations in enzyme kinetics. Biochem. J. 80: 324–332.

William, P. J. L. & L. R. Muir, 1981. Diffusion as a constraint on the biological importance of microzones in the sea. In J. C. J. Nihoul (ed.), Ecohydrodynamics, Elsevier: 209–218.

Received 27 June 1989; in revised form 27 November 1989; accepted 26 February 1990

Hydrobiologia **207**: 319–331, 1990.
D. J. Bonin & H. L. Golterman (eds), Fluxes Between Trophic Levels and Through the Water–Sediment Interface.

Biogeochemistry of organic matter in Lake Geneva: I – particulate hydrocarbons as biogenic and anthropogenic molecular markers

Pierre Scribe[1], Jean-Sulpice Ngoumbi-Nzouzi[1], Christine Fuché[1], Claude Pèpe[2] & Alain Saliot[1]
[1]*Laboratoire de Physique et Chimie Marines, Université Pierre et Marie Curie, UA CNRS 353, Tour 24, 4 place Jussieu, 75252 Paris Cedex 05, France;* [2]*Laboratoire de Spectrochimie Moléculaire, Université Pierre et Marie Curie, Batiment F, 4 place Jussieu, 75252 Paris Cedex 05, France*

Key words: Lake Geneva, particles, organic matter, alkanes, alkenes, seasonal variations

Abstract

A study of the biogeochemistry of organic matter in Lake Geneva at a central station (SHL2), was carried out at key periods of the biological cycles from November 1985 to September 1986. The seasonal variability of particulate hydrocarbons was studied in the epilimnion and the hypolimnion, focusing on specific periods such as high phytoplankton productivity and rising herbivorous zooplankton activity (May 1986), clearness of waters (June 1986) and intense mixing of waters (February 1986).

Analysis of non-aromatic hydrocarbons (NAH) indicated that their major source is the autochtonous phytoplankton. Year 1986 was characterized by a spring bloom of nanoplankton (chlorophyceae and cryptophyceae) which remained predominant during summer and fall. These biological events were correlated with an increase of biogenic NAH, particulary n-alkane and alkene concentrations; the latter increased by one order of magnitude in May. n-Alkane distributions maximized at n-C_{17} or n-C_{29}, and the n-C_{17}/n-C_{29} ratio varied in surface water from 2.8 in May to 0.06 in November of the preceding year, reflecting phytoplankton dominant species. Values of NAH/POC, n-alkanes/POC, alkenes/POC and pristane/phytane ratios underwent seasonal and vertical variations which could be related to grazing and/or settling of particles. NAH, n-alkane and alkene enrichment in hypolimnion particles was interpreted to be a result of zooplankton migrations.

A series of 37 monoenes and polyenes having from 15 to 31 carbon atoms, originating from planktonic sources, were recognized in the particulate matter by gas chromatography/mass spectrometry (GC/MS). Their distribution was described in 3 sized particles (>200 μm, 200–60 μm and <60 μm, identified as adult zooplankton, young zooplankton and nanoplankton respectively) at two different periods of the biological cycle, May and June 1986. Low level inputs originating from terrestrial higher plants were evident in particles; the value of the Carbon Preference Index (CPI) in the carbon range 23–35 was less than 2.8.

In all samples analyzed by GC, an unresolved complex mixture of hydrocarbons (UCM) was present and was interpreted, at least in May, as partly originating from the degradation of organic matter by microorganisms. Typical anthropogenic PAH, such as benzofluoranthene and dibenzo (a) anthracene, were found at a low level (9 and 3 ng l^{-1} respectively). The absence of hopane-type hydrocarbons, except in November 1985 and in September 1986, and typical combustion PAH associated with particles indicated high temperature pyrolytic contaminants rather than petroleum-derived compounds.

Introduction

Lipid analysis can provide information on the trophic status (Cranwell, 1974; 1981; Cranwell *et al.*, 1987; Nishimura, 1977), as well as on the origin of the sedimentary organic matter (Cardoso *et al.*, 1983). However, further interpretation of biomarker data is limited by the lack of chemotaxonomic information and of complete knowledge of the changes that lipids undergo *via* biochemical processes, such as assimilation, excretion, cell death and bacterial degradation. Several studies have shown that microbial and planktonic successions can greatly affect the lipid distribution and composition of particles through biological reworking, as they sink to the sediment.

Lake Geneva has been the subject of several limnological studies in the past 30 years (Burkard, 1984a). It is still at present under supervision by a committee for water quality: 'Commission Internationale pour la Protection des Eaux du Lac Léman contre la pollution' (CIPEL), which carries out bimonthly sampling for planktonic and microbiological studies, and physical and chemical parameter measurements since 1957. Lake Geneva organic matter has also been the object of several investigations using the biomolecular marker approach. Fatty acids and sterols have been determined in phytoplankton (Mermoud *et al.*, 1981) and surface sediment (Mermoud, 1982), in order to relate lipids present in the benthos to their possible sources. These previous studies have revealed a small allochtonous input to the sediment and the marked aquatic character of the sedimentary organic matter. Therefore, it would be of interest to focus on a biomarker approach in order to study the settling particles and the effect of biological processes upon the quantity and characteristics of organic materials over the period of a seasonal cycle.

This paper deals with a study of non-aromatic hydrocarbons (NAH) and polycyclic aromatic hydrocarbons (PAH) associated with particles in both epilimnion and hypolimnion of the lake at key-seasonal biological and hydrological periods, from November 1985 to September 1986.

Description of the site studied

Lake Geneva is the largest freshwater reservoir in Europe (582 km^2–89 km^3). Spring thermal stratification prevents the water column from mixing and partitioning into two physically distinct compartments. According to winter temperature (Januar–February) and wind strength, lake water undergoes more or less complete mixing, leading to partial oxygenation of the deep water, particulary in the area called 'Plaine centrale', the deepest part of Lake Geneva. Most of the carbon is fixed by epilimnetic phytoplankton. This primary production is efficiently cropped by zooplankton, mainly herbivorous (Balvay *et al.*, 1984; Pelletier *et al.*, 1987). Organic matter undergoes aerobic mineralization by heterotrophic bacteria. Seasonal variations of bacterial biomass and activity follow those of the phytoplankton (Dufour *et al.*, 1987).

The phytoplankton in Lake Geneva showed marked variations over the period from November 1985 to September 1986 and can be characterized as follows (Fig. 1): 1) in early November 1985, the biomass is on the decreasing slope of the fourth peak of the year and is accompanied by a drastic decrease of dinophyceae and an increase of cyanophyceae expressed as a percentage of the total biomass. 2) The whole year 1986 is characterized by the development of chlorophyceae specially in the surface layer in May. The maximum abundance of *Chlorella vulgaris* was observed on May 20th and reached extreme values in surface water that have never been observed before (Pelletier *et al.*, 1987). 3) The nannoplankton (chlorophyceae and cryptophyceae) remained unusually predominant throughout the summer and fall seasons (yearly average: 84% of the total biomass; it reached 99% in May).

Materials and methods

Sample handling

Water samples were collected at the central station of Lake Geneva (SHL2) at three depths:

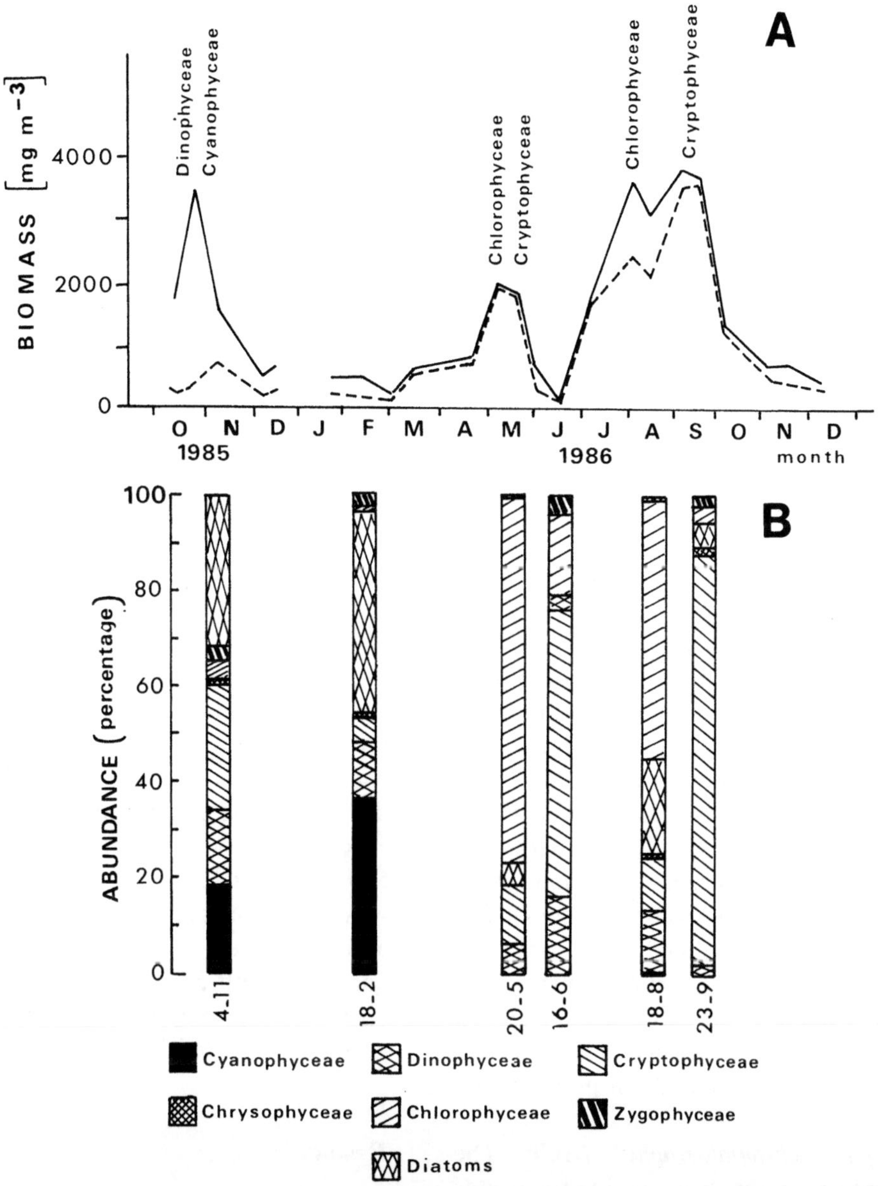

Fig. 1. A. Phytoplankton biomass (mg m^{-3}) and dominating classes in 1986 at station SHL2 (full line: total biomass; dotted line: nanoplankton biomass). B. Average abondances of the main phytoplankton classes in percentage over the period indicated (after Pelletier *et al.*, 1987).

5 m, 50 m, 100 m, from early November 1985 to mid-September 1986.

Water samples (20 liters) were filtered through solvent pre-extracted glass fibre filters having a nominal retention size of 0.7 μm (Whatman GF/F) and a diameter of 145 mm. Net tows were performed vertically, in order to collect large size particles, >200 μm and particles between 64 μm and 200 μm. The content of the collector was recovered on a glass fibre filter.. The filters were kept in the dark in a freezer ($t < -20$ °C) until laboratory analysis.

Analysis

Lipid isolation and fractionation: Filters containing the suspended particles were extracted according to the Bligh & Dyer procedure (1959). Extracts were dried over $CaCl_2$ and evaporated under vacuum and under a stream of pure nitrogen. Each extract was spiked with a known quantity of docosane or deuterated tetracosane as internal standards, in order to evaluate the efficiency of the next analytical steps. The extract was saponified under argon in a glass tube (SVL type, Sovirel, France) with 3 ml of 1.5 N MeOH-KOH and 3 ml of toluene, at refluxing temperature over a period of 2 hours.

After addition of 2 ml of bidistilled water, the reaction mixture was acidified to pH 2, and both unsaponifiable fraction (hydrocarbons, alcohols, sterols) and fatty acids were extracted 4 times with 1 ml of a mixture of hexane-ether (9:1, v/v). Extracts were evaporated under a stream of pure nitrogen. Lipid fractionation into non-aromatic and aromatic hydrocarbons was performed by adsorption chromatography on a small column (5 cm long and 7 mm internal diameter) filled with 1 g of SiO_2 Merck, type G_{60}. The first fraction was eluted with 6 ml of hexane contained non-aromatic hydrocarbons (n-alkanes, alkenes), and the second fraction was eluted with 3 ml of a mixture of hexane/ethyl acetate (50:1 v/v) contained aromatic hydrocarbons (>3 rings). The next eluted fraction contained the fatty acids, which are the subject of a study in preparation.

Capillary gas chromatography (GC): The fractions containing non-aromatic hydrocarbons were analyzed using a Girdel 3000 gas chromatograph equipped with flame ionization detector and Ross injector. The temperature of injector and detector were held at 250 °C and 320 °C respectively. A fused silica capillary column (25 m × 0.32 mm i.d.) coated with SE 54 was used. The temperature of the column was programmed from 100 °C to 280 °C at a rate of 2 °C min^{-1} for NAH, and from 50° to 280° at a rate of 2 °C min^{-1} for PAH. Quantification was based on calibrated flame ionization response to internal standard (n-C_{22} or deuterated n-C_{24} for NAH). The unresolved complex mixture (UCM) was quantified using the calibration factor determined for C_{22}. PAH was determined using an external standard composed of a mixture of PAH (naphthalene, fluorene, fluoranthene, pyrene, chrysene, benzofluoranthene, benzo (a)pyrene, perylene, dibenzoanthracene, benzoperylene). Recovery experiments, using the standard mixture and the same extraction procedure as for natural samples, showed total recovery ranging from 80 to 95% for individual components, with major losses for phenanthrene. Special care was taken during the entire analytical procedure (use of bidistilled solvents, clean glassware). Under these conditions, satisfactory procedural blanks were obtained routinely, with a detection limit for individual PAH around 0.005 ng l^{-1}.

Capillary gas chromatography-mass spectrometry (GC/MS): Analysis was performed using an R10-10C Nermag quadrupole mass spectrometer coupled with a Girdel 32 gas chromatograph in the conditions previously described. The mass spectrometer was scanned repeatedly from 31 to 550 amu at one scan per second. Electron energy was 70 eV; mass spectrometry data were acquired and processed using on-line PDP 11/23 plus computer with Sidar III data system. The mass spectra were compared with those of standards and data reported in literature.

Results

NAH, UCM, n-alkanes, and alkenes associated with particles were quantified from November 1985 to September 1986 at a central station of Lake geneva. POC was simultaneously determined and seasonal variations of relevent ratios are shown in Table 1.

A representative gas chromatogram of particulate non-aromatic hydrocarbons is displayed in Figure 2. Chromatograms are generally characterized by a uniform distribution pattern of n-alkanes from n-C_{15} to n-C_{35}, most of the time

Table 1. Concentrations ($\mu g\, l^{-1}$) of particulate organic carbon (POC), non-aromatic hydrocarbons (NAH), n-alkanes, alkenes and unresolved complex mixture (UCM) and their ratios related to POC (percentage). The upper and lower numbers correspond to the epilimnion (5 m depth) and the hypolimnion (50 m depth) values respectively.

	4 Nov. 1985	18 Feb. 1986	20 May 1986	16 June[1] 1986	18 Aug.[1] 1986	23 Sept. 1986
POC		100 141	1007 281	323 15	570 89	510 63
NAH	3.11 1.40	4.00 5.65	16.94 3.42	1.68 0.93	0.92 0.75	2.95 3.74
n-Alkanes	0.38 0.17	0.40 0.47	1.72 2.64	0.34 0.14	0.11 0.08	0.16 0.48
Alkenes	0.04 0.02	0.03 0.05	1.34 0.02	0.11 0.07	0.20 0.04	0.09 0.05
UCM	1.72 1.08	2.83 4.48	12.65 0.52	1.00 0.65	0.42 0.53	1.43 2.83
NAH/POC $\times 10^{-2}$		4.0 4.0	1.7 1.2	0.5 6.2	0.1 0.8	0.6 6.0
n-Alkanes/POC $\times 10^{-5}$		0.4 0.3	0.2 0.9	0.1 1.0	<0.1 0.1	<0.1 0.8
Alkenes/POC $\times 10^{-5}$		30 35	133 7	34 460	35 44	17 79
UCM/POC $\times 10^{-2}$		2.8 3.2	1.3 0.1	0.3 4.3	0.1 0.6	0.3 4.5
n-C_{17}/POC $\times 10^{-5}$		7.0 0.7	30.7 1.4	6.8 14	0.3 3.5	2.0 4.7
n-C_{17}/n-C_{29}	0.06 0.53	0.25 0.02	2.81 0.01	0.81 0.15	0.15 0.24	0.75 0.03
Pristane/Phytane	1.04 1.00	0.77 0.80	nd (Pris.)	0.65 (Pris.)	(Phyt.) 0.93	0.87 0.95
U/R	1.2 0.6	2.4 3.8	3.0 0.1	1.5 2.2	0.8 2.7	0.9 3.7
CPI_{23-35}	1.0 1.0	1.2 1.2	0.9 1.0	2.8 1.3	1.5 1.4	1.2 2.7

[1] Hypolimnion sampled at 100 m depth; (Pris.) and (Phyt.) indicate that only pristane or phytane was detected; U/R ratio = UCM/total GC resolved hydrocarbon peaks; CPI: Carbon Preference Index = 1/2 (23–35 odd/24–36 even + 23–35 odd/22–34 even); nd: not detected.

maximizing at n-C_{27} or n-C_{29}, more rarely at n-C_{17} or n-C_{15}. n-Alkanes from n-C_{21} to n-C_{35} showed a very slight odd/even carbon chain predominance, as indicated by the Carbon Preference Index (CPI), values varying from 1 to a maximum values of 2.8 (Table 1). A series of alkene homologues and isomers has been identified and listed in Table 2 (Scribe *et al.*, 1988). Chromatograms are also characterized by a hump representing an unresolved complex mixture of hydrocarbons (UCM), which belong mainly to the branched cyclic naphthene series.

Table 2. Concentrations (ng l^{-1}) of mono- and poly-enes associated with particles collected in the epilimnion (5 m depth). In the first column, the first number indicates the number of carbon atoms, the second the number of double bond and the third (in brackets), the location of the double bond in the carbon chain. RI = Retention Index on a SE-54 GC column.

Alkenes	RI SE-S4	4 Nov. 1985	18 Feb. 1986	20 May 1986	16 June 1986	18 Aug. 1986	23 Sept. 1986
15:1	1492			44.0		2.0	11.0
16:1	1593			11.0			5.0
17:1(7)	1689			40.0		30.0	
17:1(5)	1691			70.0		34.0	
17:1	1692	8.0	30.0	70.0	14.0	11.0	30.0
17:1(3)	1694			5.0	4.0	3.0	
17:1(1)	1696	0.4	6.0	350.0	1.0	50.0	50.0
18:1	1792	4.0	20.0	30.0	5.0		10.0
19:1	1894			30.0	3.0		2.0
20:1	1995			2.0	4.0	1.0	
21:1	2057		2.0	30.0	15.0	9.0	8.0
22:1	2119				5.0		
23:1	2295			270.0	84.0	8.0	25.0
24:1	2394				3.0		
25:1	2496			160.0	13.0	11.0	21.0
27:1	2621				17.0	5.0	4.0
28:1	2763					0.8	
29:1	2877					47.0	
31:1	3078				48.0	14.0	
17:2	1648	10.0		180.0			
18:2	1786	4.0	2.0	10.0			
19:2	1883	5.0	20.0	30.0			10.0
21:2	2044				4.0	8.0	0.1
27:2	2693	20.0		3.0		3.0	
28:2	2724					6.0	
31:2	3068					1.0	
19:3	1871		10.0	41.0			11.0
21:3	2076					2.0	
28:3	2765					4.0	
19:4	1863		30.0				
21:4	2021					1.0	
22:4	2171						1.6
25:4	2444						1.5
28:4	2745					4.0	
21:5	2079					3.0	
28:5	2749					1.0	
21:6	2074					10.0	

Discussion

Hydrocarbons enter the lacustrine environment both by endogenic natural sources: planktonic production (Balvay *et al.*, 1984) and by exogenic inputs: terrestrial vegetal detritus mainly from the Rhone and the Dranse rivers and petroleum hydrocarbons from waste water and sewage inputs (Burkard, 1984b; Burkard & Thelin, 1984).

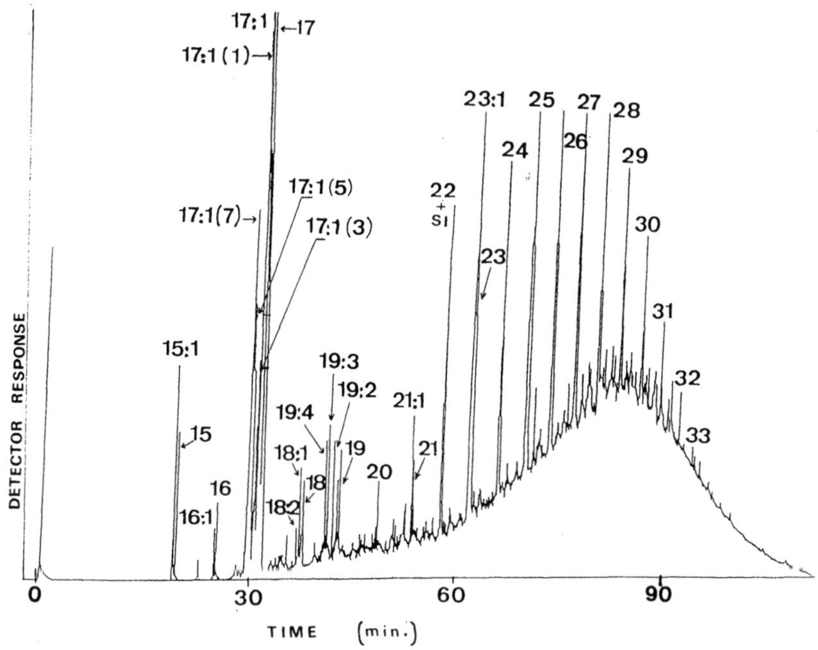

Fig. 2. Gas chromatogram of particulate non-aromatic hydrocarbon fraction in May 1986 at station SHL2. Simple numbers indicate the n-alkanes (number of carbon atoms); composite numbers indicate alkenes (number of carbon atoms, number and position of the double bonds).

Natural sources

Saturate hydrocarbons: Particulate NAH concentrations vary by a factor of 4 to 18, maximizing in the epilimnion, in May and by a factor not exceeding 6 in the hypolimnion throughout the year. Similarly, the maximum level of POC was observed at the same period as NAH, although with smaller variations (Table 1).

Most of the time, n-alkanes show a rather uniform monomodal distribution pattern ranging from n-C_{25} to n-C_{40}, maximizing at n-C_{27}, n-C_{29}, n-C_{31}. However, a bimodal distribution pattern with prominent peaks at n-C_{15}, n-C_{17}, n-C_{18}, n-C_{19}, n-C_{20}, and n-C_{25}, occurs variably in terms of space (water column), and time (seasons).

Values of the ratio of n-C_{17}/POC indicate a clear increase in epilimnion in May (Table 1). Chlorophyceae frequently produce short alkanes (n-C_{15}-C_{19}) with a predominance of heptadecane (n-C_{17}) and lesser amounts of extended long alkanes (C_{21}-C_{36}), whereas cryptophyceae produce alkanes over a broad range (n-C_{14}-C_{32}) maximizing at n-C_{29} (Weete, 1976). These two phytoplankton species strongly dominated the plankton biomass in May, and decreased rapidly from the end of May to middle of June.

The n-C_{17}/n-C_{29} ratio decreased from 2.81 to 0.81 during the same period and probably resulted from selective grazing of phytoplankton species producing n-C_{17} as the major component. In other words, the n-C_{17}/n-C_{29} ratio in Lake Geneva may reflect changes in phytoplankton species rather than differences in aquatic-versus-land hydrocarbon inputs, as found by Meyers *et al.* (1983) and Gardner *et al.* (1989) in Lake Michigan. It is noteworthy that n-C_{17}/n-C_{29} values in the hypolimnion are 2 order of magnitude lower than in the epilimnion during the phytoplankton bloom, and can be seen as a sedimentation rate effect on the settling particles. This effect can also lead to enrichment of n-C_{17} in the lower layer, as in November (Table 1). n-C_{17}/n-C_{29} values also depend on the relative degradation rate which is not equivalent *a priori*. From this particular point of view, the pristane/

phytane ratio was close to 1 most of the time, except in the hypolimnion in May and June, when pristane was found without phytane. Widely distributed in lake particulate hydrocarbons, pristane and phytane have been shown to be produced biosynthetically *via* phytol, their presumed precursor released from the side-chain of chlorophyll. In the limnic environment, pristane of recent biogenic origin was identified in zooplankton diet (Clark & Blumer, 1967). Pristane without phytane found in the hypolimnion (Table 1) suggests that active transfer of hydrocarbons and consequently organic matter may occur by way of zooplankton vertical migration. This process may explain perceptible enrichment in biogenic hydrocarbons associated with the hypolimnion particles observed in August and September (Table 1). n-Alkane distributions from n-C_{23} to C_{35} dominated by odd carbon-numbers are indicative of contributions from higher plants (Eglinton & Hamilton, 1967; Saliot, 1981). However, the n-alkane distribution lacking odd-carbon predominance occurred at the central station SHL2, where CPI in the range of C_{21} to C_{35} were close to unity, whatever the season except in June, suggesting slight hydrocarbon allochtonous terrestrial inputs, as mentioned by Mermoud *et al.* (1981) for fatty acids, and sterols in the sediment of station SHL2 (Mermoud, 1982).

Unsaturated hydrocarbons: A series of mono- and polyenes ranging from n-C_{15} to n-C_{32} associated with particles were identified. Their distribution in the epilimnion underwent large qualitative and quantitative seasonal variations (Table 2). Linear alkenes are common in freshwater (Weete, 1976), since they are biosynthesized by many algae (Gelpi *et al.*, 1968). Alkenes associated with particles have been reported in fluvial (Scribe *et al.*, 1990) and coastal waters (Albaigès *et al.*, 1984). Long chain n-alkenes ($>C_{22}$) have been attributed to higher plants (Giger *et al.*, 1980). However, the alkene/POC ratio reached highest values in surface water (5 meters depth) in May, underlining the fact that alkenes with a high structural diversity are found in the nanoplankton during the bloom period. The high values of the ratio alkene/POC in June was due to the extremely low POC content in the hypolimnion. Furthermore, two sizes of particles (>200 μm and 64–200 μm) were collected with vertical net tows between 50 meters and the surface. In May, the >200 μm sized net content was essentially constituted by adult herbivorous cyclopoïda, cladocera (*Bosmina*), whereas the 64–200 μm content was a mixture of juvenile zooplankton of small size. A slight differences in alkene distribution are observed between these two different sized particles (Fig. 3). Monoalkenes were dominated by heptadecene for

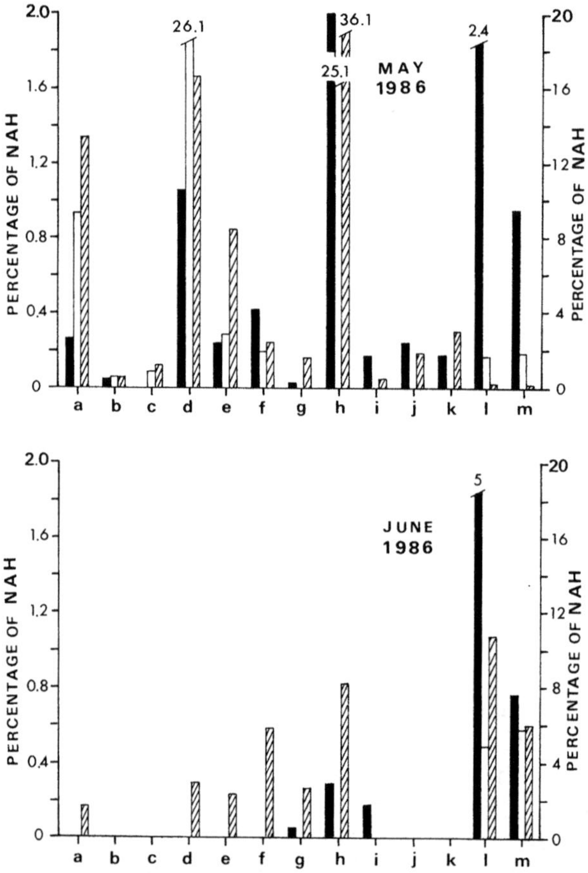

Fig. 3. Monthly variation of alkenes associated with sized particles in May and June 1986 at SHL2 station. Weight percentage of total non-aromatic hydrocarbons (NAH). ■ <64 μm (scale on the left); □ 64–200 μm (scale on the right); ▨ >200 μm (scale on the right). a) 15 : 1(1), b) 16 : 1(1), c) 16 : 1(1), d) 17 : 2, e) 17 : 1(7), f) 17 : 1(5), g) 17 : 1(3), h) 17 : 1(1), i) 18 : 1, j) 19 : 3, k) 19 : 2, l) 23 : 1(1), m) 25 : 1(1).

which 4 isomers were identified, and heptadecadiene. The distribution of n-alkenes found in both sizes of zooplankton reflects fairly well the alkene composition of the particles collected by filtration of the water sampled with the 10 liter Niskin bottle, which consisted of nanoplankton. The similarity in alkene composition of the three samples may be interpreted as the result of grazing; alkenes being one of the chemically fairly stable components of zooplankton diet. In June, considerable changes occurred in both size classes of particles: 1. Alkenes were not found in the 64–200 μm particles. 2. Polyenes had completely disappeared from both size classes reflecting the lack of this particular type of component in the nannoplankton sample (Table 2). 3. A clear decrease of the relative abundance of monoalkenes relative to the total NAH, except tricosaene ($C_{23:1}$) and pentacosaene ($C_{25:1}$) which were found in both classes.

The origin of $C_{23:1}$ and $C_{25:1}$ has been attributed to higher plants (Giger *et al.*, 1980). This origin is very unlikely here in May because the range of homologs is limited to 23 : 1 and 25 : 1, and is usually very similar to the n-alkane distribution. However, this hypothesis may be considered during the summer, when the number of alkene homologs increased and ranged from $C_{21:1}$ to $C_{31:1}$ (Table 2). This parallels a slight increase of the CPI of the n-alkanes n-C_{21}-C_{35} from 1.0 in May to 2.5 in June, and 1.9 in August (Table 1).

Anthropogenic sources

Identification of pollution sources is the second objective of the present paper. The assessment of multi-source and long term contamination is always a challenge in environmental analysis. In lacustrine environment, NAH may not only might originate from several anthropogenic inputs, as well as biosynthesis, but they might have been subject more or less recently to various weathering processes (Muller, 1987).

Non-aromatic hydrocarbons: Possible sources of the petroleum type of residues would include atmospheric precipitation of fossil fuel derivatives from shipping, street run-off and other industrial and/or domestic sources. n-Alkanes in the C_{14}-C_{36} range without even or odd carbon number predominance may also be representative of petrogenic inputs, in particular when this distribution pattern coincides with an unresolved complex mixture of hydrocarbons (UCM) (Farrington & Tripp, 1977; Muller, 1987). UCM may also be produced with modal distribution of n-alkanes ranging from C_{20}-C_{28} without carbon number preference by bacterial activity (Johnson & Calder, 1973). Are petrogenic paraffins significantly represented in the hydrocarbons associated with the particles? If we compare two typical situations in the epilimnion in May 86 (intensive bloom) and June (clearness of the water), no clear tendencies are observed in the n-alkane distribution, except the appearance of a prominent n-C_{17} (planktonic marker) and, to a lesser extent, n-C_{25} and n-C_{20} (of possible bacterial origin) in May.

Separation of the diastereoisomers of hopanoids, ubiquitous in organisms as well as in crude oils (Dastillung & Albrecht, 1976), was attempted. Except in November 1985 and September 1986, when samples collected in the epilimnion exhibited a rather high terpane content (13% and 19% of total NAH, respectively), the major part of the samples exhibited extremely low or undetectable levels of cyclic hydrocarbons by GC/MS. However, in November and September 1986, significant amounts of pentacyclic triterpanes (m/z = 191), such as $17\alpha(H)\,21\beta(H)\,C_{30}$ hopane were identified as petrogenic contaminents.

Polycyclic aromatic hydrocarbons: PAH fractions of crude oils contain high percentages of 2 and 3 rings, with a predominance of alkylated over unsubstituted homologs. Thus, in order to confirm the low level of petroleum contamination of the surface water particles, GC/MS analysis on a limited number of PAH was carried out on the same samples. This is a limited evaluation, which does not claim to be an exhaustive and quantitative determination of particulate PAH in Lake Geneva. Our objective was to compare the PAH concentration at a central station of the lake with those observed in the Rhone river, at different levels of petroleum contamination for specific components.

Quantitative results for the PAH analysis of surface particles in February, May and September and for hypolimnion in February are shown in Table 3 and compared with those obtained for particles collected by filtration (Whatman GF/F) of riverine water sampled at two stations, upstream and downstream from Port-Saint-Louis respectively (Lipiatou, 1989). Except for fluoranthene in September (150 ng l^{-1}) and perylene in May and September (37 ng l^{-1} and 44 ng l^{-1} respectively), the concentration of listed PAH was 2 times higher than the less contaminated upstream station, and at least one order of magnitude under the level of the station downstream from Port-Saint-Louis. However, the ratios of the listed PAH/POC show that its values are of the same order of magnitude as the particles collected at the station upstream from Port-Saint-Louis.

In the case of Lake Geneva, a low level of contamination by petroleum products was found. A study of ions m/z = 192, 206 and 216 corresponding to alkylated phenanthrene, anthracene and pyrene, shows that these compounds were below the GC/MS detection limit (0.005 ng l^{-1}), whereas their parent hydrocarbons were found in substantial quantity (Table 3). A wide variety of hydrocarbon compounds is formed during combustion of crude and fuel oil, and the proportion of alkylated compounds in PAH mixtures correlates closely with their temperature of formation.

Table 3. Concentrations (ng l^{-1}) of PAH associated with particles and ratios relative to POC (percentage × 10^{-3}).

Compounds	$M^{+\cdot}$ (¹)	Rhone(³)		18 Feb. 1986		20 May 1986		23 Sept. 1986	
		ng l^{-1}	% × 10^{-3}	ng l^{-1} (²)	% × 10^{-3}	ng l^{-1}	% × 10^{-3}	ng l^{-1}	% × 10^{-3}
Phenanthrene	178	19 50	16 104	nd 1.6	– 11	2.7	3	nd	–
Fluoranthene	202	22 73	19 152	6.1 5.0	61 35	5.9	6	150.0	294
Pyrene	202	21 43	18 89	9.9 3.0	99 21	4.7	5	5.0	10
Benzofluoranthene	228	22 148	19 308	9.1 8.0	91 57	9.0	9	3.8	7
Pyrelene	252	4 26	3 54	nd 19.0	– 135	37.0	37	44.0	86
Benzo(a)pyrene	252	11 68	9 142	nd nd	– –	nd	–	nd	–
Dibenzo(a)anthracene	252	9.0 49	8 102	nd nd	– –	3.0	3	nd	–
Benzoperylene	276	11 68	9 142	nd 1.8	– 13	nd	–	nd	–

(¹) M^+: molecular ion (m/z); (²) The upper and lower numbers correspond to the epilimnion (5 m depth) and the hypolimnion (100 m depth) values respectively; (³) The upper and the lower numbers correspond to values obtained with riverine particles sampled at 2 stations upstream and downstream from Port-Saint-Louis respectively (Lipiatou, 1989); nd: not detected.

On the basis of this hypothesis, several investigators have presented a method for the discrimination of PAH mixtures from petroleum and combustion sources by analysis of the alkyl homolog distribution (Youngblood & Blumer, 1975; Sporstøl *et al.*, 1983).

Perylene is thought to originate from higher plants (Aizenshtat, 1973) or from phytoplankton (Wakeham *et al.*, 1979). This compound was identified in surface waters in May and September only, during the phytoplankton blooms, indicating a possible autochthonous origin.

In our case, we have a very limited number of PAH as anthropogenic markers, so we must be very prudent in our diagnosis concerning the interpretation of the UCM, which most of the time represent the major part of the hydrocarbon fraction (more than 50% of the total NAH). It is noteworthy that no relation is observed between the petroleum and/or combustion contaminent and UCM (ratio U/R), nor between biogenic hydrocarbons and UCM. But according to phenanthrene, pyrene, benzofluoranthene, dibenzo(a)anthracene concentrations (Table 3), which indicate a low background of contamination of the particles by combustion products, it seems more likely that the strong increase of the UCM concentration should be, at least in May, predominantly due to biological processes such as zooplankton excretion and/or microbial reworking of the feces and the whole organic matter pool. On May 20th, zooplankton production as function of the season is on the rising slope (the abundance of the total planktonic crustaceans is 3.0×10^6 ind m^{-2}). The zooplankton production was still close to its yearly lowest level (around 0.5×10^6 ind m^{-2}) one month before, and reached the first yearly maximum one week later (4×10^6 ind m^{-2}) (Pelletier *et al.*, 1987).

Furthermore, Dufour *et al.* (1987) have shown that the greater part of bacterial activity occurs in the surface layer (0–10 m depth) in non-active phytoplankton cells and probably in relation with excreted and lyse products. Bacterioplankton seasonal variations follow those of the phytoplankton, with the difference that the maximum number of bacterial cells and the yearly peak of bacterial biomass and activity occur precisely on the May 20th. The composition of NAH associated with particles (n-alkanes, alkenes, UCM and individual compounds) shows large variations, particularly in May during a high level of biological activity, probably responsible for the particulate hydrocarbon distribution pattern.

A slight imprint of petroleum contamination is obviously present in September and November, and possibly in February, but according to limited PAH analysis, the essential part of the anthropogenic hydrocarbon input is probably constituted by compounds formed from burning of fossil fuels in combustion engines (land-based and/or pleasure craft).

Conclusion

Non-aromatic and aromatic hydrocarbons associated with particles were analyzed in the epilimnion and the hypolimnion of a central station of Lake Geneva at key-periods of the biological cycles and hydrological events. NAH and PAH data suggest slight exogenic terrestrial inputs and low level of anthropogenic contribution, due to high temperature pyrolytic contaminants rather than to petroleum derived compounds.

In May 1986, during the first phytoplankton bloom, a large proportion of the UCM seems to be related to reworking of endogenous organic matter.

The major part of the particulate non-aromatic hydrocarbons such as n-alkanes and alkenes is produced by phytoplankton. Variation of the $n\text{-}C_{17}/n\text{-}C_{29}$ ratio seems to follow phytoplanktonic succession, illustrated this year by the chlorophyceae-cryptophyceae succession in May and June. Alkenes ranging from C_{15} to C_{31} originating from nanoplankton are found in juvenile and adult zooplankton, and are thought to be valuable candidates as molecular tracers of primary producers to study the trophic relationship between phytoplankton and zooplankton in the lacustrine environment.

Acknowledgements

We thank the GRECO-LACS, research group of the Centre National de la Recherche Scientifique (CNRS), for financial support and the Institut de Limnologie de Thonon-les-Bains (France) for laboratory and sampling facilities. We thank Drs. J. P. Pelletier, G. Balvay, N. Angeli and P. Dufour from the Station d'Hydrobiologie Lacustre de Thonon (INRA) for fruitful discussions and suggestions.

References

Aizenshtat, A., 1973. Perylene and its geochemical significance. Geochim. Cosmochim. Acta 37: 559–567.

Albaigès, J., J. Grimalt, J. M. Bayona, R. Risebrough, B. De Lappe & W. Walker II, 1984. Dissolved, Particulate and sedimentary hydrocarbons in deltaic environment. Org. Geochem. 6: 237–248.

Balvay, G., J. C. Druart, J. Pelletier, E. Pongratz & R. Revaclier, 1984. Plancton. In Commission Internationale pour la Protection des Eaux du Léman contre la Pollution (ed.), Le Léman, synthèse 1957–1982: 261–313.

Bligh, E. & W. Dyer, 1959. A rapid method of total lipid extraction and purification. Can. J. Biochim. Physiol. 37: 911–917.

Burkard, P., 1984a. Bilan général du Léman et conclusions. In Commission Internationale pour la Protection des Eaux du Léman contre la Pollution (ed.), Le Léman, synthèse 1957–1982: 503-575.

Burkard, P., 1984b. Apports au Léman: Apports par les affluents. In Commission Internationale pour la Protection des Eaux du Léman contre la Pollution (ed.), Le Léman, synthèse 1957–1982: 387–422.

Burkard, P. & L. Thelin, 1984. Apports par les stations d'épuration. In Commission Internationale pour la Protection des Eaux du Léman contre la Pollution (ed.), Le Léman, synthèse 1957–1982: 423–435.

Clark, R. C. & M. Blumer, 1967. Distribution of n-paraffins in marine organisms and sediment. Limnol. Oceanogr. 12: 79–87.

Cardoso, J., S. Gaskell, M. Quirk & G. Eglinton, 1983. Hydrocarbon and fatty acid distributions in the Rostherne Lake sediment (England). Chem. Geol. 38: 107–128.

Cranwell, P. A., 1974. Monocarboxylic acids in lake sediments: indicators derived from terrestrial and aquatic biota, of paleoenvironmental trophic levels. Chem. Geol. 14: 1–14.

Cranwell, P. A., 1981. Diagenesis of free and bound lipids in the terrestrial detritus deposited in a lacustrine sediment. Org. Geochem. 3: 79–81.

Cranwell, P. A., G. Eglinton & N. Robinson, 1987. Lipids of aquatic organisms as potential contributions to lacustrine sediments. II. Org. Geochem. 6: 513–527.

Dastillung, M. & P. Albrecht, 1976. Molecular tests for oil pollution in surface sediments. Mar. Poll. Bull. 7: 13–15.

Dufour, P., S. Stroffeck & M. Colon, 1987. Le bactérioplancton, campagne 1986. In Commission Internationale pour la Protection des Eaux du Léman (ed), Rapport sur les études et recherches entreprises dans le bassin lémanique, campagne 1986: 69–75.

Eglinton, G. & R. J. Hamilton, 1967. Leaf epicuticular waxes. Science 156: 1322–1325.

Farrington, J. W. & B. W. Tripp, 1977. Hydrocarbons in western North Atlantic surface sediments. Geochim. Cosmochim. Acta 41: 1627–1641.

Gardner, W. S., B. J. Eadie, J. F. Chandler, C. C. Parrish & J. M. Malczyk, 1989. Mass flux and 'nutritional composition' of settling epilimnetic particles in Lake Michigan. Can. J. Fish. aquat. Sci. 46: 1118–1124.

Gelpi, E., J. Oro, H. J. Schneider & E. O. Bennett, 1968. Olefins of high molecular weight in two microscopic algae. Science 161: 700–702.

Giger, W., C. Schaffner & S. G. Wakeham, 1980. Aliphatic and olefinic hydrocarbons in recent sediment of Greifensee, Switzerland. Geochim. Cosmochim. Acta 44: 119–129.

Lipiatou, E., 1989. Contribution à l'étude géochimique des hydrocarbures aliphatiques et aromatiques dans le delta du Rhône et le Golf du Lion. Thèse Doct. de l'Université Pierre et Marie Curie, Paris, 195 pp.

Mermoud, F., 1982. Géochimie des stérols et des acides gras dans les sédiments du lac Léman et du Voua de la Motte. Thèse Doct. n° 2060, Univ. Genève. 159 pp.

Mermoud, F., O. Clerc, O. Gülacar & A. Buchs, 1981. Analyse des acides gras et stérols dans le plancton du Léman. Arch. Sci. Geneve 34: 367–382.

Meyers, P. A., M. J. Leenheer, B. J. Eadie & S. J. Maule, 1983. Organic geochemistry of suspended and settling particulate matter in Lake Michigan. Geochim. Cosmochim. Acta 48: 443–452.

Muller, H., 1987. Hydrocarbons in the freshwater environment. In H. J. Elster & W. O. Plön (eds), Advances in Limnology. E. Schweizerbart'sche Verlagsbuchhandlung. (Nägele U. & Obermiller J.) Stuttgart: 1–69.

Nishimura, M., 1977. The geochemical significance in early sedimentation of geolipids obtained by saponification of lacustrine sediment. Geochim. Cosmochim. Acta 41: 1817–1823.

Pelletier, J., G. Balvay & J. C. Druart, 1987. Evolution du plancton du Léman: campagne 1986. In Commission Internationale pour la Protection des Eaux du Léman (ed.) Rapport sur les études et recherches entreprises dans le bassin lémanique, campagne 1986: 47–68.

Saliot, A., 1981. Natural hydrocarbons in sea water. In E. K. Duursma & R. Dawson (eds), Marine Organic Chemistry. Elsevier Amsterdam: 327–374.

Scribe, P., J. Guezennec, J. Dagaut, C. Pepe & A. Saliot, 1988. Identification of the position and the stereochemistry of the double bond in monounsaturated fatty acid methyl esters by GC/MS of dimethyldisulfide derivatives: application to recent marine sediment. Anal. Chem. 60: 928–931.

Scribe, P., A. Barouxis, J. Tronczynski & A. Saliot, 1990. Application of hydrocarbon biochemical markers to the study of particulate matter in a high turbidity estuary. In J. Berthelin (ed.), Diversity of Environmental Biogeochemistry, Elsevier Science Publishers (in press).

Sporstøl, S., N. Gjos, R. G. Lichtenthaler, K. O. Gustavsen, K. Urdal, F. Oreld & J. Skel, 1983. Source identification of aromatic hydrocarbons in sediments using GC/MS. Envir. Sci. Technol. 17: 282–286.

Wakeham, S. G., C. Schaffner, W. Giger, I. J. Boon & J. W. De Leeuw, 1979. Perylene in sediments from the Namibian shelf. Geochim. Cosmochim. Acta 43: 1141–1144.

Weete, J., 1976. Algal and fungal waxes. In P. Kolattukudy (ed.), Chemistry and Biochemistry of Natural Waxes, Elsevier Amsterdam: 349–418.

Youngblood, W. & M. Blumer, 1975. Polycyclic aromatic hydrocarbons in the environment: homologous series in soils and recent sediments. Geochim. Cosmochim. Acta 39: 1303–1314.

Received 19 July 1989; in revised form 15 March 1990; accepted 15 May 1990

Hydrobiologia **207**: 333–339, 1990.
D. J. Bonin & H. L. Golterman (eds), Fluxes Between Trophic Levels and Through the Water–Sediment Interface.

On the discrepancies between a colorimetric and isotopic method for measuring nitrate utilization in nutrient-depleted waters: implications for the design of experimental protocols in new production studies

Gerd Slawyk, Patrick Raimbault & Valérie Gentilhomme
Centre d'Océanologie de Marseille, Faculté des Sciences de Luminy, Case 901, 13288 Marseille Cédex 9, France

Key words: nitrate uptake, nanomolar-nitrate determination, ^{15}N methodology

Abstract

Colorimetric analyses of nitrate disappearance from seawater have been compared with isotopic analyses of ^{15}N-labelled nitrate incorporation into particulate matter. The slope (1.41) of a regression line calculated from 19 sample pairs gathered during 6 time-series experiments and 2 single end-point incubations showed that nitrate incorporation is positively related to changes in nitrate concentration but that it accounts only for 71% of nitrate disappearance. ^{15}N-isotope dilution as a consequence of nitrification, if any, would not fully explain discrepancies between the two analytical procedures. A further possible mechanism responsible for the imbalance between nitrate-incorporation and -disappearance rates suggests losses of ^{15}N label from plankton biomass to an unanalyzed pool (dissolved organic nitrogen?) which increase (up to 65%) with incubation time. The lack of ^{15}N-mass balance calls for the need to consider additional nitrogen pools in ^{15}N budgets of isotope experiments and not only substrate and biomass pools as has been done so far.

Introduction

Field data on nitrate-uptake rates obtained from chemical analyses of nitrate disappearance are rather scarce and experiments have mostly been conducted in nutrient-rich environments (Price *et al.*, 1985; Dugdale & Wilkerson, 1986) where changes in nitrate concentration were detectable with conventional colorimetric methods (Strickland & Parsons, 1972). Eppley & Renger (1986) were the first to measure nitrate uptake from chemical differences in nutrient-depleted waters using a chemiluminescent reaction (Garside, 1982) and to compare this chemical method with the ^{15}N-tracer approach.

Materials and methods

In this study an improved automatic colorimetric procedure for the determination of nitrate and nitrite in the nanomolar range (Oudot & Montel, 1988; Raimbault *et al.*, 1990) has been used in parallel with the ^{15}N method (Dugdale & Goering, 1967) to estimate nitrate uptake in surface waters of 8 stations (Table 1) in the north-east Atlantic off Cape São Vincente (Portugal) during the Medatlante I cruise in January 1989 aboard the R.V. *Jean Charcot*.

Water from the 5-m depth was collected with a 30-l Niskin bottle. Two-liter aliquots were poured into a series of polycarbonate bottles, inoculated

Table 1. Environmental and experimental conditions of seawater collected for 6 time-series and 2 single end-point -experiments.

Sta	Location	Starting time of incubation (local time)	Initial $N-NO_3^-$ (nM)	^{15}N-tracer addition (% of ambiant $N-NO_3^-$)	Final PON (μM)
401	36.3417N,10.4180W	09.20	120.0	7.0	3.13
412	36.3318N,10.3952W	08.45	97.5	11.9	2.39
436	36.2879N,10.4131W	10.45	90.8	43.9	1.81
456[1]	36.3923N,10.5342W	09.15	339.8	13.8	0.90
476	36.5570N,10.4304W	09.30	33.0	12.6	1.14
491	36.3503N,10.3010W	09.45	57.8	8.6	1.57
505[1]	36.1979N,10.1114W	09.00	206.9	11.4	0.60
541	36.1387N,11.1974W	10.30	146.0	12.8	1.19

[1] Single end-point incubation.

with $^{15}N-NO_3^-$ (<14% of ambient concentration, except for station 436; Table 1) and incubated under natural light in a surface water cooled deck incubator where incident light was attenuated to 50% by perforated nickel screening. Particulate matter of the aliquots was collected by filtration (pre-combusted Whatman GF/F) at time 'zero' (starting time of incubation) and at different time intervals (2–4 h) thereafter for about 10 h. In order to avoid damage of cells during filtration (Goldman & Dennett, 1985) samples were gently filtered using very low vacuum of ≪100 mm Hg. This pressure differential is distinctly lower than the upper limit used sometimes by others performing this type of experiment (Kokkinakis & Wheeler, 1987). The filters were saved for Dumas preparation to mesure incorporation of ^{15}N label into the particulate fraction and particulate organic nitrogen (PON) concentrations as described by Slawyk *et al.* (1988). Prior to filtration, 50 ml of seawater were withdrawn from each aliquot for triplicate determination of nitrate concentration with a Technicon Autoanalyzer II following the working procedure outlined by Raimbault *et al.* (1990).

Nitrate-incorporation rates have been calculated from ^{15}N accumulation in the particulate matter according to Dugdale & Wilkerson (1986) and nitrate-disappearance rates have been computed from the change in concentration of nitrate per unit time. Both rates are reported for the time interval over which they have been calculated (nM $^{15}NO_3^-$). The precision of the two analytical procedures was almost identical with a coefficient of variation (C.V. = $s/\bar{x} \times 100$) of 16.9 and 16.3% for mass spectrometry and colorimetry respectively at levels of about 20 nM $^{15}NO_3^-$ reciprocal time.

Results and discussion

Results of a comparison of nitrate uptake measured during 6 time-series experiments and 2 single end-point incubations by the colorimetric method

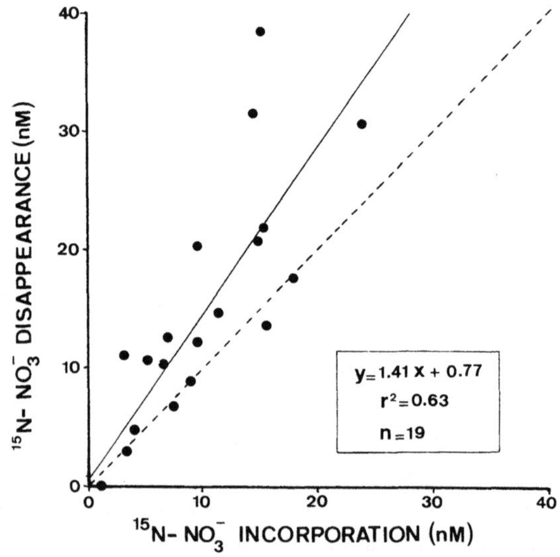

Fig. 1. Linear regression of $^{15}N-NO_3^-$ disappearance *vs* $^{15}N-NO_3^-$ incorporation. Dashed line represents a 1:1 correspondence.

vs the ^{15}N-tracer method are depicted in Fig. 1. A linear regression calculated from 19 sample pairs shows that nitrate incorporation is positively related to changes in nitrate concentration during incubation. However, the slope value of 1.41 indicates a discrepancy between the two methods, *i.e.*, only 71% of the nitrate that disappeared from the dissolved fraction was met by ^{15}N incorporation into the particulate fraction.

Our results are in contrast to those obtained by Eppley & Renger (1986) in nitrate-depleted waters off California who found that all nitrate utilized appeared in the particulate phase. They are in agreement with data reported recently by Ward *et al.* (1989) from the same study area who examined the general relationship between nitrate-concentration changes and isotope-dilution model calculations. More recently, very large discrepancies have been found by Eppley & Renger (1990) in the equatorial Pacific. The rate of nitrate removal from seawater measured with the sensitive chemiluminescence method (Garside, 1982) averaged 7.2 times the rate of ^{15}N-labelled nitrate incorporation into filterable particles. Nitrate-uptake measurements made by Dugdale & Wilkerson (1986) in nutrient-rich waters of the Peru upwelling behaved in the same way as ours with similar percentages of nitrate disappearance accounted for by the ^{15}N incorporation (76%).

Rates of ^{15}N-labelled nitrate uptake from long incubations (> 10 h) could be in error (underestimated) as a consequence of substrate depletion. However, this analytical problem did not arise in our experiments since measurable nitrate levels were always detected in the samples at the end of incubation.

Discrepancies between the two analytical procedures tested may be indicative of *in vitro* production (recycling) of unlabelled substrate (nitrate) and consequent isotope dilution. Isotope dilution, if ignored, is expected to lead to underestimation of ^{15}N-incorporation rates. Such isotope-dilution effects have been convincingly demonstrated in studies of ^{15}N-labelled ammonium uptake by microplankton (Glibert *et al.*, 1982; Harrison & Harris, 1986).

In our data set, nitrification that creates isotope dilution is rather unlikely and may not cause the observed discrepancy between ^{15}N incorporation and nitrate disappearance for at least two reasons. First, the disappearance: ^{15}N-incorporation ratio (slope of the regression line) is >1; ratios <1 would only give clear indication of nitrification (for details compare with the DU ratio – the ratio of disappearance of NH_4^+ to $^{15}NH_4^+$ uptake – for assessing the magnitude of ammonification; Dugdale & Wilkerson 1986, their Fig. 5). Second, studies on depth distribution of nitrification (Ward, 1987; Ward *et al.*, 1989) showed that nitrification occurs preferentially at nitracline depths (6% surface-light intensity) and not in near-surface waters where it is strongly inhibited by light. Of course, we cannot assert that isotope dilution has actually not occurred during the incubation period since direct measurements of isotope enrichment in the substrate at the end of incubation are not available. However, if any nitrification occurred, corresponding rates must have been lower than nitrate-uptake rates since we always observed changes (nitrification ≠ uptake) but never an increase in nitrate concentration in our incubation bottles during the time course. An indirect evaluation of isotope dilution applying the calculation procedure of Kanda *et al.* (1987) is not possible with our data because the basic assumption in their method, 'that ^{15}N which disappears from the substrate pool is equal to its increase in the particulate matter pool', is not fulfilled as we will see hereunder.

Let us assume that nitrate uptake occurs with no nitrification (thus, no isotope dilution), *i.e.*, the ^{15}N-atom% enrichment in the nitrate pool remains constant. We can then evaluate our data for ^{15}N-mass balance (Laws, 1984) and check if the ^{15}N removed from the dissolved fraction is recovered in the particulate fraction by calculating the following ratio (^{15}N-loss ratio) as:

$$^{15}\text{N loss} = [(R_o \times S_o - R_t \times S_t) - (R' \times \text{PON})] \times [(R_o \times S_o - R_t \times S_t)]^{-1} \qquad (1)$$

where R_0 and R_t are the atom% excess ^{15}N in the

nitrate fraction at time 'zero' and time t ($R_0 = R_t$, assuming no isotope dilution), S_0 and S_t the nitrate concentrations at time 'zero' and time t and R' the atom% excess ^{15}N in the PON fraction. A ratio of zero would indicate no ^{15}N loss from the experimental system whereas a value of 1 would correspond to a 100% ^{15}N loss (negative values mean experimental artifact).

Inspection of the ^{15}N-loss ratio *vs* incubation time relationship (Fig. 2) shows that losses of ^{15}N actually occur and that they increase with the length of incubation period. Closer examination of data from individual stations revealed that with short incubation periods (<6 h) most of the ^{15}N which disappeared from the dissolved fraction (75–100%) was balanced by the amount of ^{15}N which appeared in the particulate fraction. Beyond this period (6–11 h) losses of ^{15}N ranged from 30 to 65% (except station 541) and were highest at stations with high PON concentrations (stations 401 and 412; Table 1). Note that including a possible isotope-dilution effect would further drive the value of the ^{15}N-loss ratio upward. Thus, the latter results give evidence that losses (transfer) of an unknown form of nitrogen from the experimental system to a pool not considered in the mass balance are responsible, at least to some degree, for the discrepancies observed between the colorimetric and isotopic method.

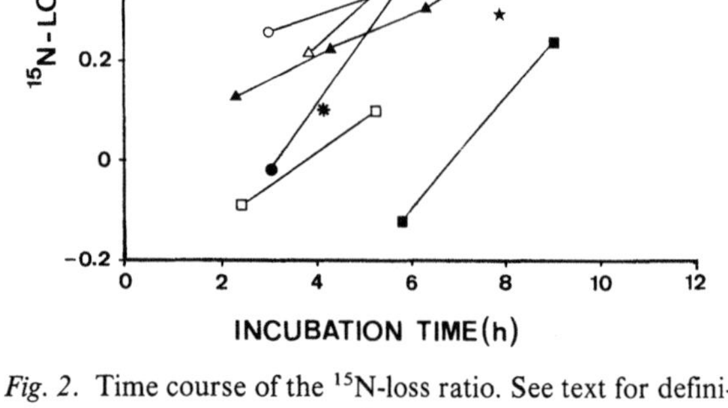

Fig. 2. Time course of the ^{15}N-loss ratio. See text for definition.

Our observation of the missing ^{15}N in the particular phase is not new and has already been reported from mass-balance calculations for ammonium (Glibert *et al.*, 1982; Laws, 1984) and urea (Hansell & Goering, 1989) as well. As in the case of the two latter regenerated nitrogen forms, mechanisms responsible for the sink of ^{15}N suggest losses of substrate (nitrate in our system) to another pathway (denitrification, *i.e.*, dissimilatory nitrate reduction by bacteria) and/or losses of some form of nitrogen from the microplankton by processes subsequent to nitrate incorporation.

If the ^{15}N loss was caused by denitrification, then some of the label from the original nitrate pool would have been transferred to the nitrite pool and not incorporated into the particulate matter, *i.e.*, not used as a nitrogen source by the organisms. Dissimilatory nitrate reduction is a more rapid process than assimilatory nitrate reduction (nitrate assimilation) which results in a correspondingly rapid extracellular accumulation of nitrite (Cole, 1987). However, significant changes in nitrite concentration were not detected during the incubation. In any case, denitrification is very unlikely in our well-oxygenated surface samples (Ward *et al.*, 1989).

Possible avenues for nitrogen lost from the particulate phase, similar to those suggested by Laws (1984) and Price *et al.* (1985) for ammonium uptake, by Hansell & Goering (1989) for urea uptake, and by Eppley & Renger (1990) for nitrate uptake, include (1) release of cellular nitrate and/or dissolved organic nitrogen (DON) through cell lysis during filtration, (2) bacterial and/or photosynthetic (microparticulate chlorophyll *a*, Taguchi & Laws 1988) uptake not retained on the GF/F filters, (3) physiological release (excretion) of nitrite and/or DON (amino acids for example; Myklestad *et al.*, 1989) by phytoplankton and (4) grazer-induced losses, such as ammonium excretion by zooplankton and leakage of nitrogen containing phytoplankton-cell contents (nitrate, DON) through 'sloppy' feeding.

Among these avenues excretion of nitrite by phytoplankton may be safely eliminated as a cause of the ^{15}N loss, since no significant changes (increase) in nitrite were evident in our incubation bottles during the time-course experiment.

Given the low pressure differential used during filtration, losses of some form of nitrogen through cell lysis are also most unlikely. Moreover, our data show that ^{15}N losses occurred essentially in the latter stage of the time-course experiment when the internal pool of low-molecular weight nitrogen should be already incorporated into proteins which are less subjected to leakage upon cell breakage (LaRoche & Harrison, 1987).

Studies of Cuhel *et al.* (1983) have shown that bacterial passage through glass-fiber filters is small. Nevertheless, it is possible that bacterial or other picoplankton uptake of nitrate has not been assessed with our GF/F filters. However, although nitrate uptake by heterotrophic bacteria has been clearly demonstrated in continuous-culture studies (Horrigan *et al.*, 1988), direct evidence from field studies that bacteria are effective in competing with phytoplankton for nitrate utilization (nitrate assimilation) is still lacking. Furthermore, ^{15}N-uptake experiments with fractionated natural microbial assemblages (Wheeler & Kirchman, 1986) have shown that the <1-μm size microorganisms utilized mainly ammonium and amino acids and only negligible amounts of nitrate. Assimilatory nitrate reduction may only occur when reduced nitrogen forms are lacking since in the presence of NH_4^+ (as in our study area) nitrate assimilation by bacteria is strongly repressed (Cole, 1987). Therefore, it would be desirable in future ^{15}N nitrate mass-balance studies to perform differential filtration through GF/F and smaller pore-size filters (<0.7 μm) compatible with mass spectrometry in order to examine to what extent bacteria and ultraphytoplankton pass GF/F filters and if so, whether this passage of microbes explains fully the ^{15}N loss.

Anyway, in our case the sink of nitrate due to incomplete recovery of PON during filtration is not consistent with the observation that the ^{15}N loss changed with incubation time unless one assumes that the uncollected picoplankton fraction grew faster than the larger cell size phytoplankton community retained on the GF/F filters as a result of bottle confinement (Ferguson *et al.*, 1984).

In view of the time dependence of ^{15}N loss (increase with incubation time) in our data set, we speculate that DON is the most likely nitrogen pool not accounted for in the mass balance and to which the missing ^{15}N has been transferred. Accumulation of ^{15}N labelled DON in phytoplankton cells, *i.e.* ^{15}N-nitrate assimilation, does not immediately catch up with nitrate uptake since it depends on nitrate reduction which is the slowest enzymatic reaction of nitrate assimilation and hence the rate-limiting step of nitrate utilization (Dortch, 1982). Consequently, the decreasing effect of ^{15}N-labelled DON losses on total ^{15}N recovery in the PON pool would increase with incubation time as observed in our data. In other words, the probability of regeneration of ^{15}N-labelled DON, contained in the PON pool, increases with rising ^{15}N enrichment of the phytoplankton as a result of prolonged uptake (Glibert *et al.*, 1985).

Although release of ^{15}N as DON may explain the lack of mass balance and, thus, discrepancies between the two analytical methods tested, we cannot state precisely if it results from physiological release by nitrate-utilizing organisms or reflects grazer-induced losses. However, the high microzooplankton abundance encountered in our study area (Gasser, personal communication) suggests that DON losses after ingestion of nitrate-assimilating cells by grazers are the most possible avenue for the fraction of ^{15}N removed from the nitrate pool which is unaccounted for in the particulate pool.

Our results raise the question whether nitrate uptake is more accurately estimated from nitrate disappearance (colorimetric method) or ^{15}N incorporation (isotopic method). Nitrate-disappearance rates represent net fluxes of nitrate since changes in concentration may encompass uptake, regeneration (nitrification) and possible fluxes to other pathways (denitrification). Depending on the relative importance of the latter processes uptake rates would be over- or underestimated. If such multiple nitrogen transfers occur in an experimental system, then the more correct approach for measuring uptake should be ^{15}N incorporation into particulates. If however, as we have suggested from these data, particulates leak significant amounts of ^{15}N during incubation, then nitrate-uptake rates and f-ratios (Eppley & Peterson, 1979) based upon these rates are underestimated by the isotopic method.

Measurements of nitrate-disappearance rates using our convenient colorimetric method are less labor-intensive than those of ^{15}N incorporation unless changes in concentration of nitrate are detectable. On the other hand, the ^{15}N methodology offers the advantage over the colorimetric method in that precisely it obviates the need of measurable changes in concentration of substrate to assess nitrogen transfers between different pools. However, when using the ^{15}N isotope for the estimation of nitrate-nitrogen fluxes (new production) an experimenter should foresee the possibility to locate missing ^{15}N by considering in addition to substrate and planktonic biomass pools a further possible nitrogen pool such as DON (LaRoche & Harrison, 1987). Exchange of label with this additional pool, which may explain the lack of mass balance, could be assessed by analyzing, in parallel with substrate, DON (amino acids, for example) in the dissolved fraction (filtrate) for changes in ^{15}N enrichment. Unless a methodological artifact is the culprit of the vanishing ^{15}N, as Dugdale & Wilkerson (1986) put it, 'it appears that nature, not science, is the problem'.

References

Cole, J. A., 1987. Assimilatory and dissimilatory reduction of nitrate to ammonia. In J. A. Cole & S. J. Ferguson (eds). Soc. gen. Microbiol., Cambridge Univ. press: 282–329.

Cuhel, R. L., H. W. Jannach & C. D. Taylor, 1983. Microbial growth and macromolecular synthesis in the northwestern Atlantic ocean. Limnol. Oceanogr. 28: 1–18.

Dortch, Q., 1982. Effect of growth conditions on accumulation of internal nitrate, ammonium, amino acids, and protein in three marine diatoms. J. exp. mar. Biol. Ecol. 61: 243–264.

Dugdale, R. C. & J. J. Goering, 1967. Uptake of new and regenerated forms of nitrogen in primary productivity. Limnol. oceanogr. 12: 196–206.

Dugdale, R. C. & F. P. Wilkerson, 1986. The use of ^{15}N to measure nitrogen uptake in eutrophic oceans, experimental considerations. Limnol. Oceanogr. 31: 673–689.

Eppley, R. W. & B. J. Peterson, 1979. Particulate organic matter flux and planktonic new production in the deep ocean. Nature 282: 677–680.

Eppley, R. W. & E. H. Renger, 1986. Nitrate-based primary production in nutrient-depleted waters off California. Océanogr. trop. 21: 229–238.

Eppley, R. W. & E. H. Renger, 1990. Nitrate utilization by plankton in the equatorial Pacific, March 1988 along 150 ° W. J. geophys. Res., in press.

Ferguson, R. L., E. N. Berckley & A. V. Palumbo, 1984. Response of marine bacterioplankton to differential filtration and confinement. Appl. envir. Microbiol. 47: 49–55.

Garside, C., 1982. A chemiluminescent technique for the determination of nanomolar concentrations of nitrate and nitrite in seawater. Mar. Chem. 11: 159–167.

Glibert, P. M., F. Lipschultz, J. J. McCarthy & M. A. Altabet, 1982. Isotope dilution models of uptake and remineralization of ammonium by marine plankton. Limnol. Oceanogr. 27: 639–650.

Glibert, P. M., F. Lipschultz, J. J. McCarthy & M. A. Altabet, 1985. Has the mystery of the vanishing ^{15}N in isotope dilution experiments been resolved? Limnol. Oceanogr. 30: 444–447.

Goldman, J. C. & M. R. Dennett, 1985. Susceptibility of some marine phytoplankton species to cell breakage during filtration and post-filtration rinsing. J. exp. mar. Biol. Ecol. 86: 47–58.

Hansell, D. A. & J. J. Goering, 1989. A method for estimating uptake and production rates for urea in seawater using [^{14}C] and [^{15}N] urea. Can. J. Fish. aquat. Sci. 46: 198–202.

Horrigan, S. G., Å. Hagström, I. Koike & F. Azam, 1988. Inorganic nitrogen utilization by assemblages of marine bacteria in seawater culture. Mar. Ecol. Prog. Ser. 50: 147–150.

Harrison, W. G. & L. R. Harris, 1986. Isotope dilution and its effects on measurements of nitrogen and phosphorus uptake by oceanic microplankton. Mar. Ecol. Prog. Ser. 27: 253–261.

Kanda, J., E. A. Laws, T. Saino & A. Hattori, 1987. An evaluation of isotope dilution effect from conventional data sets of ^{15}N uptake experiments. J. Plankton Res. 9: 79–90.

Kokkinakis, S. A. & P. A. Wheeler, 1987. Nitrogen uptake and phytoplankton growth in coastal upwelling regions. Limnol. Oceanogr. 32: 1112–1123.

LaRoche, J. & W. G. Harrison, 1987. Compartmental models of nitrogen cycling in tropical and temperate marine environments. Mar. Ecol. Prog. Ser. 38: 137–149.

Laws, E., 1984. Isotope dilution models and the mystery of the vanishing ^{15}N. Limnol. Oceanogr. 29: 379–386.

Myklestad, S., O. Holm-Hansen, K. M. Vårum & B. E. Volcani, 1989. Rate of release of extracellular amino acids and carbohydrates from the marine diatom *Chaetoceros affinis*. J. Plankton Res. 11: 763–773.

Oudot, C. & Y. Montel, 1988. A high sensitivity method for the determination of nanomolar concentrations of nitrate and nitrite in seawater with a Technicon AutoAnalyzer II. Mar. Chem. 24: 239–252.

Price, N. M., W. P. Cochlan & P. J. Harrison, 1985. Time course of uptake of inorganic and organic nitrogen by phytoplankton in the Strait of Georgia: comparison of frontal and stratified communities. Mar. Ecol. Prog. Ser. 27: 39–53.

Raimbault, P., G. Slawyk, B. Coste & J. Fry, 1990. Feasibility of using an automated colorimetric procedure for the determination of seawater nitrate in the 0–100 nM range: examples from field and culture work. Mar. Biol. 104: 347–351.

Slawyk, G., S. L'Helguen, Y. Collos & H. Freije, 1988. Quantitative determination of particulate organic N and C in marine-phytoplankton samples using mass-spectrometer signals from isotope-ratio analyses in ^{15}N- and ^{13}C-tracer studies. J. exp. mar. Biol. Ecol. 115: 187–195.

Strickland, J. D. & T. R. Parsons, 1972. A practical handbook of seawater analysis, 2nd ed. Bull. Fish. Res. Bd Can. 167: 1–310.

Taguchi, S. & E. A. Laws, 1988. On the microparticles which pass through glass fiber filter type GF/F in coastal and open waters. J. Plankton Res. 10: 999–1008.

Ward, B. B., 1987. Nitrogen transformations in the Southern California Bight. Limnol. Oceanogr. 34: 785–805.

Ward, B. B., K. A. Kilpatrick, E. H. Renger & R. W. Eppley, 1989. Biological nitrogen cycling in the nitracline. Limnol. Oceanogr. 34: 493–513.

Wheeler, P. A. & D. L. Kirchman, 1986. Utilization of inorganic and organic nitrogen by bacteria in marine systems. Limnol. Oceanogr. 31: 998–1009.

Received 2 September 1989; in revised form 12 March 1990; accepted 29 March 1990

Index

Page numbers are the first page of a paper in which the entry is discussed.

GPSR Compliance

The European Union's (EU) General Product Safety Regulation (GPSR) is a set of rules that requires consumer products to be safe and our obligations to ensure this.

If you have any concerns about our products, you can contact us on ProductSafety@springernature.com

In case Publisher is established outside the EU, the EU authorized representative is:

Springer Nature Customer Service Center GmbH
Europaplatz 3
69115 Heidelberg, Germany

Batch number: 10762213

Printed by Printforce, the Netherlands